THE FIRST STROMLO SYMPOSIUM: THE PHYSICS OF ACTIVE GALAXIES

COVER ILLUSTRATION

Optical and radio images of NGC 4261. Radio and ground-based optical data on the left and Hubble Space Telescope optical data on the right, taken by Walter Jaffe (Leiden Observatory) and Holland Ford (John Hopkins University, Space Telescope Science Institute, NASA).

A SERIES OF BOOKS ON RECENT DEVELOPMENTS IN ASTRONOMY AND ASTROPHYSICS

Printed by BookCrafters, Inc.

First published 1994

Library of Congress Catalog Card Number: 94-70287
ISBN 0-937707-73-2

D. Harold McNamara, Managing Editor of Conference Series
408 ESC Brigham Young University
Provo, UT 84602
801-378-2298

A SERIES OF BOOKS ON RECENT DEVELOPMENTS IN ASTRONOMY AND ASTROPHYSICS

Vol. 1-Progress and Opportunities in Southern Hemisphere Optical Astronomy: The CTIO 25th Anniversary Symposium
ed. V. M. Blanco and M. M. Phillips ISBN 0-937707-18-X

Vol. 2-Proceedings of a Workshop on Optical Surveys for Quasars
ed. P. S. Osmer, A. C. Porter, R. F. Green, and C. B. Foltz ISBN 0-937707-19-8

Vol. 3-Fiber Optics in Astronomy
ed. S. C. Barden ISBN 0-937707-20-1

Vol. 4-The Extragalactic Distance Scale: Proceedings of the ASP 100th Anniversary Symposium
ed. S. van den Bergh and C. J. Pritchet ISBN 0-937707-21-X

Vol. 5-The Minnesota Lectures on Clusters of Galaxies and Large-Scale Structure
ed. J. M. Dickey ISBN 0-937707-22-8

Vol. 6-Synthesis Imaging in Radio Astronomy: A Collection of Lectures from the Third NRAO Synthesis Imaging Summer School
ed. R. A. Perley, F. R. Schwab, and A. H. Bridle ISBN 0-937707-23-6

Vol. 7-Properties of Hot Luminous Stars: Boulder-Munich Workshop
ed. C. D. Garmany ISBN 0-937707-24-4

Vol. 8-CCDs in Astronomy
ed. G. H. Jacoby ISBN 0-937707-25-2

Vol. 9-Cool Stars, Stellar Systems, and the Sun. Sixth Cambridge Workshop
ed. G. Wallerstein ISBN 0-937707-27-9

Vol. 10-The Evolution of the Universe of Galaxies. The Edwin Hubble Centennial Symposium
ed. R. G. Kron ISBN 0-937707-28-7

Vol. 11-Confrontation Between Stellar Pulsation and Evolution
ed. C. Cacciari and G. Clementini ISBN 0-937707-30-9

Vol. 12-The Evolution of the Interstellar Medium
ed. L. Blitz ISBN 0-937707-31-7

Vol. 13-The Formation and Evolution of Star Clusters
ed. K. Janes ISBN 0-937707-32-5

Vol. 14-Astrophysics with Infrared Arrays
ed. R. Elston ISBN 0-937707-33-3

Vol. 15-Large-Scale Structures and Peculiar Motions in the Universe
ed. D. W. Latham and L. A. N. da Costa ISBN 0-937707-34-1

Vol. 16-Atoms, Ions and Molecules: New Results in Spectral Line Astrophysics
ed. A. D. Haschick and P. T. P. Ho ISBN 0-937707-35-X

Vol. 17-Light Pollution, Radio Interference, and Space Debris
ed. D. L. Crawford ISBN 0-937707-36-8

Vol. 18-The Interpretation of Modern Synthesis Observations of Spiral Galaxies
ed. N. Duric and P. C. Crane ISBN 0-937707-37-6

Vol. 19-Radio Interferometry: Theory, Techniques, and Applications, IAU Colloquium 131
ed. T. J. Cornwell and R. A. Perley ISBN 0-937707-38-4

Vol. 20-Frontiers of Stellar Evolution, celebrating the 50th Anniversary of McDonald Observatory
ed. D. L. Lambert ISBN 0-937707-39-2

Vol. 21-The Space Distribution of Quasars
ed. D. Crampton ISBN 0-937707-40-6

Vol. 22-Nonisotropic and Variable Outflows from Stars
ed. L. Drissen, C. Leitherer, and A. Nota ISBN 0-937707-41-4

Vol. 23-Astronomical CCD Observing and Reduction Techniques
ed. S. B. Howell ISBN 0-937707-42-4

Vol. 24-Cosmology and Large-Scale Structure in the Universe
ed. R. R. de Carvalho ISBN 0-937707-43-0

Vol. 25-Astronomical Data Analysis Software and Systems I
ed. D. M. Worrall, C. Biemesderfer, and J. Barnes ISBN 0-937707-44-9

Vol. 26-Cool Stars, Stellar Systems, and the Sun, Seventh Cambridge Workshop
ed. M. S. Giampapa and J. A. Bookbinder ISBN 0-937707-45-7

Vol. 27-The Solar Cycle
ed. K. L. Harvey ISBN 0-937707-46-5

Vol. 28-Automated Telescopes for Photometry and Imaging
ed. S. J. Adelman, R. J. Dukes, Jr., and C. J. Adelman ISBN 0-937707-47-3

Vol. 29-Workshop on Cataclysmic Variable Stars
ed. N. Vogt ISBN 0-937707-48-1

Vol. 30-Variable Stars and Galaxies, in honor of M. S. Feast on his retirement
ed. B. Warner ISBN 0-937707-49-X

Vol. 31-Relationships Between Active Galactic Nuclei and Starburst Galaxies
ed. A. V. Filippenko ISBN 0-937707-50-3

Vol. 32-Complementary Approaches to Double and Multiple Star Research, IAU Colloquium 135
ed. H. A. McAlister and W. I. Hartkopf ISBN 0-937707-51-1

Vol. 33-Research Amateur Astronomy
ed. S. J. Edberg ISBN 0-937707-52-X

Vol. 34-Robotic Telescopes in the 1990s
ed. A. V. Filippenko ISBN 0-937707-53-8

Vol. 35-Massive Stars: Their Lives in the Interstellar Medium
ed. J. P. Cassinelli and E. B. Churchwell ISBN 0-937707-54-6

Vol. 36-Planets and Pulsars
ed. J. A. Phillips, S. E. Thorsett, and S. R. Kulkarni ISBN 0-937707-55-4

Vol. 37-Fiber Optics in Astronomy II
ed. P. M. Gray ISBN 0-937707-56-2

Vol. 38-New Frontiers in Binary Star Research
ed. K. C. Leung and I. S. Nha ISBN 0-937707-57-0

Vol. 39-The Minnesota Lectures on the Structure and Dynamics of the Milky Way
ed. Roberta M. Humphreys ISBN 0-937707-58-9

Vol. 40-Inside the Stars, IAU Colloquium 137
ed. Werner W. Weiss and Annie Baglin ISBN 0-937707-59-7

Vol. 41-Astronomical Infrared Spectroscopy: Future Observational Directions
ed. Sun Kwok ISBN 0-937707-60-0

Inquiries concerning these volumes should be directed to the:
Astronomical Society of the Pacific
CONFERENCE SERIES
390 Ashton Avenue
San Francisco, CA 94112-1722
415-337-1100

SCIENTIFIC ORGANIZING COMMITTEE

G. V. Bicknell, M. A. Dopita, R. D. Ekers, R. W. Hunstead, G. K. Miley, R. P. Norris, C. A. Norman, D. E. Osterbrock, R. A. Perley, P. J. Quinn, E. M. Sadler, P. A. G. Scheuer

ASTRONOMICAL SOCIETY OF THE PACIFIC
CONFERENCE SERIES

Volume 54

THE FIRST STROMLO SYMPOSIUM: THE PHYSICS OF ACTIVE GALAXIES

Becker House, Australian Academy of Science, Canberra, Australia
27 June - 2 July 1993

Edited by
Geoffrey V. Bicknell, Michael A. Dopita, and Peter J. Quinn

CONTENTS

Contributed Papers

THE X-RAY AND Γ-RAY BACKGROUND

Invited Paper

Contributed Papers

UNIFIED SCHEMES

Invited Papers

Contributed Papers

SEYFERT GALAXIES & LINERS

Invited Papers

Contributed Papers

RADIO GALAXIES

Invited Papers

Contributed Papers

HIGH REDSHIFT RADIO GALAXIES

Invited Papers

STARBURSTS, MERGERS & ACTIVE GALAXIES

Invited Papers

Contributed Papers

SUMMARIES

Introduction

The idea for *The First Stromlo Symposium* began with a discussion with colleagues at a conference on Radio Galaxies held at Ringberg Castle in Bavaria in 1991. A number of people at that meeting were aware of new developments in Australian Astronomy - the most recent being, of course, the construction of the Australia Telescope and the funding of a Special Research Centre for Theoretical Astrophysics at Sydney University. In view of these developments and the established interests in Radio Galaxies at Mount Stromlo it was therefore thought to be a good idea that we organise an International Symposium on this topic to be held in Australia. We discussed this amongst ourselves and the proposal emerged that a symposium, not simply on the subject of Radio Galaxies, but on the general subject of Active Galaxies would have wide-ranging appeal and interest to the Australian and International astronomical communities. Professor Alex Rodgers, then director of Mount Stromlo & Siding Spring Observatories, was approached and he provided unequivocal and enthusiastic support. We then formed Scientific and Local Organising Committees and thus began arrangements for a symposium on *The Physics of Active Galaxies.* At the same time we realised that it would be a good idea, not to hold this meeting as a one-off event, but to make it the first of a series. Hence *The First Stromlo Symposium* is intended to be the initial meeting of a biennial series devoted to front-line topics in Astrophysics. We are pleased to say that the idea for this meeting quickly attracted enthusiastic support from the Australian Astronomical Community and we are happy to acknowledge generous financial assistance from The Anglo-Australian Observatory, The Australia Telescope National Facility, The Department of Applied Mathematics at Monash University and The Sydney University Research Centre for Theoretical Astrophysics - many of the key centres for astrophysical research in this country. The International Science Foundation (based in New York) provided support for a participant from the former Soviet Union. We also enjoyed generous assistance from our corporate sponsors: IBM, SUN Microsystems and Thinking Machines Corporation. Last but not least, we should acknowledge the assistance of the Australian Academy of Science who, recognising the quality of this symposium, provided Becker House of the Academy for us at a minimal cost. Academy staff were extremely helpful in assisting us to make the best use of the facilities.
What did we hope to achieve scientifically with this meeting? Essentially our plan was to hold a meeting on the Physics of Active Galaxies, to ascertain if we could collectively obtain some idea as to what it is which unites Active Galaxies and what it is which divides them - What physics is the same in Radio Galaxies and Seyfert Galaxies, for instance, and what Physics is necessarily different?

Obviously this aim encompasses the ideas central to *Unified Schemes* which are directed at unifying various sub-classes of active galaxies, mainly in terms of aspect. Certainly, *Unified Schemes* as you can see from a large number of the papers which follow, encompass a large part of the program material. However, our aims for this meeting went beyond that (if you like to the concept of *Grand Unification*) and possibly this meeting was, to some extent, unique in the mix of astrophysicists that we attracted to Canberra. For example, there was significant interaction between Radio astronomers, on the one hand, and X-Ray and Optical astronomers, on the other, two groups which have tended to diverge over the years since the first Active Galaxies were discovered.

The overwhelming response to the meeting was that it was a great success. Some reactions which particularly pleased us were the scientific stimulus that participants enjoyed, the mix of speakers (from graduate students through postdocs to acknowledged experts), the quality of the invited talks, the valuable discussion time after each talk and the high standard of the contributed papers. (Not to mention the *Great Active Nucleus Wine Tasting* and the *Great Aussie Nite* at the Gundaroo Pub, outside of Canberra.) Thus, what we have to offer in the following pages does not capture everything that happened during the main symposium and the ensuing workshop. However, we believe that you will find most of the papers here of great interest. For us, it is an enduring record of an unique occasion.

Finally, we would like to thank all of those who participated in this meeting, our local administrative staff (in particular Ian Sharpe and Vince O'Connor), and our students who willingly assisted with numerous symposium duties. We should also like to make special mention of Fiona Aplin for her invaluable assistance in the preparation of this volume. We are also grateful to the Astronomical Society of the Pacific for agreeing to publish what we trust will be an extremely useful book in the field of Active Galaxies for some years to come.

Geoffrey V. Bicknell
Michael A. Dopita
Peter J. Quinn
Canberra, December, 1993

List of Participants

M.W.B. Anderson
Australia Telescope
PO Box 76, Epping
NSW 2121, Australia
manderso@atnf.csiro.au

M. Anderson
Dept Theoretical Physics
Research School of Physical Sciences
& Engineering
Australian National University
ACT 2600, Australia
mraios@anu.edu.au

R. Antonucci
UCSB Physics Dept
Santa Barbara
California, USA
antonucci@voodoo.physics.ucsb.edu

M. Arnaboldi
Mount Stromlo & Siding Spring
Observatories
Private Bag, Weston P.O.
ACT 2611, Australia
magda@mso.anu.edu.au

J.A. Bailey
Anglo Australian Observatory
PO Box 296, Epping
NSW 2121, Australia
jab@aaoepp.oz.au

J.C. Baker
Dept. of Astrophysics
School of Physics
University of Sydney
Sydney NSW 2006, Australia
jo@astrop.physics.su.oz.

L. Ball
Research Centre for Theoretical
Astrophysics
University of Sydney
Sydney NSW 2006, Australia
ball@physics.su.oz.au

P.D. Barthel
Kapteyn Astronomical Institute
PO Box 800 NL-9700 AV Groningen
The Netherlands
pdb@astro.rug.nl

S.A. Baum
Space Telescope Science Inst
3700 San Martin Drive
Baltimore MD 21218, USA
sbaum@stsci.edu

S. Beaulieu
Mount Stromlo & Siding Spring
Observatories
Private Bag, Weston P.O.
ACT 2611, Australia
beaulieu@mso.anu.edu.a

M.C. Begelman
Joint Institute for Laboratory
Astrophysics
University of Colorado
Boulder CO 80309-0440, USA
mitch@jila.colorado.edu

J. Bell
Mount Stromlo & Siding Spring
Observatories
Private Bag, Weston P.O.
ACT 2611, Australia
bell@mso.anu.edu.au

G. V. Bicknell
Mount Stromlo & Siding Spring Observatories
Private Bag, Weston P.O.
ACT 2611, Australia
geoff@mso.anu.edu.au

R.D. Blandford
130-33 Caltech
Pasadena CA 91125, USA
rdb@tapir.caltech.edu

D.L. Blank
Department of Astrophysics
School of Physics
University of Sydney
Sydney NSW 2006, Australia
dblank@physics.su.oz.au

M. Blietz
Max-Planck-Institut für
Exterrestrische Physik
Giessenbach Str D-8046
Garching, Germany
blietz%mpei3@mpe.mpe-garching.mpg.de

E. Boldt
Code 666.0 NASA/Goddard Space Flight Center
Greenbelt MD 20771, USA
boldt@lheavx.gsfc.nasa.gov

A.M. Burgess
Dept. of Astrophysics
School of Physics
University of Sydney
Sydney NSW 2006, Australia
burgess@physics.su.oz.au

J. O. Burns
Department of Astronomy
Box 30001/Dept 4500
New Mexico State University
Las Cruces NM 88003, USA
jburns@nmsu.edu

M. Cameron
Max-Planck-Institut fur
Extraterrestrische Physik
Giessenbach Str D-8046
Garching, germany
cameron%mpei3@mpe.mpe-garching.mpg.de

R.D. Cannon
Anglo-Australian Observatory
PO BOx 296, Epping
NSW 2121, Australia
rdc@aaoepp.oz.au

A. Celotti
Institute of Astronomy
Madingley Road
Cambridge CB3 0HA, UK
celotti@mail.ast.cam.ac.uk

M. Cohen
Astronomy Department
Caltech 105-24
Pasadena CA 91125, USA
mhc@deimos.caltech.edu

M. Colless
Mount Stromlo & Siding Spring Observatories
Private Bag, Weston P.O.
ACT 2611, Australia
colless@mso.anu.edu.au

A. Comastri
Max-Planck-Institut fur
Extraterrestrische Physik
Giessenbach Str D-8046
Garching, Germany
amc@mpeu25.mpe-garching.mpg.de

M. A. Dopita
Mount Stromlo & Siding Spring Observatories
Private Bag, Weston P.O.
ACT 2611, Australia
mad@mso.anu.edu.au

M.J. Drinkwater
Anglo-Australian Observatory
Coonabarabran
NSW 2357, Australia
mjd@aaocbn.oz.au

A. Efstathiou
Division of Physical Sciences
University of Hertfordshire
Hatfield
Hertfordshire AL109AB, UK

J.A. Eilek
Physics Dept.
New Mexico Tech
Socorro NM 87801, USA
jeilek@aoc.nrao.edu

R.D. Ekers
Australia Telescope National Facility
PO Box 76, Epping
NSW 2121, Australia
rekers@atnf.csiro.au

J. English
Mount Stromlo & Siding Spring
Observatories
Private Bag, Weston P.O.
ACT 2611, Australia
english@mso.anu.edu.au

I. N. Evans
Space Telescope Science Institute
3700 San Martin Drive
Baltimore MD 21218, USA
evans@fos.stsci.edu

R. Fanti
Istituto di Radioastronomia
via Irnerio 46-40126
Bologna, Italy
rfanti@astbo1.bo.cnr.it

C. Fanti
Istituto Radioastronomia
via Irmerio 46
Bologna 40126, Italy
cfanti@astbo1.bo.cnr.it

D. J. Faulkner
Mount Stromlo & Siding Spring
Observatories
Private Bag, Weston P.O.
ACT 2611, Australia
djf@mso.anu.edu.au

H. H. Fink
Max-Planck Institute für
Extraterrestrische Physik
Giessenbach Str D-8046
Garching, Germany
hhf@ibma.ipp-garching.mpg.de

D. A. Forbes
Lick Observatory
University of California
Santa Cruz CA 95064, USA
forbes@helios.ucsc.edu

K. C. Freeman
Mount Stromlo & Siding Spring
Observatories
Private Bag, Weston P.O.
ACT 2611, Australia
kcf@mso.anu.edu.au

D. Fullager
Mount Stromlo & Siding Spring
Observatories
Private Bag, Weston P.O.
ACT 2611, Australia
dave@mso.anu.edu.au

S. C. B. Gascoigne
3 Anstey Street
Pearce ACT 2607, Australia

I. George
Code 668 NASA/Goddard Space
Flight Center
Greenbelt MD 20723, USA
george@lheavx.gsfc.nasa.gov

G. Giovannini
Istituto di Radioastronomia
Via Irnerio 46-40126
Bologna, Italy
ggiovannini@astbo1.bo.cnr.it

J. Guichard
Instituto Nacional de Astrofis Optica
y Electronica
Ap Do Postal 51 Puebla Pue
Mexico CP 72000
jguich@tonali.inaoep.mx

V.V. Gvaramadze
Krasin Str 19 Apt.
81 Moscow 123056, Russia
ochkheti%esocl.bitnet@vm.gmd.de

J.F. Hawley
PO Box 3818 University Station
Charlottesville VA 22903, USA
jh8h@astsun3.astro.virginia.edu

A. Hemming
Dept. of Astrophysics
School of Physics
Sydney University
Sydney NSW 2006, Australia
ablanch@physics.su.oz.au

R. Hes
Kapteyn Astronomical Institute
Zernikegebouw
Landleven 12 PO Box 800
9700 av Groningen, The Netherlands
rhes@kapteyn.astro.rug.nl

R. Hola
Mount Stromlo & Siding Spring
Observatories
Private Bag, Weston P.O.
ACT 2611, Australia
hola@mso.anu.edu.au

R.W. Hunstead
Astrophysics Department
School of Physics
University of Sydney
Sydney NSW 2006, Australia
rwh@physics.su.oz.au

P.A. Jones
Faculty of Science and Technology
Univesity of Western Sydney Nepean
PO Box 10
Kingswood NSW 2747, Australia
pjones@st.nepean.uws.edu.au

L. Kay
Physics Dept
Barnard College
Columbia University
3009 Broadway
New York NY 10027, USA
kay@sulis.barnard.columbia.edu

W. Kedziora-Chudczer
Research Centre for Theoretical
Astrophysics
School of Physics
University of Sydney
Sydney NSW 2006, Australia
lkedzior@robin.atnf.csiro.au

A. Kinney
Space Telescope Science Institute
3700 San Martin Drive
Baltimore MD 21218, USA
kinney@stsci.edu

D. Koehne
Mount Stromlo & Siding Spring
Observatories
Private Bag, Weston P.O.
ACT 2611, Australia
dfk@mso.anu.edu.au

A.M. Koekemoer
Mount Stromlo & Siding Spring
Observatories
Private Bag, Weston P.O.
ACT 2611, Australia
anton@mso.anu.edu.au

A. Königl
Department of Astronomy and Astrophysics
University of Chicago
5640 South Ellis Avenue
Chicago IL 60632, USA
arieh@jets.uchicago.edu

J.K. Kotalainen
European Southern Observatory
Karl–Schwarzschild Str 2
D-8046 Garching, Germany
jkotilai@eso.org

G. A. Kriss
Department of Physics & Astronomy
Johns Hopkins University
Baltimore MD 21218, USA
gak@jhufos.pha.jhu.edu

R.A. Laing
Royal Greenwich Observatory
Madingley Road
Cambridge CB3 OEZ, UK
rl@mail.ast.cam.ac.uk

J. Li
Research Centre for Theoretical Astrophysics
School of Physics
Sydney University
Sydney NSW 2006, Australia
ljk@physics.su.oz.au

H. Liang
Mount Stromlo & Siding Spring Observatories
Private Bag, Weston P.O.
ACT 2611, Australia
haida@mso.anu.edu

C. Lidman
Mount Stromlo & Siding Spring Observatories
Private Bag, Weston P.O.
ACT 2611, Australia
lidman@mso.anu.edu.au

C. Loken
Dept of Astronomy
Box 30001/Dept 4500
New Mexico State University
Las Cruces, NM 88003 USA
cloken@nmsu.edu

M.S. Longair
Royal Greenwich Observatory
Madingley Road
Cambridge CB3 OEZ, UK
msl@ras.phy.cam.ac.uk

Q. Luo
Research Centre for Theoretical Astrophysics
School of Physics
University of Sydney
Sydney NSW 2006, Australia
luo@physics.su.oz.au

S. Maddison
Dept. of Mathematics
Monash University, Clayton
VIC 3168
maddison@widget.maths.monash.edu.au

M. Malkan
Department of astronomy
UCLA, Los Angeles
CA 90024-1562, USA
malkan@bonnie.astro.ucla.edu

F. Masci
School of Physics
University of Melbourne, Parkville
VIC 3068, Australia
fjm@tauon.ph.unimelb.edu.au

W.G. Mathews
Lick Observatory
University of California
Santa Cruz, CA 95064, USA
mathews@helios.ucsc.edu

M. Mazzolini
School of Physics
University of Melbourne
Parkville
VIC 3052, Australia
marg@tauon.ph.unimelb.edu.au

C.F. McCain
Mount Stromlo & Siding Spring Observatories
Private Bag, Weston P.O.
ACT 2611, Australia
claudia@mso.anu.edu.au

P.J. McGregor
Mount Stromlo & Siding Spring Observatories
Private Bag, Weston P.O.
ACT 2611, Australia
peter@mso.anu.edu.au

K. Meisenheimer
Max-Planck-Institut für Astronomie
Königstuhl 17 D-69117
Heidelberg, Germany
meise@mpia-hd.mpg.de

D.B. Melrose
Dept. of Theoretical Physics
School of Physics
University of Sydney
Sydney
NSW 2006, Australia
melrose@physics.su.oz.au

P.F. Michelson
Dept. of Physics
Stanford University
Stanford CA 94305, USA
e7.f69@forsythe.stanford.edu

G.K. Miley
Leiden Observatory
Post Bus 9513, 2300 RA Leiden
The Netherlands
miley@rulhl1.leidenuniv.nl

J.S. Miller
Lick Observatory
University of California
Santa Cruz, CA 95064, USA
miller@helios.ucsc.edu

J.J. Monaghan
Department of Mathematics
Monash University, Clayton
VIC 3168, Australia
jjm@vaxc.cc.monash.edu.au

R. Morganti
Istituto di Radioastronomia
via Irnerio 46-40126
Bologna, Italy
morganti@astbo1.bo.cnr.it

J.R. Mould
Mount Stromlo & Siding Spring Observatories
Private Bag, Weston P.O.
ACT 2611, Australia
jrm@mso.anu.edu.au

J. Murray
Dept. of Mathematics
Monash University, Clayton
VIC 3168, Australia
bigred@mobius.maths.monash.edu.au

J.A. Nicholls
Research Centre for Theoretical Astrophysics
School of Physics
University of Sydney
Sydney, NSW 2006, Australia
jan@physics.su.oz.au

R.P. Norris
Australia Telescope National Facility
PO Box 76, Epping
NSW 2121 Australia
rnorris@atnf.csiro.au

J.E. Norris
Mount Stromlo & Siding Spring Observatories
Private Bag, Weston P.O.
ACT 2611, Australia
jen@mso.anu.edu.au

C.P. O'Dea
Space Telescope Science Institute
3700 San Martin Drive
Baltimore MD 21218, USA
odea@stsci.edu

R.V. Ojik
Leiden Observatory
PO Box 9513, 2300 RA Leiden
The Netherlands
ojik@reusel.leidenuniv.nl

T. Osterloo
Australia Telescope National Facility
PO Box 76, Epping
NSW 2121, Australia
tosterlo@atnf.csiro.au

F.N. Owen
NRAO, PO Box 0
Socorro NM 87801, USA
fowen@aoc.nrao.edu

P. Parma
Istituto di Radioastronomia
Via Irnerio 46-40126
Bologna, Italy
parma@astbo1.bo.cnr.it

J.J. Perry
Institute of Astronomy
University of Cambridge
Madingley Road
Cambridge CB3 OHA, UK
jjp@mail.ast.cam.ac.uk

M. Pope
Dept. of Theoretical Physics
School of Physics
University of Sydney
Sydney NSW 2006, Australia
pope@physics.su.oz.au

R.J. Protheroe
Dept. of Physics
University of Adelaide
GPO Box 498, Adelaide
SA 5001, Australia
rprother@physics.adelaide.au

P.J. Quinn
Mount Stromlo & Siding Spring Observatories
Private Bag, Weston P.O.
ACT 2611, Australia
pjq@mso.anu.edu.au

S. Rawlings
Dept. of Astrophysics
Nuclear Astrophysics Labs
Keble Road
Oxford OX1 3RM, UK
sr@astrophysics.oxford.ac.uk

J. Reynolds
Australia Telescope National Facility
PO Box 76, Epping
NSW 2121, Australia
jreynold@atnf.csiro.au

W. Roberts
Mount Stromlo & Siding Spring Observatories
Private Bag, Weston P.O.
ACT 2611, Australia
bill@merlin.anu.edu

J.G. Robertson
Dept. of Astrophysics
School of Physics
University of Sydney
Sydney NSW 2006, Australia
jgr@astrop.physics.su.oz.au

A.W. Rodgers
Mount Stromlo & Siding Spring Observatories
Private Bag, Weston P.O.
ACT 2611, Australia
alex@merlin.anu.edu.au

A. Roy
Australia Telescope National Facility
PO Box 76, Epping
NSW 2121, Australia
aroy@atnf.csiro.au

S. Ryder
Mount Stromlo & Siding Spring
Observatories
Private Bag, Weston P.O.
ACT 2611, Australia
ryder@mso.anu.edu.au

E.M. Sadler
Dept. of Astrophysics
School of Physics
University of Sydney
Sydney NSW 2006, Australia
ems@physics.su.oz.au

P.A.G. Scheuer
Cavendish Laboratory
Madingley Road
Cambridge CB3 QHE, UK
pags@phoenix.cambridge.ac.uk

A. Schroeder
Mount Stromlo & Siding Spring
Observatories
Private Bag, Weston P.O.
ACT 2611, Australia
anja@mso.anu.edu.au

S.M. Simkin
Dept. of Physics and Astronomy
Michigan State University
East Lansing MI 48824-1116, USA
simkin@grus.pa.msu.edu

A.K. Singal
NFRA
Postbus 2
7990 AA Dwingeloo
The Netherlands
singal@nfra.nl

A. Szabo
Dept. of Physics & Mathematical
Physics
University of Adelaide
GPO Box 498
Adelaide SA 5001, Australia
aszabo@physics.adelaide.edu.au

L.J. Tassie
Dept. of Theoretical Physics
Research Scool of Physical Sciences &
Engineering
Australian National University
ACT 2600, Australia

S. Tingay
Mount Stromlo & Siding Spring
Observatories
Private Bag, Weston P.O.
ACT 2611, Australia
tingay@mso.anu.edu.au

J. Turner
Code 668 NASA/Goddard Space
Flight Center
Greenbelt MD 20771, USA
turner@lheavx.gsfc.nasa.gov

A. Tzioumis
Australia Telescope National Facility
PO Box 76, Epping
NSW 2121, Australia
atzioumi.atnf.csiro.au

C.M. Urry
Space Telescope Science Institute
3700 San Martin Drive
Baltimore MD 21218, USA
cmu@stsci.edu

F. Vagnetti
Dipartimento di Fisica
Università di Roma “Tor Vergata”
Via della Ricerca Scientifica
I-00133 Roma, Italy
vagnetti@roma2.infn.it

S. Veilleux
KPNO/NOAO
950 N Cherry Avenue, Tucson
AZ 85726-6732, USA
veilleux@noao.edu

T. Venturi
Istituto di Radio Astronomia
Via Irnerio 46-40126
Bologna, Italy
tventuri@astbo1.bo.cnr.it

N. Visvanathan
Mount Stromlo & Siding Spring
Observatories
Private Bag, Weston P.O.
ACT 2611, Australia
vis@mso.anu.edu.au

M.A. Walker
Research Centre for Theoretical
Astrophysics
School of Physics
Sydney University
Sydney NSW 2006, Australia
maw@physics.su.oz.au

R. Webster
School of Physics
University of Melbourne
Parkville VIC 3068, Australia
webster@tauon.ph.unimelb.edu.au

D.W. Weedman
525 Davey Lab
Astronomy and Astrophysics Dept.
Pennsylvania State University
University Park
PA 16801, USA
weedman@astro.psu.edu

B. Wells
Mount Stromlo & Siding Spring
Observatories
Private Bag, Weston P.O.
ACT 2611, Australia
wells@mso.anu.edu.au

J. Whiteoak
Australia Telescope National Facility
PO Box 76, Epping
NSW 2121, Australia
jwhiteoa@atnf.csiro.au

B.J. Wilkes
Smithsonian Astrophysical
Observatory
60 Garden Street
Cambridge MA 01833, USA
belinda@cfa.harvard.edu

A.S. Wilson
Space Telescope Science Institute
3700 San Martin Drive
Baltimore MD 21218, USA
awilson@stsci.edu

K. Wu
Research Centre for Theoretical
Astrophysics
School of Physics
University of Sydney
Sydney NSW 2006, Australia
kinwah@physics.su.oz.au

THE CENTRAL ENGINE AND ITS ENVIRONS

Invited Papers

The First Stromlo Symposium: The Physics of Active Galaxies
ASP Conference Series, Vol. 54, 1994
G. V. Bicknell, M. A. Dopita, and P. J. Quinn (eds.)

HUBBLE SPACE TELESCOPE OBSERVATIONS OF ACTIVE GALAXIES

I. N. EVANS
Space Telescope Science Institute, 3700 San Martin Drive, Baltimore, MD 21218

H. C. FORD, G. A. KRISS, AND Z. TSVETANOV
Center for Astrophysical Sciences, Department of Physics and Astronomy, Johns Hopkins University, Baltimore, MD 21218

ABSTRACT Observations of the nuclei of nearby active galaxies using the *Hubble Space Telescope* have produced optical and ultraviolet images of the nuclear emission line regions with unprecedented spatial resolution. These imaging data, together with spatially resolved spectrophotometry of the nuclear emission line regions, have provided new observational constraints for the current generation of unified models for active galactic nuclei. *Hubble Space Telescope* observations of the optical and ultraviolet emission associated with extragalactic radio jets have revealed remarkable similarities between the optical and radio structures of a number of jets. These observations are placing stringent new constraints on models for particle acceleration, confinement, and diffusion processes.

INTRODUCTION

In the three years since its launch, the *Hubble Space Telescope* (*HST*) has yielded new and fundamental observational data in the field of active galaxies. *HST* observations of a range of phenomena have been obtained. Examples of the focus of such studies include (a) narrow and broad emission line regions in Seyfert, LINER, and starburst galaxies; (b) disks and tori in the nuclei of nearby active galaxies; (c) optical counterparts associated with extragalactic radio jets; (d) ultraviolet emission and polarization of active galactic nuclei (AGN) and quasars, BL Lac objects, and blazars, (e) Lyman α absorption systems in quasars; (f) gravitationally lensed quasars; and many others. Even though interesting new and fundamental results in each of these areas have resulted from *HST* observations, it is clearly impractical to discuss all of these topics in the space available. Therefore, the authors have elected to restrict the discussion here to some of the recent results from *HST* observations of (a) the nuclei of nearby active galaxies (Seyferts, LINERS, and starburst galaxies); and (b) the optical and ultraviolet counterparts of extragalactic radio jets.

NEARBY ACTIVE GALACTIC NUCLEI

In recent years, the now commonly accepted unified model of AGN invokes anisotropic emission from the nucleus in order to explain the observational differences between various types of objects. The primary hypothesis of this model is that the optical, UV, and probably soft X-ray radiation is emitted in two oppositely directed ionization cones, presumably through geometric shadowing by an optically thick torus or disk that collimates the ionizing radiation field. An immediate consequence of this model is that many of the observed properties of these objects should depend on our line of sight to the central source. Those objects for which our line of sight lies interior to one of the ionization cones should afford us a direct view of the broad line region (BLR), and possibly the ionizing continuum source, whereas the BLR and ionizing continuum source should be obscured in the case of those objects for which our line of sight falls outside the ionization cones.

Spectropolarimetric observations of the prototypical narrow line Seyfert 2 galaxy NGC 1068 (Antonucci and Miller 1985) revealed the presence of a broad component to the Balmer lines, visible only in polarized light. Antonucci and Miller (1985) interpreted this as emission scattered into our line of sight from a Seyfert 1 nucleus, and proposed that an optically thick absorber is responsible for obscuring the nucleus from direct observation. A model for NGC 1068 invoking a very optically thick molecular torus to obscure the nucleus has been developed theoretically by Krolik and Begelman (1986). Morphological studies also support the hypothesis that the primary difference between Seyfert 1 and Seyfert 2 nuclei results from orientation. Pogge (1988) produced emission line excitation maps from ground based narrow band imaging data to demonstrate that the ionizing radiation from the obscured nucleus of NGC 1068 is observable in a single sided cone.

The Faint Object Spectrograph Investigation Definition Team have used the Wide Field and Planetary Camera (*WF/PC*) on board the *HST* to study the sub-arcsecond morphology of the narrow line region (NLR) of NGC 1068 (Evans et al. 1991) as part of an intensive imaging and spectrophotometric program to use the *HST* to investigate the physics of the AGN phenomenon. The narrow band image obtained in the [O III] λ5007 emission line reveals a single sided conical morphology opening towards the northeast. The opening angle and position angle of the sub-arcsecond [O III] λ5007 cone is similar to that found at larger radii from ground based imaging. If the [O III] λ5007 emission is the visible signature of the material ionized directly by the nuclear radiation, then the nuclear ionizing continuum source must be located at the apex of the cone. Although the cone does not appear to be completely filled with material, the astrometry of Evans et al. (1991) places the apex of the cone very near the southernmost component of the radio triple observed by Ulvestad, Neff, and Wilson (1987), and consistent with the location of the water megamaser detected by Claussen and Lo (1986).

Narrow band [O III] λ5007 imaging of NGC 1068 by Boksenberg (1992) using the Faint Object Camera (*FOC*) aboard the *HST* reveals a very similar morphology to that identified by Evans et al. (1991), but shows more extended and fainter emission. In particular, Boksenberg (1992) identifies emission projecting further to the southwest than is seen on the *WF/PC* image. This emission may arise from a southwest counter-cone on the opposite side of the nucleus to the strong emission detected by Evans et al. (1991). Boksenberg (1992) suggests that the reason for the difference between the [O III] λ5007 morphology observed using the *FOC* and the [O III] λ5007 morphology observed using the *WF/PC* arises due at least partly to differences between the filter passbands and the presence of significant velocity structure in the emission regions. This interpretation is borne out by Faint Object Spectrograph (*FOS*) spectrophotometry of several of the NLR clouds that indicates that a strong velocity gradient is present along the cone axis from the north east of the nucleus to the south west of the nucleus.

Of considerable interest is the location of the electron scattering "mirror" that is responsible for reflecting the light from the BLR into our line of sight. Miller, Goodrich, and Mathews (1990) note that the relative narrowness of the Balmer lines in polarized flux requires that the temperature of the electron scattering plasma not exceed $\sim 3\times10^5$ K, and that this is consistent with all of the other constraints only if the electron scattering region is extended and located several pc from the nucleus. *HST* imaging of the nucleus in the 5500Å continuum (Lynds et al. 1991) reveals a bright resolved structure with a FWHM of $\sim 0''.15$ (11 pc for an assumed Hubble constant $H_0 = 75\,\mathrm{km\,s^{-1}\,Mpc^{-1}}$) superposed on an extended cloudlike structure. An astrometric comparison of the emission-line and continuum images places the continuum peak $\sim 0''.09$ (7 pc; Evans et al. 1991) to the northeast of the apex of the [O III] λ5007 cone, suggesting that this continuum source may be associated, at least in part, with the electron scattering region.

FOS observations centered on the nuclear continuum source were obtained by Caganoff et al. (1991). Although a broad component to the Balmer lines with a FWHM of $\sim 2200\,\mathrm{km\,s^{-1}}$ is clearly visible in the spectra obtained through the $0''.3$ diameter aperture used for the observations, the same component is detected on the strong forbidden lines also. This argues that the $2200\,\mathrm{km\,s^{-1}}$ component is associated with the emission arising from a bright [O III] λ5007 cloud that was included in the aperture, rather than with the scattering region. Caganoff et al. (1991) conclude that they do not detect a component with the FWHM of the Balmer lines seen in polarized flux, and therefore that either the continuum peak is not the scattering region or that the scattering region is extended with respect to the $0''.3$ aperture. Comparison of the *FOS* continuum and broad emission line fluxes with large aperture International Ultraviolet Explorer data (Kinney et al. 1991) implies that the scattering region is extended on the arcsecond scale (Kriss et al. 1993), in agreement with the model of Miller, Goodrich, and Mathews (1990).

The optical spectrum of the bright Seyfert 1.5 galaxy NGC 4151 includes both strong broad permitted lines and prominent narrow forbidden lines. This object has attracted considerable attention in recent years because of ground based imaging observations that have revealed the presence of both an unresolved nuclear point source responsible for the broad lines, and an extended region of narrow line emission that is elongated in the NE–SW direction. Because of this dichotomy of observational characteristics, NGC 4151 was a prime candidate for early *HST* imaging studies.

A narrow band [O III] λ5007 image and several UV and optical continuum images of NGC 4151 were obtained by Boksenberg (1992) using the *FOC*. The images show an unresolved nuclear point-source when viewed in continuum light, but reveal a clear biconical structure with several discrete clouds centered on the (saturated) nucleus in the emission line image.

The sub-arcsecond morphology of the NLR of NGC 4151 has been studied in detail by Evans et al. (1993) using the *WF/PC*. The [O III] λ5007 image resolves the NLR into a number of emission line clouds with a "stringy" morphology. They are distributed in a biconical structure with apices coincident with the unresolved nuclear point source. The opening angle projected onto the plane of the sky of the cone is $\sim 75°$, which is much wider than the distribution of the emission line gas seen at larger radii from the nucleus. The observed decrease in cone opening angle as distance from the nucleus increases is consistent with the hypothesis that the ionization cone is filled only partially with gas, and that either the gas density, or the volume filling factor, or both, decreases with distance from the nucleus.

Comparison of the imaging data of Evans et al. (1993) with ground based observations suggest a geometry for the ionizing cones that is *incompatible* with a simple model that invokes a very optically thick torus surrounding the BLR to shadow geometrically the nuclear continuum source. The *HST* data indicate that for feasible model parameters our line of dight must be exterior to the ionization cones, and therefore a model that invokes pure geometric shadowing would predict that the BLR should not be visible directly, contrary to observation.

Alternative models of obscuration that have an effective neutral hydrogen column density along our line of sight of $\lesssim 10^{21}\,\mathrm{cm}^{-2}$ are compatible with the data. Such a medium would be optically thick to the ionizing UV and soft X-ray, but optically thin in the visible. If this material were arranged in a toroidal or disk-like geometry surrounding the central source, then the location of the ionization cones would be determined by absorption of the ionizing UV and soft X-ray. However, our line of sight to the BLR in the visible would be relatively unobscured. This hypothesis is supported strongly by recent observations of UV absorption in NGC 4151 (Kriss et al. 1992). Objects such as NGC 4151 do not readily fall into the current unified scheme that depends solely on orientation to define the nature of the nucleus; while a simple modification to the standard model allows the incorporation of AGN that possess strong UV absorption, their observational characteristics are sufficiently distinct that we should consider them as a different class of object from their Seyfert 1 or Seyfert 2 siblings.

Antonucci and Miller (1985) proposed to explain the obscuration of the BLR in NGC 1068 by invoking a torus of absorbing material surrounding the active nucleus. This model is very appealing, because it explains naturally not only the absorption, but also through geometric shadowing the projected shapes of the observed ionization cones. Theoretical models of the physical properties of the torus have been developed by Krolik and Begelman (1986) and Krolik and Lepp (1989). Recently, the *HST* has yielded images that place the importance of the torus on a much firmer observational footing.

Observations of the LINER galaxy M51 obtained using the VLA have resolved a bidirectional radio jet centered on the nucleus (Crane and van der Hulst 1992). Ford et al. (1985) detected in emission a pair of "bubbles," apparently inflated by the bidirectional jet. The bright rim of the southern bubble shows considerable evidence for shock excitation at the location where the jet is decelerated by the surrounding interstellar medium (ISM). The *WF/PC* observations of Ford et al. (1992) indicate that the [N II] λ6584/Hα ratio is very high in this rim of the bubble, supporting strongly this interpretation. Ford (1993) also discovered the presence of a dark bar, $\sim 1''$ long, and approximately perpendicular to the jet axis. A plausible explanation for these observations is that the dark bar is the result of a dusty disk seen nearly edge on. The dimensions of the disk are similar to the outer dimensions of the molecular torus proposed by Krolik and co-workers, and the bar may actually be the first clear observational evidence for the presence of a molecular torus surrounding the active nucleus.

Additional observational evidence for the presence of disk like structures in AGN comes from *HST* imaging of NGC 4261 by Jaffe et al. (1993). This bright elliptical galaxy in the Virgo cluster has a pair of radio jets emanating from the nucleus. An optical continuum *WF/PC* image of the region reveals a smooth absorbing disk-like structure with well-defined edges and an unresolved point source nucleus. The plane of the disk is approximately perpendicular to the radio axis, and, assuming that the structure is circular, has an inclination of $\sim 64°$ relative to the plane of the sky. Assuming that the distance to NGC 4261 is the same as that of the Virgo cluster (14.7 Mpc), then the diameter of the disk is of order 120 pc, which is somewhat larger than expected from theoretical models for the nuclear molecular torus. Although the exact nature of this disk is still somewhat uncertain, it is clear that the structure provides a natural mechanism for dissipating the angular momentum of material accreted towards the active nucleus.

In addition to the objects discussed above, the *HST* has been used to observe a number of other nearby active galaxies of the Seyfert and LINER types to attempt to further understand the nature of the active nucleus. For example, as part of the Faint Object Camera Investigation Definition Team program, Boksenberg (1992) has detected a biconical structure in Mkn 78 that is apparently devoid of emission at the location of the cone apices, suggesting the possible presence of an absorbing structure along our line of sight to the nucleus in that object. Boksenberg (1992) finds strong extended emission and a resolved nucleus in both the Seyfert 2 galaxies in his sample, and a lack of strong extended emission and an unresolved nucleus in each of the six Seyfert 1 objects in his sample.

Similar statistics are found using *HST* images obtained as part of the Faint Object Spectrograph Investigation Definition Team observing program (see Table 1). As can be seen from the table, the broad line objects tend to have unresolved nuclei or compact ($\lesssim 0''.3$ diameter) morphologies, whereas the narrow line objects tend to have resolved structures with dimensions $\sim 0''.7$ or greater. Apparently conical or quasi-linear morphologies are prevalent for the narrow line objects, whereas the compact morphologies tend to be more approximately symmetric. These correlations appear to be consistent with the hypothesis that our line of sight is interior to the ionizing cone in the case of the broad line objects, but exterior to the cone for the narrow line objects.

The *HST* imaging data of nearby active galaxies suggest that *the AGN paradigm must be modified to include the notion that the angle between the preferred axis of the central engine (the normal to the accretion disk and the plane of the torus) and the plane of the host galaxy is important in determining the extent of the NLR.* If the ionizing cones intersect the plane of the galaxy, then the ionizing radiation and jets will produce significant extended emission in the plane of the galaxy, and the extended emission line region will be probably ionization bounded. An example of this is M51, where the nuclear engine apparently lays on its side, perhaps because of the potential of the triaxial nucleus. On the other hand, if the ionizing cones are more nearly normal to the plane of the galaxy, the extent of the emission will be limited by the scale height of the disk, and the extended emission line region will be probably density bounded. M81 is an example of this. Although in M81 our line of sight falls inside the ionization cone, and thus we see directly the BLR, there is probably very little extended emission because there is very little gas above the disk and along the axis of the ionization cones to be ionized and shocked. In this case, where our line of sight is interior to the ionization cone, the emission measure along our line of sight through the outer regions of the ionization cone will be significantly less than through the core of the cone. Any statistical study of size and orientation will need to consider the effects of reduced surface brightness in these cases.

TABLE I Imaging of Nearby Active Galactic Nuclei

Object	Broad Lines?	Structure	Features
3C 120[a]	Yes	Unresolved	
3C 390.3[a]	Yes	Unresolved	
Akn 120[a]	Yes	Unresolved	
NGC 1275[a]	Yes	Unresolved	
NGC 1566	Yes	Unresolved	
NGC 3031	Yes	Unresolved	
NGC 3783	Yes	Unresolved	
NGC 4051	Yes	Compact	
NGC 4151	Yes	Extended	Cones
NGC 6814	Yes	Unresolved	
NGC 5548[a]	Yes	Unresolved	
NGC 7469[a]	Yes	Unresolved	
Mkn 3	No	Extended	"Jet"?
Mkn 78[a]	No	Extended	Cones
Mkn 463	No	Extended	"Jet"
Mkn 477	No	Compact	
NGC 1052	No	Compact	
NGC 1068	No	Extended	Cones
NGC 3079	No	Extended	
NGC 4258	No	Compact	
NGC 5194	No	Extended	Cones

[a]Boksenberg (1992)

While the conversion of gravitational energy through accretion onto a massive black hole is the most popular scenario for the power source for high luminosity Seyfert and other AGN, at lower luminosities the situation is not so well defined. A nuclear starburst, rather than mass accretion, could be responsible for the energy output from low luminosity active nuclei. Terlevich and Melnick (1985) have demonstrated that a cluster of massive stars may produce a continuum spectral distribution that resembles the power law spectrum seen in AGN, and Terlevich, Tenorio-Tagle, Franco, and Melnick (1992, and references cited therein) have argued that Type II supernovae evolving in a dense, metal rich ISM can reproduce the observed properties of the BLR in high luminosity Seyfert 1 nuclei and QSOs.

If the starburst hypothesis applies to the BLRs of AGN, then the least intrinsically luminous Seyfert 1 galaxy, NGC 4395, would seem to be a very likely candidate for this explanation (Filippenko and Sargent 1989). To test this hypothesis, Filippenko, Ho, and Sargent (1993) have obtained *HST* optical images and ultraviolet spectrophotometry of NGC 4395. Their spectrum of this object reveals narrow emission lines covering a wide range of ionization states, and permitted lines with broad wings, resembling very closely Seyfert 1 nuclei with higher luminosities. The absence of stellar absorption lines in the UV spectrum of the nucleus is particularly noticeable. Filippenko, Ho, and Sargent (1993) argue that such lines should be present if the active nucleus is energized by a starbursting cluster of massive stars with the appropriate initial mass function necessary to produce the observed continuum spectral shape. The lack of stellar absorption line features in the UV spectrum is also in agreement with the result of optical and near infrared spectrophotometry. The optical images obtained using the *WF/PC* resolve the nucleus in [O III] λ5007 emission, but in continuum light the nucleus is unresolved at the scale of $\sim 0\farcs7$ pc (for an assumed distance to NGC 4395 of 2.6 Mpc). Together, these observations constitute strong evidence against the applicability of the starburst hypothesis of Terlevich et al. (1992) to NGC 4395, and Filippenko, Ho, and Sargent (1993) conclude that this object is likely powered by nonstellar phenomena, consistent with the standard model of mass accretion onto a black hole.

JETS

Another area that has attracted considerable interest among researchers using the *HST* is the field of extragalactic jets. For the first time, the *HST* allows observers to study the optical and UV emitting plasmas associated with these jets with a spatial resolution similar to that of the radio observations. Comparison of the data from the optical and radio emitting plasmas places additional constraints on the mechanisms responsible for powering the energetic particles in the the radio plasma.

The nearby giant elliptical M87 is a well studied active galaxy that hosts a prominent FR-I radio jet and a double lobed radio structure. Ground based optical imaging has revealed that the optical structure of the jet closely matches the complex structure visible in the radio. The similarity between the morphological structure of the jet seen at very different wavelengths, together with the observed high polarization, argues that non-thermal synchrotron emission is responsible for the jet emission.

As part of their program to study extragalactic jets, the Faint Object Camera Investigation Definition Team has imaged extensively the M87 jet at optical and UV wavelengths using the *FOC*. From these data, Boksenberg et al. (1992) find a remarkable similarity between their UV images of the jet and radio images of comparable resolution, suggesting that the spatial distributions of the old and young electrons are similar. This, together with their observation that the ultraviolet spectral index is remarkably constant and shows no steepening with distance along the jet out to a radius of $\sim 1.5\,\text{kpc}$, support strongly the model of Owen, Hardee, and Cornwell (1989) in which there is no *in situ* particle acceleration, but in which energetic electrons are transported along the low loss inner core of the jet and the synchrotron emission arises primarily from the interface region between the jet and the surrounding ISM (Boksenberg et al. 1992).

Although the radio and *HST* optical images are remarkably similar, they are not identical. According to Macchetto (1992), the knots visible in the jet have intrinsically higher contrast in the optical and UV when compared to the radio data, with the knots being more centrally condensed in the optical/UV wavebands. The radio-to-optical spectral index of the inter-knot regions is significantly steeper than the radio-to-optical spectral index of the knots themselves. Sparks, Biretta, and Macchetto (1992) have considered the implications of the new observational constraints that the *FOC* observations impose on the various models for the energy transport in the M87 jet, and suggest that an overall spectral steepening, rather than lifetime/diffusion effects, are responsible for producing the observed steepening of the radio-to-optical spectral index between the knots (Macchetto 1992).

Besides M87, the *HST* has been used to observe the optical emission associated with the radio jets in several other objects. Thomson, Wright, and Disney (1992) used the *FOC* to obtain a high resolution B band image of 3C 273 in linearly polarized light. They find that the position angle of the electric vector is perpendicular to the jet along most of its length. Their images resolve the outermost tip of the jet into a double arc structure, and the position angle of the electric vector rotates to become parallel to the arc at that point. This is consistent with a synchrotron origin for the optical as well as the radio emission, but is not conistent with the scattering model of Morrison and Sadun (1992) for the optical emission. *FOC* images of the optical counterparts of the radio jets in PKS 0521−36 (Macchetto et al. 1991a) and 3C 66B (Macchetto et al. 1991b) have been obtained also. For the former object, these imaging data apparently resolve completely the jet structure, and reveal the presence of a component that may be an inner extension of the jet that is not visible in the VLA radio data (Keel 1986), and which could be a site where electron acceleration is taking place. In 3C 66B, Macchetto et al. (1991b) find that the optical jet has a double stranded filamentary morphology that may be due to edge brightening of the jet/ISM interface. Both PKS 0521−36 and 3C 66B have observed optical and radio morphologies, separated by five decades in frequency, that are very similar. This implies that the old and young electrons have similar spatial distributions (Macchetto et al. 1991b), even though there are significant variations in physical conditions along the jet. Clearly, these observations impose new and stringent constraints on models for particle acceleration in jets.

HST imaging observations using both the *FOC* and the *WF/PC* have resulted in serendipitous discovery of optical jet structures in nearby active galaxies. In the elliptical galaxy 3C 264 (NGC 3862), Crane et al. (1993) report the existence of a hitherto unidentified, $\sim 0\farcs65$ long, optical jet emanating from the nucleus. Morphologically, the optical jet is similar to the synchrotron jets seen in M87 and 3C 66B (Crane et al. 1993). The total optical jet flux, optical spectral index, and metric scale are all similar to the values found in other synchrotron jets. Since 3C 264 is a typical radio loud elliptical, Crane et al. (1993)

have suggested that many observed nonthermal jets in ellipticals in the radio continuum may have optical counterparts that can be observed using the *HST*.

Uomoto et al. (1993) also report serendipitous discovery of an optical jet in the Seyfert 2 galaxy Mkn 463 using the *WF/PC*. The $0''.84$ long jet is directed radially from the Eastern nucleus of this double nucleus active galaxy, and terminates at a radio point source $1''.2$ from the nucleus. Unlike 3C 264, the optical jet in Mkn 463 is not synchrotron in nature, although the jet is morphologically very similar to the optical jet in 3C 273. Although *FOS* spectrophotometry is yet to be obtained, Uomoto et al. (1993) were able to estimate some of the jet characteristics directly from comparison of the jet fluxes in their observed optical filter passbands. They conclude that the jet spectrum is likely continuum dominated, with weak line emission. This description fits the polarized spectrum of the Eastern nucleus of Mkn 463 observed by Miller and Goodrich (1990), and suggests that at least some of the jet radiation is scattered light from the hidden Seyfert 1 nucleus.

CONCLUSIONS

We have briefly reviewed some of the areas of interest in the field of active galaxies that observations using the *HST* have contributed to in the past three years. It is clear that the excellent spatial resolution of the *HST* is producing new observational data for the innermost regions of active galaxies, and consequently imposing new constraints on our theories for the energy formation mechanisms in these objects.

Based on observations with the NASA/ESA *Hubble Space Telescope*, obtained at the Space Telescope Science Institute, which is operated by the Association of Universities for Research in Astronomy, Inc., under NASA contract NAS5-26555. This work was supported in part by NASA grant NAG5-1630.

REFERENCES

Antonucci, R. R. J., and Miller, J. S. 1985, *ApJ*, **297**, 621

Boksenberg, A. 1992, in *Science with the Hubble Space Telescope*, P. Benvenuti and E. Schreier, Munich: European Southern Observatory, 61

Boksenberg, A., et al. 1992, *A&A*, **261**, 393

Caganoff, S., et al. 1991, *ApJ*, **377**, L9

Claussen, M. J., and Lo, K.-Y. 1986, *ApJ*, **308**, 592

Crane, P. C., and van der Hulst, J. M. 1992, *AJ*, **103**, 1146

Crane, P., et al. 1993, *ApJ*, **402**, L37

Evans, I. N., Ford, H. C., Kinney, A. L., Antonucci, R. R. J., Armus, L., and Caganoff, S. 1991, *ApJ*, **369**, L27

Evans, I. N., Tsvetanov, Z., Kriss, G. A., Ford, H. C., Caganoff, S., and Koratkar, A. P. 1993, *ApJ*, **417**, 82

Filippenko, A. V., Ho, L. C., and Sargent, W. L. W. 1993, *ApJ*, **410**, L75

Filippenko, A. V., and Sargent, W. L. W. 1989, *ApJ*, **342**, L11

Ford, H. C. 1993, *NASA Press Release, ST ScI-PRC92-17*

Ford, H. C., Caganoff, S., Kriss, G. A., Tsvetanov, Z., and Evans, I. N. 1992, *BAAS*, **24**, 818

Ford, H. C., Crane, P. C., Jacoby, G. H., Lawrie, D. G., and Van Der Hulst, J. M. 1985, *ApJ*, **293**, 132

Jaffe, W., Ford, H. C., Ferrarese, L., Van Den Bosch, F., and O'Connell, R. 1993, *Nature*, **363**, 213

Keel, W. C. 1986, *ApJ*, **302**, 296

Kinney, A. L., Antonucci, R. R. J., Ward, M. J., Wilson, A. S., and Whittle, M. 1991, *ApJ*, **377**, 100

Kriss, G. A., et al. 1992, *ApJ*, **392**, 485

Kriss, G. A., et al. 1993, in *First Light in the Universe: Stars or QSO's*, B. Rocca-Volmerange, M. Dennefeld, Thanh Van, Paris: Editions Frontieres, 379

Krolik, J. H., and Begelman, M. C. 1986, *ApJ*, **308**, L55

Krolik, J. H., and Lepp, S. 1989, *ApJ*, **347**, 179

Lynds, R., et al. 1991, *ApJ*, **369** L31

Macchetto, F. 1992, in *Science with the Hubble Space Telescope*, P. Benvenuti and E. Schreier, Munich: European Southern Observatory, 73

Macchetto, F., et al. 1991a, *ApJ*, **369**, L55

Macchetto, F., et al. 1991b, *ApJ*, **373**, L55

Miller, J. S., and Goodrich, R. W. 1990, *ApJ*, **355**, 456

Miller, J. S., Goodrich, R. W., and Matthews, W. G. 1991, *ApJ*, **378**, 47

Morrison, P., and Sadun, A. 1992, *MNRAS*, **254**, 488

Owen, F. N., Hardee, P. E., and Cornwell, T. J. 1989, *ApJ*, **340**, 698

Pogge, R. W. 1989, *ApJ*, **345**, 730

Sparks, W. B., Biretta, J. A., and Macchetto, F. 1992, *ApJ*, in press

Terlevich, R., and Melnick, J. 1985, *MNRAS*, **213**, 841

Terlevich, R., Tenorio-Tagle, G., Franco, J., and Melnick, J. 1992, *MNRAS*, **255**, 713

Thompson, R. C., Wright, A. E., and Disney, M. J. 1992, in *Science with the Hubble Space Telescope*, P. Benvenuti and E. Schreier, Munich: European Southern Observatory, 83

Ulvestad, J. S., Neff, S. G., and Wilson, A. S. 1987, *AJ*, **93**, 22

Uomoto, A., et al. 1993, *AJ*, **105**, 1308

The First Stromlo Symposium: The Physics of Active Galaxies
ASP Conference Series, Vol. 54, 1994
G. V. Bicknell, M. A. Dopita, and P. J. Quinn (eds.)

HIGH ENERGY GAMMA RAY EMISSION FROM ACTIVE GALAXIES: EGRET OBSERVATIONS AND IMPLICATIONS

P.F. MICHELSON
Department of Physics and W.W. Hansen Experimental Physics Laboratory, Stanford University, Stanford, CA 94305, USA

ABSTRACT The Energetic Gamma Ray Experiment Telescope (EGRET) on the Compton Observatory has so far detected more than 24 active galaxies (AGN) above 100 MeV. The majority of the sources are quasars and five are usually classified as Bl Lac objects. A common characteristic is that they are all radio-loud, flat spectrum radio sources. Many of them are seen as superluminal sources as well. A summary of the EGRET observations is presented, including discussion of the energy spectra, temporal variability of the sources, and some of the implications for emission models.

INTRODUCTION

Before the launch of the Energetic Gamma Ray Experiment Telescope (EGRET) on the Compton Gamma Ray Observatory, high energy gamma-ray emission had been detected from only one AGN, the relatively nearby quasar 3C 273 (Swanenburg et al. 1978; Bignami et al. 1981). EGRET has so far detected, with a high degree of certainty (significance $> 5\sigma$), 22 active galactic nuclei in high energy gamma-rays ($E > 100$ MeV) during Phase I (6/91 - 11/92) and the early part of Phase II of the Compton Observatory mission (Bertsch et al. 1991; Hartman et al. 1992a; Fichtel et al. 1992a; Michelson et al. 1992; Hartman et al. 1992b; Hartman et al. 1992c; Lin et al. 1992; von Montigny et al. 1993a; Kanbach et al. 1992). There are marginal detections (significance between 3 and 5 σ) of 10 other blazars. One of these sources (Mkn 421) has subsequently been observed to be a TeV gamma-ray source (Punch et al. 1992). A common characteristic of these sources is that they are all radio-loud, flat or complex radio spectrum sources. Many of them are seen as superluminal radio sources as well. They all appear to be members of the blazar class of AGN (Bl Lac objects, HPQ and OVV quasars), exhibiting one or more of the characteristics of this source class (non-thermal continuum spectrum, flat radio spectrum, strong variability, optical polarization). In addition, a number of bright, unidentified high-latitude gamma-ray sources have been detected. These sources may be members of a new class or uncataloged blazars.

Among the most interesting results obtained thus far from analysis of the EGRET observations are the following: (1) the gamma-ray energy flux in many of the sources is dominant over the flux in lower energy bands; (2) the "apparent" gamma-ray luminosity, assuming isotropic emission which is almost certainly incorrect, ranges from less than 3×10^{44} erg s^{-1} to more than 10^{49} erg s^{-1} with values in the range 3×10^{47} erg $s^{-1}$to 3×10^{48} erg s^{-1} being typical; (3) the sources for which redshifts have been measured are distributed over a wide range of redshifts (0.03 to 2.17); (4) the photon spectra in the EGRET energy range are generally well-represented by power laws in energy with a wide range of photon number spectral indiçes (-1.6 to -2.6); (5) many of the sources exhibit variability in the gamma-ray band on timescales from days to months; (6) many active galaxies relatively close to earth, including Seyfert galaxies and some notable blazars, have not been detected in high energy gamma rays (von Montigny et al. 1993b; Lin et al. 1993).

In the remainder of this paper, examples of the spectra and time variability observed in the EGRET sources will be presented. The implications these observations have for emission models is discussed with emphasis on the need for relativistic beaming in order to avoid absorption in the source due to gamma-gamma interactions. A brief description of the EGRET instrument is given for the benefit of readers who may not be familar with techniques for detecting high energy gamma-rays.

All of the work described here is a result of the efforts of members of the EGRET science team from NASA Goddard Space Flight Center, Max Planck Institut für Extraterrestrische Physik, Grumman Aerospace Corporation, Stanford University, and Hampden-Sydney College. Dr. Carl Fichtel (Goddard) and Prof. Klaus Pinkau (MPE) are the co-principal investigators for the EGRET instrument.

THE EGRET INSTRUMENT

The EGRET telescope is one of four instruments on the Compton Gamma Ray Observatory that together cover the energy range from 0.06 MeV to 30 GeV. EGRET is sensitive to gamma-rays in the energy range from about 30 MeV to 30 GeV. In the mode used for most of the observations of active galaxies discussed here the effective area of the telescope is about 1.0×10^3 cm^2 at 150 MeV and 1.5×10^3 cm^2 around 0.5-1 GeV for targets near the center of the field of view. EGRET's effective area is maximum when the target is on axis and falls to approximately 50% of this value when the angular offset reaches 20°. The large field-of-view (FOV) of EGRET is important for the study of time variable sources such as blazars. The instrument, shown schematically in Figure I, has components typically used in high energy gamma-ray telescopes; an anticoincidence system to discriminate against charged particle radiation, a multilevel thin-plate spark chamber system to convert gamma-rays and determine the trajectories of the secondary electron-positron pair, a triggering telescope that detects the presence of the pair with the correct direction of motion, and an energy measuring calorimeter, which in the case of EGRET is a NaI(Tl) crystal. Descriptions of the instrument are given by Hughes et al. (1980) and Kanbach et al. (1988). Details of the instrument calibration, both before and after launch, are given by Thompson et al. (1993) and Nolan et al. (1992). The instrument is designed to be free of internal background and its calibration tests have verified that the internal background is at least an order of magnitude below the extragalactic diffuse gamma radiation. EGRET observations are carried

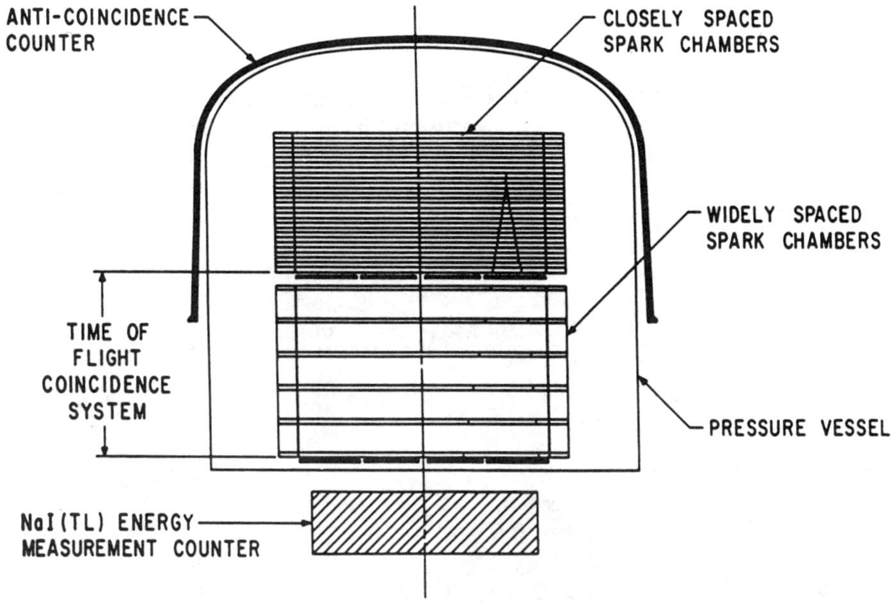

FIGURE I   Schematic diagram of the EGRET instrument showing the four major subsystems; the spark chamber, the time-of-flight system, the anticoincidence dome, and the NaI(Tl) crystal calorimeter.

out as pointed observations that last from a few days to a few weeks with 3 weeks being the duration of a typical viewing period.

EGRET BLAZAR DETECTIONS AND IDENTIFICATIONS

Table I lists active galaxies that were detected above 100 MeV with high significance during one or more viewing periods during Phase I and the early part of Phase II of the Compton mission. Also listed are the observed fluxes above 100 MeV, the photon spectral index, the redshift if known, and the luminosity. Here "high significance" means that the probability of false detection of a point source is less than 6×10^{-7} (corresponding to 5 σ) for a single trial. Since there are about 10^4 independent resolution elements on the full sky, assuming a typical resolution element has a 1° radius, the expected number of false source detections at this confidence level is less than 6×10^{-3} for a uniform full-sky survey. This conclusion is based on statistical considerations alone. The systematic uncertainties are believed to be small enough that there is very little doubt about about positive detections at this level. With respect to source identification, all of the 22 identifications in Table I are from a list of approximately 500 radio loud AGNs found largely in the Kühr catalog (Kühr et al. 1981), which was used as the basis of source identification to avoid searching for counterparts "after-the-fact". Considering that there are about 30 gamma-ray sources detected above 5 σ (excluding the galactic plane (ie., $-10° \leq b \leq 10°$), but including unidentified sources above 5 σ), the a priori expected number of chance associations with radio loud AGNs is ~0.27. Thus it is likely that all of the sources in Table I have been correctly identified.

TABLE I Characteristics of Active Galaxies detected by EGRET. The source luminosity (0.1 GeV < E < 5 GeV), computed using the known redshift assuming H_0 = 75 km s^{-1} Mpc^{-1} and q_0 =1/2, is equal to the relative luminosity shown in the table times (10^{48} erg s^{-1}) f, where f is an unknown beaming factor. Typically f is thought to be in the range from 10^{-2} to 10^{-3}.

source		flux (10^{-6} cm^{-2} s^{-1}) E > 100 MeV	photon spectral index	z	relative luminosity
0202+149	(4C+15.05)	1.0 ± 0.3			
0208-512	PKS	1.0 ± 0.1	−1.68 ± 0.04	1.003	2
0219+428		0.17 ± 0.04		0.444	
0234+285		0.16 ± 0.04		1.207	
0235+164	(OD+160)	0.8 ± 0.2	−1.9 ± 0.2	0.94	2.0
0420-014	(OA 129)	0.5 ± 0.1	−1.9 ± 0.3	0.915	0.7
0446+112	PKS	0.2 to 0.95			
0454-234		0.62 ± 0.15		0.890	
0454-463	PKS	0.25 ± 0.1		0.858	0.3
0528+134	PKS	0.4 to 1.6	−2.56 ± 0.1	2.06	4 to 13
0537-441	PKS	0.3 ± 0.1	−2.0 ± 0.2	0.894	0.4
0716+714		0.20 ± 0.06	−1.8 ± 0.1	...	...
0836+710	(4C+71.07)	0.15 ± 0.04	−2.4 ± 0.3	2.17	2.6
1101+384	(Mrk 421)	0.11 ± 0.03	−2.0 ± 0.14	0.031	0.0003
1226+023	(3C 273)	0.24 ± 0.06	−2.4 ± 0.2	0.158	0.008
1253-055	(3C 279)	0.6 to 4.9	−1.9 ± 0.1	0.536	0.3 to 2
1606+106	(4C+10.45)	0.35 to 0.56	−2.2 ± 0.3	1.23	...
1633+382	(4C+38.41)	0.8 ± 0.1	−1.9 ± 0.1	1.814	6
2022-077		6.7 ± 1.2	−1.5 ± 0.2		
2052-474	PKS	0.45 ± 0.06		1.489	
2230+114	(CTA 102)	0.24 ± 0.07	−2.6 ± 0.2	1.037	0.5
2251+158	(3C 454.3)	0.8 ± 0.1	−2.2 ± 0.1	0.859	0.5

There have also been a number of active galaxies with marginal detections ($3\sigma \leq$ significance $< 5\sigma$) by EGRET. However these identifications are also affected more by systematic uncertainties, for example, in the background model. Hence we characterize them as "marginal" detections. These sources are discussed in more detail elsewhere (Michelson et al. 1993).

SOURCE CHARACTERISTICS

The background corrected differential gamma-ray energy spectra for PKS 0528 +134, PKS 0208-512, and 3C 279 are shown in Figure II. In all three cases, the spectra are well represented by a power law. This is generally true of all of the sources listed in Table I. Note that there is a wide range of variation of the power law index from -1.6 to about -2.6.

Many of the EGRET detected blazars show significant time variability of the gamma-ray flux on timescales ranging from days to months or longer. The high energy gamma-ray flux variations observed for PKS 0528+134, PKS 0208-512, and 3C 279 are shown in Figure III.

The continua spectra of almost all AGN span more than 12 decades in frequency and often exhibit significant temporal variability in many of these energy bands. Simultaneous multiwaveband spectra of blazars are important for testing and constraining specific blazar emission models. For example, the correlation (or lack of) of variability across these various bands from radio to high energy gamma-rays can provide important clues to the geometry of the source. Unfortunately, the typical observational situation is that there are usually large frequency gaps across which there is no data and the data in different frequency bands are often contemporaneous rather than simultaneous. This is especially true of gamma-ray observations to date. The situation may improve substantially during Phase III of the Compton Observatory mission with several multiwavelength observing campaigns being planned around EGRET observations of likely high energy blazar candidates.

Lacking extensive simultaneous multiwavelength data for the EGRET sources, it is very useful to construct multiwavelength spectra of these sources using archival data when available. Several examples are shown in Figure IV. A more complete set of spectra for the EGRET sources is in preparation (Michelson et al. 1993).

DISCUSSION AND CONCLUSIONS

Gamma-ray absorption via photon-photon pair production, the EGRET observations of high fluxes above 1 GeV, and short timescale variability (at least in some sources) appear to rule out emission models in which the gamma radiation is generated in the same volume as the observed X-ray radiation, unless the radiation is beamed towards us (Jelly 1966). This conclusion is also consistent with the interpretation of blazars as objects in which the non-thermal radiation arises predominantly from a "jet" of relativistic particles directed close to the observer's line of sight (Blandford & Rees 1978). However, the gamma-ray emission in most of the EGRET sources cannot come from radii as large as VLBI sources because of variability constraints. Thus the gamma emission is an important link between large radio structure and the central engine (Blandford 1993).

The requirement that the source be transparent to photon-photon collisions and the large apparent gamma-ray luminosities and rapid variability observed in many of the EGRET sources, provide constraints on emission models in which the gamma-rays arise from a relativistic jet, in particular a lower limit on the

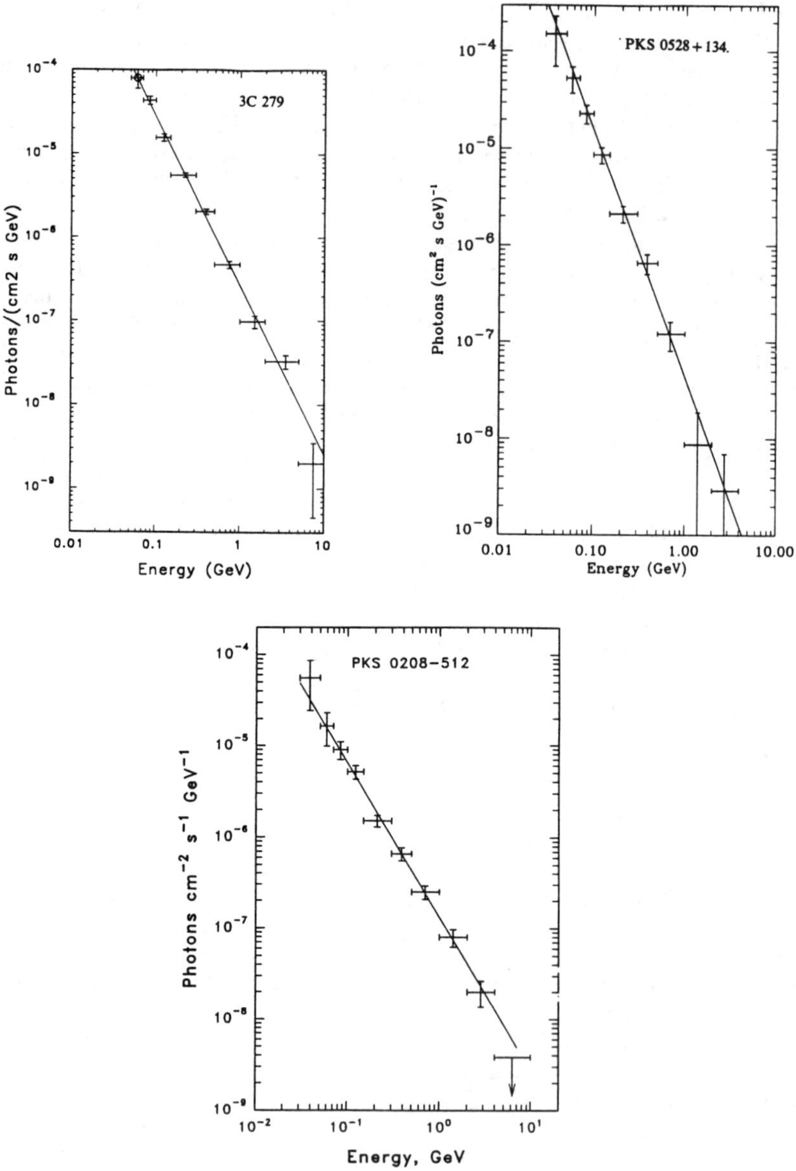

FIGURE II High-energy gamma-ray spectrum of (a) 3C 279, (b) PKS 0528+134, and (c) PKS 0208-512. The figures are adapted from Hartman et al. (1992), Hunter et al. (1993), and Bertsch et al. (1993), respectively.

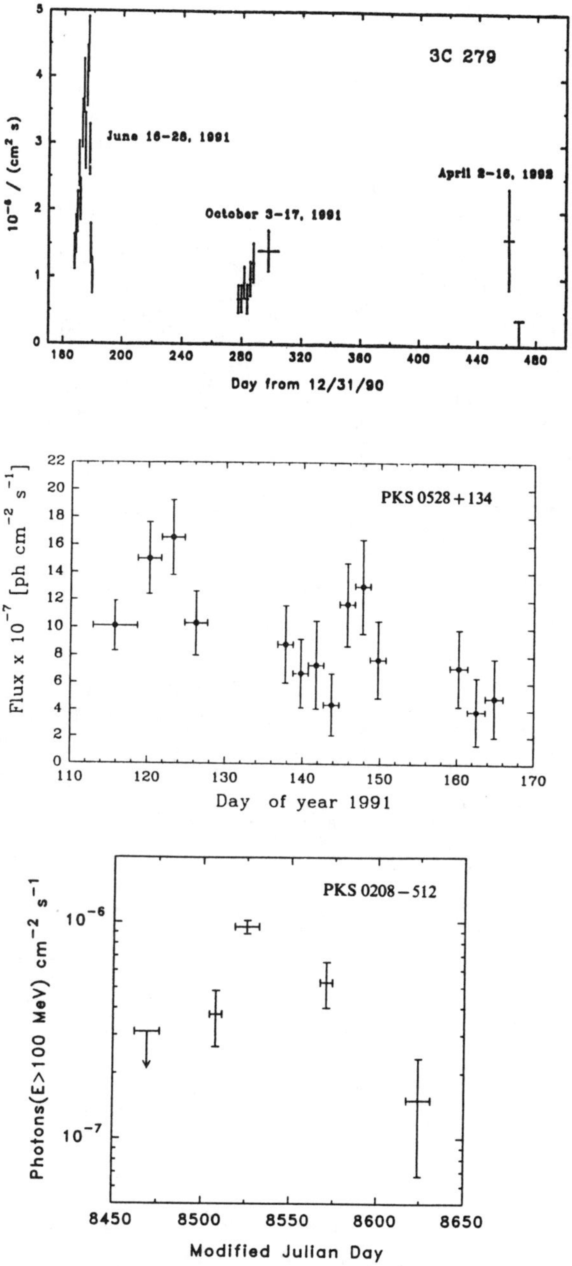

FIGURE III Observed temporal variations of the high-energy gamma-ray flux (above 100 MeV) from (a) 3C 279, (b) PKS 0528+134, and (c) PKS 0208-512 The figures are adapted from Kniffen et al. (1993), Hunter et al. (1993), and Bertsch et al. (1993), respectively.

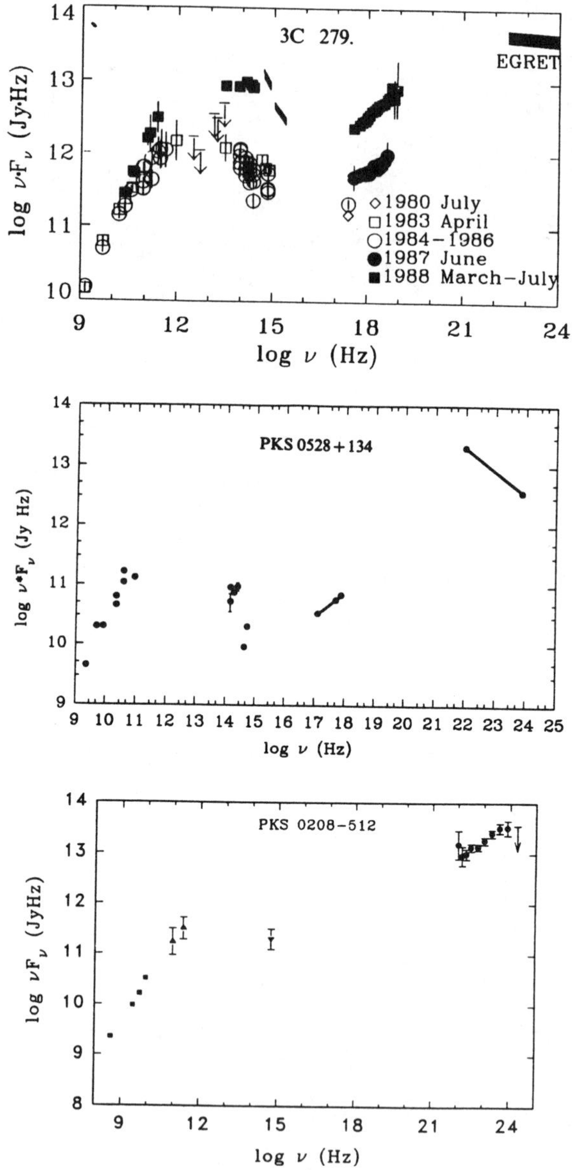

FIGURE IV Multifrequency spectra of (a) 3C 279, (b) PKS 0528+134, and (c) PKS 0208-512 The figures are adapted from Hartman et al. (1992a), Hunter et al. (1993), and Bertsch et al. (1993), respectively. See these papers for references to the observations in energy bands below the EGRET band.

relativistic Doppler factor (Mattox et al. 1993; Maraschi et al. 1992; Sikora, Begelman, & Rees 1993b). For example, Mattox et al. (1993) find that the minimum Doppler factors required, if bulk relativisic motion is responsible for beaming, are 7.6 and 4, respectively, for 3C 279 and PKS 1633+382. These limits are consistent with VLBI observations of superluminal motion in blazars.

The apparent integrated luminosity between 100 MeV and 10 GeV for 3C 279, observed during a flaring episode in 1991 June, was 2×10^{48} erg s^{-1} (Hartman et al. 1992a). With the minimum Doppler factor inferred above from optical depth considerations, this corresponds to an intrinsic integrated luminosity of less than 7.8×10^{45} erg s^{-1}. For 1633+382, the apparent luminosity is 9×10^{48} erg s^{-1} (Mattox et al. 1993) and the intrinsic luminosity is less than 2.7×10^{45} erg s^{-1}. Thus beaming effects in blazars can reduce the total gamma-ray luminosity problem by several orders of magnitude.

The now almost standard explanation of the nonthermal radio to infrared emission from the blazar class of AGNs involves a well-collimated jet of relativistic plasma and magnetic field flowing out of the nucleus at speeds near that of light (Blandford & Königl 1979). Synchrotron radiation is produced, beamed in the forward direction into a narrow cone cone. The opacity considerations outlined above suggest that the gamma-ray emission from blazars is also beamed.

Models of how the jets actually produce gamma-ray emission mostly involve inverse Compton scattering of lower energy photons up to gamma-ray energies by the beamed relativistic electrons. The low energy photons come from either a disk (Melia & Königl 1989; Dermer, Schlickeiser, & Mastichiadis 1992) or are self-generated by the electrons as synchrotron radiation ("synchrotron self-Compton" emission) (Jones, O'Dell, & Stein 1974; Marscher 1980; Königl 1981; Marscher & Gear 1985; Ghisellini & Maraschi 1989; Marscher & Bloom 1992; Maraschi et al. 1992). It has recently been proposed (Blandford 1993; Sikora, Begelman, & Rees 1993a, 1993b) that UV or IR thermal radiation from an accretion disk which has been scattered or reprocessed in a halo (eg., broad-line region clouds, an accretion disk wind, dust) may be more effectively Compton scattered than thermal radiation coming directly from the disk. This is essentially because in the frame of the gamma-ray producing plasma in the outflowing relativistic jet the unscattered disk radiation from the central source is strongly redshifted while diffuse reprocessed radiation is strongly blueshifted as the jet passes through the reprocessing region. Details of several of these models have been discussed by R. Blandford, M. Begelman, and A. Königl at this conference. The interested reader is referred to their papers for details.

REFERENCES

Bertsch, D.L., et al. 1991, IAU Circ. 5311

Bertsch, D.L., et al. 1993, *ApJ*, **405**, L21

Bignami, G.F., et al. 1981, *A&A*, **93**, 71

Blandford, R.D., 1993, in M. Friedlander and N. Gehrels (eds) *Proceedings of the Compton Symposium*, New York: AIP, in press

Blandford, R.D. and Königl, A. 1979, *ApJ*, **232**, 34

Blandford, R. and Rees, M.J., 1978, in A.N. Wolfe (ed) *Pittsburgh Conference on Bl Lac Objects, University of Pittsburgh Press*, 328

Dermer, C.D., and Schlickeiser, R. 1992, Science, **257**, 1642

Fichtel, C.E., et al. 1992a, IAU Circ. 5460

Hartman, R.C., et al. 1992a, *ApJ*, **385**, L1

Hartman, R.C., et al. 1992b, IAU Circ. 5477

Hartman, R.C., et al. 1992c, IAU Circ. 5519

Hughes, E.B., et al. 1980, IEEE Trans. Nucl. Sci., **NS-27**,364

Hunter, S.D., et al. 1993, *ApJ*, **409**, 134

Jelly, J.V., 1966, Nature, **211**, 472

Jones, T.W., O'Dell, S.L., and Stein, W.A. 1974, *ApJ*, **188**, 353

Kanbach. G., et al. 1988, Space Sci. Rev., **49**, 69

Kanbach, G., et al. 1992, IAU Circ. 5431

Kniffen, D.A., et al. 1993, *ApJ*, **391**, 659

Königl, A., 1981, *ApJ*, **243**, 700

Kühr, H., et al. 1981, *A&A*, **45**, 367

Lin, Y.C., et al. 1992, *ApJ*, **401**, L61

Lin, Y.C., et al. 1993, *ApJ*, submitted

Maraschi, L., Ghisellini, G., and Celotti, A. 1992, *ApJ*, **397**, L5

Marscher, A.P., 1980, *ApJ*, **235**, 386

Marscher, A.P., and Gear, W.K., 1985, *ApJ*, **298**, 114

Marscher, A.P., and Bloom, S.D., 1992, in *Proc. of the CGRO Science Workshop*, NASA Conf. Pub. **3137**, 346

Mattox, J.R., et al. 1993, *ApJ*, **410**, 609

Melia, F., and Königl, A., 1989, *ApJ*, **340**, 162

Michelson, P.F., et al. 1992, IAU Circ. 5470

Michelson, P.F., et al. 1993, in preparation

von Montigny, C., et al. 1993a, *A&A*Suppl., **411**, 133

von Montigny, C., et al. 1993b, in M. Friedlander and N. Gehrels (eds) *Proceedings of the Compton Symposium*, New York: AIP, in press

Nolan, P.L., et al. 1992, IEEE Trans. Nucl. Sci., **NS-39**, 993

Punch, M., et al. 1992, Nature, **358**, 477

Sikora, M., Begelman, M.C., and Rees, M.J., 1993a, in M. Friedlander and N. Gehrels (eds) *Proceedings of the Compton Symposium*, New York: AIP, in press

Sikora, M., Begelman, M.C., and Rees, M.J. 1993b, preprint

Swanenburg, B.N., et al. 1978, Nature, **275**, 298

Thompson, D.J., et al. 1993, *ApJ*, Suppl., **86**, 629

The First Stromlo Symposium: The Physics of Active Galaxies
ASP Conference Series, Vol. 54, 1994
G. V. Bicknell, M. A. Dopita, and P. J. Quinn (eds.)

HOLES, DISKS, STARS AND JETS IN ACTIVE GALACTIC NUCLEI

R. D. BLANDFORD
130-33 Caltech, Pasadena, CA 91125, USA

ABSTRACT A brief overview is presented of the four basic components – black holes, accretion disks, stars and jets – whose interplay is believed to account for the variety of galactic nuclear activity. It is proposed that relativistic torques caused by a misaligned central black hole may induce strong shocks in accreting gas which form the O-UV-X spectrum of AGN. It is proposed that radio-loud AGN are collimated by large scale hydromagnetic torques and that excessive stellar mass loss precludes this in a radio-quiet object. Recent EGRET observations of core-dominated, radio-loud AGN are interpreted in terms of a model in which an electromagnetic energy flux, powered by a spinning black hole is steadily transformed into a relativistic electron-positron jet which inverse Compton scatters ambient soft X-rays into γ-rays.

INTRODUCTION

A contemporary discussion of the *physics* of Active Galactic Nuclei (AGN) inevitably focuses upon their three special constituents, black holes, accretion disks and outflowing jets. To these must be added stars, which are surely present in abundance in galactic nuclei and whose influence it seems as foolish to ignore as it is to imagine that they are responsible for all activity. The evidence in favor of black holes remains circumstantial although it has improved steadily with time. Disks are sometimes observed on scales $\gtrsim$ 100 pc, though there is increasing doubt of their existence on much smaller scales. Phenomenologically, jets are in the strongest position as they are generally observed in radio-loud AGNs and may also be present in a minority of radio-quiet objects. Improvements in both resolution and dynamic range are enabling us to trace these jets closer to their putative sources. The relationship between these four components is a current focus of AGN research.

There have been several recent summaries of the basic properties of holes, disks, stars and jets together with critiques of the circumstantial evidence that has been adduced in their favor. These include articles to be found in Blandford, Netzer & Woltjer 1991, Duschl & Wagner 1991, Holt, Neff & Urry 1991, where extensive bibliographies can be found. There seems little purpose in recapitulating this material in a conference proceedings (although it did constitute the majority of the oral presentation). Instead, I shall concentrate on some more recent developments and introduce some highly speculative new ideas.

BLACK HOLES

A massive black hole is completely characterized by its mass (m in geometrical units), and its spin $\mathbf{S}$ (limited in magnitude to m^2 in geometrical units), or equivalently its angular velocity given by

$$\Omega_H = \frac{m - (m^2 - S^2/m^2)^{1/2}}{2S} \tag{1}$$

Gas that accretes onto the black hole will liberate binding energy with an efficiency that depends upon the mode of accretion and the spin of the hole but which is generally estimated to be ~ 10 percent. When the rate of mass accretion is large enough that the luminosity approaches the Eddington luminosity, $\sim 10^{46}(M/10^8 \mathrm{M}_\odot$ erg s^{-1}, the radiation pressure may inhibit the flow (assuming that it is fully ionised) (*e.g.* Rees 1984).

A hole possesses an *irreducible* mass m_0 which cannot be extracted by classical processes. It is related to the total mass m through the familiar-looking formula

$$m = \frac{m_0}{(1 - \beta^2)^{1/2}} \tag{2}$$

where $\beta = 2\Omega_H m_0/c < 2^{-1/2}$ is a dimensionless rotational speed. The spin energy that is extractable in principle is $m - m_0 < 0.29m$. Magnetic flux Φ that threads the black hole horizon and is supported by external currents may tap a large fraction (of order 50 percent) of this spin energy in the form of a large scale Poynting flux(*e.g.* Thorne *et al.* 1986). The extracted power can be approximated by $L_{EM} \sim \Omega_H^2 \Phi^2$. The associated current flowing through the hole is $I \sim 3 \times 10^{18} (L_{EM}/10^{46} \mathrm{ergs}^{-1})^{1/2}$ A and the total potential difference induced across the horizon of the hole satisfies $V \sim 3 \times 10^{20} (L_{EM}/10^{46} \mathrm{ergs}^{-1})^{1/2}$ V. For a rapidly spinning hole, the power released through the electromagnetic channel can match the rate of binding energy release by accreting gas prior to its plunge through the horizon.

The best dynamical evidence for the presence of large central masses in AGN comes from stellar kinematic studies of their supposed fossil remnants, local, high mass galaxies (*e.g.* Kormendy, Evans & Richstone 1993). Viral velocities $\sim 1 - 3 \times 10^{-3} c$ are measured and so we are necessarily only probing radii $\sim 10^5 - 10^6$ times the radius of a central black hole. It is to be hoped that a refurbished Hubble Space Telescope will have improved angular resolution and produce more certain mass determinations.

One longshot for improving the observational status of AGN black holes is that an X-ray spectroscopic black hole binary may be discovered in a nearby Seyfert galaxy. In order for this to occur, the relative velocity of the two holes V must be large enough to be detectable, while the orbital period P is short enough to be measurable and the lifetime to gravitational radiation energy loss t_{GR} is long enough for this to have a reasonable chance to observe this phenomenon in a small sample. It is straightforward to show that

$$V = 3760Q \left(\frac{10^7 P}{t_{GR}} \right)^{1/5} \mathrm{kms}^{-1} \tag{3}$$

where $Q = (m_1 + m_2)^{2/5}(m_1 m_2)^{-1/5}$ is a model-dependent mass factor that lies in the range $\sim 1.3 - 3$. Future X-ray spectroscopic missions may attain this spectral resolution.

DISKS

There are several good reasons for believing in accretion disks in AGN. The specific angular momentum of gas in the galaxy is much larger than that associated with the hole and so it ought to become rotationally supported in its plane of origin, indicative of the overall galaxy potential. In addition, we observe direct evidence for this flow in the form of dust lanes and disks (eg Quillen *et al.* 1993). Jets and ionization cones provide indirect evidence for a durable symmetry axis, presumably associated with the angular momentum of accreting gas. The infrared emission from radio-quiet AGN can be attractively interpreted as re-radiation of a large fraction of the UV continuum by a dusty molecular disk or torus with the longer wavelengths originating at larger radii (*e.g.* Sanders *et al.* 1989). This molecular gas, may be the raw material for the broad emission line clouds (*e.g.* Emmering , Blandford & Shlosman 1992).

It has commonly been assumed that the hole's spin axis is aligned with the orbital angular momentum of the accreting gas. The rationale for this view (*e.g.* Bardeen & Petterson 1972, Rees 1982) has been that Lense-Thirring torques will cause inclined rings of gas to precess with an angular frequency $\boldsymbol{\Omega}_{LT} = 2S/r^3$. When the precessional period becomes comparable with the inflow timescale, it is supposed that dissipation will cause the gas to settle into the equatorial plane of the black hole, although the details remain obscure (*e.g.* Kumar 1990). If a disk does form, it is further commonly assumed that it is optically thick and radiates the photoionizing UV continuum from its surface as well as excite a corona where a non-thermal pair plasma produces a Comptonized X-ray "reflection" spectrum (*e.g.* Begelman these proceedings).

There are, however, significant observational problems with the notion that these disks remain thin and coplanar at small radii (*e.g.* Kinney, these proceedings). The first is that the observed optical-UV spectrum is not what would be expected from a thin disk. Secondly, there is little or no evidence for Lyman edges or polarisation which would both be expected to be produced by a disk atmosphere. Finally the suggestion that there is "microlensing" in the gravitationally lensed quasar 2237+0305 appears to be inconsistent with a simple disk spectrum (Rauch & Blandford 1991). There are specific remedies that have been proposed for each of these difficulties, but the simplest solution of all is that the inner flow is not disk-like.

One possible alternative is that, sufficiently close to the hole, the non-Newtonian forces cause an inclined disk to thicken in an unsteady fashion. In addition to the Lense-Thirring precession, general relativity introduces two more corrections to the motion of a fluid element. The first is that an inclined orbit cannot be circular due to quadrupolar corrections to the metric $O(S^2/m^2r^2)$. The second is that spacetime curvature will cause a non-circular orbit to precess in its orbital plane with an angular frequency $O(m^{3/2}/r^{5/2})$. Inclined orbits will not close and we expect streams of gas to approach one another with relative speeds $\sim (m/r)^{5/2}$. When this speed exceeds the sound speed of the accreting gas, shocks will form and the streams can be deflected out of their orbital planes. Differential Lense-Thirring precession then guarantees that gas streams from different radii will collide and that a thickened, though highly inhomogeneous gas, distribution be created.

It is very hard to model this flow. To the extent that it can be idealized as a small number of discrete flows, we might be tempted to use the techniques of smooth particle dynamics (SPH) as a conceptual guide. Individual particles can be endowed with a cross section and allowed to collide with relative speeds $V \sim 10^4$ km s^{-1}. The ensuing shocks are likely to be radiative so that colliding "particles" will coalesce. However, many of the collisions will be glancing and tidal forces, applied post-impact will tear apart these "particles". Overall, we expect that the particles will fragment and this is easy to incorporate into a numerical simulation. A thickened distribution should be produced with a small filling factor for the shocked gas. There is a natural feedback. If the torus is too thin, then the collision mean free path for the gas streams will be quite short and the disk will quickly thicken. There will consequently be a natural covering factor $C \sim 0.1$.

These shocks may account for some of the gross radiative properties of AGN. Let us make a Lorentz transformation into the frame in which the shock is perpendicular and consider a one-dimensional flow of gas and radiation both ahead of and behind the shock front. The immediate post-shock ion temperature will be $\sim 10^9(V/10^4\mathrm{kms}^{-1})^2$ K. Although there may be some acceleration of relativistic electrons by the Fermi process, the post-shock gas should quickly cool to the Compton temperature, $T_C \sim 10^7$ K, determined by the local radiation spectrum. The post-shock gas will continue to cool through free-free emission at more or less constant pressure. The radiation that propagates back toward the shock front will be responsible for the Compton cooling; the oppositely-directed flux will be absorbed by cooler and denser gas and eventually re-radiated close to the effective black body temperature $\sim 3 \times 10^5$ K. The shocked gas will therefore re-emit a fraction of the the X-rays in the UV at least until tidal forces reduce its density or a second collision takes place. Both of these are likely to occur within a dynamical time $\sim r^{3/2}/m^{1/2}$, of order several days. The effective cooling times evaluate to $\sim (P_{gas}/10^7\mathrm{dyne\ cm}^{-2})^{-1}$ s, the associated column densities are $\sim 10^{24}$ cm^{-2}. Most of the radiant energy will be produced close to the shock front as Comptonized free-free emission in the hard X-ray band. However, roughly half of this will be re-processed in to the ultraviolet. A minority of the radiation will appear in the soft X-rays, and here we might expect there to be some spectral curvature produced by reflection effects. Rapid X-ray variability can be associated with individual shocks whose emission may be boosted by Doppler beaming (*cf.* the recent analysis of variability data by Lawrence & Papadakis 1993). These features are in qualitative accord with observations of integrated spectra of radio-quiet AGN. Clearly, a more careful calculation must be carried out to test the viability of this model.

STARS

Stars that pass close by a black hole may either collide or be torn apart by tidal stresses. The subsequent evolution has been the subject of much speculation.

The prospect for observing individual stars in AGN received a boost with the report of a $\sim$ 3 hour periodicity in the Seyfert galaxy NGC 6814. Unfortunately, it now appears that this variation is attributable to a nearby cataclysmic variable, not the AGN (Rosen *et al.* 1993). Despite this setback, the various attempts to explain an AGN period of three hours at least highlighted some of the possible signatures of stars near black holes that might be sought in other sources. Firstly the period itself is (in both an AGN and a CV) very roughly what might be expected from a low mass star that is being stripped tidally and therefore supplying accretable gas (*i.e.* $\sim \pi(G\bar{\rho})^{-1/2} \sim 3(\bar{\rho}/\rho_{\odot})^{-1/2}$hr). Secondly, although an equatorial disk might be expected to force a star to corotate in its plane (Syer *et al.* 1992), there is no guarantee that the gas flow will be this simple, as we have argued above, and a large quasi-periodic modulation of the emitted X-rays seems a distinct possibility. Sudden flaring in nearby, otherwise dormant, galactic nuclei associated with tidal capture of a passing star on a near parabolic trajectory might also be sought, particularly in supernova searches (Rees 1990).

A quite different (although rather improbable) way of detecting stars is if the black hole acts as a very strong gravitational lens. A sudden double-peaked increases in the flux might then be measured as a star passes through the caustic surfaces formed by the black hole. This is most likely to be possible at X-ray energies (Rauch & Blandford 1993).

JETS

The most striking, recent observational development is the discovery, made using the EGRET telescope on the Compton Gamma Ray Observatory, that bright, compact, flat-spectrum radio sources frequently are associated with unexpectedly large, hard ($\sim$ 100 MeV-10GeV) γ-ray fluxes (*e.g.* Michelson these proceedings). (TeV γ-rays have been reported from one extragalactic source, MKN 421, Punch *et al.* 1992.) To date, 26 such identifications have been made. (No such identifications have been made with radio-quiet quasars or Seyfert galaxies.) Many of the EGRET sources exhibit superluminal expansion.

The most powerful γ-ray fluxes reported correspond to powers $\sim 10^{49} f$ erg s^{-1}, where f is the beaming factor. As these fluxes are not observed from the supposed unbeamed counterparts of blazars, it is reasonable to suppose that the beaming fractions are $f \sim \gamma^{-2} \sim 10^{-2} - 10^{-3}$. Even so, the γ-ray powers of these objects are still as large, or sometimes larger than, the total powers from the rest of the electromagnetic spectrum. The best studied example is the quasar 3C279 (Hartmann *et al.* 1992), which, appropriately, was the first quasar to exhibit superluminal motion. The observed γ-ray flux ($\sim 1 - 6 \times 10^{48}$ erg s^{-1}), is as much as 50 times the optical flux and it seems to radiate roughly equal power in each octave of γ-ray energy and so this could be an underestimate. Furthermore it has shown variability on a timescale of $\sim$ 2 d presumably originating from a source significantly smaller than the smallest resolved radio feature.

The simplest interpretation of these γ-ray sources is that they involve inverse Compton scattering of ultraviolet and X-ray soft photons by relativistic electrons and perhaps positrons accelerated within the outflowing jet. That positrons are involved is made plausible by the observation that the ambient soft photon number density at the maximum radius for γ-ray emission, based on the variability, is so large that the γ-rays can barely avoid creating electron-positron pairs by combining with the soft photons. This absorption process motivates us to introduce the idea of a "γ-ray photosphere" which is the minimum radius from which a γ-ray of a given energy can escape. It is expected to *increase* with E_γ. (This contrasts with the expected properties of the radio jets where the radius of the radio photosphere is expected to *decrease* with increasing radio frequency.)

On this interpretation, the active constituents of these jets are pairs and γ-rays and their observed spectra are formed principally by balancing particle acceleration with inverse Compton scattering and pair production. As long as we concentrate upon these processes and ignore secondary effects like pair annihilation, synchrotron emission and thermal Comptonisation, it is simplest to work in the frame of the AGN and measure the pair and photon energies in this frame. In practice, we expect there to be a well-defined fluid rest frame in which the electrons and positrons are isotropically distributed as they gyrate about a tangled magnetostatic field and that this frame moves with Lorentz factor Γ through the AGN frame. In the AGN frame, individual particles will not conserve energy on the gyrational timescale and will spend a small fraction of their orbits moving radially inward. However, we can define an average energy distribution in the AGN frame and ignore those particles that, at a given instant, are not moving within an outwardly-directed cone with opening angle $\sim \Gamma^{-1}$. This can be shown to be a good approximation provided that $\Gamma >> 1$. In particular this formalism allows us to deal with variable Lorentz factors such as are to be expected to arise as a result of radiative deceleration and γ-ray absorption.

Let us introduce the symbol $I_X(E_X, \mu, r)$ to denote the dimensionless soft photon intensity. This is the energy flux per unit photon energy $E_X << 1$ and steradian at radius r propagating at an angle $\cos^{-1}\mu$ to the outward radial direction in units of $c/\sigma_T r$. (All energies are measured in units of $m_e c^2$.) Also, let $N_\gamma(E_\gamma, r)$ denote the number of hard γ-rays per unit energy $E_\gamma >> 1$ and per unit jet radius and let $N_e(E_e, r)$ be the corresponding quantity for the pairs. We can now write down a pair of coupled, linear integro-differential equations for the evolution of N_γ, N_e.

$$\frac{\partial N_\gamma}{\partial \ln r} = -\tilde{\kappa}_{PP} N_\gamma + \int_{E_\gamma} dE_e \eta_{C\gamma} N_e \tag{4}$$

$$\begin{aligned} \frac{\partial N_e}{\partial \ln r} &= -\tilde{\kappa}_C N_e + \int_{E_e} dE'_e \eta_{Ce} N_e(E'_e, r) \\ &\quad + \int_{E_e} dE_\gamma \eta_{Pe} N_\gamma + S_e(E_e, r) \end{aligned} \tag{5}$$

We have introduced the dimensionless pair production and inverse Compton opacities

$$\tilde{\kappa}_{PP}(E_\gamma, r) = 2\pi \int_{-1}^{1} d\mu (1-\mu) \int d\ln E_X I_X \left(\frac{\sigma_{PP}}{\sigma_T}\right) \tag{6}$$

$$\tilde{\kappa}_C(E_e, r) = 2\pi \int_{-1}^{1} d\mu (1-\mu) \int d\ln E_X I_X \left(\frac{\sigma_C}{\sigma_T}\right) \tag{7}$$

incorporating the pair production and Compton cross sections, σ_P, σ_C. In addition we must define three redistribution functions

$$\eta_{C\gamma}(E_e, E_\gamma, r) = 2\pi \int_{-1}^{1} d\mu(1-\mu) \int d\ln E_X I_X \frac{1}{\sigma_T}\frac{d\sigma_C}{dE_\gamma}(E_X, E_e; E_\gamma) \quad (8)$$

$$\eta_{Ce}(E'_e, E_e, r) = 2\pi \int_{-1}^{1} d\mu(1-\mu) \int d\ln E_X I_X \frac{1}{\sigma_T}\frac{d\sigma_C}{dE_e}(E_X, E'_e; E_e) \quad (9)$$

$$\eta_{Pe}(E_\gamma, E_e, r) = 2\pi \int_{-1}^{1} d\mu(1-\mu) \int d\ln E_X I_X \frac{1}{\sigma_T}\frac{d\sigma_P}{dE_e}(E_X, E_e; E_\gamma) \quad (10)$$

and a source function of freshly accelerated pairs, $S_e(E_e, r)$. If we specify the soft photon background, then we can solve for the γ-ray output at large r for a δ-function (in both energy and radius) injection of high energy pairs. The total output can be obtained by integrating over this Green's function response.

To proceed further, we need to make some assumptions about the origin of the soft photons. Three possibilities have been explored in the literature, the inner accretion disk near the base of the jet (*e.g.* Dermer & Schlickheiser 1992, the jet itself, *e.g.* Ghisellini *et al.* 1993 and nuclear photons Thomson-scattered into the jet by the accreting gas *e.g.* Blandford 1993, Sikora, Begelman & Rees 1993). There is a good chance that future γ-ray observations, combined with simultaneous ground-based monitoring, will enable us to distinguish between these three general classes of model.

An analysis of this model as well as some alternative approaches will be given in Blandford & Levinson (1993, in preparation). However, some general qualitative conclusions can be reported. Firstly, as the pair production and Compton scattering cross sections are so similar, there is a natural balance between these two processes. It can be shown that, under a quite broad range of conditions, the electrons and positrons automatically radiate much of their energy near the photospheres for γ-rays of similar energy. Ongoing particle acceleration is mandated and synchrotron emission is presumably responsible for the infrared-ultraviolet polarised emission from blazars. Secondly, the approximate energy to radius mapping for the photosphere is likely to be reflected in an increase of variability timescale with increasing γ-ray energy. Thirdly, at small radii, the electrons and positrons can cool to subrelativistic energy in an outflow timescale and will consequently annihilate. Thermal Comptonisation may also be important for the lowest energy γ-rays.

How are these jets powered and collimated? The γ-ray variability on timescales of $\lesssim 2$ day in 3C279 indicates that the outflows are already well collimated and moving relativistically by radii ~ 0.1 pc, significantly smaller than the smallest resolved radio structure and perhaps no more than a few hundred times the radius of an associated central black hole that is massive enough to supply all the energy associated with this source. (If "intraday variability", as reviewed by Melrose in these proceedings, is caused by yet faster outflows moving with Lorentz factors $\Gamma \sim 30 - 100$, then even more stringent conditions are imposed upon the sources of jets.) Although we can argue observationally that these γ-ray jets are observed close to their γ-ray photospheres where pairs are freely created, and simple theoretical models are consistent with the notion that most of the observed emission is produced at this location, jets are not likely to originate in a form mainly comprising electrons and positrons because the associated radiative drag would quickly decelerate them. There must then be another cache of momentum. Protons are a possibility, though if electron-proton jets are accelerated to bulk Lorentz factors $\Gamma \gtrsim 10$, at small radii, then they too will be subject to radiative deceleration (*e.g.* Phinney 1987). A more attractive possibility is that the momentum is carried, at least initially in an electromagnetic form and is extracted from the central spinning black hole.

It is unlikely that power extracted from a spinning hole can be self-collimated to the impressive degree required by radio observations and suggested by the EGRET detections. It is also not credible that external gas pressure can be large enough to effect this collimation as it would be subject to catastrophic radiative losses if it were dense enough. Again an alternative possibility is that there is an orbiting accretion disk and that this launches a bipolar, hydromagnetic wind. The natural speed of this wind is the orbital speed at its base and so we might expect this to diminish inversely with cylindrical radius. This wind is unlikely to be ultrarelativistic. However, it might collimate the relativistic ouflow derived from the black hole, particularly through the action of its toroidal magnetic field.

DISCUSSION

The most important division in the taxonomy of AGN is between radio-loud and radio-quiet objects and this is the first challenge to any general theory of AGN. We know that this characteristic must be stable for the lifetimes of extended radio sources, $\gtrsim 10^7$ yr. There is also general evidence that the radio quiet Seyferts have spiral hosts and that the radio galaxies are associated with ellipticals, (although the extent to which this is also true of quasars is still unclear). One possible explanation, consistent with the model outlined above, involves the torques that act on disks. It now appears to be more widely accepted that these are primarily magnetic, (*e.g.* Hawley, these proceedings). Perhaps in most disks, these torques are internal. However, suppose that in a minority of AGN, those that form jets, strong stresses are also applied externally and much of the angular momentum is extracted by the bipolar hydromagnetic wind. Suppose further that spiral galaxies contain a dense central cluster of young stars losing mass through stellar winds, supernovae and planetary nebulae and that ellipticals contain a lower density of an older population. Perhaps the hydromagnetic wind can only form in the latter case. We might guess that a necessary condition for the formation of collimating MHD flows is that they be able to pass through an intermediate critical point before they are appreciably loaded with gas. If r_i is the radius of the critical point and r_f the radius of the associated magnetic footpoint, then if the stellar mass loss rate exceeds a fraction $\sim (r_f/r_i)^2$ of the mass accretion rate through the disk, then the wind will not be able to form. This would prevent flux being concentrated on the black hole and any outflow from being collimated. The AGN would then be radio-quiet for the lifetime of the star cluster.

There is a secondary dichotomy involving radio-loud AGN and this is the separation into low power, apparently slow FR1 sources and high power, apparently fast FR2 sources (Owen, these proceedings). This too can be rationalised. As Laing and Bicknell (these proceedings) have argued, there is now good evidence that both types of radio source start with jets that are at least mildly relativistic. Perhaps what happens here is that the high power sources contain rapidly spinning central holes that create powerful relativistic cores which carry more linear momentum than the surrounding non-relativistic and collimating, hydromagnetic sheaths. So, when these two components interact, the sheaths end up being accelerated to the speed of the core. Just the opposite may occur in the low power FR1 sources and it is possible to incorporate this in an evolutionary scheme. Of course, any quantitative assessment of this explanation must take into account the momentum apparently radiated by relativistic jets in the form of γ-rays and this may have some observational implications.

ACKNOWLEDGMENTS

I thank the Mount Stromlo and Siding Springs Observatory for financial support to attend this conference. Support under NSF contract AST 92-23370 and NASA contracts NAGW 2816, 2372 is gratefully acknowledged.

REFERENCES

Blandford, R. D. 1993, in *Compton Gamma-Ray Observatory*, M. Friedlander, N. Gehrels & D. J. Macomb, New York: AIP Press, 533

Blandford, R. D., Netzer, H. & Woltjer, L. 1991, *Active Galactic Nuclei*, Berlin: Springer-Verlag

Dermer, C. & Schlikheiser, R. 1993, *ApJ*, in press

Duschl, W. & Wagner, S. (ed.) 1991, *Physics of Active Galactic Nuclei*, Berlin: Springer-Verlag

Emmering, R. T., Blandford, R. D. & Shlosman, I. 1992, *ApJ*, **385**, 460

Ghisellini, G. Padovani, P., Celotti, A. & Maraschi,L. 1993, *ApJ*, **407**, 65

Hartmann, R. C. *et al.* 1992, *ApJ*, **385**, L1

Holt, S., Neff, S. & Urry, C. (ed.) 1991, *Testing the AGN Paradigm*, New York: American Institute of Physics

Kormendy, J., Evans, A. S. & Richstone, D. O. 1993, in preparation

Kumar, S. 1990, *ApJ*, **245**, 670

Lawrence, A. & Papadakis, I. 1993, *ApJ*, **414**, L85

Phinney, E. S. 1987, in *Superluminal Radio Sources*, J. A. Zensus & T. J. Pearson, Cambridge: Cambridge University Press, 301

Punch, C. W. *et al.* 1992, *Nature*, **358**, 477

Quillen, A., Graham, J. R. & Frogel, J. A. *ApJ*, in press

Rauch, K. P. & Blandford, R. D. 1991, *ApJ*, **381**, L39

Rauch, K. P. & Blandford, R. D. 1993, *MNRAS*, in press

Rees, M. J. 1984, *Ann Rev. Ast. Ap.*, **22**, 471

Rees, M. J. 1990, *Science*, **247**, 817

Rosen, S., Done, C., Watson, M. & Madejski, G. 1993, *IAU Circ. No. 5850*

Sikora, M., Begelman, M. C. & Rees, M. J. 1993, *ApJ*, in press

Syer, D., Clarke, C. J. & Rees, M. J. 1991, *MNRAS*, **250**, 205

Thorne, K. S., Price, R. M. & MacDonald, D. 1986, *Black Holes: The Membrane Paradigm*, New Haven: Yale University Press

The First Stromlo Symposium: The Physics of Active Galaxies
ASP Conference Series, Vol. 54, 1994
G. V. Bicknell, M. A. Dopita, and P. J. Quinn (eds.)

PAIR CASCADES AND HIGH-ENERGY EMISSION IN ULTRARELATIVISTIC OUTFLOWS FROM ACTIVE GALACTIC NUCLEI

ARIEH KÖNIGL
Department of Astronomy & Astrophysics, University of Chicago, 5640 South Ellis Avenue, Chicago, IL 60637, U.S.A.

ABSTRACT The high-energy γ-rays detected in blazars most likely represent soft photons that are Compton-upscattered by a directed relativistic outflow. The general properties of the interaction between a relativistic particle beam and an ambient (possibly inhomogeneous and anisotropic) radiation field have been studied with the help of a state-of-the-art kinetic code. Specific scenarios for the origin of the γ-ray emission are considered in the context of the radiative deceleration of an ultrarelativistic proton beam in the vicinity of a nuclear accretion disk.

INTRODUCTION

As reported in this Symposium by P. Michelson, there are now over 30 AGNs that have been detected at $\gtrsim 100$ MeV energies by the CGRO/EGRET experiment. All of these objects are associated with compact radio sources and are classified as blazars, and many show evidence for apparent superluminal motions. In the energy regime ~ 10 MeV – 10 GeV the measured spectra are well approximated by power laws with a mean spectral index $\alpha \approx 1$ ($F_\nu \propto \nu^{-\alpha}$), although values in the range $\sim 0.8 - 1.6$ have been observed. There are also indications of strong variability: for example, in 3C 279, the flux (which translates into an isotropic luminosity $L \gtrsim 10^{48}$ ergs s^{-1}) was observed to vary by a factor ~ 4 over a period $\Delta t \lesssim 2$ days (Kniffen et al. 1993). The high luminosity and short variability time scale imply that the high-energy γ-rays would be degraded by collisions with low-energy (X-ray) photons in the vicinity of the central black hole unless either (1) L is overestimated and Δt is underestimated because of relativistic Doppler boosting, or (2) both the γ-rays and the X-rays are beamed, or (3) the γ-rays are produced sufficiently far from the center, implying an efficient mechanism of energy transport away from the nucleus. All of these alternatives point to a relativistic jet as the origin of the high-energy emission, a possibility that is clearly consistent with the blazar nature of these sources.

The most likely explanation of the observed high-energy spectrum is that it arises in Compton upscattering of a soft radiation field by relativistic electrons. A quasi-isotropic (fan-shaped) ultrarelativistic electron beam injected with Lorentz factor $\gamma_{e,i} < \gamma_{\rm KN} \equiv m_e c^2/2.7kT = 2 \times 10^4 T_5^{-1}$ into a homogeneous and isotropic radiation field that is characterized by a blackbody temperature $T = 10^5 T_5$ K and a large absorption optical depth ($\tau_{\gamma\gamma}$) will be decelerated by

Thomson scatterings and will give rise to an observed power-law spectrum with spectral index $\alpha_T = 0.5$ that will break at $x_{max} \approx \gamma_{e,i}^2/\gamma_{KN}$, where $x \equiv h\nu/m_e c^2$ (Zdziarski 1988). If the distribution of the initial Lorentz factors is of the form $dN(\gamma_{e,i})/d\gamma_{e,i} \propto \gamma_{e,i}^{-\beta}$, then the composite spectrum will be a power-law with index $\alpha = \beta/2$. Now, an attractive way of producing the requisite high electron injection energies is Fermi acceleration in shocks, and, as this mechanism typically yields $\beta \gtrsim 2$ (which translates into $\alpha \gtrsim 1$, in accord with the observations), this idea has been adopted in some of the models for the origin of the high-energy γ-rays. The proposed models appeal to the bulk relativistic motion of the jet to account for the inferred Doppler beaming but they differ in their kinematic details as well as in whether the soft radiation field is attributed to the nonthermal emission in the jet itself or to an ambient (thermal or nonthermal) source of photons (such as the continuum emission from the central source, impinging directly on the jet or being scattered by ambient material). M. Begelman and R. Blandford have discussed specific proposals along these lines in this Symposium (see, e.g., Henri & Pelletier 1991, Maraschi et al. 1992, and Dermer & Schlickeiser 1993 for additional examples).

In this contribution I describe recent work (Coppi & Königl 1993) that attempts to elucidate the dynamical and spectral details of the interaction between ultrarelativistic particles and an ambient radiation field. To be specific, it is assumed that the particles are injected in the close vicinity of the central black hole and interact with the thermal radiation field of a nuclear accretion disk (cf. Melia & Königl 1989). This setting allows one to explore additional manifestations of the Compton-upscattering scenario for the high-energy emission in blazars as well as such issues as the jet's composition and the distribution of terminal Lorentz factors in the flow.

RADIATIVE DECELERATION AND PARTICLE SHOWERS

If the Lorentz factor of the injected electron exceeds γ_{KN}, an e^+e^- pair–photon cascade will be initiated. For $\gamma_e \lesssim 250\gamma_{KN}$ the cascade will be governed by the $\gamma\gamma$ pair-production and inverse-Compton scattering processes (e.g., Burns & Lovelace 1982), but for higher Lorentz factors triplet pair production ($e^\pm + \gamma \rightarrow e^\pm + e^- + e^+$) will dominate scattering and must also be taken into account (e.g., Mastichiadis 1991). If $\tau_{\gamma\gamma} \gg 1$, $\sim \gamma_{e,i}/\gamma_{KN}$ electrons and positrons will be produced: by the time these pairs are decelerated by Thomson scattering, most of the injected energy will reside in the upscattered radiation, with only a fraction $\sim \gamma_{KN}^{-1} \ll 1$ converted into pair rest mass.

Because of their larger mass, protons are more readily accelerated to high Lorentz factors by the Fermi process and are less strongly decelerated by the interaction with the ambient radiation field. Consequently, they can attain Lorentz factors as high as $\gamma_{p,i} \approx 10^8$ (e.g., Sikora et al. 1987) and, through their interaction with the ambient radiation field, produce electrons with $\gamma_{e,i} \lesssim 3 \times 10^{10}$. For $\gamma_p > \gamma_\pi \approx 142\gamma_{KN}$ this interaction is dominated by photomeson production, for $\gamma_{KN} < \gamma_p < \gamma_\pi$ inverse-Compton scattering and pair production are the main processes, and for $\gamma_p < \gamma_{KN}$ Thomson scattering is the relevant proton-deceleration mechanism (e.g., Begelman et al. 1990).

Although the radiative deceleration of ultrarelativistic protons is an efficient

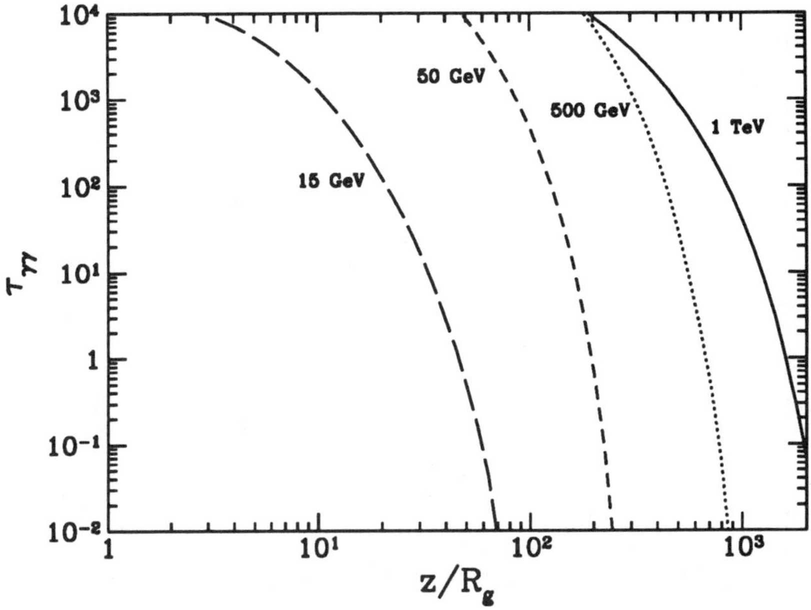

FIGURE I Optical depth for photon-photon pair production off disk photons as a function of height above the disk. The disk is characterized by inner and outer radii $R_{\rm in} = 10R_g$ and $R_{\rm out} = 200R_g$, a maximum temperature $T_{\rm max} = 2 \times 10^5$ K, and a sum-of-blackbodies spectrum. Here $R_g = 1.5 \times 10^{13}$ cm, corresponding to a black hole mass of $10^8\,M_\odot$. The different curves are labeled by the incident photon energy.

means of generating copious e^+e^- pairs, it is unlikely that the composition of the relativistic jet component that emerges from the interaction region is dominated by pairs produced in this way. This follows from the evidence (e.g., Celotti & Fabian 1993) that the radiated power $L_{\rm rad}$ of relativistic jet sources does not usually exceed the kinetic power $L_{\rm kin}$ of the associated jets. If this constraint is combined with the condition that the number of generated positrons greatly exceed the number of protons moving with the bulk Lorentz factor $\gamma_{\rm j}$ of the large-scale jet, then it follows that $\gamma_{\rm j}$ should be $\gg \gamma_{\rm KN} m_e/m_p \approx 5(\gamma_{\rm KN}/10^4)$, whereas the observations imply $\gamma_{\rm j} \lesssim 10$. This reinforces the arguments (Celotti & Fabian 1993) for parsec-scale radio jets being composed primarily of protons and electrons. Consider a beam of ultrarelativistic protons of energy $\gamma_{p,i} m_p c^2$ that is injected along the axis of a nuclear accretion disk at a height z_i above the central black hole. Owing to the decrease in the number density and mean energy of the disk-produced soft photons with distance z along the axis and to the anisotropy of the radiation field, the optical depths for the different radiatiative interaction processes vary strongly with z (see Fig. I). Consequently, the

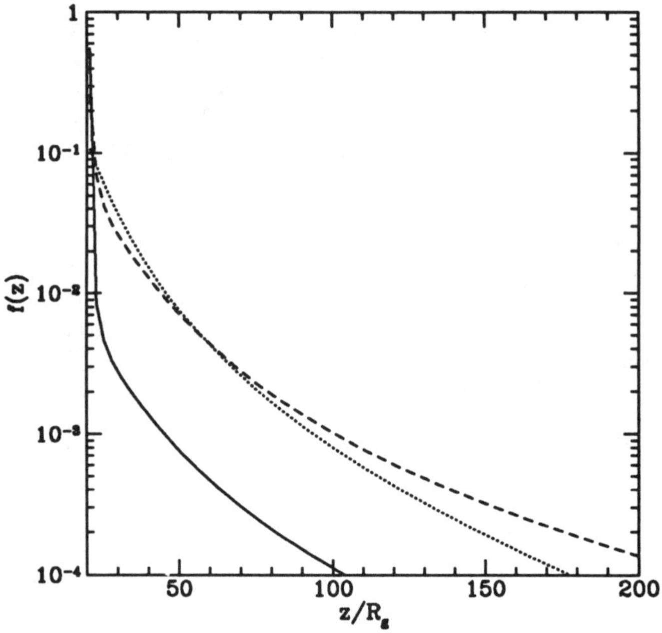

FIGURE II Evolution of unneutralized proton beams that are injected with $\gamma_{p,i} = 10^8$, 10^7, and 10^6 (*solid*, *dashed*, and *dotted* curves, respectively) at a height $z_i = 20R_g$ above a nuclear accretion disk with the same parameters as in Fig. I (except that $R_{\rm out} = 2 \times 10^5 R_g$). Shown is the fraction f of the protons' initial energy that is lost to photons and pairs by proton–photon meson- and pair-production processes as a function of height above the disk.

efficiency of radiative deceleration and the magnitude of the terminal Lorentz factor $\gamma_{p,f}$ depend sensitively on the values of z_i and $\gamma_{p,i}$. Figure II illustrates this dependence for an *unneutralized* proton beam. The results shown in this figure (as well as the spectral results discussed below) were obtained using a first-order, implicit kinetic code that incorporates all the relevant interactions and that follows the spatial evolution of both the particles and the photons in the given anisotropic radiation field. In the case of a neutralized beam (where the Compton-scattering cross section is that of an electron) the deceleration below γ_π is considerably more efficient than in the unneutralized case, and, for the disk and black-hole parameters of this example, $\gamma_{p,f} \approx 10^2 - 10^3$ for $z_i \approx 10 - 30R_g$, independent of $\gamma_{p,i}$ (cf. Melia & Königl 1989).

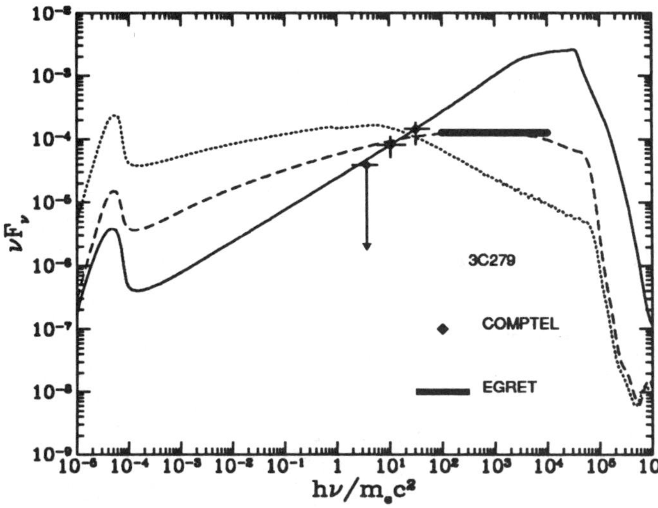

FIGURE III Spectra produced by the radiative deceleration of an unneutralized proton beam characterized by $\gamma_i = 10^7$, $z_i = 60R_g$, and the disk/black-hole parameters of Fig. II. The *solid* curve shows the original spectrum, whereas the *short-dashed* and *long-dashed* curves show the modifications induced by a further interaction with X-ray photons of spectral index $\alpha_X = 0.5$ and 1, respectively. The X-rays were assumed to originate in a corona modeled as a flat, uniformly emitting disk of radius $80R_g$. The EGRET and COMPTEL data are from Hartman et al. (1992) and Hermsen et al. (1993), respectively. The normalizations of the curves are arbitrary and were chosen so as to make the curves go through the COMPTEL data points.

SPECTRAL CHARACTERISTICS

The emergent high-energy spectrum for a radiatively decelerated, unneutralized proton beam is shown by the solid curve in figure III. As expected, the superposition of the contributions from all the cascade orders in the quasi-isotropic beam produces a spectrum characterized by $\alpha \approx 0.5$ over most of its range. The spectrum cuts off at $\gtrsim 10$ GeV, a value that exceeds the simple estimate $x_{\rm max} \approx \gamma_{\rm KN}(T_{\rm max})$ obtained without taking into account the dilution and anisotropy of the radiation field. Since the energy deposition of the injected protons occurs, in this case, over a distance Δz that is $\gtrsim z_i$ (cf. Fig. II), the upscattered spectrum could be further modified by absorption on nuclear X-ray photons if the latter permeate the interaction region. In particular, if the X-ray spectral index is α_X, then the emergent spectrum will steepen to $\alpha = \alpha_T + \alpha_X$, reflecting the frequency dependence of the optical depth ($\tau_{\gamma\gamma}[x] \propto x^{\alpha_X}$). This scenario is consistent with the observations of 3C 279, where $\alpha_X \approx 0.5$ (Makino et al. 1989) and $\alpha \approx 1$ in the EGRET energy regime. In this picture, the observed range of α's may be at least partially due to different underlying X-ray

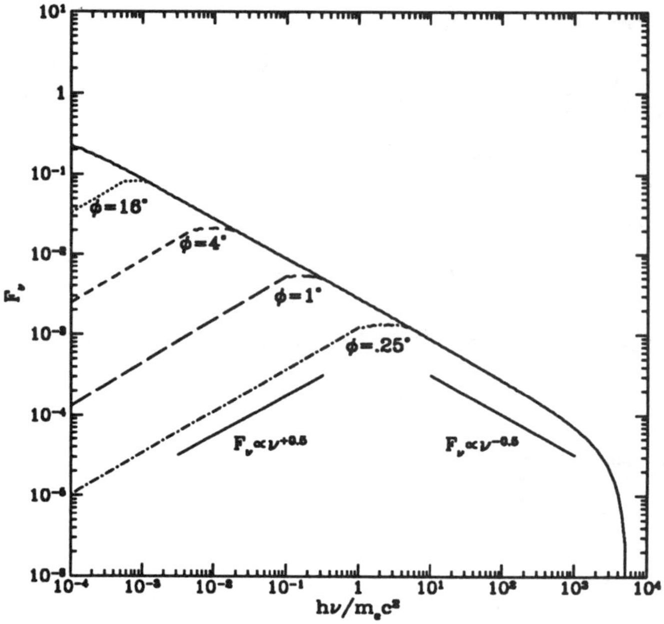

FIGURE IV Effect of decreasing the opening half-angle ϕ of the jet, illustrated for an electron beam of $\gamma_{e,i} = 10^4$ that is decelerated in a homogeneous and isotropic blackbody radiation field of temperature $T = 2 \times 10^5$ K. The observer is assumed to look down the jet axis.

spectra (see *dashed* curves in Fig. III). The apparent hardening of the 3C 279 spectrum below $\sim$ 10 MeV could be attributed, in this scheme, to the beam having a relatively small half-opening angle ϕ. In that case the upscattered radiation could no longer be approximated as originating in an isotropic particle distribution, and the spectral index α_{T} would change from 0.5 to -0.5 below a break frequency that increases with decreasing ϕ (see figure IV).

Even if the injected protons are rapidly neutralized and decelerated, so that $\Delta z \ll z_i$ and the radiative interaction region can no longer be approximated as a slab of uniform emissivity, it may be possible to account for the observed spectra with this scheme by considering a (neutralized) proton distribution of the form $dN(\gamma_{p,i})/d\gamma_{p,i} \propto \gamma_{p,i}^{-\beta}$, with $\gamma_{p,\mathrm{min}} < \gamma_{p,i} < \gamma_{p,\mathrm{max}}$. For $\beta > 2$ one can, for simplicity, equate $\gamma_{p,\mathrm{min}}$ with the bulk Lorentz factor γ_{j} of the asymptotic flow, as this distribution will satisfy the constraint $L_{\mathrm{rad}} \lesssim L_{\mathrm{kin}}$. Such a distribution could in principle arise in the course of the initial acceleration of the jet (e.g., through field reconnection processes; cf. Romanova & Lovelace 1992), or else it can form through Fermi acceleration in oblique shocks that are produced, e.g., after an electromagnetically accelerated jet converges toward the axis not far from the origin (cf. Pelletier & Pudritz 1992; Li et al. 1992). As discussed in the Introduction, such a distribution would give rise to a γ-ray spectrum with $\alpha = \alpha_T + (\beta - 1)/2$ and an upper cutoff x_{max} at the effective value of γ_{KN} (assuming $\gamma_{p,max} > \gamma_{KN}$). The lower cutoff (actually, a break $\Delta\alpha \approx -0.5$) would be determined in this case by the terminal Lorentz factor of the radiatively decelerated protons and would occur at $x_{min} \approx \gamma_{p,f}^2/\gamma_{KN}$. The CGRO/COMPTEL measurements of 3C 279 indicate a break of this type below $\sim$ 10 MeV (see Fig. III), which is consistent with the value $\gamma_{p,f} \gtrsim 10^2$ (corresponding to $\gamma_{KN} \gtrsim 10^4$) that we obtain in our model. It is interesting to mention in this connection that bulk Lorentz factors $\gtrsim 10^2$ may be required to explain the intraday variability measured in certain blazars (e.g., Quirrenbach et al. 1991).

This scheme cannot account for the apparent extension of the γ-ray spectrum to $\sim$ 1 TeV indicated in Mkn 421 (Punch et al. 1992): for typical disk parameters, Fig. I shows that such emission must be produced on scales $\gtrsim 10^{16}$ cm if it is to avoid being absorbed by disk-generated photons. One conceivable way of preventing ultrarelativistic protons that are accelerated near the nucleus from being rapidly decelerated by the ambient radiation field (so that they can produce γ-rays further out) is to shield them from the effect of the external photons by means of an e^+e^- sheath of transverse Thomson optical depth $\tau_\perp \gtrsim 1$. The sheath could in principle form by a pair–photon cascade at the periphery of the beam and would have an annihilation length $\sim (32\sigma_T/3\pi m_e^2 c^3)(\gamma_{p,f}^2 L_{beam}/\gamma_{KN}\tau_\perp^2) \approx 10^{16}\tau_\perp^{-2}(L_{beam}/10^{46}\,\mathrm{ergs\,s^{-1}})(\gamma_{p,f}/10)^2$ $(\gamma_{KN}/10^4)^{-1}$ cm (where σ_T is the Thomson cross section and L_{beam} is the kinetic power of the beam). Since a nonnegligible $\tau_\perp$ may have a feedback effect on the pair creation processes, one would need to adapt the kinetic code to the treatment of *nonlinear* cascades in order to explore this possibility. Even if relativistic particles can be transported to a distance of $\gtrsim 10^{16}$ cm, further shielding may be needed to forestall absorption of TeV γ-rays in the radiation field of the broad-line region (see the articles by Begelman and by Blandford in these Proceedings). Very-high-energy γ-rays that escape the AGN would interact with the microwave and infrared/optical background photon fields and could give rise to a circumgalactic ($\gtrsim$ 1 Mpc), γ-ray–luminous pair halo (Aharonian et al. 1993).

ACKNOWLEDGMENTS

I am grateful to Paolo Coppi, my collaborator on this project, for his invaluable input and for his assistance in producing this contribution. I also thank John Kartje for his help and the organizers of this Symposium for their encouragement and generosity. This work was supported in part by NASA grant NAGW-1636.

REFERENCES

Aharonian, F. A., Coppi, P. S., & Völk, H. J. 1993, in preparation
Begelman, M. C., Rudak, B., & Sikora, M. 1990, *ApJ*, **362**, 38
Burns, M. L., & Lovelace, R. V. E. 1982, *ApJ*, **262**, 87
Celotti, A., & Fabian, A. C. 1993, *MNRAS*, **264**, 228
Coppi, P. S., & Königl, A. 1993, in preparation
Dermer, C. D., & Schlickeiser, R. 1993, *ApJ*, **416**, 458
Hartman, R. C., et al. 1991, *ApJ*, **385**, L1
Henri, G., & Pelletier, G. 1991, *ApJ*, **383**, L7
Hermsen, W., et al. 1993, *A&AS*, **97**, 97
Kniffen, D. A., et al. 1993, *ApJ*, **411**, 133
Li, Z., Chiueh, T., & Begelman, M. C. 1992, *ApJ*, **394**, 459
Makino, F., et al. 1989, *ApJ*, **347**, L9
Maraschi, L., Ghisellini, G., & Celotti, A. 1992, *ApJ*, **397**, L5
Mastichiadis, A. 1991, *MNRAS*, **253**, 235
Melia, F., & Königl, A. 1989, *ApJ*, **340**, 162
Pelletier, G., & Pudritz, R. E. *ApJ*, **394**, 117
Punch, M., et al. 1992, *Nature*, **358**, 477
Romanova, M. M., & Lovelace, R. V. E. 1992, *A&A*, **262**, 26
Quirrenbach, A., et al. 1991, *A&A*, **372**, L71
Sikora, M., Kirk, J. G., Begelman, M. C., & Schneider, P. 1987, *ApJ*, **320**, L81
Zdziarski, A. A. 1988, *ApJ*, **335**, 786

The First Stromlo Symposium: The Physics of Active Galaxies
ASP Conference Series, Vol. 54, 1994
G. V. Bicknell, M. A. Dopita, and P. J. Quinn (eds.)

AGN IN X- AND γ-RAYS: A VIEW OF THE HEART

BELINDA J. WILKES
Smithsonian Institution, Astrophysical Observatory, 60 Garden St., Cambridge, MA 02138

ABSTRACT Active Galaxies are strong X-ray sources, emitting a comparable amount of energy per unit frequency throughout the infrared – soft X-ray regimes. Successive generations of X-ray satellites have revealed that their X-ray spectra are complex, containing a number of components originating in distinct sites within the quasar and with different mechanisms. In this paper, the evidence for, characteristics of and current interpretation of these various components is reviewed. The paper closes with a brief look at the γ-ray emission from these objects.

INTRODUCTION

The first X-ray satellites demonstrated the importance of X-ray emission for the study of active galactic nuclei (AGN, Tananbaum *et al.* 1978, Elvis *et al.* 1978). They are strong X-ray sources, with X-rays contributing a significant fraction of their emitted luminosity. X-ray surveys are the most efficient way to discover AGN given their higher surface density in this waveband. The combination of the strong X-ray flux and rapid variability in the X-rays is currently our best evidence for a central compact object.

The spectrum of the X-ray emission is complex, including several emission and absorption components originating in distinct sites in the active galaxy. These components include: a radio-linked emission component which dominates the X-ray spectrum of radio-loud objects and is probably beamed; a soft excess component at energies $\lesssim$0.6 keV, present in both radio-loud and radio-quiet objects; a reflection component at energies $\gtrsim$10 keV; absorption due to ionised (warm) gas; and absorption due to neutral (cold) gas. These various components will be reviewed in an order which is roughly chronological in terms of their discovery.

THE ROLE OF RADIO EMISSION

Einstein observations in the soft (0.1–3.5 keV) band showed that radio-loud quasars are brighter than radio-quiet by a factor of $\sim$ 3 (Zamorani *et al.* 1981). This was explained by an additional, synchrotron self-Compton emission component in the X-rays linked to the radio emission in radio-loud objects. Spectral studies of a smaller number of bright quasars with the *Einstein* Imaging Proportional Counter (IPC) confirm this picture as the X-ray slopes for radio-loud objects are significantly flatter ($\alpha_E \sim 0.5$; $F_\nu \propto \nu^{-\alpha_E}$) than those for radio-quiet

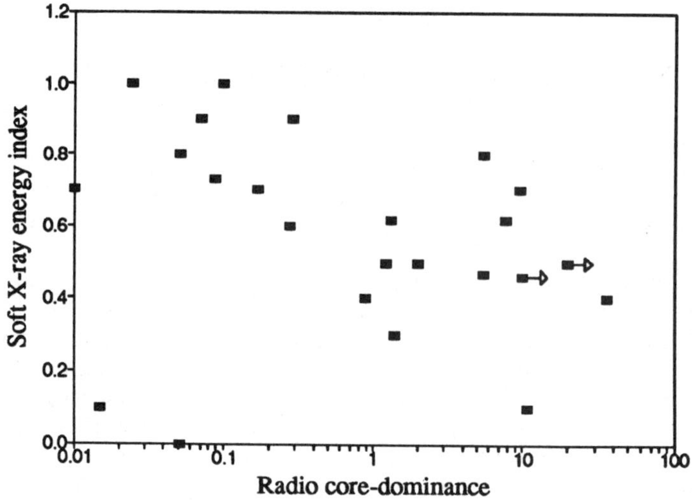

FIGURE I The correlation of radio core dominance (log R) with soft X-ray slope, from Shastri *et al.* 1993

($\alpha_E \sim$ 1.0, Wilkes and Elvis, 1987, Shastri *et al.* 1993).

In addition, the X-ray slope of core and lobe-dominated radio-loud quasars is different (Boroson, 1989, Canizares and White, 1989) with slope increasing as core dominance (R) decreases (Fig I). This implies beaming of the radio-linked, $\alpha_E \sim$ 0.5, X-ray component which then dominates the spectrum of the core-dominated objects. In the lobe-dominated objects the X-ray emission would then be the same as that in the radio-quiet objects although their slightly flatter slopes imply some contribution from the beamed component.

This trend of slopes with core dominance is also confirmed in the softer energy band of the ROSAT PSPC (0.1–2 keV, Brunner *et al.* 1992 and see below).

THE SOFT EXCESS

Both radio-loud and radio-quiet quasars and Seyfert 1 galaxies are frequently observed to have excess low energy emission ($\lesssim$1 keV) above the extrapolation of the higher energy power law (Turner and Pounds, 1989, Masnou *et al.* 1992). It is often variable but persists over periods of at least several years (Elvis, Wilkes and McDowell, 1991). A popular interpretation is as the high energy tail of the optical/ultraviolet blue bump component, since the latter must turn over in the extreme UV in order to meet the X-ray emission (Fig II). Models of the blue bump, such as accretion disk or free-free emission, can reproduce the soft X-ray emission (Fig II), though generally some reprocessing is required to avoid exceptionally high temperatures in the emitting regions (Czerny and Elvis, 1987, Sun and Malkan 1989, Laor, 1990, Barvainis, 1993).

As expected, the soft excess is more dominant in ROSAT's softer band (0.1-

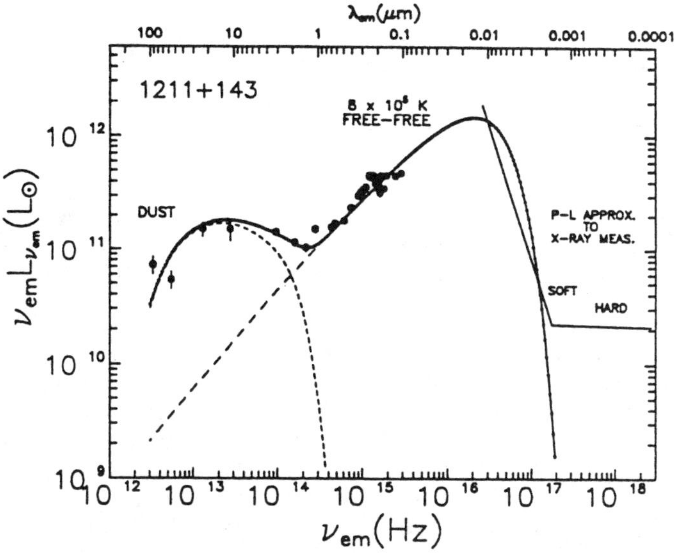

FIGURE II The infrared(IR)–X-ray spectral energy distribution of PG1211+143 compared with a simple free-free plus warm dust continuum model, taken from Barvainis, 1993

2 keV). Initial results from the all sky survey confirm that slopes are steeper by $\Delta\alpha \sim 0.5$ in all types of quasar and Sy1 but with the systematic difference between radio-loud and radio-quiet discussed earlier still present (Brunner *et al.* 1992). This indicates that although the soft excess is more dominant, the higher energy components still contribute significantly.

The soft excess is variable on short timescales ($\sim$ hours) in the few cases where data are available (Turner and Pounds, 1988, Elvis *et al.* 1991), suggesting its origin is a small region close to the core of the active galaxy, such as the inner edge of an accretion disk.

RADIO-QUIET OBJECTS

Early spectral studies of Seyfert 1 galaxies indicated a surprisingly constant slope $\alpha_E = 0.65 \pm 0.15$ (2–10 keV, Mushotzky, 1984, Rothschild *et al.* 1983). EXOSAT results over a similar energy band, confirm this mean spectral index but indicate that a range is present (Turner and Pounds, 1989). However, in the softer band of *Einstein*, slopes for radio-quiet quasars ($\alpha_E \sim 0.9$ Wilkes and Elvis, 1987) and Seyfert 1 galaxies ($\alpha_E \sim 0.8$, Kruper, Urry and Canizares 1990) appear steeper. This result holds both on average in the different samples and for individual Sy1 galaxies which are common to both *Einstein* and EXOSAT samples (Kruper, Urry and Canizares, 1990). From this comparison alone it is not clear whether the difference is due to the luminosity of the objects or the energy range over which they are observed. A correlation between X-ray slope and luminosity (Kruper, Urry and Canizares, 1990, Turner and Pounds, 1989) and a lack of correlation between slope and redshift for the quasars (Canizares

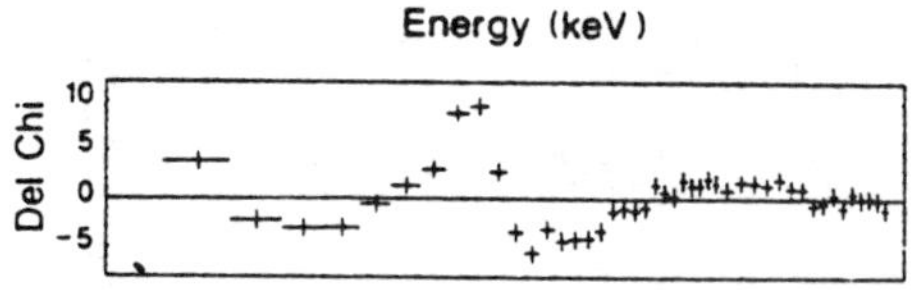

FIGURE III The residuals to a simple power law fit to the mean of 12 Sy1 *Ginga* spectra showing systematic deviations including an emission line at 6.4 keV and a flattening at energies $\gtrsim$10 keV, from Pounds *et al.* (1990)

and White, 1988) imply that both effects must contribute.

This is confirmed and so the dichotomy resolved by *Ginga* observations of Seyfert 1 galaxies in the 1–20 keV energy band which show a spectrum more complex than a simple power law. Residuals to a power law fit show a strong emission line $\sim$ 6.4 keV, probably due to Fe fluoresence in cold gas, and a flattening of the spectral index above $\sim$ 10 keV (Fig III, Pounds *et al.* 1990). These features agree remarkably well with those predicted by Compton scattering in dense neutral gas (Guilbert and Rees, 1988, Lightman and White, 1988). With this component included, the underlying X-ray spectrum, *i.e.* that input to the Compton scattering gas, has a spectral index closer to 0.9 than 0.7, in good agreement with the IPC results. The IPC's lower energy range is not contaminated by the reflection component and thus is dominated by the input power law. In addition, *Ginga* observations of quasars show a weak or negligible contribution from the scattered component (Williams *et al.* 1992). Thus the difference between the *Einstein* and harder X-ray results is a combination of both the different energy bands and the source luminosities. For the lower luminosity AGN (Sy 1) the presence of the reflection component at higher energies causes the apparent spectrum to be flatter if modelled by a simple power law. For the high luminosity AGN (quasars) the reflection component is weak/absent and thus a steeper slope is observed throughout.

The obvious, unresolved question in our understanding of the X-ray emission in the radio-quiet quasars is the source of the input ($\alpha_E \sim$ 0.9) power law. Possible suggestions include: pair production combined with down-scattering of the high energy photons; direct synchrotron (which would require a very wide range of electron energies); and synchrotron self-Compton emission from the lower energy (IR?) photons. This latter model predicts a close relationship between the X-ray (0.6-10 keV) and IR emission which, although seen in X-ray selected samples (Carleton *et al.* 1987, Brissenden, 1989, Green *et al.* 1992), is not universally present (Sanders *et al.* 1989, Elvis *et al.* 1993a). In addition, there are an increasing number of mm measurements for IR bright, low luminosity AGN which show a far-IR turnover too steep ($\alpha_{IR} \lesssim -2.5$) to be synchrotron self-absorption (except in an extremely homogeneous source) but consistent with a cool dust model (T$\sim$ 30 K, Chini *et al.* 1989, Lawrence *et al.* 1991, Barvainis, 1990, Hughes *et al.* 1993). In this case the IR emission would originate in the outer regions of the host galaxy and would not be available to Compton scatter the synchrotron emission from the core to generate the X-rays. No direct relation between IR and X-ray emission would then be expected.

The resolution of these conflicting results remains of prime importance in understanding the origin of the X-ray emission. The number of objects in the X-ray selected and mm measured samples is small, with no objects in common, and both samples are strongly biased. In order to determine the relative importance of thermal and non-thermal processes in various AGN, these selection effects must be understood or removed, along with any contaminating effects of the beamed radio/X-ray component.

Multi-wavelength monitoring of variable AGN including both X-ray and IR (or optical) observations are of prime importance, though unfortunately very difficult to carry out. A search for correlated variations will show directly the presence/absence of an X-ray–IR relation. An excellent example of this is the dramatic observation by Done *et al.* (1990) of NGC4051 showing no relation between the X-ray emission (2-9 keV), which varies rapidly by factors of 2-3, and the optical B magnitude which is accurately steady. This places extremely strong constraints on the presence of an underlying IR–X-ray power law arguing in favour of the pair production model in this object.

X-RAY VARIABILITY

X-ray monitoring of AGN is available mostly for the lower luminosity AGN (Sy 1) rather than quasars, due to their relative proximity and resulting higher fluxes. Most AGN (including quasars) vary on timescales of weeks and longer. Close to half the Sy1s monitored on shorter timescales vary, with no preferred timescale (power density spectrum $\sim f^{-1}$) and generally with no detected cut-off at high frequency, on the shortest detectable timescales ($\lesssim$hours, Grandi *et al.* 1992). This indicates that the X-rays originate in a very small region in a substantial fraction of AGN.

An example is NGC6814 in which variations on all timescales down to ~250 secs are observed in both the X-ray continuum and the Fe Kα emission line (Kuneida *et al.* 1990). In addition a strong, persistent (over ~ 6 years,, 1985-1991) periodic variation (P = 12100 secs) has been reported (but see Madejski *et al.* 1993). A number of models to explain this observation have been proposed and are discussed elsewhere in these procedings.

Spectral monitoring is generally difficult but where data exist the spectrum generally steepens as the flux increases. The large number of components present in the spectra confuse the interpretation of spectral variations. They are generally consistent with one component varying more than another rather than an intrinsic change in spectrum.

LINE EMISSION

As mentioned above, a strong line (W_λ = 50–350 eV) is frequently observed at ~ 6.4 keV in the spectra of Sy 1 and Sy 2 galaxies. The energy of the line implies an origin in cold (neutral) gas, consistent with being part of the reflection component.

There is also evidence for a blend of line features at ~ 0.8 keV (Turner *et al.* 1991, Marshall *et al.* 1993) probably dominated by Fe L emission. The observed strength implies the presence of ionised gas in addition to the neutral gas already

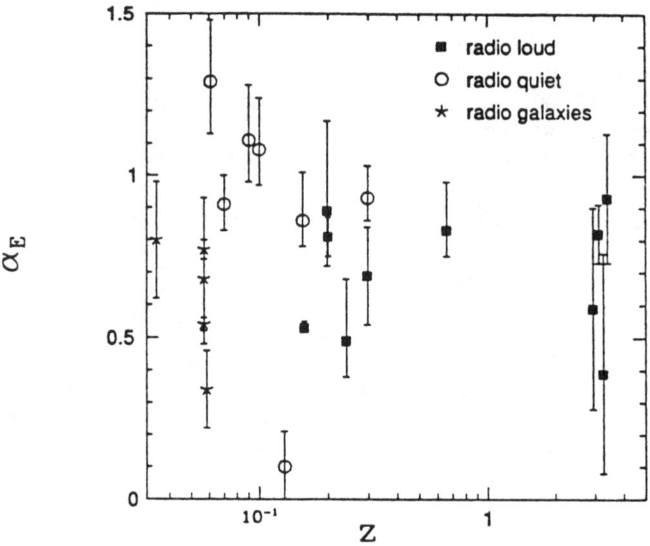

FIGURE IV Soft X-ray slopes of AGN as a function of their redshifts showing no correlation, from Elvis *et al.* (1993b)

invoked in these objects. Few data are available on extended emission in the X-rays around these objects, but ROSAT HRI observations of the obscured Sy1 NGC1068 show that 50% of the soft X-ray emission is extended similarly to the optical image (Wilson *et al.* 1992). This is interpreted (though not uniquely) in terms of a hot, outflowing wind which would provide sufficient FeL line emission to explain that observed (Turner *et al.* 1991). Although for only one object so far, the detection of extended emission suggests an even more complex picutre for the origin of the X-ray emission in AGN in which a substantial fraction does not originate in the core. It is clearly extremely important to make high spatial resolution observations of nearby AGN.

HIGH REDSHIFT QUASARS

Although a number of high redshift quasars were detected with *Einstein* (Wilkes *et al.* 1993), only ROSAT has obtained sufficient signal-to-noise to allow spectral studies. Observations of several quasars with z$\sim$ 3 show spectral indices somewhat flatter than their lower-redshift counterparts in the same observed energy band (0.1-2.0 keV, Elvis *et al.* 1993b). However, at z$\sim$ 3 the rest frame energy range in which these quasars are being observed is comparable to that observed by the EXOSAT ME and HEAO1-A2 experiments (*i.e.* $\sim$2–10 keV) facilitating a comparison in a constant rest frame energy range as a function of redshift. This shows (Fig IV) that there is no significant change in the spectral index for radio-loud quasars (the only ones for which we currently have observations at high redshift).

In addition, the initial results have shown a surprisingly high incidence of soft X-ray absorption in excess of that expected due to the column density of the Galaxy (Wilkes *et al.* 1992, Elvis *et al.* 1993b). At low redshift, although lower luminosity AGN frequently show strong, intrinsic absorption, quasars do not (Wilkes and Elvis, 1987). In our current high-redshift sample of 6 quasars, we have strong evidence for excess absorption in two and suggestive results for a third. The low resolution of the PSPC does not allow determination of the location of this absorbing material, *i.e.* along the line-of-sight or intrinsic to the quasar. Line-of-sight absorption is a heavily studied phenomenon in the optical and ultraviolet spectra of quasars, although the measured neutral hydrogen column densities are far too low to be detectable in the X-rays ($\sim 10^{13-17}$cm^{-2} for Lyman α forest and metal line systems). There remains a substantial uncertainty in the ionisation conditions of these absorbers which prevents metal line systems from being ruled out as the X-ray absorber. However the ionisation would be very high: (N(HI)/N(HII) $\sim 10^{-6}$) in order to match the observed X-ray column densities ($\sim 10^{21-22}$cm^{-2} depending on the redshift of the absorber, Wilkes *et al.* 1992). The most likely line-of-sight absorbers are the high column density damped Lyα systems which have neutral hydrogen column densities up to $\sim 10^{21}$ cm^{-2}. None of the X-ray absorbed quasars have a damped Lyα absorption system in their optical spectra, but given the number density of these systems, the chance of one occuring at a lower redshift is $\sim$ 20% (Elvis *et al.* 1993b). However data at lower redshift are sparse and the number density is thought to decrease with decreasing redshift (Lanzetta *et al.* 1991) which would decrease the probabilities quoted here.

In terms of intrinsic absorbers, all the objects in our sample are GPS radio-loud quasars (O'Dea, Baum and Stanghellini, 1991). These are compact radio sources with a curved radio spectrum rather than the characteristic flat spectrum. They are thought to be confined by a dense surrounding medium ($\sim$ 100 pc) rather than beamed. Perhaps this dense material is absorbing the X-rays (Elvis *et al.* 1993b). If so the size and column densities of the medium are reminiscent of cooling flows. The increase in incidence of GPS, cooling flows and soft X-ray absorption with increasing redshift provides additional, circumstantial evidence for a connection. Clearly these are early days and a good deal more data on soft X-ray absorption are required to positively identify the absorbing material.

WARM ABSORBERS

The presence of a warm (ionized) absorber in low luminosity AGN has frequently been inferred to explain apparent soft excesses in the EXOSAT observations. With ROSAT it is now possible to detect individual absorption features. For example the quasar 3C351 (Fig V) has an OVII edge at 0.76$\pm$0.08 keV (Fiore *et al.* 1993). Similar observations have been reported for MCG6-30-15 (Nandra and Pounds, 1992) and NGC5548 (Nandra *et al.* 1993).

The identity of the warm absorber has not yet been determined. For 3C351, the possibility that it is a cloud in the broad emission line region has been investigated (Mathur *et al.* 1993). The ionisation parameter for the warm absorber is around U = 6.7 $\pm$ 0.3 (Fiore *et al.* 1993). Using the observed OVIλ1034 and CIVλ1549 emission line strengths from IUE and HST measurements and the observed continuum of 3C351 the ionisation parameter deduced is lower:

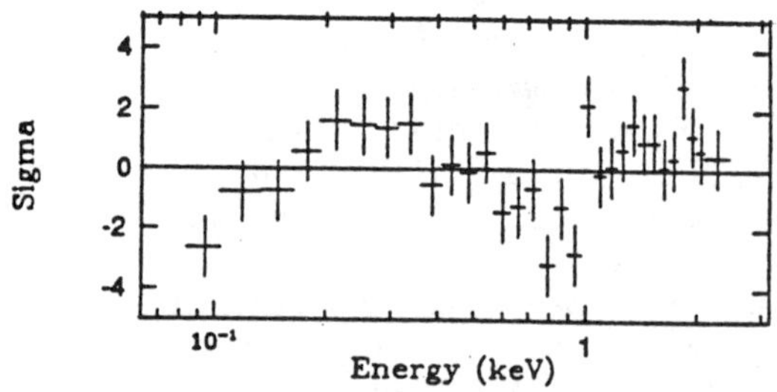

FIGURE V Residuals to a simple power law fit to the ROSAT PSPC spectrum of 3C351 showing the presence of an absorption feature at 0.76±0.8 keV, from Fiore *et al.* (1993)

U$\sim$0.3 (Mathur *et al.* 1993). This does not rule out the emitting and absorbing material being from BLR clouds at different radii in a stratified BLR. The HST spectrum of 3C351 also shows a strong, metal-line absorption system including features of CIVλ1549, NVλ1240 and OVIλ1034. The ionisation conditions inferred from these lines are consistent with origin in the same gas as the X-ray absorption suggesting that the warm absorber and the associated metal line absorber are one and the same in this object (Mathur *et al.* 1993). This link opens up the possibility of determining the ionisation conditions in the metal-line systems since the X-rays constrain the column density. In 3C351 the ionisation fraction of hydrogen would be $\sim 10^{-5}$.

γ-RAY EMISSION FROM QUASARS

This topic is covered elsewhere in these proceedings (Michelson, this volume), only the main results here will be summarised here for completeness. EGRET on GRO has now detected 24 quasars in the 50–2000 MeV range and a number of others at lower significance ($< 5\sigma$). All these quasars are compact, and radio-loud, many with blazar characteristics, implying that beaming plays an important role in the observation of γ-ray emission. Perhaps the most exciting new observation is the detection at TeV energies with the Whipple Observatory of the closest, GRO-detected AGN MKN 421 (Punch *et al.* 1992).

REFERENCES

Barvainis, R., 1993, *ApJ*, submitted

Barvainis, R., 1990, *ApJ*, **353**, 419

Boroson, T. A., 1989, *ApJ*, **343**, L9

Brissenden, R. J. V., 1989, Ph.D. Thesis, The Australian National University

Brunner, H., Friedrich, P., Zimmerman, H-U. and Staubert, R., 1992, in "X-ray Emission from Active Galactic Nuclei and the X-ray Background" eds. W. Brinkmann, and J. Truemper, MPE Report **235**, 1981,

Canizares, C. R. and White, J. L., 1989, *ApJ*, **339**, 27

Carleton, N. P, Elvis, M., Fabbiano, G., Wilner, S. P., Lawrence, A. and Ward, M., 1987, *ApJ*, **318**, 595

Chini, R., Kreysa, E. and Biermann, P. L. 1989 *A&A*, **219**, 87

Czerny, B. and Elvis, M., 1987, *ApJ*, **321**, 305

Done, C., Ward, M. J., Fabian, A. C., Kunieda, H., Tsuruta, S., Lawrence, A., Smith, M. G. and Wamstekker, W., 1990, *MNRAS*, **243**, 713

Elvis, M., *et al.* 1978, *MNRAS*, **183**, 129

Elvis, M., Fiore, F., Wilkes, B. J., McDowell, J. and Bechtold, J., 1993b, *ApJ*, in press

Elvis, M, Giommi, P., Wilkes, B. J. and McDowell, J., 1991, *ApJ*, **378**, 537

Elvis, M., Wilkes, B. J. and McDowell, J. C., 1991, in *Extreme Ultraviolet Astronomy*, R. F. Malina and S. Bowyer, 238

Elvis, M., Wilkes, B. J., McDowell, J. C., Green, R. F., Bechtold, J., Willner, S. P., Cutri, R., Oey, M, S., and Polomski, E., 1993a, *ApJS*, submitted

Fiore, F., Elvis, M., Mathur, S., Wilkes, B. and McDowell, B. J., 1993, *ApJ*, in press

Grandi, P., Tagliaferri, G., Giommi, P., Barr, P. and Palumbo, G. G. C., 1992, *ApJS*, **82**, 93

Green, P. J., Anderson, S. F. and Ward, M. J., 1992, *MNRAS*, **254**, 30

Guilbert, P. W. and Rees, M. J., 1988, *MNRAS*, **233**, 475

Hughes, D. H., Robson, E. I., Dunlop, J. S. and Gear, W. K. 1993 *MNRAS*, **263**, 607

Kruper, J. S., Urry, C. M. and Canizares, C. R., 1990, *ApJS*, **74**, 347

Kuneida, H., Turner, T. J., Awaki, H., Koyama, K., Mushotzky, R. F. and Tsusaka, Y., 1990, Nature **345**, 786

Laor, A., 1990, *MNRAS*, **246**, 369

Lanzetta, K. M., Wolfe, A. M., Turnshek, D. A., Lu, L., McMahon, R. G. and Hazard, C., 1991, *ApJS*, **77**, 1

Lawrence, A., Rowan-Robinson, M., Efstathiou, A., Ward, M. J., Elvis, M., Smith, M. G., Duncan, W. D. and Robson, E. I. 1991 *MNRAS*, **248**, 91

Lightman, A. P. and White, T. R., 1988, *ApJ*, **335**, 57

Madejski, G. M. *et al.* 1993, *Nature*, in press

Marshall, F. E. *et al.* 1993, *ApJ*, **405**, 168

Masnou, J-L, Wilkes, B. J., Elvis, M., Arnaud, K. A. and McDowell, J. C., 1992, *A&A*, **253**, 35

Mathur, S. *et al.* 1993, in preparation

Mushotzky, R. F., 1984, *Adv. Sp. Res.* **3**, 157

Nandra, K. and Pounds, K. A., 1992, *Nature*, **359**, 215

Nandra, K. *et al.* 1993, *MNRAS*, **260**, 504

O'Dea, C. P, Baum S. A. and Stanghellini, C., 1991, *ApJ*, **380**, 66

Pounds, K. A., Nandra, K, Stewart, G. C. George, I. M., and Fabian, A. C., 1990, *Nature*, **344**, 132

Punch, M. *et al.* 1992, *Nature*, **358**, 477

Rothschild, R. E., Mushotzky, R. F., Baity, W. A., Gruber, D. E., Matteson, J. L. and Peterson, L. E., 1983, *ApJ*, **269**, 423

Sanders, D., Phinney, E. S., Neugebauer, G., Soifer, B. T. and Matthews, K., 1989, *ApJ*, **347**, 29

Shastri, P., Wilkes, B. J.,Elvis, M. and McDowell, J. C. 1993 *ApJ*, **410**, 29

Sun, W-S and Malkan, M., 1989, *ApJ*, **346**,68

Tananbaum, H., Peters, G., Forman, W., Giacconi, R., Jones, C. and Avni, Y., 1978, *ApJ*, **223**, 74

Turner, T. J. and Pounds, K. A., 1988, *MNRAS*, **232**, 463

Turner, T. J. and Pounds, K. A., 1989, *MNRAS*, **240**, 833

Turner, T. J, Weaver, K. A., Mushotzky, R. F., Holt, S. S. and Madejski, G. M., 1991, *ApJ*, **381**,85

Wilkes, B. J. and Elvis, M., 1987, *ApJ*, **323**, 243

Wilkes, B. J., Elvis, M., Tananbaum, H., McDowell, J. C. and Lawrence, A., 1992, *ApJ*, **393**, L1

Wilson, A. S., Elvis, M., Lawrence, A. and Bland-Hawthorne, J., 1992, *ApJ*, **391**, L75

Williams, O.R., *et al.* 1992, *ApJ*, **389**, 157

Zamorani, G. *et al.* 1981, *ApJ*, **245**, 357

ACKNOWLEDGEMENTS

This work was supported in part by NASA contract NAS5-30934 and grants NAG5-1771 and NAG5-1724.

The First Stromlo Symposium: The Physics of Active Galaxies
ASP Conference Series, Vol. 54, 1994
G. V. Bicknell, M. A. Dopita, and P. J. Quinn (eds.)

THERMAL AND NONTHERMAL PROCESSES IN ACTIVE GALACTIC NUCLEI

MITCHELL C. BEGELMAN[1]
Joint Institute for Laboratory Astrophysics, University of Colorado and National Institute of Standards and Technology, Boulder, CO 80309-0440

ABSTRACT I review current ideas about the mechanisms responsible for broadband continuum radiation from active galactic nuclei. Both thermal and nonthermal mechanisms may be important, possibly with one or the other dominating in a given spectral band. Reprocessing, which seems to be important at all energies, greatly complicates the interpretation.

INTRODUCTION

Can we distinguish between thermal and nonthermal radiation mechanisms in AGNs? Historically, there has been a temptation to regard AGNs primarily as nonthermal objects because their radiation energy outputs are spread fairly evenly over as many as eight or more decades in frequency, and to a very crude approximation resemble near-power laws over large stretches of photon energy. However, as observations become more sensitive in each band, the spectra which originally looked like featureless continua often display an increasing complexity of bumps and wiggles, curvature, and breaks in slope. These developments have made such an impression that, in recent years, the vogue has been to interpret as much of the spectrum as possible in terms of the superposition of thermal spectra. Yet we know that some AGNs produce a large fraction of their energy outputs in the highly nonthermal form of jets and winds. Below I will try to characterize the current status of arguments for and against the different kinds of radiation mechanism in each spectral band.

SUB-MILLIMETER → INFRARED

The debate here centers on whether this portion of the spectrum is dominated by nonthermal synchrotron radiation, or by the superposition of thermal spectra radiated by dust with a wide range of temperatures. Over the past few years "public opinion", which started out favoring the nonthermal interpretation, has swung overwhelmingly to the dust side. This seems to be a strong case for efficient thermal reprocessing of the radiation from the central source.

---

[1]Also at Department of Astrophysical, Planetary, and Atmospheric Sciences, University of Colorado, Boulder, CO 80309-0391.

Pro-Synchrotron

Three arguments have been used to support the synchrotron interpretation of the sub-mm–IR spectrum. First, the spectra of many AGNs, from the far-IR through the optical, vaguely "look" nonthermal, i.e., roughly power-laws with energy spectral index $\alpha \equiv -d \log F_\nu / d \log \nu \approx 1$ (Ward et al., 1987; Carleton et al., 1987). Second, there seemed to be a correlation between the IR intensity and the X-ray intensity at a few keV (Kriss 1988), suggestive of a single power-law connecting the bands. Third, the sharp turnovers observed at frequencies $\sim 10^{12} - 10^{13}$ Hz occurred roughly where one would predict synchrotron self-absorption to set in, if the far-IR radiation were produced close to the putative black hole (Begelman 1989). The discovery of spectra steeper than $\nu^{2.5}$ longward of the turnovers (Engargiola et al., 1988; Edelson et al., 1988; Chini, Kreysa, and Biermann, 1989; Barvainis, Antonucci, and Coleman 1992) rules out self-absorbed synchrotron radiation from a *power-law* electron distribution, although it may be consistent with other plausible electron distributions (de Kool and Begelman 1989; however, see Hughes et al., 1993).

Pro-Dust

However, it has been harder to explain away the extreme steadiness of the infrared emission. If the IR comprised merely the low-frequency end of the same synchrotron spectrum which produces the optical and even X-ray emission, then one would expect it to vary on similar timescales. Yet no convincing variation has ever been seen in a non-blazar AGN, despite careful attempts to detect it (Done et al., 1990). Moreover, systematic studies of the IR spectra of AGNs have shown that they are far from simple power-laws. The characteristic 3 μm "bump" and the trough at $\sim$ 1 μm can be associated with dust sublimation (Sanders et al., 1989), and a sub-mm cutoff with a slope as steep as $\sim \nu^3$ is natural if the dust has a minimum temperature.

Caveats and Tests

On balance, it seems more "natural" to ascribe the sub-mm–IR spectrum to reprocessing by dust, but some unanswered questions remain. To yield roughly comparable fluxes per decade of frequency, as observed, the dust distribution in each temperature range must intercept and reprocess the right amount of radiation, a significant fraction of the total radiation output. Sanders et al. propose that this occurs as the result of a generic "warped disk" geometry, but the reasons that such a geometry should develop are not entirely clear. (The recent analysis of the warped disk in Centaurus A [Quillen et al., 1992] does support this general picture, however.) The alternative would be some kind of a thick dusty torus (Pier and Krolik 1992b), but it is not clear how such a torus could be supported against gravity (Krolik and Begelman 1988; Pier and Krolik 1992a).

The dust picture makes one extremely robust prediction about the variability of the far-IR emission: there should not be any, on any accessible timescale. Detection of far-IR variability would rule out a thermal origin.

OPTICAL → ULTRAVIOLET

This portion of the spectrum is home to the "UV Bump", but to confuse matters there are also features which can be ascribed to reprocessing in the broad emission-line region, such as Balmer continuum and a mess of Fe II lines. The consensus seems to be that the bump is thermal emission, possibly from a region close to the black hole, but there are some nagging doubts. Even if it is thermal, there is *no solid evidence* that it comes from a gas distribution that resembles an accretion disk (but there is no evidence against such a structure, either).

Pro-Thermal

The principal arguments for the thermal nature of the UV Bump are that it "looks" thermal and fits in (more or less) with theoretical expectations. The inner portion of an accretion disk around a $10^8\ M_\odot$ black hole, radiating at $\gtrsim 1\%$ of the Eddington limit, is expected to produce thermal radiation with a temperature $\sim 10^5$ K — but, then, so is *any* cool, optically thick slab of gas subjected to the radiation density thought to be present close to the black hole! There is strong evidence that "cool" ($T \lesssim 10^6$ K), optically thick gas does intercept and reprocess a large fraction of the hard X-ray continuum in many Seyfert nuclei, imprinting a "Compton reflection" component, iron K-shell fluorescence lines, and an iron K-shell absorption edge on the spectrum (e.g., Pounds et al., 1990). However, it is difficult to tell whether this gas lies in close proximity to the source of the hard radiation, or much further out.

Shields (1978) was the first to point out that thermal radiation from an accretion disk could dominate the optical–UV continua of quasars, and the decomposition of AGN spectra into a power-law plus a thermal (Malkan and Sargent 1982) or accretion disk spectrum (Malkan 1983) lent credence to this idea. Since then, increasingly sophisticated models for accretion disk spectra have been fit to observations of AGNs (Czerny and Elvis 1987; Wandel and Petrosian 1988; Sun and Malkan 1989; Laor 1990; among others). One problem with making these fits is that the peak of the inferred thermal component falls in the extreme ultraviolet range of frequencies, which is seldom observable. Second, there are now so many dials to be turned on these models (black hole mass, accretion rate, disk inclination, black hole angular momentum, etc.) that it would be surprising if the relatively featureless continua could not be fit in a continuum of ways (Laor 1990).

Pro-Nonthermal

Questions concerning the viability of a thermal accretion disk model for the optical–UV radiation were raised by Rauch and Blandford (1991), based on an analysis of the apparent microlensing event observed in the quasar Q2237+0305 (Irwin et al., 1989; Corrigan et al., 1991). They found that a thermal source would have to be too large to be consistent with the time scale and amplitude of the microlensing event. It should be emphasized that their result, taken at face value, would not rule out the presence of a thermal source, only that the accretion disk provided more than a certain fraction of the optical flux. More recent analyses (Jaroszyński, Wambsganss, and Paczyński 1992; Czerny, Jaroszyński, and Czerny 1993) appear to have relaxed this constraint somewhat.

Cold Opaque Gas ≠ Accretion Disk

It is important to keep in mind that the cool gas responsible for a thermal UV bump need not take the form of an accretion disk. Any distribution of gas with a sufficient covering factor, which is effectively optically thick to UV, could reprocess incident X-rays into thermal UV radiation (Ferland and Rees 1988). In order to avoid heating up to the inverse Compton temperature ($\sim 10^7$ K), the gas pressure would have to exceed about 10% of the radiation energy density (Guilbert and Rees 1988), corresponding to a particle density $\gtrsim 10^{16}(T_{eff}/10^5 \text{ K})^3$ cm^{-3}. An advantage of a disk configuration is that the vertical component of gravity can easily confine the gas at the required pressure, given a modest column density. Moreover, such a configuration affords a large covering factor for reprocessing. Note, however, that the gas can resemble a disk geometrically without being what we normally think of as a "viscous accretion disk" (Shakura and Sunyaev 1973), in which gravitational binding energy is dissipated internally and radiated through the disk atmosphere. Instead, the disk might be a "passive reprocessor" (Begelman and de Kool 1991). Alternatively, the gas could have a chaotic spatial distribution, taking the form of filaments confined by a magnetic field (Rees 1987). In this case, the reprocessor is more likely to be "mixed in" with the primary source of hard radiation. To avoid excessive modification of the X-ray spectrum by Compton effects, one might envisage a spray of tiny, extremely dense ($n \gtrsim 10^{19}$ cm^{-3}) cloudlets, with large free-free optical depths but small depths to Thomson scattering (Celotti, Fabian, and Rees 1992).

In principle, the thermally emitting gas need not be optically thick, and the UV need not come from reprocessing. Optically thin bremsstrahlung at roughly a single temperature between 10^5 and 10^6 K could explain the optical-to-near UV slopes observed in many AGNs (Antonucci and Barvainis 1988; Ferland, Korista, and Peterson 1990) without violating size and Compton reprocessing constraints (Barvainis 1993), provided that the emitting gas takes the form of very small, dense clouds (i.e., similar to the Celotti et al. model).

On the other hand, there is no evidence that any cool thermal gas present in the nucleus of an AGN is *not* in the form of an accretion disk. Often-quoted observational arguments against disks are relatively easy to circumvent. For example, the large scattering opacities of disk atmospheres suggest that the emergent radiation should have polarizations of at least a few per cent, yet AGNs are often observed to be very weakly polarized. However, simple scaling arguments suggest that Faraday rotation due to the diffuse gas in an AGN could depolarize the radiation up to frequencies in the UV band. A "rough" disk surface, caused, e.g., by turbulence at the disk–corona interface, could also reduce the level of polarization substantially (Coleman and Shields 1990). Another argument suggests that disks should show Lyman edges, yet these are seldom if ever seen (Antonucci, Kinney, and Ford 1989; Koratkar, Kinney, and Bohlin 1992). However, theoretical modeling shows that the prominence and properties of such edges are likely to be highly model-dependent (Czerny and Pojmański 1990); moreover, they are readily smeared out by kinematic and radiative transfer effects in the nucleus (Czerny and Zbyszewska 1991).

X-RAYS → GAMMA-RAYS

Observations in this part of the spectrum have seen the greatest development in recent years. EXOSAT and Ginga revealed features associated with "Compton reflection" and iron K-shell fluorescence, indicating the presence of "cool" gas (Guilbert and Rees 1988; Lightman and White 1988) in many AGNs. At much higher energies, the Compton Gamma-Ray Observatory (CGRO) is starting to produce some dramatic surprises, but as yet no resolution of the thermal vs. nonthermal controversy.

Pre-CGRO

For several years a major goal in X-ray studies of AGNs was to explain the "canonical" $\alpha_X \simeq 0.7$ spectral index in the 2–10 keV band (Turner and Pounds 1989). Nonthermal pair cascades had been invented (Zdziarski and Lightman 1985; Fabian et al., 1986), in part, to do this, but were having trouble explaining α_X in conjunction with other observational constraints. The discovery of Compton reflection (Pounds et al., 1990) showed that, in fact, 0.7 was really only the slope of the composite (incident + reflected) spectrum. When the reflected component was subtracted, the slope of the "intrinsic" spectrum was closer to 0.9.

Proponents of pair cascades were quick to point out that $\alpha_X \simeq 0.9$ could be obtained readily in a "saturated" pair cascade (Zdziarski et al., 1990). Unfortunately, $\alpha \approx 1$ turns out to be easily explained by a variety of other mechanisms as well. For example, Haardt and Maraschi (1991) used a simple "reflection feedback" mechanism to obtain this spectrum. In their model, the disk serves mainly to reprocess $\gtrsim 50\%$ of the radiation emitted by an active corona. The corona, in turn, Comptonizes the reprocessed radiation from the disk, closing the feedback loop. Another way to get $\alpha \simeq 1$ is to arrange for a hot corona to Comptonize a very soft input spectrum (e.g., far-IR photons from a self-absorbed synchrotron spectrum). If the amplification creates a power-law tail extending over several decades in frequency, the spectral index will approach 1 provided that most of the energy goes into heating the corona.

The lesson of this is that an $\alpha \sim 1$ spectral index tells us little about the nature of the radiation mechanism, not even whether it involves thermal or nonthermal (i.e., relativistic) Comptonization. Note, however, that all three of the models discussed above, including the pair cascade, involve most of the energy being injected first into a hot corona, rather than diffusing out from the interior of an accretion disk.

Implications of OSSE Results

Before the advent of CGRO, we knew from HEAO-1 observations that AGN spectra often continued in power-law fashion up to ten's of keV (Rothschild et al., 1983). Observations with the OSSE instrument on CGRO are now showing that the spectra of type 1 Seyfert galaxies steepen considerably beyond a break energy $E_b \sim 100$ keV (Cameron et al., 1993; Maisack et al., 1993). The break can be interpreted as a Wien exponential cutoff due to Compton upscattering of soft photons by nonrelativistic electrons (or pairs) with $kT \sim E_b/3$. One can tentatively reject a model in which the cutoff results from *down*scattering of a much harder (nonthermal) power-law by colder electrons, since the observed E_b implies an optical depth τ which is too small to explain the steepness of the cutoff. According to the Sunyaev and Titarchuk (1980) model for spectrum formation due to unsaturated Comptonization, a spectral index in the range $0.5 - 1$ and a spectral break at ~ 100 keV requires a Thomson optical depth $\tau \sim 3$. One could have a cool disk or opaque filaments producing UV "seed" photons, which are Comptonized by a strongly heated corona. Strong heating of the corona is necessary because the required coronal temperature is much higher than the inverse Compton temperature, which is the equilibrium temperature for low-density gas in the absence of extra heating. Note that the active corona must not completely cover the UV source, if the UV bump is to stand above the X-ray continuum.

Still a Role for Pairs?

Evidence for a Wien cutoff caused by nonrelativistic electrons does *not* imply that nonthermal processes are unimportant. They may be responsible for injecting particles into the corona in the first place. In fact, the corona may be composed primarily of cooled electron–positron pairs created by a nonthermal cascade. In this case, the Thomson optical depth is fixed by the balance between pair production and annihilation and is given by $\tau \sim (Y\ell_{nt})^{1/2}$, where Y is the "pair yield", the fraction of nonthermal energy going into rest mass of pairs, and ℓ_{nt} is the "nonthermal compactness", a measure of the power going into nonthermal processes. Zdziarski, Lightman, and Maciołek-Niedźwiecki (1993) have shown how such a model can reproduce the high-energy spectrum of NGC 4151; however, several parameters must be adjusted to fit the cutoff energy. If *thermal* pair production is taken into account, it may be possible to explain why the break energies fall in the observed range. The temperature sensitivity of the thermal pair production rate can produce a thermostatic effect, limiting E_b to $50 - 100$ keV when the power input into the corona is large (Svensson 1984).

Thus, the recent observational results from OSSE, while extremely valuable, do not reveal all of the basic processes shaping the X- and gamma-ray spectra of AGNs. The spectral features produced by pair annihilation might well be smeared beyond recognition by Comptonization in the corona, although one should not give up hope of detecting a weak annihilation feature and/or high-energy tail (extending up to ~ 1 MeV).

Gamma-rays from Blazars

At the higher gamma-ray energies observed with the EGRET instrument on CGRO, all of the detected sources belong to the blazar class of AGNs, i.e., BL Lac objects and optically violently variable quasars. It is thought that the dominant emission from these objects comes from relativistic jets pointed nearly towards us. The observed rapid (day–week timescale) variability of GeV gamma-rays (e.g., Hartman et al., 1992; Fichtel et al., 1993), which cannot be produced too close to the nucleus because of the opacity to pair production, provides some of the strongest evidence for bulk relativistic outflows with Lorentz factors $\gtrsim 10$. One BL Lac object, Mrk 421, has also been detected at TeV energies using the ground-based Whipple Telescope (Punch et al., 1992).

The X-ray emission from blazars has generally been interpreted using synchrotron-self-Compton (SSC) models (Königl 1981; Marscher and Gear 1985; Ghisellini and Maraschi 1989). SSC models possess the elegant feature that the same population of relativistic electrons both produces the low-energy seed photons and Comptonizes them, but it has proven difficult to fit the observed high-energy gamma-ray spectra using these models (Bloom and Marscher 1993; Maraschi, Ghisellini, and Celotti 1992). An attractive alternative hypothesis is that the gamma-rays are produced by Comptonization of a radiation field which originates outside the jet, e.g., UV radiation from the central accretion flow which is scattered or reprocessed by gas in the region through which the jet is passing (Dermer, Schlickeiser, and Mastichiadis 1992; Dermer 1993; Sikora, Begelman, and Rees 1993, 1994; Blandford 1993). Sikora et al. (1993, 1994) have shown that the broad-band spectral properties of 3C 279 can be understood within the context of a "one-zone" model of this type.

CONCLUSIONS

If there is any conclusion to be drawn from the arguments presented above, it is that the radiation processes operating in AGNs are not nearly as well-constrained as one might like. In part this is due to a lack of concrete clues about the source geometry. Curiously, the geometric clues are strongest for the most exotic classes of AGN, the blazars, since the weight of evidence overwhelmingly favors Doppler-boosted radiation from relativistic jets. However, even here such basic parameters as the opening angles and bulk Lorentz factors of the jets are poorly constrained.

What are the prospects for gaining a better handle on the structure of the central regions of AGNs? At the time we met in Canberra, we thought we had a very promising candidate, the Seyfert galaxy NGC 6814. Ginga and EXOSAT observations of this object showed a clear 12,100 s periodicity (Mittaz and Branduardi-Raymont 1989; Done et al., 1992), which was best explained as that of a star orbiting the black hole at $\sim$ 20 Schwarzschild radii (Syer, Clarke, and Rees 1991; Sikora and Begelman 1992). Moreover, the very sharp "dropouts" in the X-ray flux showed all the earmarks of eclipses (K. Leighly, private communication) which could provide tight geometric constraints on the primary and reprocessed components of the emission. Alas, ROSAT observations which became public just weeks after the meeting showed that the periodicity was actually associated with a previously unknown Galactic source 37 arc min away from the AGN, probably a cataclysmic variable (Madejski et al., 1993). The "Rosetta stone" of AGN structure has yet to be unearthed.

ACKNOWLEDGMENTS

My work on AGNs is supported in part by the National Science Foundation (grant AST91-20599) and the National Aeronautics and Space Administration.

REFERENCES

Antonucci, R., and Barvainis, R. 1988, *ApJ*, **332**, L13

Antonucci, R. R. J., Kinney, A. L., and Ford, H. C. 1989, *ApJ*, **342**, 64

Barvainis, R. 1993, *ApJ*, **412**, 513

Barvainis, R., Antonucci, R., and Coleman, P. 1992, *ApJ*, **399**, L19

Begelman, M. C. 1989, in *Active Galactic Nuclei*, D. E. Osterbrock and J. S. Miller, Dordrecht: Kluwer, 141

Begelman, M. C., and de Kool, M. 1991, in *Variability of Active Galactic Nuclei*, H. R. Miller and P. J. Wiita, Cambridge: Cambridge University Press, 198

Blandford, R. D. 1993, in *Compton Gamma-Ray Observatory, AIP Conf. Proc. 280*, M. Friedlander, N. Gehrels, and D. J. Macomb, New York: Amer. Inst. Phys., 533

Bloom, S. D., and Marscher, A. P. 1993, in *Compton Gamma-Ray Observatory, AIP Conf. Proc. 280*, M. Friedlander, N. Gehrels, and D. J. Macomb, New York: Amer. Inst. Phys., 578

Cameron, R. A., et al. 1993, in *Compton Gamma-Ray Observatory, AIP Conf. Proc. 280*, M. Friedlander, N. Gehrels, and D. J. Macomb, New York: Amer. Inst. Phys., 478

Carleton, N. P., Elvis, M., Fabbiano, G., Willner, S. P., Lawrence, A., and Ward, M. 1987, *ApJ*, **318**, 595

Celotti, A., Fabian, A. C., and Rees, M. J. 1992, *MNRAS*, **255**, 419

Chini, R., Kreysa, E., and Biermann, P. L. 1989, *A&A*, **219**, 87

Coleman, H. H., and Shields, G. A. 1990, *ApJ*, **363**, 415

Corrigan, R. T., et al. 1991, *AJ*, **102**, 34

Czerny, B., and Elvis, M. 1987, *ApJ*, **312**, 325

Czerny, B., Jaroszyński, M., and Czerny, M. 1993, *MNRAS*, submitted

Czerny, B., and Pojmański, G. 1990, *MNRAS*, **245**, 1P

Czerny, B., and Zbyszewska, M. 1991, *MNRAS*, **249**, 634

de Kool, M., and Begelman, M. C. 1989, *Nature*, **338**, 484

Dermer, C. D. 1993, in *Compton Gamma-Ray Observatory, AIP Conf. Proc. 280*, M. Friedlander, N. Gehrels, and D. J. Macomb, New York: Amer. Inst. Phys., 541

Dermer, C. D., Schlickeiser, R., and Mastichiadis, A. 1992, *A&A*, **256**, L27

Done, C., Madejski, G. M., Mushotzky, R. F., Turner, T. J., Koyama, K., and Kunieda, H. 1992, *ApJ*, **400**, 138

Done, C., Ward, M. J., Fabian, A. C., Kunieda, H., Tsuruta, S., Lawrence, A., Smith, M. G., and Wamsteker, W. 1990, *MNRAS*, **243**, 713

Edelson, R. A., Gear, W. K. P., Malkan, M. A., and Robson, E. I. 1988, *Nature*, **336**, 749

Engargiola, G., Harper, D. A., Elvis, M., and Willner, S. P. 1988, *ApJ*, **332**, L19

Fabian, A. C., Blandford, R. D., Guilbert, P. W., Phinney, E. S., and Cuellar, L. 1986, *MNRAS*, **221**, 931

Ferland, G. J., Korista, K. T., and Peterson, B. M. 1990, *ApJ*, **363**, L21

Ferland, G. J., and Rees, M. J. 1988, *ApJ*, **332**, 141

Fichtel, C. E., et al. 1993, in *Compton Gamma-Ray Observatory, AIP Conf. Proc. 280*, M. Friedlander, N. Gehrels, and D. J. Macomb, New York: Amer. Inst. Phys., 461

Ghisellini, G., and Maraschi, L. 1989, *ApJ*, **340**, 181

Guilbert, P. W., and Rees, M. J. 1988, *MNRAS*, **233**, 475

Haardt, F., and Maraschi, L. 1991, *ApJ*, **380**, L51

Hartman, R. C., et al. 1992, *ApJ*, **385**, L1

Hughes, D. H., Robson, E. I., Dunlop, J. S., and Gear, W. K. 1993, *MNRAS*, **263**, 607

Irwin, M. J., Webster, R. L., Hewett, P. C., Corrigan, R. T., and Jedrzejewski, R. I. 1989, *AJ*, **98**, 1989

Jaroszyński, M., Wambsganss, J., and Paczyński, B. 1992, *ApJ*, **396**, L65

Königl, A. 1981, *ApJ*, **243**, 700

Koratkar, A. P., Kinney, A. L., and Bohlin, R. C. 1992, *ApJ*, **400**, 435

Kriss, G. 1988, *ApJ*, **324**, 809

Krolik, J. H., and Begelman, M. C. 1988, *ApJ*, **329**, 702

Laor, A. 1990, *MNRAS*, **246**, 369

Lightman, A. P., and White, T. R. 1988, *ApJ*, **335**, 57

Madejski, G. M., Done, C., Turner, T. J., Mushotzky, R. F., Serlemitsos, P., Fiore, F., Sikora, M., and Begelman, M. C. 1993, *Nature*, in press

Maisack, M., et al. 1993, *ApJ*, **407**, L61

Malkan, M. A. 1983, *ApJ*, **268**, 582

Malkan, M. A., and Sargent, W. L. W. 1982, *ApJ*, **254**, 22

Maraschi, L., Ghisellini, G., and Celotti, A. 1992, *ApJ*, **397**, L5

Marscher, A. P., and Gear, W. K. 1985, *ApJ*, **298**, 114

Mittaz, J. P. D., and Branduardi-Raymont, G. 1989, *MNRAS*, **238**, 1029

Quillen, A. C., de Zeeuw, P. T., Phinney, E. S., and Phillips, T. G. 1992, *ApJ*, **391**, 121

Pier, E. A., and Krolik, J. H. 1992a, *ApJ*, **399**, L23

Pier, E. A., and Krolik, J. H. 1992b, *ApJ*, **401**, 99

Pounds, K. A., Nandra, K., Stewart, G. C., George, I. M., and Fabian, A. C. 1990, *Nature*, **344**, 132

Punch, M., et al. 1992, *Nature*, **358**, 477

Rauch, K. P., and Blandford, R. D. 1991, *ApJ*, **381**, L39

Rees, M. J. 1987, *MNRAS*, **228**, 47P

Rothschild, R. E., et al. 1983, *ApJ*, **269**, 423

Sanders, D. B., Phinney, E. S., Neugebauer, G., Soifer, B. T., and Matthews, K. 1989, *ApJ*, **347**, 29

Shakura, N. I., and Sunyaev, R. A. 1989, *A&A*, **24**, 337

Shields, G. 1978, *Nature*, **272**, 706

Sikora, M., Begelman, M. C., and Rees, M. J. 1993, in *Compton Gamma-Ray Observatory, AIP Conf. Proc. 280*, M. Friedlander, N. Gehrels, and D. J. Macomb, New York: Amer. Inst. Phys., 598

Sikora, M., and Begelman, M. C. 1992, *Nature*, **356**, 224

Sikora, M., Begelman, M. C., and Rees, M. J. 1994, *ApJ*, in press

Sun, W. H., and Malkan, M. A. 1989, *ApJ*, **346**, 68

Sunyaev, R. A., and Titarchuk, L. G. 1980, *A&A*, **86**, 121

Svensson, R. 1984, in *X-ray and UV Emission from Active Galactic Nuclei*, W. Brinkmann and J. Trümper, Munich: MPE-Report 184, 152

Syer, D., Clarke, C. J., and Rees, M. J. 1991, *MNRAS*, **250**, 505

Turner, T. J., and Pounds, K. A. 1989, *MNRAS*, **240**, 833

Wandel, A., and Petrosian, V. 1988, *ApJ*, **329**, L11

Ward, M., Elvis, M., Fabbiano, G., Carleton, N. P., Willner, S. P., and Lawrence, A. 1987, *ApJ*, **315**, 74

Zdziarski, A. A., Ghisellini, G., George, I. M., Svensson, R., Fabian, A. C., and Done, C. 1990, *ApJ*, **363**, L1

Zdziarski, A. A., and Lightman, A. P. 1985, *ApJ*, **294**, L79

Zdziarski, A. A., Lightman, A. P., and Maciołek-Niedźwiecki, A. M. 1993, *ApJ*, in press

The First Stromlo Symposium: The Physics of Active Galaxies
ASP Conference Series, Vol. 54, 1994
G. V. Bicknell, M. A. Dopita, and P. J. Quinn (eds.)

SIGNATURES OF UV AND OPTICALLY EMITTING ACCRETION DISKS

A.L. KINNEY
Space Telescope Science Institute, Baltimore, MD 21218

ABSTRACT

Observational evidence in direct support of accretion disks which dominate the optical and UV emission is reviewed. The evidence concerns temperature proportional to radius of the disk, presence of a Balmer jump due to thermal emission, double peaked line profiles, and variability.

INTRODUCTION

In this paper, I will present unambiguous observational evidence for a UV and optically emitting accretion disk. The evidence comes in a number of forms: 1) blue and UV continuum emitted when the disk is visible, with the temperature dependent on radius within the disk, 2) a prominant Balmer jump, as predicted by models of thermally emitting disks, 3) double peaked line profiles and 4) variability characteristic of an accretion disk.

The evidence that I will present is for the class of objects known as Cataclysmic Variables (CV). In the second half of this paper I will review the lack of evidence which would link UV and optical emission in AGN and quasars to an accretion disk.

A CV is a binary star system which is exchanging mass (see Horne 1990 for a review). In a typical CV, a white dwarf (the primary) is surrounded by an accretion disk. A cool star (the secondary) is filling its Roche lobe, and providing accreting material to the primary. The cartoon in Figure 1 illustrates a CV. These objects have a number of properties that make them ideal for observing accretion disks. Some CVs are eclipsing, with the inflated, cool secondary orbiting into the line of sight and obscuring the accretion disk. Some CVs have steady state accretion, with accretion disks that are always in outburst, while others have material dumped onto the disk randomly so that they can be observed while the emission from the disk is varying.

A CV is a relatively well understood system (relative to a quasar) which contains elements that astronomers believe to be present in quasars; material introduced onto an accretion disk, which radiates.

OBSERVATIONAL EVIDENCE

Thermal Emission from the Disk

The most compelling evidence for the presence of a UV and optically emitting thermal accretion disk in CVs is also the simplest. When an eclipsing cataclysmic

Figure 1. This cartoon of a CV shows the low mass secondary accreting material via the inner Lagrangian point onto an accretion disk surrounding the more massive primary.

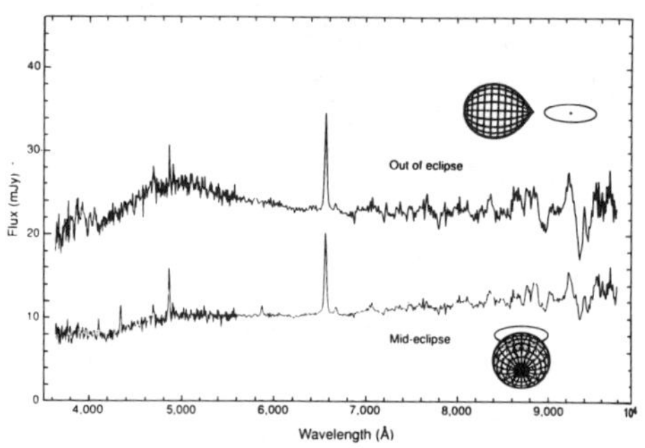

Figure 2. Rutten et al. 1993 show that the CV flux is blue when observed out of eclipse and red when observed in mid-eclipse.

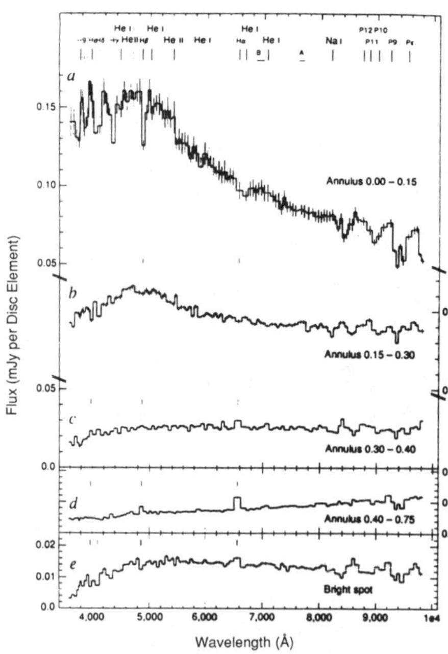

Figure 3. Applying eclipse mapping methods to spectra taken during different CV phases, Rutten et al. 1993 find that the spectra at different disk radii have different temperatures.

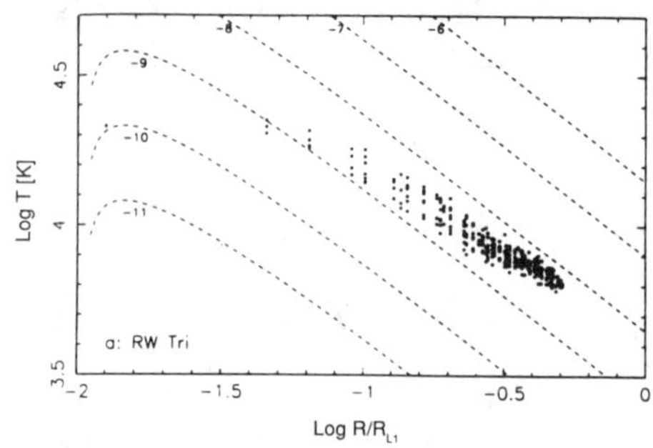

Figure 4. Rutten et al. 1992, using eclipse mapping, finds temperature as a function of radius, which agree well with accretion disk models.

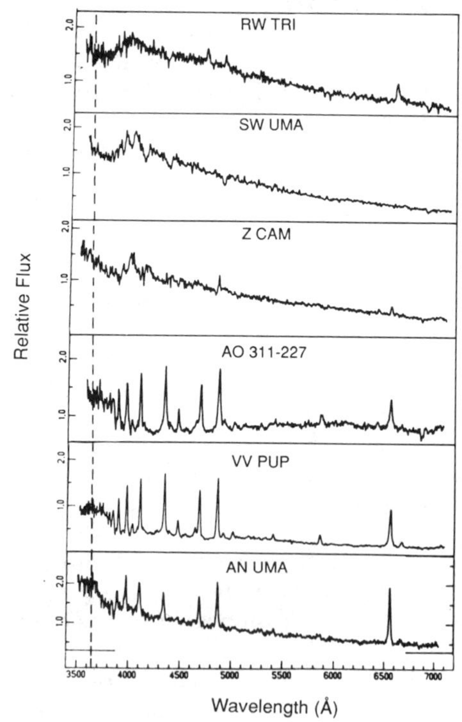

Figure 5. Balmer jumps are observed in CVs in both absorption and emission (Williams 1983).

variable is observed at different phases of the eclipse, the spectrum becomes bluer out of eclipse, and losses its blue component in mid-eclipse. This change in color with phase of eclipse is illustrated in Figure 2, from Rutten et al. (1993) in an observation of UX Ursae Majoris. The top spectrum, observed when the CV is out of eclipse, as illustrated by the cartoon above the top spectrum, has a blue bump of thermal emission. The bottom spectrum, observed when the CV is in mid-eclipse, illustrated below the bottom spectrum, is much less blue. Emission from the secondary is negligible at short wavelengths.

To further illustrate this mapping of the accretion disk, Rutten et al. (1993) have obtained light curves of the eclipse at each wavelength interval. The light curves at each wavelength have been fit independently with an eclipse mapping algorithm. The maps at each wavelength can be combined to produce spectra as a function of annulus in the accretion disk. When this is done, as shown in Figure 3, there is a clear progression of the temperature of the spectra, with the hot thermal emission coming from the inner disk, and the cool thermal emission coming from the outer disk. When the results from this method are compared with models for the temperature of a thermal accretion disk versus radius, the observations (shown in Figure 4 for RW Tri) match the models quite well.

The eclipsing CVs demonstrate that a small scale structure (presumably an accretion disk) does indeed emit thermal continuum, with the inner region being hot ($\approx 17,000^0$) and the outer region being cool ($\approx 6,000^0$).

The Balmer Jump, Signature of a Thermally Emitting Region

A geometrically thin, optically thick accretion disk such as is seen in CVs and hypothesized in AGN, should show the spectral features due to the large change in optical depth caused by the ionization of hydrogen from the ground state at 912Å (the Lyman edge) and caused by the ionization of hydrogen from the first excited state, at 3646Å (the Balmer jump). Balmer jumps are observed in both absorption and in emission in the spectra of a number of CVs observed by Williams (1983), as shown in Figure 5. The features appear at slightly longer wavelengths than the Balmer jump, due to the convergence of the high order Balmer lines into a pseudo continuum. While modeling of accretion disks by Wade (1984) predicts strong Balmer jumps in absorption, the observations show some objects with no Balmer jumps, some objects with Balmer jumps that are not deeply absorbed, and even some objects with Balmer jumps in emission instead of absorption. The presence of the Balmer jumps, however varied, does demonstrate the important point that the emission mechanism is thermal.

Double Peaked Line Profiles

Another signature of an accretion disk predicted by models is a double peaked emission line profile (Marsh, & Horne 1988). A disk that is emitting emission lines, and that is not observed face on will show velocity structure in the emission lines indicative of the rotation of material in the disk. Figure 6 is a cartoon of the most extreme case of a disk, drawn face–on with the line of sight to the object in the plane of the paper, and with lines of equal velocity shown on the disk. Above the cartoon of the disk is the line profile that results from material emitting at different velocities. The reddest and bluest velocities come from regions with the fastest Keplerian orbits, moving transverse to the line of sight, i.e. points along the inner edges of the disk, which contain little material. Material moving perpendicular to the line of sight, i.e., a line down the center of the disk, emits at zero velocity, while the peaks of the line profile are from the regions with the greatest amount of material, all moving at the same velocity.

A stunning observation of IP Peg emission lines during eclipse demonstrates the viability of this picture. Figure 7 (Marsh & Horne 1990, plot thanks to K. Horne) shows the Hβ line profile immediately after mid–eclipse, immediately before mid–eclipse, and out of eclipse. The cartoons to the left of the spectra show the configurations of the binary system after and before mid–eclipse. After mid–eclipse, the part of the disk rotating towards the viewer is seen, and the blue shifted part of the double peaked profile is observed. Before mid–eclipse, the part of the disk rotating away from the viewer is seen, while the other part of the disk is obscured by occultation of the secondary, and the red shifted peak of the double peaked profile is observed.

Variability

The complex behavior of CVs in outburst is reviewed by Szkody and Cropper (1988, and references within). In the rise to outburst, the UV is observed to lag behind the optical with a delay of about a half a day in a number of CVs. The model of mass transfering from secondary to primary, and then from outer, cool, optically–emitting disk, to inner, hot, UV-emitting disk, (Pringle, Verbunt, & Wade 1986) is able to reproduce the observations and, especially, the time lag observed between the optical and the UV. The important point here is not the details of the model, but the fact that the variability is characteristic of mass transfer in a disk.

CVs as Prototypes for Objects with Mass Accretion

The type of accretion geometry observed in CVs depends on the strength of the white dwarf magnetic field: the stronger the magnetic field, the more disrupted the accretion disk becomes. The CVs divide themselves into three categories: 1) those with $B < 10^4$, which have a well defined accretion disk ("non-magnetic"), 2) those with $10^4 < B < 10^7$, where the accretion disk is truncated as the accreting material begins to flow along the field lines ("intermediate polar", with prototype DQ Her), and 3) those with $B > 10^7$, where the accreting material flows directly from the overflowing Roche lobe of the secondary, along the field lines onto the white dwarf ("polar" type, with prototype AM Her). These types are illustrated in Figure 8. The largest majority of CVs are the non-magnetic CVs, with only a small percentage of CVs being DQ Her and AM Her type.

A CV always has Roche lobe overflow, but does not always have an accretion disk. Accretion occurs via a disk only if the magnetic field is weak. The presence of a strong magnetic field causes accretion to follow the field lines, truncating the disk at some inner radius (Figure 8). When the magnetic field becomes greater than about 10^7, the accretion occurs with material sliding down the magnetic field lines along an accretion funnel, with no accretion disk.

CVs have found at least one way to accrete material without an accretion disk, in spite of the very systematic introduction of material from the secondary to the primary, occuring in a manner that tends to increase the angular momentum of the primary. The inner Lagrangian point, from which material passes

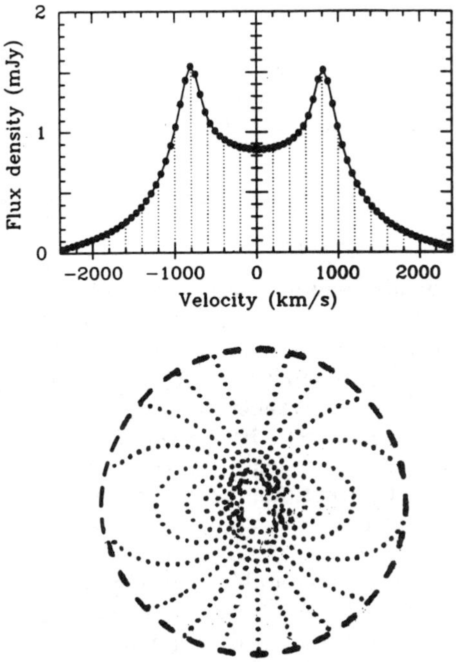

Figure 6. Top shows the profile calcuated assuming a disk rotating with Keplerian orbits. Bottom shows disk with lines of equal velocity (Marsh & Horne 1988).

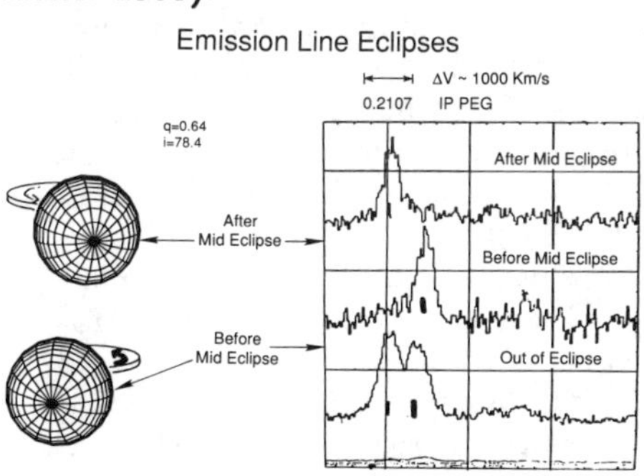

Figure 7. Line profiles observed by Marsh and Horne (1990 and 1993) as a function of phase of eclipse. Cartoons show CV phase corresponding to immediately before and after mid-eclipse.

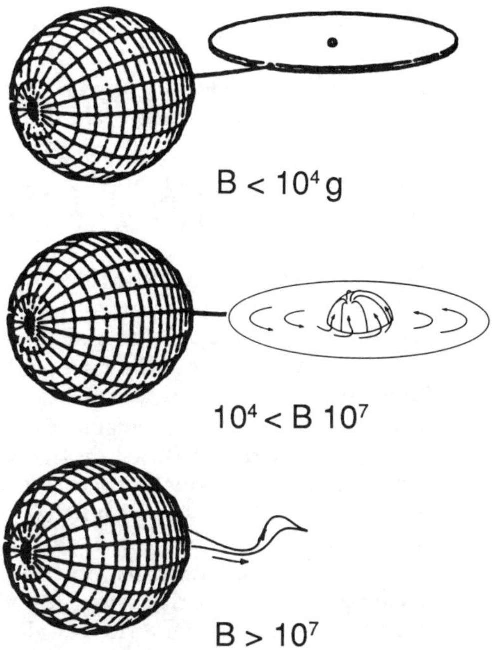

Figure 8. Accretion disks are present in CVs when the magnetic field is low. CVs with high magnetic fields accrete via accretion funnels.

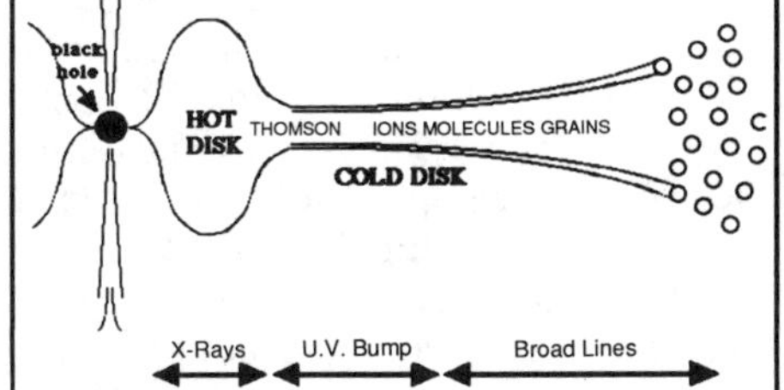

Figure 9. Accretion disks: different things to different theoreticians (Collin–Souffrin 1992). The thin disk, assumed to emit the optical and UV continuum, is the part of the disk under discussion here.

from secondary to the primary, introduces material in one plane, torquing the primary. A more random transfer of material, via a wind, for example, in the case of a high mass x-ray binary such as Vela X-1 does not increase the angular momentum of the primary, and does not require an accretion disk.

EVIDENCE FOR A UV AND OPTICALLY EMITTING ACCRETION DISK IN AGN AND QUASARS

Can one imagine an AGN without an accretion disk feeding material into a black hole? Disk structure is part of the AGN vocabulary, from the thick inner disk radiating in X-rays, to the thin disk radiating in the UV and optical, to the thick molecular torus. Figure 9, taken from Collin-Souffrin (1991) shows the larger picture of an accretion disk. I will address the smaller picture here; **Is the UV and optical spectrum of an AGN dominated by emission from an accretion disk?** Based on the points of evidence presented above for CVs, the answer to this smaller question is: 1) no, the continuum shape shows no indication of a blue component that would be the unique component from a UV and optically emitting disk, 2) no, there is no indication of a Lyman edge such as expected from thermally emitting bodies, 3) no, the double peaked profiles predicted by the velocity structure of a disk are not seen, except for a few rare possible cases, and 4) no, the variability in the UV and optical do not show the characteristics of a varying accretion disk with a hot inner region and a cool outer region.

UV and Optical Continuum: From a Disk?

Much of the original motivation for the accretion disk model came from multifrequency observations of quasars showing a big blue bump of emission that was well fit by a blue, thermal component identified with an accretion disk (Malkan 1983). On close inspection, the blueness of that bump is dependent on what is assumed to be the shape of the underlying IR emission. The original model of Malkan assumed a power law IR component, which was subtracted from the quasar spectrum, leaving a very blue UV continuum (Figure 10). However, Barvainis (1992 and references therein) has argued that the IR component is thermal and most likely due to dust, the IR emission dropping rapidly for $\lambda < 1\mu$m. Thus when the IR component is subtracted, the resulting optical and UV component is quite flat (see also Figure 1, Malkan (1991) for a fit using an IR dust component). Figure 11 demonstrates the spectral index of the UV spectrum resulting from subtracting a power law IR component compared to subtracting a dust component. The residual UV spectrum is flatter than the $F_\nu \propto \nu^{1/3}$ spectrum expected for an AGN disk model with a wide range of radii and hence disk temperatures. Malkan (1989) is still able to fit the data with the accretion disk plus thermal dust model by raising the black hole mass by a factor of a few. While the flatter UV spectral index found after removing the thermal dust component does not exclude the possibility that the UV emission arises from a thermal accretion disk, it does mean that the parameters of such a disk would be considerably different than those commonly discussed in the literature.

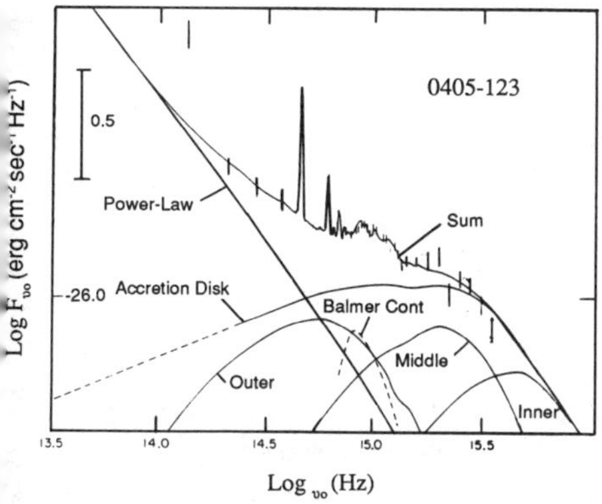

Figure 10. When an IR power law is removed from the optical plus UV spectrum of a quasar, the remainder is well fit by thermal emission, assumed to be from an accretion disk (Malkan 1983).

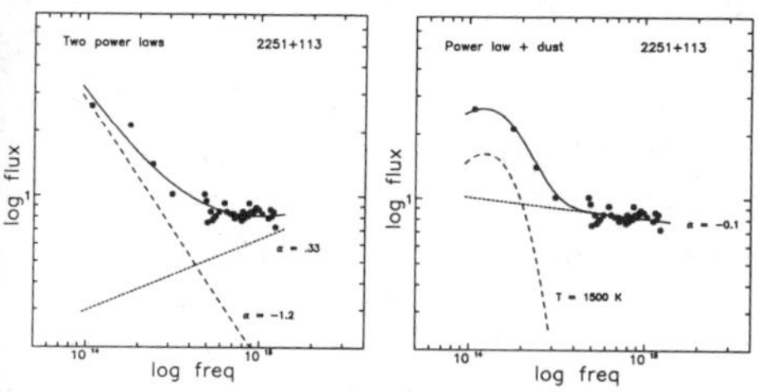

Figure 11. When the IR component is instead taken to be due to dust (Barvainis 1992 and references therein), the optical plus UV emission is not rising towards the shortest wavelengths, as required by a thermally emitting accretion disk.

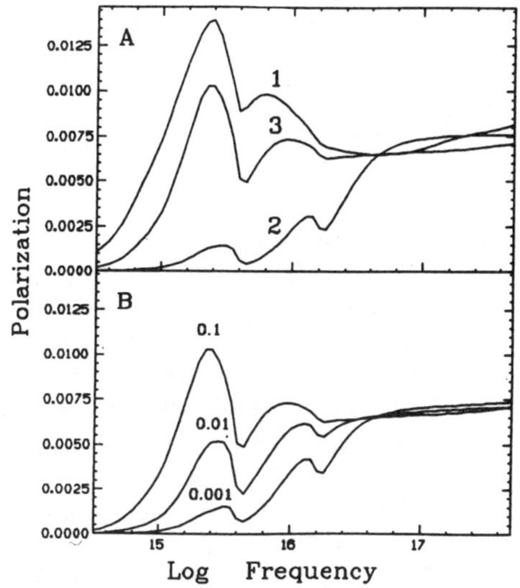

Figure 12. The results from the models of Laor, Piran, & Netzer (1990) for polarization across the Lyman edge. The polarization should increase monotonically with decreasing wavelength from 1 μm to 1000Å, and have a rapid change at the Lyman edge.

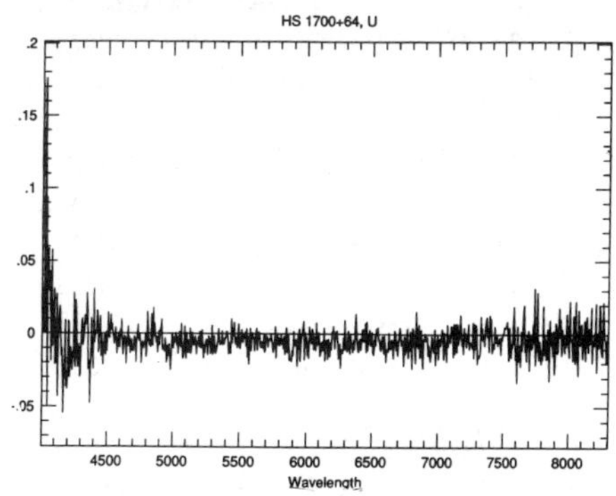

Figure 13. Polarization data from Antonucci, Hurt, & Miller (1994) for HS1700+6416(z=2.72) shows no change from rest wavelength 2230Å down to 1150Å, contrary to predictions of Laor, Piran, & Netzer (1990).

Lyman Edge Observations

Geometrically thin, optically thick accretion disks are expected to show strong Lyman edges (Kinney 1992), just as CVs are expected to show strong Balmer jumps. While the AGN spectra do show evidence of Balmer jumps, the Balmer jump lies in the region where both the host galaxy and the central source make contributions to the flux. However, the Lyman edge lies in a region where the host galaxy makes *no* contribution to the flux, and the Lyman edge originates only from the central source. Thus the Lyman edge is the more important feature to analyse in AGN. Two studies have been undertaken of the Lyman edge region in quasars. The first, by Antonucci, Kinney, & Ford (1989) concentrated on high redshift quasars, observed in the optical. The second, Koratkar, Kinney, & Bohlin (1992) concentrated on low redshift quasars observed in the UV. Neither study found evidence of Lyman edges intrinsic to the quasar. All Lyman edges that were detected were associated with intervening absorbing material, or were found to have corresponding *narrow* absorption lines, so that they could not be from a disk which has an intrinsic velocity spread. Preliminary results from an ongoing study of the Lyman edge region of quasars with the Faint Object Spectrograph of HST by Tytler and collaborators (D. Tytler and C. Davis 1993, private communication) verify the earlier results: changes in flux across the Lyman edge must be less than about 4%.

One of many ways to get rid of the Lyman edge is to invoke a scattering atmosphere over the surface of the disk: the scattering atmosphere redistributes the photons so as to wash out features such as the Lyman edge. However, a scattering atmosphere would leave another signature – high polarization (Antonucci 1992). When Laor, Netzer, & Piran (1990) model both the spectrum and the polarization of the disk with scattering atmospheres which are denser than those previously used, they find a low polarization which increases monotonically with decreasing wavelength from 1μm down to 1000Å, accompanied by a large change in polarization at the Lyman edge (see Figure 12). Polarization observations by Antonucci (1992) show no evidence of either a monotonic increase in polarization with decreasing wavelength, or of a change in polarization at the Lyman edge. The lack of a Lyman edge in absorption, and the lack of high polarization just longward of the Lyman edge provides firm evidence against the picture of the UV emission arising from a thermal accretion disk.

Double Peaked Profiles

A systematic search for double peaked Hα profiles in radio galaxies and radio loud quasars has been carried out by Eracleous and Halpern (1994). This signature of disk emission had been found in a number of individual radio loud AGN (see Eracleous and Halpern for references, and Storchi-Bergmann, Baldwin, & Wilson 1993) but has not been observed in the Hα lines of radio quiet AGN. Of the 74 objects included in their study, 12 are found to have double peaked profiles. While the double peaked profiles imply that a disk may be present, the objects with the double peaked profiles show optical continua *dominated by starlight*, in contrast to the optical continua of most AGN, which are dominated by the featureless blue continuum usually attributed to disk emission. Figure 13 demonstrates this stellar component for 3C 332 in the form of NaD and TiO absorption features. While a disk may be producing the double peaked pro-

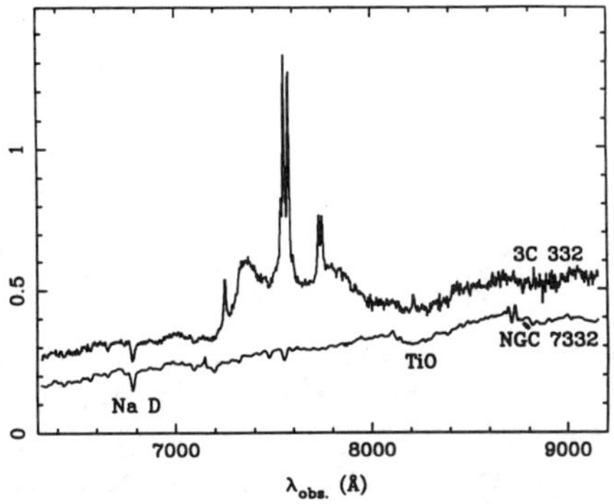

Figure 14. Double peaked Hα profile from Eracleous and Halpern (1994): note that the continuum is dominated by the galaxy, not by a featureless continuum.

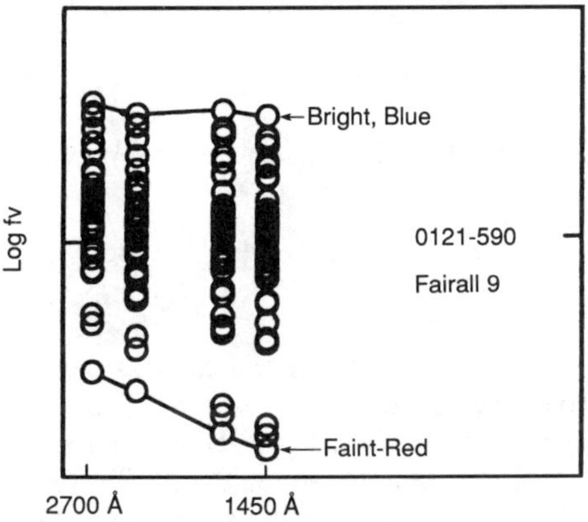

Figure 15. Variability of Fairall 9 in the UV (Kinney et al. 1991) shows that when the object is bright, it is blue, and when it is faint, it is red. This behaviour is not suggestive of an accretion disk.

files seen in these objects, the same disk is not producing appreciable optical continuum.

Variability

Variability in quasars and AGN is common. These objects vary the least in the red (Cutri et al. 1985), and the most in the ultraviolet (Kinney et al. 1991). The UV variability is so strong, that in the UV, blazars and quasars are not distinguishable by the amount they vary. Figure 14 shows averages in 100Å bins for the AGN Fairall 9 as monitored by the IUE (Kinney et al. 1991). Unlike the CVs, which become brighter first at longer wavelengths, and then at shorter wavelengths as the mass is transfered through increasingly hotter parts of the disk, the AGN vary at all wavelengths simultaneously, becoming bluer when the are bright, and redder when they are faint. There is no lag observed between the optical and the UV continuum emission when the objects vary (see Peterson 1993 for a review of AGN variability). The simultaneity of the variability is not characteristic of a thermally emitting accretion disk.

CONCLUSION

The UV and optical continuum emission in AGN and quasars does not come primarily from an accretion disk.

ACKNOWLEDGMENTS

I would like to thank Mike Eracleous, and Kirk Korista for many discussions on accretion disk properties, and would like to give a special thanks to Ski Antonucci and Keith Horne for reading and providing critical comments on the manuscript. Thanks also to R. Rutten, G. Williams, T. Marsh, K. Horne, S. Collin-Souffrin, M. Malkan, R. Barvainis, A. Laor, S. Antonucci, and M. Eracleous for permission to use their plots and diagrams.

REFERENCES

Antonucci, R.R.J., in *Testing the AGN Paradigm*, S. Holt, S. Neff, & C.M. Urry, New York: AIP, 1992, 486

Antonucci, R.R.J., Hurt, T., & Miller, J. 1994, in preparation

Antonucci, R.R.J., Kinney, A.L., & Ford, H.C. 1989, *ApJ*, **342**, 64

Barvainis, R., in *Testing the AGN Paradigm*, S. Holt, S. Neff, & C.M. Urry, New York: AIP, 1992, 129

Collin–Souffrin, S. 1992, in *Testing the AGN Paradigm*, S. Holt, S. Neff, & C.M. Urry, New York: AIP, 1992, 119

Cutri, R.M., Wisniewski, W.Z., Rieke, G.H., & Lebofsky, M.J. 1985, *ApJ*, **296**, 423

Eracleous, M., & Halpern, J.P. 1994, *ApJS*, **90**, in press

Horne, K. 1990, in *Structure and Emission Properties of Accretion Disks, Proceedings of the Sixth IAP Astrophysics Meeting, IAU Colloquium no. 129*, C. Bertout, S. Collin, J.-P. Lasota, & J. Tran Thanh Van, Gif sur Yvette Cedex: Editions Frontieres, 1990, 3.

Kinney, A.L. 1991, in *Testing the AGN Paradigm*, S. Holt, S. Neff, & C.M. Urry, New York: AIP, 1992, 139

Kinney, A.L., Bohlin, R.C., Blades, J.C., & York D.G. 1991, *ApJS*, **75**, 645

Koratkar, A.P., Kinney, A.L., & Bohlin, R.C. 1992, *ApJ*, **400**, 435

Laor, A., Netzer, H., & Piran, T. 1990, *MNRAS*, **242**, 560

Malkan, M.A, 1983, *ApJ*, **268**, 582

Malkan, M.A. 1989, in *Theory of Accretion Disks*, F. Meyer, W.J. Duschl, J. Frank, & E. Meyer-Hofmeister, NATO ASI Series Vol. 290: Kluwer Academic Publishers, p. 23

Malkan, M.A. 1991, in *The Physics of AGN* , W.J. Duschl, & S.J. Wagner, Heidelberg: Springer-Verlag, p. 113

Marsh, T., & Horne, K. 1988, *MNRAS*, **235**, 269

Marsh, T., & Horne, K. 1990, *ApJ*, **349**, 593

Peterson, B. 1993, *PASP*, **105**, 247

Pringle, J., Verbunt, F., & Wade, R.A. 1986, *MNRAS*, **221**, 169

Rutten, R.G.M., Dhillon, V.S., Horne, K., Kuulkers, E., & van Paradijs, J. 1993, *Nature*, **362**, 518.

Rutten, R.G.M., van Paradijs, J., & Tinbergen, J. 1992, *AA*, **260**, 213

Storchi–Bergmann, T., Baldwin, J.A., & Wilson, A.S. 1993, *ApJ*, **410**, L11

Szkody, P., & Cropper, M. 1987, in *Multiwavelength Astrophysics*, F. Cordova, Cambridge: Cambridge University Press, 1987, 109

Wade, R.A. 1984, *MNRAS*, **204**, 381

Williams, G. 1983, *ApJS*, **53**, 523

The First Stromlo Symposium: The Physics of Active Galaxies
ASP Conference Series, Vol. 54, 1994
G. V. Bicknell, M. A. Dopita, and P. J. Quinn (eds.)

NUMERICAL SIMULATIONS OF ACCRETION DISKS

J. F. HAWLEY, C. F. GAMMIE AND S. A. BALBUS
Astronomy Department, University of Virginia, Charlottesville, VA 22903

ABSTRACT

The nature of the anomalous viscosity in accretion disks is a central issue of physics for Active Galactic Nuclei. The presence of a powerful, general, and local magnetohydrodynamic (MHD) instability in accretion disks suggests that the cause of angular momentum transport is MHD turbulence. We review recent results from three dimensional numerical MHD simulations in a simplified rotationally-supported shearing system. These models demonstrate that the instability leads to field amplification and angular momentum transport.

INTRODUCTION

The talks and poster papers at this conference have presented an impressive, detailed set of multiwavelength observations of Active Galactic Nuclei (AGN) and their environs. The wide variety of observed phenomena suggests an equally wide variety of physical processes, reflected in the encompassing title of this conference, "The Physics of Active Galaxies." Honoring the spirit of this title in its breach, this presentation focuses on a single, important aspect of the accretion disk in the AGN central engine, namely the origin of anomalous viscosity in disks and the role of numerical simulations in its elucidation.

The current "standard model" for AGN central engines is accretion into a supermassive black hole through a time-stationary optically thick accretion disk. However, as Anne Kinney reminds us in her contribution to these proceedings, despite all the data and observations there is still no clear, *direct* spectral evidence for such accretion disks in AGN. This stands in embarrassing contrast to X-ray binary systems, where the observational signatures of disks are clearly seen. Although the lack of disk-like features in AGN spectra obviously provides no support for the standard accretion disk picture, it is not a compelling refutation of that model either. Supporters of the standard model can simply argue that a more complete understanding of the behavior of orbiting gas near a black hole is required. At present, observations provide only the broadest of constraints for theoretical disk models, although clearly the variability across the spectrum, the X- and gamma-ray fluxes, and the existence of jets are features that require more than the standard quiescent, optically thick accretion disk.

An important tool for investigating the physics of accretion disks is numerical experiment, that is, computer simulation. The advantage of time-dependent finite-differencing and related techniques is that they can study the full complex, nonlinear, multidimensional equations of motion. The drawbacks to such techniques are the limited spatial and temporal scales ("dynamic range") that they can examine. This is a particular problem for a calculation involving a turbulent fluid. At present, numerical simulations can address certain specific questions about the dynamics of accretion systems, but they cannot model an entire disk even *hydrodynamically*, much less with the radiative transfer necessary for detailed observational predictions. Although the scope of the answerable questions is limited compared to what we would all like it to be, it is nevertheless possible to make a good start at refining AGN models.

DISK INSTABILITIES

Progress in understanding the nature of accretion disks has long been stymied by the absence of a physical basis for the anomalous angular momentum transport mechanism. Molecular and radiative viscosities in a disk are tiny and cannot account for the angular momentum transport required by the mass accretion rates. However, if present, an instability can change the laminar flow in a disk into a turbulent one, leading to an effective viscosity orders of magnitude larger than the molecular value. Although turbulence has long been a leading candidate for this anomalous viscosity, extensive studies by many investigators have failed to find any simple, local, generic hydrodynamic instability in accretion disks.

This is not to say that various special circumstances cannot lead to interesting instabilities. For example, disks might develop a vertical structure that is convectively unstable. The resultant fluid motions could drive turbulence that, in turn, might transport angular momentum. This scenario is probably more applicable to protostellar than to AGN disks (Lin 1989). Even then, it is far from clear that convective instability will lead to effective angular momentum transport, since it is not shear-driven. Indeed, Goodman and Ryu (1993) have recently carried out a linear treatment of the nonaxisymmetric convective instability. They found that transport was inward, the opposite of what is needed to drive accretion. Nonlinear transport can be studied by nonaxisymmetric numerical simulations of convection in disks. Such studies can investigate whether or not significant outward angular momentum transport is produced by the convectively driven turbulence (Kley, Lin & Papaloizou 1993). But convection does not appear to be a particularly promising mechanism for AGN.

Another possibility comes from recent investigations into the global stability of the "thick disk" or accretion torus. These thick disks have been of particular interest in the study of AGNs because their novel geometry may lead to super-Eddington luminosities and radiative jet acceleration along narrow axial "funnels." The most extreme of the thick disks is the constant angular momentum torus. Since local hydrodynamic stability is determined by the Rayleigh criterion, which states that stable angular momentum configurations must increase outward ($dl/dR \geq 0$), these disks are marginally stable to local perturbations. However, Papaloizou and Pringle (1984) demonstrated that constant angular momentum tori are unstable to a *global* hydrodynamic mode, a result they subsequently extended to other angular momentum distributions (Papaloizou & Pringle 1985). Subsequent work by many authors has elucidated the general properties of the Papaloizou-Pringle instability (reviewed, e.g., by Narayan 1990).

The thick, distended structure of the torus depends upon its maintaining both a nearly constant angular momentum distribution and substantial pressure gradient. This in turn requires significant optical depths and relatively low accretion rates, i.e., low anomalous viscosities. The Papaloizou-Pringle instability seemed to be fatal to the model. However, this concern was tempered somewhat by a number of subsequent findings from analytic perturbation studies: the global modes have significant growth rates only for nearly constant angular momentum distributions, they require at least one reflecting boundary to the disk (Narayan, Goldreich & Goodman 1987), and they are stabilized by accretion flows (Blaes 1987). Ultimately the importance of an unstable mode depends upon its properties at nonlinear saturation, and this is a question beyond the ability of linear perturbation analysis to answer. Direct three-dimensional (3D) numerical simulations (Hawley 1991) confirm that only in rather special tori can the instability grow to large amplitude. Optically-thick constant angular momentum tori are ruled out, but thick disks with angular momentum distributions closer to Keplerian appear to be acceptable.

This represents progress, but leaves us no closer to understanding the anomalous transport in Keplerian disks. Since nonlinearity has long been the last refuge of the theorist, one can ask, are disks nonlinearly unstable? Here the argument about the level of molecular and radiative viscosities in disks is turned around: Since these viscosities are so low and the Reynolds number consequently so high, might not turbulence, once established, be self-sustaining? This possibility was recognized at least as early as the seminal Shakura & Sunyaev (1973) paper in which turbulent transport was parameterized with the now-familiar α, so that the net stress in the disk was proportional to αP where P is the local pressure. However, as Shakura & Sunyaev pointed out, the high Reynolds number of a disk does not guarantee that it will be turbulent in the absence of an instability in the laminar state. Although there is free energy available in the differential rotation a mechanism is required to tap it. Nonlinear instabilities are present in certain high Reynolds number laboratory fluid experiments, but whether or not disks behave the same way is difficult to determine by purely analytic techniques. The answer is, by definition, outside the realm of linear perturbation theory.

We believe that the origin of turbulence in accretion disks lies elsewhere, with magnetohydrodynamics rather than with hydrodynamics, and with *linear* rather than nonlinear instabilities. In a recent series of papers, Balbus & Hawley (1991, 1992a,b) have considered the linear stability of an accretion disk containing a weak magnetic field. Remarkably, axisymmetric perturbations in such systems are locally unstable. Their stability is determined not by the Rayleigh criterion of an outwardly increasing specific angular momentum, but by the requirement that the angular *velocity* increase outward (Balbus & Hawley 1991)

$$d\Omega^2/dR \geq 0. \qquad \text{(STABILITY)}$$

This condition is, of course, violated in accretion disks. The instability was originally analyzed in a global calculation of magnetized Couette flow by Velikhov (1959), but its relevance to the accretion disk transport problem went unrecognized for over three decades. The instability arises from the magnetic field's tendency to transfer angular momentum from a fluid element on a *low* angular momentum orbit to a fluid element on a *high* angular momentum orbit. This causes the elements to separate further and to increase the magnetic tension, leading to further angular momentum transfer. Since it feeds off the free energy of differential rotation, the instability is precisely the sort of process needed to produce an anomalous viscosity. This instability is present only if the field is weak; perturbing a disk with a strong field results in wave propagation, not instability. Of course, angular momentum transport in a strongly magnetized disk is not particularly mysterious.

What is surprising is the wide generality of the instability. Its maximum growth rate is independent of field properties, including field strength and orientation. A magnetic field is unstable for wavenumbers k such that

$$(\mathbf{k} \cdot \mathbf{v_A})/\Omega \leq \sqrt{3},$$

where $\mathbf{v_A}$ is the Alfvén velocity, and Ω is the rotational frequency. The role of the magnetic field strength is solely to determine the unstable wavenumbers. The growth rate of the most unstable wavenumber

$$(\mathbf{k_{max}} \cdot \mathbf{v_A})/\Omega = \sqrt{15/16},$$

is the local Oort A-value of the disk, equal to $\frac{3}{4}\Omega$ for a Keplerian disk (Balbus & Hawley 1992b). For weak fields, i.e., $v_A \ll c_s \ll R\Omega$, where c_s is the sound speed, there will be unstable modes with wavelength smaller than the disk scale height, $H \sim c_s/\Omega$. For stronger fields, the shortest unstable wavelength would not "fit" within one disk scale height, leading to a stable, strongly magnetized, disk.

The instability is not limited to axisymmetric modes. An initial-value time-dependent linear analysis for the nonaxisymmetric accretion disk has been carried out in Balbus & Hawley (1992a). Analysis of the nonaxisymmetric instability is somewhat complicated because the radial wavenumber k_r evolves due to the shearing background. However, the results are easy to summarize because the field itself is sheared in such a way that the combination $\mathbf{k} \cdot \mathbf{v_A}$ remains constant with time. As was true for the axisymmetric case, the magnetic field enters the linear equations only through this Alfvén parameter. In the absence of buoyancy, the only other important factor governing the instability's growth rate is the wavenumber ratio k/k_z. As the radial wavenumber of a nonaxisymmetric mode evolves from leading to trailing, a mode experiences a period of exponential growth if $(\mathbf{k} \cdot \mathbf{v_A})^2 < -d\Omega^2/d\ln R$. Since typical e-folding times are a fraction of the orbital period, this is sufficient to ensure field amplification factors of many orders of magnitude.

Under very general circumstances, therefore, disks are unstable with an e-folding time significantly less than one orbital period. With such fast growth rates the effect on a laminar Keplerian accretion disk will be both immediate and profound. However, the character of the saturated nonlinear instability is not easily obtained by analytic means. Fortunately, both the wavelength and timescale of the instability make it suitable to study by direct numerical simulation. In the remainder of this paper we will review the simulations that have been done to date, and their implications.

2D AND 3D MHD SIMULATIONS

We begin our investigation with a series of problems with simple input physics, first in two, and then in three dimensions (2D and 3D).

Hawley & Balbus (1991) considered a particularly simple set of initial conditions, that of an isolated uniform B_z flux tube in an axisymmetric uniform Keplerian flow. These 2D simulations verified that the numerical scheme could reproduce the linear modes and their growth rates as predicted by perturbation theory. This was also important for confirming that neither the gas sound speed nor the presence of a subthermal azimuthal magnetic field had a significant influence on the development of the instability. The simulations reveal many of the important nonlinear properties of the instability, including significant angular momentum transport, mode-mode coupling transferring energy from the most unstable mode to both larger and smaller wavenumbers, and reconnection as an amplitude-limiting dissipation process. Reconnection occurs in our simulations due to numerical effects, namely the advection of oppositely directed field into one zone. The present code includes no explicit resistivity.

The next step was to extend the numerical models to longer evolutionary times using a greater variety of initial field configurations. The simulations of Hawley & Balbus (1991) were limited by the use of hard-wall boundary conditions in the radial direction. To overcome this problem we use the *shearing box* approximation and periodic radial shearing boundary conditions. The shearing box is a local corotating approximation to a rotationally supported system. One selects a radius R_o, rotating with angular velocity Ω_o, and constructs pseudo-Cartesian coordinates $x = R - R_o$, $y = R_o(\phi - \Omega_o t)$, and z. The equations of motion are linearized around R_o; centrifugal, gravitational, and Coriolis forces are included but curvature effects are of higher order in the local approximation and are ignored. A background shearing velocity v_y produces the appropriate equilibrium for the system of interest. For a Keplerian accretion disk $v_y = -\frac{3}{2}\Omega_o x$. Thus, the shearing box approximation allows a well-resolved numerical grid to represent a local piece of a larger Keplerian disk.

The shearing box system retains many of the physical properties of an accretion disk that are essential to the local behavior of the MHD instability, but without many of the disk's global complexities. The shearing box is therefore a model with which we can study the nonlinear behavior of the instability in isolation. The outcome of a simulation can depend on the physical dimensions of the box, the grid resolution, and the strength and orientation of the initial magnetic field, but there is no vertical gravity, g_z, and hence no buoyancy, and the initial density and pressure are uniform. In the absence of buoyancy, and because the instability is incompressible, the sound speed of the gas plays, at most, a secondary role.

The 2D (x, z) version of the shearing box was used in the simulations of Hawley & Balbus (1992). These models examine the evolution of several different initial field geometries: uniform vertical B_z fields, uniform radial B_x fields, and radially varying B_z fields. As before, during the linear phase of the evolution the numerical growth rates agree with the analytic linear perturbation results. The subsequent nonlinear behavior of the instability can vary considerably, however, depending upon the initial field geometry. Simulations with uniform net vertical field B_z produced some of the more dramatic outcomes. This field configuration produces large-scale radial streaming motions and strong magnetic fields. Within a few orbits the modes evolve to the largest lengthscales afforded by the grid, and the flow organizes itself into two channels, one moving out and carrying excess angular momentum, the other a channel consisting of subkeplerian fluid moving in. This result requires only the presence of a background B_z field; the presence or absence of a weak toroidal field is immaterial. These "channel" solutions were obtained for a variety of initial field strengths, computational domain sizes, and grid resolutions. The rate of angular momentum transport is large, and the field strength and flow velocities grow exponentially. This surprising behavior is clearly related to the finding by Goodman and Xu (1993) of an exact *nonlinear* exponentially growing (incompressible) channel solution from an initially vertical magnetic field.

In contrast to the large-scale, exponentially growing coherent nonlinear mode obtained from a vertical field, an initial radial field model develops structures with small vertical and large radial wavelengths. These saturate at relatively modest amplitudes. Here, the size of the computational domain appears to be the predominate factor limiting growth.

Perhaps the most suggestive 2D simulations were those with an initial vertical field that varied sinusoidally with radius so that the net field on the grid is zero. This field geometry produces complex turbulent poloidal fluid motions, angular momentum transport, and field amplification. The instability increases the total magnetic field energy (B^2) by a factor of ten or more. However, since this simulation is axisymmetric with zero net field, the "anti-dynamo" theorem (Moffatt 1978) precludes unlimited field amplification. Once the turbulence is fully developed, the volume-averaged magnetic field energy declines slowly with time. The rate of the subsequent reduction of magnetic field energy is determined by numerical reconnection, which is, in turn, set by grid resolution.

The results clearly show the need for going to three dimensions if questions of sustained field growth are to be addressed. We present here a summary of recent 3D simulation results and place them in the context of the general questions that can now be addressed by numerical simulations. The complete 3D shearing box results are presented in Hawley, Gammie & Balbus (1993).

In our standard shearing box we set the vertical size of the box equal to the ratio of the sound speed to the rotational frequency, c_s/Ω_o. In an accretion disk this would correspond approximately to the disk vertical scale height; as the shearing box has no vertical gravity, this simply establishes a sound crossing time in units of Ω. The radial dimension is also set equal to c_s/Ω, and the angular (y) dimension is 2π larger. We set the initial pressure and density equal to constants, assume background Keplerian shear flow, and add magnetic fields of various orientations and strengths. Most of the simulations were run with $31 \times 63 \times 31$ physical grid zones in (x, y, z). Select test simulations were also run at both half and twice the resolution as a means of measuring the numerical truncation error.

Every magnetic field strength and geometry that is unstable in 2D remains unstable in 3D, i.e., vertical B_z initial fields, radial B_x initial fields, and zero-net initial fields. One important difference in the 3D simulations is that the axisymmetric "channel solutions" are unstable to nonaxisymmetric perturbations. This has been nicely demonstrated by Goodman and Xu (1993), who did a linear perturbation analysis of their exact channel solution and found that they are subject to several types of "parasitic" instabilities. Because of these parasitic instabilities the coherent flow breaks apart into a turbulent state that proved to be self-sustaining.

The 3D simulations are able to follow the development of the instability in uniform weak toroidal (y) fields. The instability is nonaxisymmetric in this case, and so could not be studied in 2D. Again, the results were consistent with the local, nonaxisymmetric linear perturbation study (Balbus & Hawley 1992a). Pure toroidal fields show less growth and field amplification than those that begin with an initial poloidal field. However, whether the initial field is toroidal or poloidal, the instability ultimately saturates in a highly nonisotropic turbulent state. Thus the problem of determining the anomalous viscosity in accretion disks is "reduced" to understanding nonisotropic MHD turbulence!

These simulations are too crude to provide substantial detail about the MHD turbulent state over a wide range of wavenumbers, but we can make a few observations. First, although in a time-averaged sense the magnetic and thermal energies are in steady state, they do show significant variation on an orbital timescale. The fluctuation amplitudes of the perturbed magnetic fields and velocities, e.g., δB, δv, are much larger than their mean values. (The failure to include these fluctuations, which are generally correlated, is perhaps the most significant shortcoming of the α-disk models.) A general trend is that the largest component of magnetic field is the toroidal field, followed by radial, and then vertical. This is because the toroidal field grows both by differential rotation and the instability, the radial field grows just by the action of the instability, and the vertical field is affected largely directly by the other two components. The velocity components follow a similar hierarchy. Of all the variables, only the thermal energy has a secular change; it grows steadily with time. It can be shown that the boundary conditions chosen will do work on the shearing box in proportion to the angular momentum transport. Since we have not included cooling, the box must gradually heat up. This effect can be reduced by decreasing the adiabatic $\gamma \rightarrow 1$, and it does not otherwise affect the outcome of the simulation.

By examining fourier transforms of the data we can measure the spectral energy density in the fundamental variables. In all cases most of the power is found in the lowest wavenumbers. In practical terms this means that large scale magnetic fields account for most of the magnetic energy and stress. Experiments in which the grid resolution, initial field strength, and box size are systematically varied show that in simulations with a purely vertical initial field the saturation amplitudes of the magnetic and kinetic energies are determined primarily by the size of the computational box. The field strength grows until the smallest unstable wavelength just fits within the box.

The reason for saturation is less clear for simulations with initial toroidal fields. In this case, saturation depends upon several factors, including grid resolution, computational domain size, and the strength of the initial field. In the case of a background toroidal field, perturbations are amplified as wavenumbers k_r swing from leading to trailing and pass through the range of unstable wavenumbers. The time available for this amplification process is set by the periodicity size of the box. The amplification must also contend with the competing process of dissipation, produced here primarily by numerical diffusion. Simulations with higher resolution and larger physical domain sizes should provide further insight into the toroidal field case.

Since one of our original motivations was to find the source of the anomalous viscosity in an accretion disk, can we determine an effective "alpha" value from our simulations? After the instability has established the turbulent state in the shearing box, angular momentum is transported outward at significant rates. Of the two components of local angular momentum transport, $\rho v_x v_y$ and $-B_x B_y/4\pi$, the magnetic stress is larger by a factor of $3 \sim 4$. In these simulations the rate of angular momentum transport is more directly proportional to the magnetic energy, B^2 rather than the thermal energy P. It is traditional to parameterize the anomalous viscosity by the ratio of the stress to the gas pressure P. Taking the time-averaged stress and comparing it with our initial gas pressure we can thus obtain a Shakura-Sunyaev alpha value. Our α depends on the scale we assume for the box. However, it is natural to set the box size to the disk scale height, as we have done, and this yields $\alpha \sim 0.1 - 0.4$ for the vertical field models and about a factor of 10 lower for the toroidal initial field configurations. Of course, since the shearing box is a highly oversimplified model, one should not attach too much significance to the precise values. The important point is that these transport rates are achieved with *no assumptions* other than a weak magnetic field in a Keplerian shear flow.

Largely as a control, but also as a study that is interesting in its own right, we have simulated the behavior of purely hydrodynamic perturbations in our standard shearing box. Sound and epicyclic waves are present due to the initial perturbations, but these are stable. Even when perturbed at the 10% level, no nonlinear instabilities develop. The angular momentum transport corresponds to $\alpha < 10^{-5}$, a level consistent with the amplitude of the imposed perturbations. In another test the magnetic fields were removed partway through a turbulent MHD run; the simulation continued with the hydrodynamic terms only. This allows us to compare directly the efficacy of MHD with hydrodynamic processes. In the MHD case angular momentum transport is self-sustaining. When the fields are "switched off" this angular momentum transport quickly dies out, demonstrating that it is the action of the fields that maintains the turbulence.

One should bear in mind that the simulations have numerical diffusion and resistivity that give the box effective hydrodynamic and magnetic Reynolds numbers that are much smaller than those expected in a real disk. However, while it is possible that the purely hydrodynamic results might be somewhat different if the Reynolds number were larger, the MHD instability produces and sustains turbulence *despite* these numerical effects. Hence, at better resolution the relative abilities of hydrodynamic versus magnetohydrodynamic processes to produce and maintain turbulence should remain the same. In light of this, the question as to whether or not nonlinear instabilities can maintain purely hydrodynamic turbulence appears to be moot.

CONSEQUENCES FOR DISKS

Although the shearing box is an idealized representation of an accretion disk, these are direct simulations of the three-dimensional equations of MHD, and no further simplifying assumptions have been made about the physics. Relating our results back to accretion disks places us on slightly thinner ice, of course, but we can nevertheless identify significant consequences as well as pose questions for future investigation.

First, the anomalous viscosity in accretion disks is due to MHD turbulence. This was identified as a likely candidate by Shakura & Sunyaev (1973), and now the local MHD instability provides the physical cause. The rate of angular momentum transport ("α") depends upon the magnetic energy sustained within the turbulent disk. Our results suggest that in actual disks the magnetic field strength and angular momentum transport rate will be determined by macroscopic disk properties such as scale height rather than by microscopic details such as reconnection length. If the scale height proves to be the primary determining factor, then we expect field energies to be near equipartition, corresponding to $\alpha \lesssim 1$.

The question as to whether or not significant magnetic energies can be built up in a disk in the face of the inevitable dissipative processes is essentially the disk dynamo issue. Does the instability constitute a full MHD dynamo? Certainly all the ingredients are there: the differential rotation is a source of energy, and the instability a means to extract it. Although the present results are promising, higher resolution simulations involving additional features such as buoyancy and spherical geometry are required. Buoyancy in particular is likely to be important. The vertical motions resulting from buoyant or Parker instabilities could twist toroidal field into vertical field, thereby increasing the effect of the instability. Such a scenario has been proposed by Tout & Pringle (1992). We note that the kinematic dynamo does not seem to be a viable model. A kinematic dynamo assumes field amplification from preexisting turbulence in which the back reaction from the Lorentz forces can be neglected. We have found that it is precisely those Lorentz forces, inappropriately regarded as negligible in kinematical models, that drive the turbulence in the first place.

The magnetic dynamo question also has implications for the production of MHD winds and jets. Contributions in this proceedings by Roger Blandford, Mitch Begelman and Arieh Königl describe several scenarios that involve magnetically accelerated winds originating in AGN accretion disks. These models require a large-scale net magnetic field threading the disk and extending out to large distance. Such a field could be advected in with the accreting matter, but it remains to be seen if this occurs in a turbulent MHD disk. As Roger Blandford has emphasized, an MHD wind can lift considerable angular momentum directly out of the disk in contrast to the MHD turbulence which transports angular momentum radially through the disk. Simulations of stratified MHD disks should be able to compare the efficacy of these two processes.

How and where the orbital energy is released will have a significant impact on what an accretion disk looks like. The standard picture assumes local release of energy and immediate thermalization. However, in our simulations the majority of the transport is due to magnetic fields and a good deal of the available energy (up to equipartition levels) is stored therein. This need not be released locally inside the disk but might, for example, be released by reconnection events in a disk corona. In this manner a good deal of the energy dissipated could emerge in the form of nonthermal emission.

All these points highlight the most important limitation of the present work:

the absence of vertical gravity. The models described above provide sufficient evidence on that point. The magnetic field energy varies considerably within the box; the ratio $P_{gas}/P_{mag} = \beta$ ranges over several orders of magnitude. The fluid density also has substantial spatial variation. If such a system were placed within a vertical gravitational field, buoyant and Parker instabilities would be inevitable.

What then are the prospects for future simulations employing vertical gravity? The difficulties are technical, not conceptual or fundamental. Such simulations will require three dimensions with good resolution over several scale heights. It is straightforward to incorporate vertical gravity in the shearing box model. The highly parallel architecture of computers such as the Connection Machine holds promise for providing both the memory and the time required for such simulations. This project is now underway in collaboration with James Stone of the University of Maryland.

Eventually we must graduate to global disk simulations. The local nature and high degree of symmetry in the shearing box means that we cannot assign an accretion rate to a given angular momentum transport rate; as much mass moves in as moves out. Investigations will need to include physics beyond ideal MHD, such as radiative transfer, ambipolar diffusion, and resistive dissipation. Accretion disks promise to tax the resources of both the astrophysicist and the supercomputer for some time to come.

ACKNOWLEDGMENTS

This work is partially supported by NSF grants PHY-9018251 and AST-9021348, and NASA grants NAGW-1510 and NAGW-2376. Computations were carried out on the Cray YMP system of the National Center for Supercomputing Applications.

REFERENCES

Balbus, S. A., & Hawley, J. F. 1991, *ApJ*, **376**, 214

Balbus, S. A., & Hawley, J. F. 1992a, *ApJ*, **400**, 610

Balbus, S. A., & Hawley, J. F. 1992b, *ApJ*, **392**, 662

Blaes, O. M. 1987, *MNRAS*, **227**, 975

Goodman, J., & Ryu, D. 1993, preprint

Goodman, J., & Xu, G. 1993, preprint

Hawley, J. F. 1991, *ApJ*, **381**, 496

Hawley, J. F. & Balbus, S. A. 1991, *ApJ*, **376**, 223

Hawley, J. F. & Balbus, S. A. 1992, *ApJ*, **400**, 595

Hawley, J. F. & Gammie, C. F., & Balbus, S. A. 1993, preprint

Kley, W, Lin, D. N. C., Papaloizou, J. 1993, *ApJ*, in press

Lin, D. N. C. 1989, in *Theory of Accretion Disks*, ed. F. Meyer et al., Dordrecht: Kluwer, 89

Narayan, R. 1990, in *Structure and Emission Properties of Accretion Disks*, C. Bertout et al., Cedex: Editions Frontieres, 153

Moffatt, H. K. 1978, Magnetic Field Generation in Electrically Conducting Fluids, London: Cambridge University Press

Narayan, R., Goldreich, P., and Goodman, J. 1987, *MNRAS*, **228**, 1

Papaloizou, J. C. B. & Pringle, J. E. 1984, *MNRAS*, **208**, 721

Papaloizou, J. C. B. & Pringle, J. E. 1985, *MNRAS*, **213**, 799

Tout, C. A. & Pringle, J. E. 1992, *MNRAS*, **259** , 604

Shakura, N.I. & Sunyaev, R.A. 1973, *A&A*, **24**, 337

Velikhov, E.P. 1959, *Soviet JETP*, **36**, 995

The First Stromlo Symposium: The Physics of Active Galaxies
ASP Conference Series, Vol. 54, 1994
G. V. Bicknell, M. A. Dopita, and P. J. Quinn (eds.)

A NEW MODEL OF THE AGN'S CENTRAL ENGINE

V.V.GVARAMADZE & G.Z.MACHABELI
Abastumani Astrophysical Observatory of Georgian Academy of Sciences, A.Kazbegi ave., 2-A, Tbilisi, 380060, Republic of Georgia

ABSTRACT The influence of large-scale magnetic fields on accretion disk formation is either neglected or included inconsistently in standard models of active galactic nuclei (AGNs). In this connection a new model of AGN is proposed providing a new mechanism for the formation of jet outflows.

INTRODUCTION

It is generally accepted that an Active Galactic Nucleus consists of a black hole and a Keplerian accretion disk. It is also acknowledged that jet formation and other manifestations of activity in AGN are directly related to large-scale magnetic fields. At the same time, the influence of the magnetic field on AGN formation is usually neglected (although there are indications that the gravitational and magnetic energy in an AGN are of the same order of magnitude). That is, it is supposed that the accretion disk first forms and only after that does the amplification of magnetic field occur.

It seems quite plausible that a weak large-scale homogeneous magnetic field is present in intergalactic space. The feed-back of such a regular magnetic field on AGN formation was first considered in detail by Bisnovatyi-Kogan & Ruzmaikin (1976). It was shown that field lines of initially homogeneous magnetic field frozen into plasma accreting onto the black hole acquire an "hour-glass" form. It is assumed that plasma freely moves along magnetic field lines and a thin stationary accretion disk, supported in an equilibrium formation by the poloidal magnetic field, is formed at the equatorial plane (inside the cross-piece of the "hour-glass"). This magnetic field is dipolar near the disk. In the equatorial region, outside the external boundary of the accretion disk, a ring-shaped stagnation zone forms. But this model is idealized, as it is supposed that plasma does not leave the stagnation zone and the plasma motion along the inhomogeneous magnetic field is considered to be free. A more consistent consideration of the feed-back of a magnetic field leads to the necessity of developing a principally new model of the AGNs (Gvaramadze & Machabeli, 1993). Our model of AGN is in a certain sense opposite to the above model. According to our model, in the quasiequilibrium state the inner boundary of the accretion disk is kept at a certain (relatively large) distance from the gravitating centre by a poloidal magnetic field (see Fig. I). The accretion disk corresponds to the stagnation zone in the model of Bisnovatyi-Kogan & Ruzmaikin. A part of the accreting plasma leaves the disk to fall onto the black hole in the form of filaments produced by the flute instability. Another part of the accreting plasma flows out from the

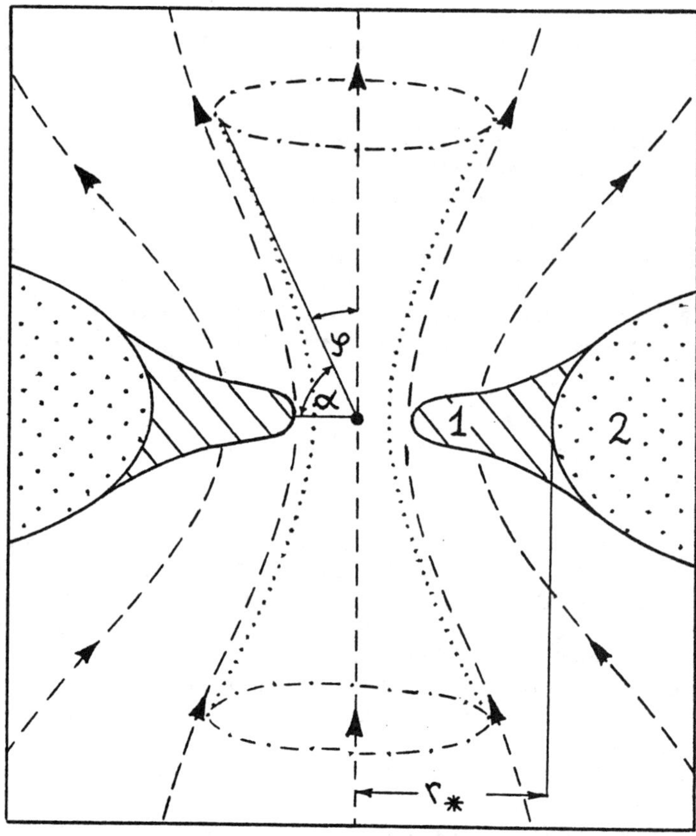

FIGURE I Our model of the AGN: 1 - the filaments; 2 - the accretion disk. The averaged biconical front of the converging shock waves is shown by dotted lines.

disk along the lines of the poloidal magnetic field.

THE AGN FORMATION

Plasma motion at the distance

$$R_a = 10^{23} \left(\frac{M}{10^9 M_\odot}\right) \left(\frac{T_\infty}{10^4 K}\right)^{-1} \text{cm} \tag{1}$$

from the black hole is basically determined by gravitation force. If the plasma does not possess angular momentum, it initially falls onto the black hole freely and isotropically at the velocity $v_{ff} = (2GM/R)^{1/2}$, with the plasma density growing as $\rho = \rho_\infty (R_a/R)^{3/2}$, where ρ_∞ is the plasma density at distance R_a. The magnetic field lines frozen into the plasma are deformed and take the form of an "hour-glass". Since the magnetic flux is conserved, the strength of the magnetic field grows as $H = H_\infty(R_a/R)^2$, where H_∞ is the homogeneous magnetic field strength at the distance R_a. The magnetic energy density grows more rapidly than the gravitational energy density and the accretion eventually becomes anisotropic. The feed-back of the magnetic field affects the plasma motion first near the black hole. Gradually the region of appreciable feed-back action expands. For the plasma moving in the equatorial plane from the distance R_a towards the black hole, the magnetic and dynamic pressures become equal, $H^2/8\pi \simeq \rho v^2/2$, at the distance r_* from the centre (see Fig. I) given by

$$r_* = \left(\frac{v_{A\infty}}{v_{ff\infty}}\right)^{4/3} R_a, \qquad (2)$$

where $v_{A\infty} = H_\infty/(4\pi\rho_\infty)^{1/2}, v_{ff\infty} = (2GM/R_a)^{1/2}$. At this distance regular accretion in the equatorial plane is halted. At smaller radii plasma motion across the magnetic field is possible only in the form of filaments produced by the flute instability (Kazbegi et al., 1993).

Plasma motion along the inhomogeneous magnetic field $\vec{H}(z)$ towards the black hole is hampered due to the conservation of the adiabatic invariant $\mu \equiv mv_\perp^2(z)/2H(z)$, where $v_\perp$ is the cyclotron rotation velocity and z is the coordinate measured along the magnetic field lines. For a power-law distribution of magnetic field, $H = H_0(z_0/z)^n$, the diamagnetic force responsible for this effect has the form:

$$|\vec{F}_\mu| = n\frac{mv^2}{z}\sin\psi\cos\psi,$$

where $v^2 = v_\perp^2 + v_\parallel^2$, ψ is pitch angle ($\tan\psi \simeq v_\perp/v_\parallel$) and $v_\parallel$ is the plasma particle's velocity along the magnetic field. It can be shown that the diamagnetic force averaged over the pitch angles is n times larger than the gravitational force ($n = 3$ for the dipolar magnetic field). As a result, the majority of the particles are reflected towards the region of weaker magnetic field.

The filaments moving across the magnetic field represent magnetic tubes loaded by plasma. These tubes are evacuated near the black hole and a gradual accumulation of magnetic field occurs at $r = r_*$. This growing magnetic field may keep more and more plasma in the quasiequilibrium state. The most plausible case is that the mass of the plasma kept by the magnetic field reaches the mass of the central black hole. In this case the strength of the magnetic field reaches its maximum possible value determined by the virial theorem:

$$\frac{H^2}{8\pi}r_*^3 \leq \frac{GM^2}{r_*}.$$

For illustration, we take $M = 10^9 M_\odot, T_\infty = 10^4 K, \rho_\infty = 10^{-29} g \cdot cm^{-3}, H_\infty = 10^{-12} G$. It follows from (2), with the help of (1), that $r_* \simeq 10^{17}$ cm. The corresponding maximum admissible value of the magnetic field strength is $H_{max} \simeq 10^5$ G. The plasma configuration with the inner boundary supported in quasiequilibrium by a poloidal magnetic field can be identified with the accretion disk (see Fig. I).

The thickness of the accretion disk near its inner boundary is determined by the characteristic scale of the poloidal magnetic field, but its estimation is a complicated problem. One can only say that the magnetic field inside the funnel of the accretion disk is nearly uniform, similar to the magnetic field inside a solenoid or a ring current. At large distances from the equatorial plane the magnetic field is dipolar.

THE CUMULATIVE FORMATION OF JETS

The filament motion can be considered to be free since the magnetic field does not increase towards the centre inside the funnel of the accretion disk. The plasma penetrates the funnel from the filaments due to diffusion. This plasma is less dense than that in the accretion disk. Plasma density perturbations produced by the filament motion propagate at the characteristic velocity $v_A \simeq H/(4\pi\rho_{fun})^{1/2}$, where ρ_{fun} is the plasma density inside the funnel. It is evident that the inequality $v_A > v_{ff}$ is fulfilled for the initial stage of filament motion, i.e. the density perturbations rapidly propagate away from a falling filament. But the plasma is distributed inhomogeneously inside the funnel. At a certain distance from the centre, the speed of the filaments v_{ff} may become larger than v_A if the plasma density decreases towards the centre slower than r^{-1} or increases. In consequence, quasiperpendicular collisionless shock waves arise, travelling ahead the filaments. The tangent to the shock front at a point far from the equatorial plane is inclined to the direction of motion of the filaments by an angle close to the Mach angle (see Fig. I): $\alpha = \arcsin(M_A^{-1})$, where $M_A \equiv v_{ff}/v_A \geq 1$ is the Alfven number.

The averaged front of converging shock waves has a biconical symmetry (see Fig. I). The closer the Alfven number M_A to unity, the smaller the bicone half-angle: $\varphi = \frac{\pi}{2} - \alpha$.

The converging biconical shock wave produces, owing to the cumulative effect, two well-collimated high-velocity jets moving in opposite directions along the bicone axis (see Sokolov, 1988). For small bicone half-angles, the jet velocity is estimated to be $v_j \simeq v_{ff}/\varphi$ and may be much larger than the velocity v_{ff} of the converging biconical shock wave. The width of the jets sharply decreases when the bicone half-angle decreases.

Note that in the case of converging shock waves with biconical symmetry there is no necessity, for jet formation, in a source of axial momentum. In that case the axial pressure gradient is built up in a natural way due to the cumulative effect that leads to plasma ejection.

Since the averaged front of the converging biconical shock wave may not have a strict axial symmetry, the jet direction can deviate, from time to time, from the bicone axis.

Furthermore, the formation of the filaments can occur in bursts, rather than in a purely random manner. Such a time correlation between filaments seems to be natural if some time is required to pile up enough plasma to form a new generation of filaments. Then the jet itself consist of plasma blobs.

FORMATION OF THE EXTERNAL ENVELOPE OF THE JETS AND ELECTRON ACCELERATION

The flow of the accreting plasma is braked by magnetic field near the inner disk edge. As a result, a jump of the magnetic field strength and plasma density, i.e. a standing circular shock wave, forms.

On the shock front, a Harris-type instability (Harris, 1959) develops due to the existence of a preferential ion velocity ($v_{\perp i} \gg v_{\parallel i}$). High-frequency potential waves are excited, with the wave vector having a considerable component along the magnetic field. These waves are effectively damped by the resonant electrons thereby transferring to the latter the greater part of the ion energy. As a result, electrons are accelerated up to relativistic velocities.

Furthermore, the feed-back on the ions of the excited waves results in quasi-linear diffusion, which redistributes the ion energy over pitch angle ($v_{\perp i} \simeq v_{\parallel i}$). This leads to a wide ($\sim 10^{17}$ cm) two-sided plasma outflow along the poloidal magnetic field lines in the region of the decelerating, accreting plasma (near the standing shock wave). This outflow can be considered as an external envelope of the well-collimated high-velocity jets produced by the cumulative effect.

CONCLUSION

The basic ideas of the proposed model may be used for the explanation of the formation of biconical nebulae with the jets directed along the bicone axis that are observed in star-forming regions (Scarrott et al., 1986).

Our model is also promising in solving the problem of hard X-ray emission of AGNs and a related problem of effective energy exchange between protons and electrons (see Kazbegi et al., 1993).

ACKNOWLEDGMENTS

The authors are grateful to G.Bisnovatyi-Kogan and A.Volokitin for helpful discussions. One of us (V.G.) acknowledges the hospitality of G. Bicknell during a visit to Mt. Stromlo Observatory.

REFERENCES

Bisnovatyi-Kogan,G.S., & Ruzmaikin,A.A. 1976, *Astrophys. Space Sci.*, **42**, 401

Gvaramadze,V.V., & Machabeli,G.Z. 1993, *MNRAS*, submitted

Harris,E.G. 1959, *Phys. Rev. Lett.*, **2**, 34

Kazbegi,A.Z., Krishan,V., Machabeli,G.Z., & Melikidze,G.I. 1993, *MNRAS*, submitted

Scarrott,S.M., Warren-Smith,R.F., Draper,P.W., & Gledhill,T.M., 1986, *Canad. J. Phys.*, **64**, 426

Sokolov,I.V. 1988, *Teplofizika Visokikh Temperatur*, **26**, 560 (in Russian)

The First Stromlo Symposium: The Physics of Active Galaxies
ASP Conference Series, Vol. 54, 1994
G. V. Bicknell, M. A. Dopita, and P. J. Quinn (eds.)

LOW-FREQUENCY VARIABILITY OF EXTRAGALACTIC RADIO SOURCES

D.B. MELROSE
Research Centre for Theoretical Astrophysics, School of Physics, University of Sydney, NSW 2006, Australia

ABSTRACT Interpretations of low-frequency variability, flickering and intra-day variability of extragalactic radio sources are reviewed with emphasis on refractive interstellar scintillations and models involving relativistic boosting. Coherent emission is severely constrained by induced Compton scattering.

INTRODUCTION

Variability of compact extragalactic radio sources was first reported by Hunstead (1972) and soon recognized to be relatively common (Shapirovskaya 1978). Flat-spectrum sources are now known to exhibit several different forms of variability (e.g., Rickett 1990): low-frequency variability, flickering, intra-day variability (IDV) and extreme scattering events (Fiedler et al. 1987). These are in addition to flaring and other forms of variability that are clearly intrinsic to the source (e.g., Marscher and Gear 1985). The properties of these phenomena may be summarized as follows. (1) Low-frequency ($\lesssim 1\,$GHz) compact ($\lesssim 10\,$mas) sources show variations of amplitude 3–30% over months to years. These variations depend on the galactic latitude of the source (e.g., Mantovani et al. 1990), which is strongly indicative of a scintillation in the interstellar medium. (2) Flickering involves a few per cent flux variations on a time scale of a few days at frequencies of a few GHz (Heeschen 1982). (3) IDV involves more rapid variability of a few percent on time scales of a few hours at higher frequencies (Heeschen et al. 1987; Witzel 1990). There is evidence for a correlation between the radio and optical variations (Wagner and Witzel 1992). (4) Extreme scattering events involve changes up to 50 percent over a few days.

A constraint on the interpretation of such variability arises from the maximum brightness temperature, $T_B \approx 10^{12}\,$K, that is determined by inverse-Compton losses in a self-absorbed synchrotron source. If observed variations on a time scale Δt are assumed to imply a source size $\sim c\Delta t$, then the implied brightness temperature can exceed the inverse-Compton limit. The inferred values for IDV, $T_B \sim 10^{17}$–$10^{19}\,$K are well in excess of the limit. Three ways of avoiding this difficulty have been suggested: (1) attribute the variability to scintillations in the interstellar medium (ISM) (Shapirovskaya 1978; Rickett, Coles and Bourgois 1984); (2) invoke relativistic boosting together with some intrinsic time-dependence in the source, such as shocks (e.g., Marscher and Gear 1985; Qian et al. 1991) or a lighthouse effect (Camenzind and Krockenberger 1992); (3) abandon the synchrotron hypothesis and invoke some coherent emission process (e.g., Colgate 1967; Baker et al. 1988).

In this paper, these interpretations of the radio variability of compact sources are reviewed. First, the theory of scintillations is summarized, emphasizing refractive interstellar scintillations (RISS), and then its application to the interpretation of the variability is discussed. Next the relativistic beaming and lighthouse models are described, and finally, coherent emission mechanisms are discussed critically, emphasizing a limitation imposed by induced Compton scattering.

THE THEORY OF SCINTILLATIONS

Scintillations are a result of fluctuations, $\delta\phi$, in the phase of a wave due to variations in the refractive index along the line of sight. In a plasma, variations in the refractive index are due to spatial variations, δn_e, in the electron density, n_e. The observer sees a phase deviation from the straight-line path

$$\delta\phi(\mathbf{r}, L) = r_e\lambda \int_0^L dz\, \delta n_e(\mathbf{r}, z), \tag{1}$$

where the z-axis lies along the line of sight, $\mathbf{r} = (x, y)$ denotes a transverse displacement, $r_e = 2.82 \times 10^{-15}\,\mathrm{m}$ is the classical radius of the electron, and λ is the wavelength of the wave. The properties of the fluctuations may be characterized by the autocorrelation function of the phase variations, usually written as the wave structure function, $D(\mathbf{r}, L)$, with

$$D(\mathbf{r}, z) = \langle[\delta\phi(\mathbf{s}, z) - \delta\phi(\mathbf{s} - \mathbf{r}, z)]^2\rangle, \tag{2}$$

where the angular brackets denote an ensemble average, intended to describe averages over nearby lines of sight. Through (1), the phase fluctuations are related to the density fluctuations, which may be described by

$$R_n(\mathbf{x}) = \langle\delta n_e(\mathbf{x}')\delta n_e(\mathbf{x}' - \mathbf{x})\rangle, \qquad Q_n(\mathbf{K}) = \int d^3\mathbf{x}\, R_n(\mathbf{x}) \exp(i\mathbf{K}\cdot\mathbf{x}), \tag{3}$$

that is, $Q_n(\mathbf{K})$ is the spatial Fourier transform of $R_n(\mathbf{x})$. The density fluctuations in the ISM are assumed to be isotropic with a power-law dependence on wavenumber,

$$Q_n(\mathbf{K}) = C_n^2 K^{-\beta}, \tag{4}$$

with cutoffs at an inner scale, $K > 2\pi/r_{\rm in}$, and an outer scale, $K < 2\pi/r_{\rm out}$, and where C_n^2 is the (density) structure constant. A Kolmogorov spectrum corresponds to $\beta = 11/3$. For such a spectrum, the phase structure function is a power law,

$$D(r) = (r/r_{\rm diff})^{\beta-2}, \tag{5}$$

where $r_{\rm diff}$ includes all normalization constants, and where cutoffs at $r < r_{\rm in}$ and $r > r_{\rm out}$ are not included explicitly. In the theory, $r_{\rm diff}$ plays the role of a coherence scale, defining the width of the visibility function (e.g., Rickett et al. 1984), with $\theta_0 = \lambda/2\pi r_{\rm diff}$ a scattering angle.

There are two important lengths in scintillation theory. One is $r_{\rm diff}$, which characterizes the size of the density inhomogeneities. The other is the Fresnel

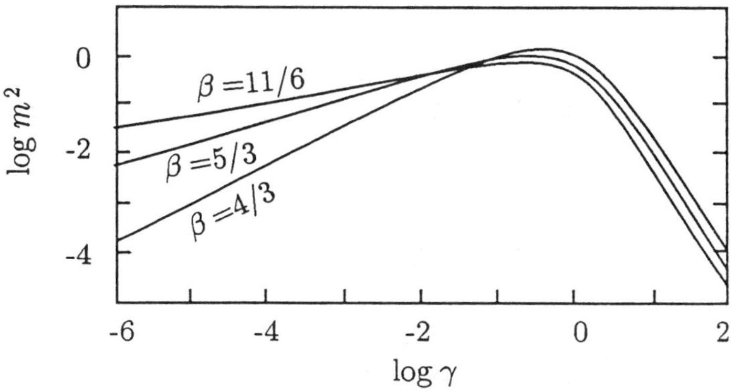

FIGURE I Plots of the square of the scintillation index versus $\gamma = r_{\rm in}/\theta_0 L$ for a source with $\theta_{\rm S} \ll \theta_0 = \lambda/2\pi r_{\rm diff}$, and for various power-law density spectra with inner scales $r_{\rm in}$. [After Coles et al. (1987).]

scale $r_{\rm F} = (L/k)^{1/2}$, $k = 2\pi/\lambda$, which characterizes the contribution, $r^2/2r_{\rm F}^2$, to the phase from diffraction, with L the distance between the scattering inhomogeneity and the observer. The *strength* of the scattering may be described using the parameter $u = r_{\rm F}/r_{\rm diff}$ (e.g., Rickett 1990). For $u \ll 1$ the scattering is said to be *weak*, in which case (5) implies $D(r) \ll 1$, so that the phase fluctuations are small. For $u \gg 1$ the scattering is said to be *strong*.

Strong scattering separates into two distinct scales: diffractive scintillations on a scale $r \lesssim r_{\rm F}$, and refractive scintillations on a scale $r \sim u r_{\rm F}$. Thus refractive scintillations occur on a characteristic scale $r_{\rm ref} = r_{\rm F}^2/r_{\rm diff}$. Estimates for $\lambda = 6\,{\rm cm}$ give $r_{\rm diff} \sim 10^6\,{\rm m}$, $r_{\rm F} \sim 10^9\,{\rm m}$, $r_{\rm ref} \sim 10^{13}\,{\rm m}$ (Mutel and Lestrade 1990). Diffractive scintillations can be observed only for a source with an angular size less than the diffractive limit, $\lambda/r_{\rm diff} \sim 0.1\,{\rm mas}$, which is not satisfied for the extragalactic sources of interest here. Only RISS are relevant for compact extragalactic sources with angular diameter $\theta_S \gtrsim 1\,{\rm mas}$. The properties of RISS may be deduced using geometric optics (e.g., Blandford and Narayan 1985), which is much simpler than the wave theory for strong scattering.

The strength of scintillations is usually described in terms of the scintillation index, m, which is related to the fluctuations in the intensity, I, by $m^2 = \langle (I - \langle I \rangle)^2 \rangle / \langle I \rangle^2$. In Figure I, m^2 is plotted for an extended source with a gaussian angular profile, $\propto \exp(-\theta^2/2\theta_{\rm S}^2)$, with $\theta_{\rm S} \gg \theta_0$. Note that the strongest scintillations occur for a certain combination of parameters that involves the inner cutoff scale and the wavelength of the radiation. It follows from Figure I that large amplitude variations ($m^2 \gtrsim 10^{-1}$) due to RISS are possible only in a relatively narrow parameter range. However, variations of a few percent ($m^2 \gtrsim 10^{-4}$) can occur over six orders of magnitude in the parameter γ.

The characteristic size scale, $r_{\rm scint}$, time scale, $t_{\rm scint}$, modulation index and decorrelation bandwidth for RISS were estimated by Narayan (1992) for a Kolmogorov spectrum ($\beta = 11/3$):

$$r_{\rm scint} \approx r_{\rm ref}\frac{\theta_S}{\theta_{\rm scatt}}, \qquad t_{\rm scint} \approx \frac{r_{\rm ref}}{v}\frac{\theta_S}{\theta_{\rm scatt}},$$
$$m \approx \left(\frac{r_{\rm diff}}{r_{\rm F}}\right)^{1/3}\left(\frac{\theta_S}{\theta_{\rm scatt}}\right)^{-7/6}, \quad \delta\lambda_{\rm dc} \approx \lambda, \tag{6}$$

with $\theta_{\rm scatt} = r_{\rm ref}/L$, and where v is the transverse speed of the inhomogeneities relative to the observer, which is thought to be of order the orbital speed of the Earth, $v \sim 50\,{\rm km\,s^{-1}}$. The decorrelation width $\delta\lambda_{\rm dc}$ is the range of wavelengths over which correlation in the fluctuations in lost.

The scaling of the quantities in (6) with wavelength follows from $r_{\rm ref} \propto \lambda^{\beta/(\beta-2)} \propto \lambda^{2.2}$, where the latter form is for $\beta = 11/3$. However, this dependence is complicated by the dependence of the source size, θ_S, on λ in (6). An intrinsic dependence of the angular size θ_S of the source on λ is implied by standard models to explain the flat spectrum (e.g., Marscher 1977; Blandford and Königl 1979). Rickett (1986) described the dependence of the various models for the flat spectrum in the form $\theta_S \propto \lambda^x$, with x in the range $0 \lesssim x \lesssim 1.25$. This gives $t_{\rm scint} \propto \lambda^x$ and $m \propto \lambda^{2-7x/6}$.

INTERPRETATION OF VARIABILITY IN TERMS OF SCINTILLATIONS

Some, but not all the variability of compact extragalactic radio sources may be explained in terms of RISS.

For the low-frequency variability, the original suggestion (Shapirovskaya 1978; Rickett et al. 1984) that it is due to RISS is generally supported by the observations (e.g., Rickett 1990). The observed dependence of the variability on galactic latitude (e.g., Mantovani et al. 1990) is strongly indicative of RISS. Some of the variability over months to years appears to be intrinsic, associated with flaring and other secular changes in the source. This type of variability does not violate the 10^{12} K limit. Nevertheless there are observed features that are not readily explained. For example, the amplitude of the variability can be higher than the theory for RISS implies. There is also the variability reported by Slee and Siegman (1988) at 80 MHz and 160 MHz which does not show the dependence on galactic latitude expected for RISS. However, scintillation theory does not explain all the phenomena satisfactorily for pulsars, where the theory is very well established. For example, Gupta, Rickett and Coles (1993) argued that RISS theory applied to pulsars at 74 MHz seemingly underestimates the strength of the actual scattering, as described by the scintillation index. These authors noted that this raises questions over the quantitative application of a simple version of RISS theory to extragalactic sources.

As an example, consider the source 1741–038, which has a compact, flat-spectrum core with an extended steeper-spectrum component. VLA observations of the variation of this source were reported by Hjellming and Narayan (1986), who found that the variations at 1.49 GHz and 4.9 GHz are independent of stronger variations at higher frequencies 15 GHz and 22 GHz. Some of the data from recent monitoring of this source at NRL and by the MOST are

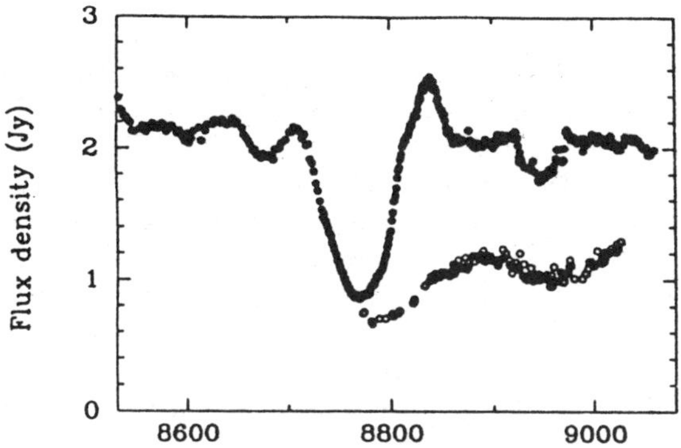

FIGURE II The variation of the flux of the source 1741–038 at 2.3 GHz (upper plot, courtesy NRL) and 843 MHz (lower plot, courtesy MOST) as a function of time (Julian days 244000).

illustrated in Figure II. At 8.3 GHz (not shown) the source shows a secular increase by about 30% over ~ 3 yr, with some fluctuations. The amplitude of the variations is very large at 2.3 GHz, and only slightly smaller at 843 MHz, with indications of a delay of about a week. It appears that at the minimum in the fluctuations at 843 MHz, the signal from the flat-spectrum core falls below that of the steeper-spectrum component (Hunstead, private communication). This would imply a scintillation index $m \approx 1$, which is higher than predicted by the existing theory for RISS.

The interpretation of flickering in terms of RISS was initially thought to be unfavorable (Hjellming and Narayan 1986; Simonetti, Cordes and Heeschen 1985). However, Blandford, Narayan and Romani (1986) showed that the flickering can be explained in terms of RISS in an extended medium provided that the intrinsic source size varies with wavelength. An additional complication in the earlier applications of RISS to flickering is that the distinction between the average image, and the ensemble-average image was not recognized. These depend on whether the observation involves an average over a time, t, in the range $t_{\rm diff} \ll t \ll t_{\rm ref}$ or in the range $t \gg t_{\rm ref}$, respectively, where $t_{\rm diff} = r_{\rm diff}/v$ is the time scale for diffractive scintillations and $t_{\rm ref} = r_{\rm ref}/v$ is the time scale for refractive scintillations. Narayan and Goodman (1989) argued that observations are in the average-image regime, rather than the ensemble-average regime as had been assumed implicitly, and Mutel and Lestrade (1990) reported observations that support this suggestion.

The scintillation index is expected to decrease strongly with increasing frequency, and the smaller amplitude of the flickering at higher radio frequencies, compared with the low-frequency variations is consistent with this. The strength of the scattering also decreases with increasing frequency, and another possibility is that at higher frequencies the scattering is weak rather than strong. Rickett (1990) noted that all possible regimes for the interpretation of the observed variabilities had not yet been explored, in particular the limit of weak scattering

(WISS).

Blandford et al. (1986) developed a model for flickering due to refractive scintillations. The main observation that these authors aimed to explain is the structure functions of the flux density fluctuations, $D^{(1)}(\tau) = \langle\delta F(t)\delta F(t)\rangle - \langle\delta F(t)\delta F(t+\tau)\rangle$, with τ the time lag between two observations. The observations suggest $D^{(1)}(\tau) \propto \tau$ throughout the observation period, which is not consistent with the predictions of simple theory. Blandford et al. (1986) assumed the inhomogeneities to be moving relative to the observer at speed v, and for simplicity assumed a power law $\beta = 4$ in (5). They argued that the intensity correlation is of the form

$$D^{(1)}(\tau) = \frac{Q_0}{2\pi k^4 \theta^2} \exp\left(-\frac{v^2\tau^2}{2L^2\theta^2}\right), \tag{7}$$

where Q_0 is defined by writing the Fourier transform of (5) in the form $D(K) = Q_0 K^{-\beta}$. In the case $\beta = 4$, the scatter image of a point source would lead to $\theta \propto \lambda^2$. Then, if the scatter broadening is negligible, the coefficient in (7) varies as λ^2. The observations suggest that the intensity fluctuations are roughly independent of λ (Simonetti et al. 1985). Moreover, for $v^2\tau^2 \ll 2L^2\theta^2$, (7) implies $D^{(1)}(\tau) \propto \tau^2$, again contrary to observation.

Blandford et al. (1986) made two assumptions to resolve these inconsistencies. First, they assumed that the scattering occurs over a large range of distances, rather than in a single screen at distance L and, second, they assumed that the intrinsic size of the source varies with wavelength as $\theta_I \propto \lambda$. The first assumption requires that Q_0 in (7) be replaced by a local value per unit length, which they took to be $dQ_0(L)/dL = (2Q_0/\sqrt{\pi}H)\,exp(-L^2/H^2)$, where H is the scale height of the Galactic disk. For $v\tau \ll H\theta$ this leads to

$$D^{(1)}(\tau) \approx \frac{Q_0 v}{\sqrt{2}\pi k^4 \theta^3}\left(\tau - \frac{\sqrt{2}v\tau^2}{H\theta}\right), \tag{8}$$

which gives the desired form $D^{(1)}(\tau) \propto \tau$. For a flat spectrum source with a constant brightness temperature core, the angular size varies as $\theta_I \propto \lambda$, and then (8) implies rms fluctuations varying $\propto \lambda^{1/2}$, in reasonable agreement with the observations. The success of this model supports the interpretation of flickering in terms of RISS.

INTERPRETATION OF THE IDV

In discussing the interpretation of IDV one needs to decide whether to take account of the suggested correlation between radio and optical variations, because such a correlation would exclude RISS. Suppose this suggested correlation is ignored. Consider refractive scintillations due to a plasma inhomogeneity of size Δr at a distance l that causes variations on a time scale $\Delta t = \Delta r/v$. For the inhomogeneity to cover a source of angular size θ_S at a distance L, requires $\Delta r \gtrsim l\theta_S$, implying $l/L \lesssim (v/c)(\theta_{\Delta t}/\theta_S)$, where $\theta_{\Delta t} = c\Delta t/L$ is the angular size estimated from the light propagation time. For the inverse-Compton limit to be satisfied, IDV then requires $l/L \lesssim 10^{-3}(v/c)$, which can only plausibly be satisfied for a screen in our Galaxy, e.g., $l/L \sim 10^{-7}$, $v/c \sim 10^{-4}$. For example,

RISS with $\Delta r \sim 10^9$ m, $l \sim 3$ kpc, $v \sim 30\,\mathrm{km\,s^{-1}}$ appears capable of accounting for IDV (e.g., Wambsganss et al. 1989).

Some emphasis has been placed on the rapid changes in polarization that sometime accompany IDV (e.g., Krichbaum, Quirrenbach and Witzel 1992). Such variations are expected due to random changes in the rotation measure associated with scintillations (Melrose 1993). The polarization fluctuations should increase rapidly with wavelength (Melrose 1993), providing a test for this suggested interpretation.

Now assuming the correlation between radio and optical variations to be real, there are at least three suggested explanations for the IDV. (1) Doppler boosting enhances the observed value of T_B compared to that in the rest frame of the source. Consider a source moving toward the observer at an angle θ to the line of sight with Lorentz factor $\gamma_S = (1-\beta_S^2)^{-1/2}$. The ratio of the observed to the intrinsic T_B equals $\mathcal{D}^3$, with $\mathcal{D} = [\gamma_S(1-\beta_S\cos\theta)]^{-1}$. For IDV the inferred T_B exceeds the inverse-Compton limit by a factor 10^6, which is equal to $\mathcal{D}^3$ for $\mathcal{D} \sim 10^2$. Thus this model requires $\gamma_S \sim 10^2$, which is larger than seems plausible for radio-emitting jets (e.g., Qian et al. 1991). (2) A model proposed by Qian et al. (1991) is a modification of a model for flaring due to shocks propagating along a jet (e.g., Marscher and Gear 1985). A shock is assumed to brighten on hitting inhomogeneities in the jet, producing variations on a time scale $\Delta t = \Delta L(1+z)/c\beta_S\mathcal{D}\gamma_S$, with z the red shift and ΔL the separation between the inhomogeneities. Suggested parameters $\Delta L = 0.16$ pc, $\gamma_S \approx 14$, $\mathcal{D} \approx 28$ can account for $\Delta t \approx 1$ day. (3) Camenzind and Krockenberger (1992) suggested a lighthouse model in which plasma bubbles are ejected at relativistic speed nearly along the axis of a magnetized jet. These bubble spiral about the axis of the jet, and the variations are attributed to the changing angle between the line of sight and the bubbles.

COHERENT EMISSION

Several different coherent emission mechanisms have been proposed to overcome the difficulty with the 10^{12} K-limit. Here these various mechanisms are reviewed briefly, and some critical comments are made on their application to extragalactic sources.

All the coherent emission mechanisms proposed for extragalactic sources (Colgate 1967; Baker et al. 1988; Sol, Pelletier and Asséo 1989; Krishnan and Wiita 1990; Weatherall and Benford 1991; Benford 1992) involve a two stage process in which some instability generates longitudinal waves which produce the escaping radiation either through nonlinear plasma processes (e.g., Melrose 1991) or through scattering by relativistic particles. For example, Baker et al. (1988) developed a model in which a relativistic electron beam injected from the central engine into the inner portion of a jet produces Langmuir turbulence. The electromagnetic radiation is assumed to be due to a form of free electron maser emission (e.g., Attwood et al. 1985), which is related to linear acceleration emission (e.g., Rowe 1992a, 1992b), with the Langmuir turbulence acting as the wiggler field. The coherence is attributed to bunching, in the sense that the total power radiated is assumed to exceed the power radiated incoherently by a factor $N \sim n_b\lambda^3$, where N is the number of particles per bunch, n_b is the number density of the beam electrons and $\lambda = 2\pi c/\omega$ is the wavelength of the Langmuir waves. (In my opinion this is not correct, and the correct factor is the number of beam electrons per coherence volume of the *emitted* radiation.) Baker et al. (1988) estimated the ratio of the power radiated by this process to the power radiated in synchrotron emission to be $G = (E^2/c^2B^2)(N^2N^*/n_e)$, where N^* is the number of bunches, and n_e is the number of electrons, and where E is the electric field strength in the waves. The other models for coherent emission differ from this model in specific details of the instability and the conversion mechanism to electromagnetic radiation.

The main motivation for invoking coherent emission is to explain radio emission brighter than the 10^{12} K-limit. However, as pointed out by Rees (1992) and Coppi, Blandford and Rees (1993), the effective optical depth for induced Compton scattering is of order T_B/m_ec^2 times the optical depth for Thomson scattering, $\sim \sigma_T n_e L$, where L is the depth of the source. The typical optical depth for a broad-line emission region in an AGN is $\sim 10^{-2}$. Although the optical depth in any postulated source of coherent emission needs to be estimated directly, it is clear that $\sim \sigma_T n_e L$ cannot be very small because a relatively high value of n_e is required for the mechanisms discussed above to operate, and a relatively large value of L is required on energetic grounds. It follows that for $T_B \sim 10^{18}$–10^{20} K, which is required in the most extreme cases, induced scattering would be a very strong effect. The expected signatures of induced Compton scattering in AGN were discussed by (Coppi et al. 1993), who argued that the absence of these signatures implies a limit on the Thomson optical depth, and hence on the electron density. For a source with $T_B = 10^{18}$ K and radius $r = 10^{15}$ cm, they estimated $n_e \leq 10\,\mathrm{cm}^{-3}$, which they noted to be ten orders of magnitude smaller than the value required in the source region to sustain the observed power.

Another physical process that can modify and prevent escape of radiation involves a collimated beam of photons generating Langmuir waves, which then scatter the beam photons. This process appears to be important in eclipsing pulsars (Gedalin and Eichler 1993), where it can prevent the beam of photons propagating through a seemingly transparent plasma. The constraints imposed by induced scattering and this photon-beam instability appear to be severe, and seemingly preclude any coherent emission mechanism producing the very bright emission postulated.

DISCUSSION AND CONCLUSIONS

(1) Low-frequency variability includes both RISS, which depends on the galactic latitude of the source, and intrinsic variations that do not violate the inverse-Compton limit.

(2) Flickering appears to be due to RISS

(3) IDV may be due to RISS if the suggested radio-optical correlation is not real. Otherwise, it must be intrinsic to the source, and can be explained in terms of relativistic boosting, with Lorentz factor $\sim 10^2$, a relativistic shock encountering inhomogeneities on a scale ~ 0.1 pc, or a specific lighthouse model involving relativistic (Lorentz factor ~ 14) motion of plasma bubbles in a twisted magnetic field.

(4) Fluctuations in polarization associated with IDV may be due to random changes in rotation measure (Melrose 1993), which suggestion can be tested through a strong predicted dependence on wavelength.
(5) Coherent emission mechanisms are significantly constrained by induced scattering, and perhaps also by photon-induced plasma turbulence in or near the source.
(6) There remain features that are not obviously consistent with the existing theory of RISS, notably the large amplitudes of some variations. It may be that enhanced refraction from specific structures needs to be invoked, e.g., as suggested by Romani, Blandford and Cordes (1988) for the extreme scattering events.

ACKNOWLEDGMENTS

I thank Ralph Fiedler for permission to show the NRL data at 2.3 GHz for Figure II, and Dick Hunstead and Duncan Campbell-Wilson for supplying the MOST data at 843 MHz for Figure II.

REFERENCES

Attwood, D., Halbach, K., and Kim, K.J. 1985 *Science*, **228**, 1265

Baker, D.N., Borovsky, J.E., Benford, G., and Eilek, J.A. 1988 *ApJ*, **326**, 110

Benford, G. 1992 *ApJ*, **391**, L59

Blandford, R.D., and Königl, A. 1979 *ApJ*, **232**, 34

Blandford, R., and Narayan, R. 1985 *MNRAS*, **213**, 591

Blandford, R., Narayan, R., and Romani, R.W. 1986 *ApJ*, **301**, L53

Camenzind, M., and Krockenberger, M. 1992 *A&A*, **255**, 59

Coles, W.A., Frehlich, R.G., Rickett, B.J., and Codona, J.L. 1987 *ApJ*, **315**, 666

Colgate, S.A. 1967 *ApJ*, **150**, 163

Coppi, P., Blandford, R.D., and Rees, M.J. 1993 *MNRAS*, **262**, 603

Fiedler, R.L., Dennison, B., Johnston, K.J., and Hewish, A. 1987 *Nature*, **326**, 675

Gedalin, M., and Eichler, D. 1993 *ApJ*, **406**, 629

Gupta, Y., Rickett, B.J., and Coles, W.A. 1993 *ApJ*, **403**, 183

Heeschen, D.S. 1982 in D.S. Heeschen and C.M. Wade (eds) *Extragalactic Radio Sources*, Dordrecht: Reidel, 327

Heeschen, D.S., Krichbaum, Th., Schalinski, C.J., and Witzel, A. 1987 *AJ*, **94**, 1493

Hunstead, R.W. 1972 *Astrophys.Lett.*, **12**, 193

Krichbaum, T.P., Quirrenbach, A. and Witzel A. 1992 in E. Valtaoja and M. Valtonen (eds) *Variability of Blazars*, 331

Krishnan, V., and Wiita, P.J. 1990 *MNRAS*, **246**, 597

Mantovani, F., Fanti, R., Gregorini, L., Padrielli, L., and Spangler, S. 1990 *A&A*, **233**, 535

Marscher, A.P. 1977 *ApJ*, **216**, 244

Marscher, A.P., and Gear, W.K. 1985 *ApJ*, **298**, 114

Melrose, D.B. 1991 *ARA&A*, **29**, 31

Melrose, D.B. 1993 *MNRAS*, (submitted)

Mutel, R.L., and Lestrade, J.-F. 1990 *ApJ*, **349**, L47

Narayan, R. 1992 *Phil.Trans.R.Soc.Lond.*, **A341**, 151

Narayan, R., and Goodman, J. 1989 *MNRAS*, **238**, 963

Qian, S.J., Quirrenbach, A., Witzel, A., Krichbaum, T.P., Hummel, C.A., and Zensus, J.A. 1991 *A&A*, **241**, 15

Rees, M.J. 1992 in W. Brinkmann and J. Trümper (eds) *X-ray Emission from Active Galactic Nuclei and the Cosmic X-ray Background*, Garching: Max-Planck-Institut, 255

Rickett, B.J. 1986 *ApJ*, **307**, 564

Rickett, B.J. 1990 *ARA&A*, **28**, 561

Rickett, B.J., Coles, W.A., and Bourgois, G. 1984 *A&A*, **134**, 390

Romani,R.W., Blandford, R.D., and Cordes, J.M., 1988 *Nature*, **328**, 324

Rowe, E.T. 1992a&b *Aust.J.Phys.*, **45**, 1&21

Shapirovskaya, N.Ya. 1978 *Soviet Ast.*, **22**, 544

Simonetti, J.H., Cordes, J.M., and Heeschen, D.S. 1985 *ApJ*, **296**, 46

Slee, O.B., and Siegman, B.C. 1988 *MNRAS*, **235**, 1313

Sol, H., Pelletier, G., and Asséo, E. 1989 *MNRAS*, **237**, 411

Wagner, S.J., and Witzel, A. 1992 in J. Roland, H. Sol and G. Pelletier (eds) *Extragalactic Radio Sources—From Beams to Jets*, Cambridge University Press, 59

Wambsganss, J., Schneider, P., Quirrenbach, A., and Witzel, A. 1989 *A&A*, **224**, L9

Weatherall, J.C., and Benford, G. 1991 *ApJ*, **378**, 543

Witzel, A. 1990 in J.A. Zensus and T.J. Pearson (eds) *Parsec-Scale Radio Jets*, Cambridge University Press, 206

The First Stromlo Symposium: The Physics of Active Galaxies
ASP Conference Series, Vol. 54, 1994
G. V. Bicknell, M. A. Dopita, and P. J. Quinn (eds.)

THE QUASAR BROAD LINE REGION AND A BOUNCING CLOUD MODEL

WILLIAM G. MATHEWS
Lick Observatory, University of California at Santa Cruz, CA 95064

ABSTRACT Most or all current theoretical models of the broad line region require that many line-emitting gas clouds move either radially (in or out) or orbit in a very *large* region around central massive black holes. In the new model discussed here line emitting clouds oscillate radially in the quasar atmosphere. This new model is consistent with recent observations in that (1) each emission line originates within a *small* spatial region around the central object and (2) the red and blue sides of the emission lines vary together. The new model for the broad line region is fully compatible with (opaque) accretion disks.

INTRODUCTION

Recent observations of the variation of broad line fluxes and profiles have revitalized interest in this important component of active galaxies, suggesting new possibilities for theoretical interpretation. Long term, well-sampled observations of the Seyfert galaxy NGC 5548 reveal that various emission lines respond to changes in the continuum after delay times related to the distance of the line-emitting plasma from the source of ionizing radiation. As a result, many physical attributes of the broad line region (BLR) formerly regarded as firmly established are now seen to be less constrained: the size of the BLR is smaller than we thought, emission lines of different ions appear to come from spatially distinct regions, the ionization parameter and plasma density may vary over a significant range within each BLR, etc.

I propose here a new dynamical model for the BLR that has many encouraging resonances with these recent observations. In this model clouds of line-emitting gas oscillate approximately radially in the quasar atmosphere about equilibrium radii established by the balance of gravity and radiation pressure. By this means the entire broad line profile is produced in a very compact region as required by variability studies. For a background perspective, I begin with a discussion of some recent or relevant observations.

In the following I adopt the paradigm that massive black holes do indeed lie at the heart of the broad line region in active galaxies. The evidence in support of massive black holes is somewhat circumstantial but nevertheless compelling: rapid x-ray and γ-ray variability, the likely presence of (inactive) massive black holes in the cores of nearby galaxies, and (for radio loud galaxies) the formation of jets, superluminal motion and regions of high brightness temperature. In addition I adopt the point of view that the BLRs and surrounding structures are qualitatively similar in Seyfert galaxies and quasars, and I refer to both as either active galactic nuclei (AGNs) or quasars. Finally, the idea that AGNs are unified in the sense that different classes of AGN are intrinsically similar but are viewed from different aspect angles, a subject of great current interest, is not the central issue of this review: my approach is to first develop a plausible model for the physical nature of the BLR, then examine the possible implications of varying the viewing angle.

RECENT AND RELEVANT OBSERVATIONS OF THE BLR

Broad emission lines are the most prominent defining features in AGN spectra. Full widths at half maximum (FWHM) of typical lines are 5000 - 20000 km s^{-1} while the spectrum is characteristic of plasma at temperatures $T \sim 1 - 2 \times 10^4$ K, indicating Mach numbers $\mathcal{M} \sim 1000$! Typical electron densities in the emitting gas can be established by the presence or absence of forbidden and intercombination lines of various ions; $n \gtrsim 10^9$ cm^{-3} (absence of broad [OIII]) and $n \lesssim 10^{11}$ (presence of CIII]λ1909, weakness of CIIIλ977. Most line ratios can be understood within factors of ~ 2 by photoionization of single plasma regions, but the observed hydrogen line ratio $F(L_\alpha)/F(H_\beta) \sim 10$ is 3 or 4 times smaller than theoretical predictions. The physical size of the BLR is small: $R_{blr} \sim 10 - 100$ light days for Seyferts and $R_{blr} \gtrsim 100$ ld for quasars. Although the total mass of line-emitting gas is small, $\sim 10 - 100$ $M_\odot$, its proximity to the source of powerful ionizing radiation and its extraordinary dynamics must, when fully understood, indicate something quite remarkable about the mysterious engine at the heart of AGN activity.

The most important observational development during the last few years has been the discovery of stratification in the BLR from variations in the delay time of line emission following those in the continuum. Each broad line has a fairly specific and spatially limited dwelling place in the quasar atmosphere. This behavior is shown in Figure 1a (from Krolik et al. 1991). Emission lines from more highly excited ions arise closer to the source of optical and UV radiation, providing further support that the BLR is photoionized by a source much smaller than R_{blr}. Notice that L_α and the Balmer lines arise largely from different clouds! This is also largely true for the semi-forbidden CIII] line and CIVλ1549, so the density in dense, highly-excited CIV-emitting clouds (at small radii) is poorly known.

Another very important and related observation is that the red and blue sides of the broad emission lines generally change together (Koratkar & Gaskell 1991; Stirpe & de Bruyn 1991; Maoz et al. 1991; 1993), indicating that the line-emitting plasma is moving in both directions throughout the BLR. This is inconsistent with models in which the line-emitting plasma is constrained to move radially either toward or away from the central source of ionizing UV radiation.

Recent variability studies also indicate that infrared emission from dust lies just beyond the BLR. For example in the luminous Seyfert galaxy Fairall 9 Koratkar & Gaskell (1989) find line delays of 124 ± 39 ld for CIVλ1549 and 167 ± 61 ld for MgIIλ2800. The delay in the IR continuum following that in the optical-UV increases with IR wavelength (Clavel, Wamsteker, & Glass 1989). The J band flux may vary in phase with the optical-UV continuum but is masked

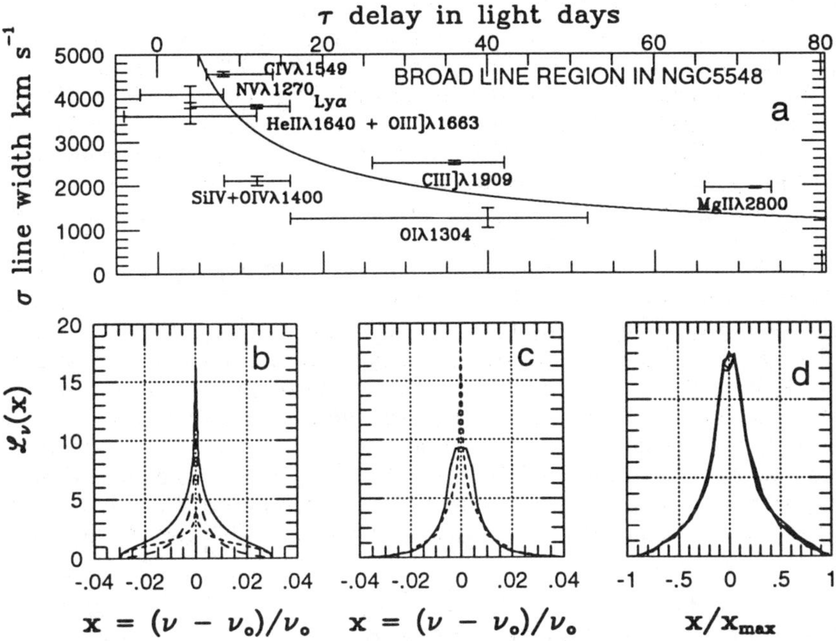

FIGURE I (a) *Vertical axis:* Characteristic width σ (in km s^{-1}) of broad emission lines in NGC 5548; *Horizontal axis:* Delay time τ (in light days) for each line to respond to changes in continuum flux, horizontal error bars show 1σ variation from fits to cross-correlation functions for the variable part of each emission line; *Curve:* Least squares fit to a purely Keplerian velocity profile $\sigma \propto \tau^{-1/2}$ as if gravity were the only force (based on Fig. 1 of Krolik *et al.* 1991). (b) *Solid Line:* Line profile for a system of clouds all having velocity amplitude u_{max} = 6000 km s^{-1} in a quasar atmosphere with $M_{bh} = 3 \times 10^8$ $M_\odot$, $n_o = 10^{10}$ cm^{-3}, and $p = 3$. *Long and short dashed lines:* Subprofiles produced while at $R > R_0$ and $R < R_0$ respectively. (c) *Dashed line:* Combined profile of three families of radially bouncing clouds having velocity amplitudes u_{max} = 3000, 6000, and 9000 km s^{-1} weighted by 2, 1, and 0.5 respectively. *Solid line:* Unnormalized profile of a system of single-amplitude (u_{max} = 6000 km s^{-1}) clouds having additional random tangential velocities $u_{tan} = u_{max}/10$ along the $R = R_0$ sphere. (d) Four profiles created by a hemispherical system of single amplitude bouncing clouds with $u_{tan} = u_{max}/10$. The emission from each cloud is isotropic but line photons incident on the central part of the accretion disk are scattered by a wind back toward the blue side of the profile. The four profiles (viewed at angles 30°, 50°, 70°, and 83° relative to the disk axis) are all very similar with a striking resemblance to observed blue-shifted and blue asymmetric CIV profiles.

by starlight and non-stellar continuum. The K band at 2.1μ exhibits a delay of 400 ± 100 ld., not far from the MgII-emitting clouds.

Profiles of lines of high excitation such as CIVλ1549 are very smooth, indicating a well-populated region of phase space for the BLR kinematics. Many CIV profiles in AGNs are somewhat more extended in the blue wings (Wilkes 1986; Wills et al. 1993), although red asymmetry has been noticed in radio-loud AGN (Corbin 1991b). Wills et al (1993) observe that CIV equivalent widths decrease as the FWHM increases and the profiles become on average less peaked. These authors suggest that this behavior results from viewing intrinsically similar AGNs from differing viewing angles.

Profiles of lines of low excitation, particularly the Balmer lines, show more striking differences from one AGN to another; this variation is evident in the Hα profiles recently published by Eracleous & Halpern (1993). For Hβ Sulentic (1989) finds all possible permutations of asymmetries and shifts to both the red and blue. Neither the profile shift nor asymmetry correlates with other obvious observable parameters. Clearly, there is a significant stochastic component in the kinematics of those parts of the BLR that emit lines of low excitation. Finally, we know from Miley and Miller (1979) that broad profiles are widened and complicated by the presence of extended, steep-slope radio sources. This implies that an additional source of broad line plasma is associated with radio jets.

SINGLE FORCE MODELS FOR THE BLR

In constructing theoretical models of the BLR, most authors have assumed that a single outward or inward force dominates the dynamics of the line-emitting plasma. The emitting gas can be regarded as either moving mostly in the radial direction or, when gravity dominates, in orbits largely perpendicular to the local radius vector. In models involving radial motion the emitting plasma is accelerated either inward or outward (by radiation), first emitting radiation in the core of the line and later generating the wings as the plasma crosses the broad line region. In models involving orbital (transverse) motions, the line-emitting gas can reside at each radius in the BLR for a longer time. Generally in these orbiting, gravity-dominated BLR models the adopted gravitational potential has been that of a single central massive object (black hole), ignoring any additional mass that could be present (stars, accretion disk, etc.). If the central mass dominates the potential throughout the BLR, the wings of the lines are created by plasma orbiting at small radii and the line cores are formed at large radii.

The important point I wish to emphasize here is that for either class of single force models – involving either radial or transverse (orbiting) motion – the range in radius required to generate the entire line profile far exceeds that indicated by recent observations of line variability.

Line variability (reverberation) studies of normal single-peaked BLR profiles are approximately consistent with gravity-dominated Keplerian (or free fall) velocity fields in which the mean velocity at each radius $(\langle u^2 \rangle)^{1/2} \propto (GM_{bh}/r)^{1/2} \propto r^{-1/2}$ (as in Fig. 1a). Many authors have assumed that these observations can be used with the virial theorem to estimate the mass of the central object M_{bh} for either orbital motion or free fall. However, a most important attribute of the results shown in Fig. 1a is that a significant fraction of the emission from *each* prominent broad line arises from a rather small range in radius ΔR represented by the horizontal error bars. This last result, when considered with the profiles, is inconsistent with BLR models involving line-emitting disks or clouds in radial free fall which require a much larger dynamical range in radius than the ΔR indicated in Fig. 1a. Clouds moving at velocity u within a spherical shell of thickness δR in free fall produce a square profile of width $\delta\nu = \nu_o(u/c)$ while orbiting clouds produce saucer-shaped profiles of the same width; in either case it is necessary to integrate these elemental profiles over all BLR radii to generate the complete profile observed. For single-peaked line profiles the line-emitting gas must have velocities that extend from the the full width of the observed lines $\sim$ 10000 km s^{-1} to at least the spectral resolution at the line core, $\sim$ 200 km s^{-1}. Since $R \propto u^{-2}$ for gravity-dominated models, this velocity range translates into an enormous dynamic range in radius, $R_{in}/R_{out} \approx (10000/200)^2 = 2500$. This large range in radius $R_{max} - R_{min}$ required to generate the observed profile is inconsistent with the smaller ΔR found in variability studies in which the *most of the profile*, from the core to the wings, varies approximately at the same time following changes in the (ionizing) continuum. Similar conclusions follow if the single dominant force in BLR models is assumed to be outward (radiation pressure, ram pressure, etc.) unless the outward radial forces are $\gtrsim$ 100 times stronger than the local gravity (an unlikely possibility).

In any case, there is now strong observational evidence against models in which the dynamics of the BLR are controlled by a single force, directed either in or out.

CLOUD MODELS FOR THE BLR

Cloud models for the BLR are by far the most popular, but there is no convincing proof that such clouds exist. The argument for clouds is usually based on the small size of the emitting regions r_{cl} to the observed size of the BLR, $R_{blr} \sim 10^{17} - 10^{18}$ cm. The Stromgren column is only $N_s = cU/\alpha_B \approx 10^{23} U$ cm^{-2} where α_B is the Case B recombination coefficient and $U \sim 0.01 - 1$ is the ionization parameter. The size r_{cl} is roughly comparable to the physical depth of the largest fully-ionized region, $r_s = N_s/n = 10^{13}(U/n_{10})$ cm, although photoionization models of optically thick clouds required to produce lines of low excitation have sizes $r_{cl} \approx (3 - 30) r_s$. Since $r_{cl} \ll R_{blr}$ cloud models are plausible.

However, three major difficulties confront cloud models: (i) confinement by external pressure or gravity is essential, (ii) randomly moving clouds collide frequently, and (iii) non-selfgravitating clouds are subject to a large number of hydrodynamic instabilities.

Confinement is desirable or essential since the time for sound to cross a cloud $t_{sc} \approx r_s/c_s \approx 0.2U/n_{10}$ yr can be less than the dynamical (or orbital) time for the cloud to cross the BLR, $t_{dy} \approx R_{blr}/\langle u\rangle \approx 2(R/10\ \mathrm{ld})(5000\ \mathrm{kms}^{-1}/\langle u\rangle)$ yr. If clouds are confined by an intercloud medium at the Compton temperature $T_{icm} = T_c \approx 10^7$ K (Mathews and Ferland 1987), the moving clouds will be slowed appreciably and suffer Rayleigh-Taylor instabilities in times $\lesssim t_{dy}$. This difficulty can be circumvented in several ways: (i) $T_{icm} \approx T_c$ but requiring the intercloud medium to move with the clouds, (ii) $T_{icm} \gg T_c$ due to additional sources of heating or cosmic ray pressure, (iii) the clouds are confined by magnetic stresses, etc.

Randomly moving clouds collide in times $t_{coll} = (n_{cl}\langle u\rangle \pi r_{cl}^2)^{-1}$ where $n_{cl} = 3\mathcal{N}_{cl}/4\pi R_{blr}^3$ is the space density of $\mathcal{N}_{cl}$ clouds. In terms of the covering factor of the BLR $f_c = \mathcal{N}_{cl} r_{cl}^2/4R_{blr}^2$, the cloud-cloud collision time is only $t_{coll} = R_{blr}/3f_c\langle u\rangle \approx (1-10)t_{dy}$. When clouds collide they are heated to temperatures $\gtrsim T_{icm}$ and can no longer emit broad line radiation. The time for new clouds to form by thermal instabilities from the intercloud medium can exceed t_{dy} (Mathews and Doane 1990). Although collisions of ambient stars with BLR clouds do not disrupt the clouds on short time scales, if the clouds *are* illuminated stars, then the the small collision time derived above indicates that the stars and the BLR would be rapidly destroyed.

If the BLR clouds are non-selfgravitating and appropriately confined their existence is still precarious because of a large number of instabilities (Mathews and Capriotti 1986). Orbiting clouds are vulnerable to disruption on times $\sim t_{dy}$ by Keplerian or Poynting-Robertson shear. Further disruption can result from (i) derimming, the selective radial dislocation by radiation forces of the outer regions of clouds having lower column densities, (iii) lateral flow instabilities driven by internal overpressures in clouds, (ii) boundary layer loss; for clouds confined by cosmic rays, a layer of thickness of the mean Larmor radius on the periphery of the cloud will have lower densities and experience less radiation force, (iv) radiation driven shock waves occur for ionization parameters $U \gtrsim 0.3$ (Mathews 1976), etc.

In spite of these concerns, I shall nevertheless adopt the cloud model here.

BOUNCING CLOUDS!

In order to produce most of the broad line profiles in a relatively small radial region ΔR surrounding the source of UV radiation, it is necessary to consider the possibility that line-emitting clouds *oscillate* about an equilibrium radius established by two opposing radial forces (Mathews 1993). This can be arranged quite naturally, for example, if the pressure in the intercloud medium P_{icm}, and therefore also the cloud density n, varies as R^{-p} with $p > 2$, a variation expected for any intercloud wind with adiabatic index $\gamma > 1$. The equation of motion for a cloud of mass m is

$$m\frac{du}{dR} = \left[\frac{n\alpha_B\langle h\nu\rangle}{Mc} - \frac{GM_{bh}}{R^2}\right] m$$

where M_{bh} is the central mass. For simplicity the cloud is assumed to be fully ionized with density $n = n_o(R/R_O)^{-p}$; the outward radiative force is provided by the momentum received by Lyman continuum absorption $\langle h\nu\rangle/c$ following each recombination. Provided $p > 2$ there is a very special radius, $R_O = (cGM_{bh}M/n_o\alpha_B\langle h\nu\rangle)^{1/2} \approx 14(M_{bh}/3\times 10^8\ M_\odot)^{1/2}(n_o/10^{10})^{-1/2}$ light days, where the outward radiation force balances gravity and about which clouds can oscillate radially. Although clouds can experience some drag force as they oscillate, it is assumed that drag does not dominate. The equilibrium radius R_O has just the correct magnitude for reasonable cloud densities and central masses.

The central masses need to be ~ 10 times larger than the "virial mass" naively determined from observed delay distances R_{blr} and line widths.

One of the great advantages of this model is that the clouds can attain velocities similar to those observed in the BLR without moving very far from the equilibrium radius; i.e. for $M_{bh} = 3 \times 10^8\ M_\odot$ and $n_o = 10^{10}$ a cloud of maximum velocity $u_{max} = 6000$ km s^{-1} produces the entire line profile within $0.65 < R/R_0 < 1.9$; for $u_{max} = 9000$ it is still only $0.55 < R/R_0 < 3.6$, i.e. $\Delta R \lesssim R_0$ as implied by Fig. 1a.

A line profile for a spherical system of radially oscillating clouds having a single amplitude u_{max} and constant, isotropic emissivity is shown (for $p = 3$) in Fig. 1b; the total profile is the sum of two complete subprofiles generated on both sides of the equilibrium radius R_0. The profile for this system of single-amplitude clouds bears a remarkable resemblance to observed profiles. An even better fit in the wings results when clouds of several amplitudes are combined (Fig. 1c). The core can easily be made less peaked by allowing the clouds to have small $\sim 0.1u_{max}$ random transverse velocities along the spherical surface R_0. The solid curve in Fig. 1c shows the profile of a spherical system of clouds with single amplitudes for radial and (small) transverse motions; this profile has almost exactly the characteristic shape of observed broad lines. In general the cloud system will contain a continuous distribution of radial and transverse velocity amplitudes (Mathews 1993).

The excitation of clouds away from the equilibrium radius R_0 is easily accomplished by imagining that random pressure fluctuations occur in the intercloud medium so that n_o and therefore R_0 can occasionally change on timescales shorter than the radial oscillation times (5 - 10 yrs). More optically thick clouds can also be excited by fluctuations in the non-stellar continuum.

There are many additional advantages of this model: (i) relatively few parameters and assumptions: concentrated central mass, cloud confinement with moderate or low drag, and $p > 2$; (ii) profiles are insensitive to the exact value of the index p; (iii) for fully-ionized clouds the profile is independent of cloud mass and shape; (iv) clouds can slowly disrupt as they interact with the intercloud medium without affecting the basic profile; (v) the entire line profile is spatially compact, $\Delta R \lesssim R_0$; (vi) both red and blue sides of the profile change together following changes in the external pressure or ionizing continuum.

Even more interesting, the bouncing cloud model is fully compatible with the existence of opaque accretion disks since the largely radial orbits of the clouds do not intersect the disk and furthermore: (vii) by symmetry the profile produced by clouds moving in a *single hemisphere* is exactly the same as the profile from a completely spherical BLR, and (viii) the single-hemisphere line profile is *independent* of the (viewing) angle between the line of sight and the normal to the disk, accounting for the general similarity of observed profiles.

The problem of frequent cloud-cloud collisions is greatly lessened in this model since the collision rate is reduced for largely radial cloud motions. Clouds having large transverse velocities (along the R_0 sphere) preferentially collide and are removed from the line-emitting system. Provided new cloud material is continuously added to the ensemble of bouncing clouds, collisional cloud destruction of transversely moving clouds may in fact quite naturally limit the covering factor to the values observed, $f_c \approx 0.1 - 0.3$. In preparing this review I have written a Monte Carlo program to explore how line profiles are influenced by several additional effects such as (i) anisotropic emission from individual clouds, (ii) conical-shaped broad line regions, and (iii) scattering of line photons from ambient material. The results of these preliminary explorations are most interesting and encouraging.

Ferland et al. (1992) have shown that most lines are emitted anisotropically from photoionized clouds either because of continuum absorption or thermalization within the clouds. Even CIVλ1549 from optically thin clouds can be significantly anisotropic due to thermalization. I have therefore computed line profiles created in a single hemisphere BLR by a system of (single-amplitude) bouncing clouds emitting anisotropically with emissivity $\epsilon(\mu) = \epsilon_{tot}\frac{1}{2}(1 - \frac{1}{2}\mu)$, where $\cos^{-1}\mu$ is the angle between emission direction and the local radius vector. Line photons incident on the disk are assumed to be either totally absorbed or reprocessed elsewhere in the spectrum. Again by symmetry, the profiles are independent of viewing angle. However, if the BLR is *conical* (as indicated by observations of Pogge 1988a,b and many others) with opening angle $\theta_c = 60^o$, approximately pole-on BLR have larger FWHM and less peaked profiles than those of nearly equator-on BLRs. This is precisely the observational result recently discussed by Wills et al. (1993) in which the equivalent width of CIVλ1549 anticorrelates with FWHM and also agrees with Corbin's (1991a) finding that CIV is broader in radio core-dominated (and presumably nearly pole-on) AGNs!

Finally, I have computed approximate CIV profiles (with isotropic and anisotropic $\epsilon(\mu)$) in which some inward-emitted line photons are electron-scattered back toward the observer from a wind (of temperature $T \lesssim 10^6$ K) flowing from the central parts of the accretion disk (Fig. 1d). These profiles are slightly blue shifted and blue-asymmetric as expected. Clearly such a wind must be shown to be fully consistent with observations of soft x-ray absorption and Fe Kα emission in quasar continua. It is possible, however, that the wind required to blue shift CIV arises from the disk, not the central core, so that it would interfere less with our line of sight to the central source. Dust-scattering by the wind is another possibility and this would allow a wider range in electron temperature, but the scattered CIV image would then be broadened by disk rotation.

Can broad lines of *low* excitation (H_β, MgIIλ2800, FeII, etc.) also be formed by bouncing clouds having equilibrium radii R_1 somewhat larger than R_0 (as in Fig. 1a)? The answer to this question seems to be yes, but the outward pressure force on these optically thick clouds must be supplied by radiation pressure on dust rather than by atomic absorption of UV or x-rays (Mathews 1993). The possibility that dust exists within the outer parts of the BLR has been discussed for many years (Netzer 1990) and has received recent support from variability observations of Fairall 9 as discussed above. Netzer & Laor (1993) provide a convincing argument that narrow line clouds exist all the way down to the BLR but the *innermost* clouds in the NLR are faint in line emission because of absorption by internal dust. My hypothesis here is a corollary to this idea in that the *outermost* parts of the BLR are also assumed to contain dust.

The ratio of radiation force on dust to gravity force, $Le^{-\tau_d}\bar{\sigma_d}/4\pi cMGM_{bh} \approx 120\, L_{45}e^{-\tau_d}/M_{bh8}$, depends on the cloud optical depth for dust absorption $\tau_d(R)$ and can be very large if the mean grain area per hydrogen is similar to that in the local interstellar medium, $\bar{\sigma_d} \approx 10^{-21}$ cm^2. This estimate shows that it is indeed possible to sustain equilibrium radii at $R_1 \sim (2 - 20)R_0$ if only a tiny fraction ~ 0.01 of normal dust abundance is present. For clouds of low ionization to bounce the net restoring force must be directed toward R_1. This is possible if the mass M_{bh} of the black hole is supplemented by a small additional mass (e.g. the mass of the accretion disk or local stars) so that the gravitational acceleration declines slightly less steeply than R^{-2} in the outer BLR. If the restoring force is somewhat weaker for low ionization clouds than for CIV-emitting clouds, to achieve approximately the same line widths Balmer and MgII-emitting clouds should oscillate in a thicker shell; this is supported by variablity studies. The dynamical requirement for dust that I propose is natural and entirely consistent with the spatially disjoint emission of L_α and H_β observed since L_α is preferentially absorbed by dust within low ionization, H_β-emitting clouds in the outer BLR.

I therefore speculate that dusty neutral or molecular clouds are introduced into the BLR at some radius $\sim 10 - 100R_0$. More distant and therefore more dusty clouds may be pushed outward by radiation on dust and/or intercloud wind to become part of the narrow line region (Mathews & Veilleux 1989). Less distant and dusty clouds oscillate about an equilibrium radius R_1 that slowly decreases as the internal dust grains are evaporated (Laor & Draine 1993). Since these distant BLR clouds are closer to the source of cloud material, their line profiles are more irregular and stochastically variable (Sulentic 1989). Recently introduced optically thick clouds collide and the orbits of surviving clouds become more radial; a variety of instabilities are expected which will tend to reduce the average mass and column density of the clouds as they oscillate with decreasing R_1, losing radiative support from the evaporating dust. Eventually when the dust is fully evaporated, the remaining gas in these clouds settles down to the innermost oscillation radius R_0; the same gas can then produce lines of high excitation until destroyed by instability or collision. The bouncing cloud model provides – for the first time – a plausible joint dynamic explanation for both the narrow and broad line regions and explains how gas evolves through each of these regions.

Because of the many instabilities that clouds experience, a single cloud is not likely to survive intact in its journey from the outer parts of the BLR to R_0. The column density of *spherical*, optically thick clouds of *constant mass* decreases with BLR radius, $N \propto n r_{cl} \propto n^{2/3} \propto R^{-2p/3}$ if $p > 2$, so the ionization level is lower at small R. However, variability studies indicate that optically thick clouds dominate at larger radii. Therefore it is necessary (and natural) to assume that optically BLR clouds at large radii break up due to numerous instabilities so that the average cloud mass and column density decrease with radius R in the BLR. In this picture, although the ionization parameter *incident* on the clouds increases with radius, $U \propto R^{p-2}$ when $p > 2$, its *average* value inside the clouds $\langle U \rangle$ is larger in clouds of lower mass close to the central UV source.

What is the original source of the dusty gas in the BLR? Obviously this is a more difficult question, and somewhat beyond the objectives of this review, but some likely sources of gas are: (i) gas of low angular momentum accreted during a galactic merging event or a gas cloud that ventures near the AGN nucleus, (ii) gas ejected from the accretion disk by bow shocks generated by stars passing through the disk; if the angular momentum of the stellar system is small compared to that of the disk material, most of this ejected gas will also have low angular momentum; (iii) gas liberated following star-star collisions which may form dust, as in some supernovae.

This theoretical work on the broad line region has been supported by the California Space Institute.

REFERENCES

Clavel, J., Wamsteker, W., & Glass, I. S., 1989, *ApJ*, **337**, 236

Corbin, M. R., 1991a&b, *ApJ*, **371**, L51; **375**, 503

Eracleous, M., & Halpern, J. P., 1993, *ApJ*, **409**, 584

Ferland, G. J., Peterson, B. M., Horne, K., Welsh, W. F., & Nahar, S. N., 1992, *ApJ*, **387**, 95

Koratkar, A. P., & Gaskell, C. M., 1989, *ApJ*, **345**, 637; 1991, *ApJS*, **75**, 719

Krolik, J. H., Horne, K., Kallman, T. R., Malkan, V. A., Edleson, R. A., & Kriss, G. A., 1991, *ApJ*, **371**, 541

Laor, A., & Draine, B. T., 1993, *ApJ*, **402**, 441

Maoz, D., *et al.*, 1991, *ApJ*, **367**, 493

Maoz, D., *et al.*, 1993, *ApJ*, **404**, 576

Mathews, W. G., 1986, *ApJ*, **305**, 187

Mathews, W. G., 1976, *ApJ*, **207**, 351

Mathews, W. G., 1993, *ApJ*, **412**, L17

Mathews, W. G. & Capriotti, E. R., 1986, in *Astrophysics of Active Galaxies and Quasi-stellar Objects* (Mill Valley:University Science Books) ed. J. S. Miller, p. 185

Mathews, W. G., & Doane, J. S., 1990, *ApJ*, **352**, 423

Mathews, W. G., & Ferland, G. J. 1987, *ApJ*, **323**, 456

Mathews, W. G. & Veilleux, S. 1989, *ApJ*, **336**, 93

Miley, G. K., & Miller, J. S., 1979, *ApJ*, **228**, L55

Netzer, H., 1990, in *Active Galactic Nuclei: Saas-Fee Advance Course 20*, eds. T. J. -L. Courvoisier and M. Mayor (Springer-Verlag:Berlin), p. 57

Netzer, H., & Laor, A., 1993, *ApJ*, **404**, L51

Pogge, R. W., 1988a&b, *ApJ*, **328**, 519; **332**, 702

Stirpe, G. M., & de Bruyn, A. G. 1991, *A&A*, **245**, 355

Sulentic, J. W., 1989, *ApJ*, **343**, 54

Wilkes, B. J., 1986, *MNRAS*, **218**, 331

Wills, B. J., Brotherton, M. S., Fang, D., Steidel, C. C., & Sargent, W. L. W. 1993, *ApJ*, (in press)

The First Stromlo Symposium: The Physics of Active Galaxies
ASP Conference Series, Vol. 54, 1994
G. V. Bicknell, M. A. Dopita, and P. J. Quinn (eds.)

REVERBERATION MAPPING OF AGN BROAD LINE REGIONS

GERARD A. KRISS AND JULIAN H. KROLIK
Department of Physics & Astronomy, Johns Hopkins University, Baltimore, MD 21218

ABSTRACT The time variable continua of active galaxies can be used to unravel the structure of the broad line region in active galactic nuclei by observing the time delays in the response of the illuminated broad line clouds to changes in the ionizing continuum. We review previous results from the 1988-1989 IUE campaign to monitor the Seyfert 1 galaxy NGC 5548, and compare transfer functions derived using a new regularized linear inversion technique to those from maximum entropy methods. We conclude with a presentation of preliminary results from the recently completed follow-up campaign to monitor NGC 5548 using the Hubble Space Telescope.

INTRODUCTION

The broad emission lines that are one of the defining characteristics of quasars and Seyfert 1 galaxies are formed in dense clouds photoionized by the high energy photons emitted by the central engine. Arguments based on single-zone photoionization models lead to sizes of order $100L_{44}^{1/2}$ lt-d. Even for the apparently brightest active galactic nuclei (AGN), the corresponding angular sizes are on the order of 10 micro-arcseconds. While crude photoionization modeling can give an order-of-magnitude estimate of the size of the broad-line region (BLR), determining if the actual structure is spherical or disk-like, smooth or lumpy, requires more information. Interferometric techniques that might resolve the BLR through direct observation are years in the future, but analysis of the delay in the response of the photoionized clouds to continuum fluctuations, reverberation mapping, can probe structures on these scales right now.

In addition to determining the shape of the BLR, reverberation mapping can also tell us about the dynamics. The response of different portions of the broad-line profiles are sensitive to the velocity field, and we can test whether motions are predominantly inflowing, outflowing, or even chaotic. Understanding the velocity field may lead to an understanding of the forces acting on the broad-line clouds, perhaps the gravity of a central black hole, or the radiation pressure of the ionizing continuum.

An independent understanding of the structure of the BLR removes one of the prime degrees of freedom in photoionization models, the ionization parameter. Thus another benefit of reverberation mapping is a more complete picture of the physical conditions in the BLR — its density, temperature, column density, and ionization structure.

Finally, reverberation mapping campaigns are beginning to span typical dynamical times in the BLR. Motions of 3000 $km\ s^{-1}$ cover a typical radius of 10 lt-days in only three years, so with repeated campaigns we might expect to see real changes in the structure of the BLR that may be associated either with matter being blown out of the central regions, or flowing inwards to feed the black hole.

As in many fields, the theory of reverberation mapping was far ahead of the ability of observers to make the requisite measurements. Bahcall, Kozlovsky & Salpeter (1972) first suggested using light travel time delays to study the structure of the photoionized BLR, followed ten years later by Capriotti, Foltz & Peterson (1982) and Blandford & McKee (1982). The formalism used by Blandford & McKee (1982) to describe the problem is the basis of much of today's work on reverberation mapping.

One of the key ingredients in a successful reverberation experiment is well calibrated photometric data, and this required the advent of modern electronic detectors. The first reverberation experiments (e.g., Antonucci & Cohen 1983; Ulrich et al. 1984; Peterson et al. 1985), however, were lacking another essential factor that has led to the success of more recent efforts — regular, intensive sampling over long periods of time. Nevertheless, these attempts gave the first inkling that the BLR might extend to smaller radii than suggested by single-zone photoionization models. More recently, long intervals of regularly spaced observations have been obtained of NGC 4151 (Maoz et al. 1991, ground-based), NGC 5548 (Clavel et al. 1991, IUE; Peterson et al. 1991, Peterson et al. 1992, and Peterson et al. 1993, ground-based), NGC 3783 (Reichert et al. 1993, IUE; Stirpe et al. 1993, ground-based), and NGC 3516 (Wanders et al. 1993, ground-based). Long, successful campaigns require the cooperation of many investigators and telescope time-assignment committees, and these social factors have been the final major ingredient of the more recent reverberation mapping experiments. Peterson (1993) gives an excellent comprehensive review of reverberation mapping theory, experiments, and results.

UNRAVELING THE STRUCTURE OF THE BLR

The earliest and simplest size estimates of the BLR were made using single-zone photoionization models. These implied nucleon densities of roughly $n = 1 \times 10^{10}$ cm^{-3} and ionization parameters of $\Xi \sim 0.1 - 1.0$, where Ξ is the ratio of the ionizing radiation pressure to the nucleon density in a sample of broad-line material — $\Xi = L_{ion}/(4\pi r^2 cnkT)$. For an ionizing luminosity typical of Seyfert galaxies, 10^{44} erg s^{-1}, this implies a radius for the BLR of 40-135 light days.

By using more of the information available in a spectrum, one could attempt to map the structure of the BLR on constant velocity surfaces by using the line profiles (e.g., Espey et al. 1989; Kallman et al. 1993). Unfortunately, the shape of the isovelocity surfaces are unknown, and they may overlap spatially if there is a random component.

Until very long baseline optical interferometry is practical, reverberation mapping appears to be our best tool for analyzing the structure of the BLR. It has the advantage over the above two techniques that the isodelay surfaces are known, effectively paraboloids given by $c\tau = r(1 - \cos\theta)$. A disadvantage, however, is that a lot of data and a little bit of luck are required to ensure that there are line and continuum variations on the timescales of interest.

Basics of Reverberation Mapping

Implicit assumptions in the reverberation mapping technique are (1) that the continuum source is point-like, or much smaller than the delays being measured, (2) that the continuum is radiated isotropically, (3) that the photoionized clouds respond to changes in the ionizing continuum on timescales much shorter than the delays of interest, and (4) that the volume of the BLR is effectively optically thin so that line and continuum radiation propagate freely at the speed of light. Formally we can express the emission line flux as a function of time as an integral of the time-dependent, nonlinear line emissivity over each isodelay surface for all delays. By subtracting the time-averaged line and continuum fluxes and performing a first-order series expansion for the line emissivity in terms of the continuum fluctuations, the line variations are given by the convolution of the transfer function $\Psi_l(\tau)$ with the continuum variations:

$$\delta F_l(t) = F_l(t) - \bar{F}_l = \int d\tau \, \Psi_l(\tau) \left[F_c(t - \tau) - \bar{F}_c \right] , \qquad (1)$$

where the transfer function is an integral over an isodelay surface of the "responsivity" of the emitting material.

$$\Psi_l(\tau) = \int dS_\tau \frac{\partial j_l[p(r), ...]}{\partial F_c} . \qquad (2)$$

It is here that most of the physics in the problem is buried since the response depends on the physical conditions in the photoionized material and its sensitivity to the ionizing flux. As qualitative examples of transfer functions for some simple geometries, consider a spherical shell and an inclined planar ring, both assuming isotropically radiating clouds. Important differentiating characteristics of these transfer functions are the strong response at zero lag for the spherical distribution in contrast to the totally absent signal for the ring, and the double-horned structure of the ring transfer function (e.g., see Peterson 1993). Whole families of transfer functions for a wide variety of geometries and cloud characteristics are illustrated by Robinson & Pérez (1993).

While reverberation studies of the integrated line flux can reveal the structure of the BLR, examining the response of different portions of a given line *profile* promises to yield insights into the dynamics of the BLR. As a simple example, consider a spherical distribution of infalling clouds radiating isotropically. Clouds on the near side of the AGN produce the red wing of the line, and clouds on the far side emit the blue wing. The far side lies on isodelay surfaces with a larger time delay than the near side, so one would expect the response of the blue wing of the line to lag behind the red wing. Welsh & Horne (1991) and Pérez, Robinson, & de la Fuente (1992) compute a variety of two-dimensional transfer functions for commonly anticipated dynamical structures in the BLR.

Solving for the Transfer Function

The first step in solving for the transfer function is accumulating a large, high signal-to-noise ratio data set, preferably evenly sampled in time. To search for structure in $\Psi(\tau)$ on timescales of τ, the data must be sampled at intervals $\Delta t < \tau/2$ for a time interval $T >> \tau$. Samples obtained at regular intervals greatly enhance the ease of recovering the transfer function from the data for a variety of restoration techniques.

A direct linear solution of $\delta F_l(t) = \int d\tau\ \Psi_l(\tau)\ \delta F_c(t-\tau)$ for $\Psi_l(\tau)$ is impractical since the large degeneracy of the problem renders the solution unstable. Blandford & McKee (1982) use a Fourier transform solution in their development of the problem, but this is overwhelmed by noise at high frequencies, and the solution is unstable in most practical applications. A useful, approximate solution to the problem, and one that has the longest history in the analysis of AGN monitoring data (first applied by Gaskell & Sparke 1986), is the cross-correlation function $C(\tau)$ defined by

$$C(\tau) = \frac{1}{\sigma_l \sigma_c} \int dt\ \delta F_l(t)\ \delta F_c(t-\tau) = \frac{1}{\sigma_l \sigma_c} \int df\ e^{-2\pi f \tau}\ \hat{\Psi}(f)\ |\delta \hat{F}_c(f)|^2. \quad (3)$$

The peak in $C(\tau)$ is an estimator of the delay to a given region of the BLR, but it is essentially only a first moment of the transfer function, not a full description, and it is sensitive to the characteristics of the autocorrelation function of the *continuum*, something that is *not* a feature of the BLR.

One method for obtaining a full solution for the transfer function uses maximum entropy model fitting to extract $\Psi(\tau_k)$ as free parameters in a χ^2 fit to the data set. These solutions are stable, but they suffer from the significant drawbacks that they are model-dependent and that they force the transfer function to be positive at all lags. At first glance the latter assumption seems physically reasonable, but Sparke (1993) has illustrated some reasonable cases where one could naturally expect $\Psi < 0$. An easy example to visualize is a highly ionized region where most of the carbon is ionized to C IV or higher. An increase in the ionizing flux will heat the region further, raising the emissivities of most lines, but it will also decrease the population of C III, leading to a net decrease in the flux of C III] λ1909.

A new technique which removes the constraint on the sign of Ψ, regularized linear inversion, has additional promising features. The method developed by Done & Krolik (1993) minimizes the function $(\underline{C} \cdot \vec{\Psi} - \vec{L})^2 + \lambda \vec{\Psi} \cdot \underline{R}^T \cdot \underline{R} \cdot \vec{\Psi}$, where $C_{ij} = \delta F_c(t_i - t_j)/\epsilon_i$, $L_i = \delta F_l(t_i)/\epsilon_i$, and $\underline{R}$ is a differencing operator that measures differences from smoothness in $\Psi(\tau)$. Minimization leads to

$$(\underline{C}^T \cdot \underline{C} + \lambda \underline{R}^T \cdot \underline{R}) \cdot \vec{\Psi} = \underline{C}^T \cdot \vec{L}\,. \quad (4)$$

This solution is stable, permits all signs in Ψ, and gives clear propagation of errors from the observations to the solution for Ψ. Drawbacks are that the Lagrange multiplier λ and the differencing operator $\underline{R}$ are free parameters, and, more seriously, the continuum time series in the time interval directly preceding the experiment is unknown, so that only a portion of the line flux time series can be used. Maximum entropy avoids this problem by considering the unmeasured continuum points as free parameters. The free parameters can be determined by requiring a good fit to the data using a statistical test such as χ^2.

THE 1988-1989 AGN WATCH CAMPAIGN ON NGC 5548

The combined IUE and ground-based campaign conducted by the International AGN Watch consortium in 1988-1989 on the Seyfert 1 galaxy NGC 5548 is the most comprehensive reverberation mapping experiment performed to date. At regular sampling intervals of four days, IUE short- and long-wavelength UV spectra were obtained over an 8 month period from December 1988 to August 1989 (Clavel et al. 1991). The simultaneous ground-based campaign (Peterson et al. 1991) concentrated on the optical continuum and the Hβ line, and it involved an extensive, world-wide network of observers whose observations are still on-going after 4 years (Peterson et al. 1992; Dietrich et al. 1993; Peterson et al. 1993).

Over the 8 months of the campaign, the UV continuum of NGC 5548 was observed to vary by a factor of ~4, with three broad maxima in flux occuring at intervals of roughly 90 days. Key results from this campaign were (1) high ionization lines (such as He II λ1640) showed larger amplitude fluctuations than did lower ionization lines (such as C III] λ1909; (2) the emission line variations followed the continuum fluctuations with lags that were longer for lower ionization lines; (3) the continuum fluctuations were stronger at shorter wavelengths so that the continuum was bluer when brighter; and (4) the continuum varied synchronously at all wavelengths from the UV to the optical within the limits of the 4-day sampling.

The cross-correlation functions presented by Clavel et al. (1991) show lags ranging from < 4 days for He II λ1640, to 8 days for C IV λ1549 and 34-72 days for the weakly varying Mg II λ2798 line. Peterson et al. (1991) found that the Hβ line had a 20 day lag comparable to that of C III] λ1909 22 days. Maoz et al. (1993) studied the response of the Fe II lines and the Balmer continuum comprising the "small UV bump" and found a lag of 6 days. The qualitative picture of a BLR with a highly stratified ionization structure stands out clearly in the simple cross-correlation results, but the transfer functions computed by Krolik et al. (1991) and Horne, Welsh & Peterson (1991) give an even more detailed picture of the BLR (see Figure I).

The transfer functions for the high ionization lines all peak near zero lag and gradually fall to zero for delays of $\sim$ 30 lt-days. This indicates that the high ionization lines can be formed in a roughly spherical region surrounding the continuum source extending from radii of less than 4 lt-days to ~15 lt-days. In contrast, the low ionization lines exemplified by C III] λ1909 and Hβ have transfer functions with low amplitude at short lags and positive peaks at delays of 15-30 lt-days. The low response at zero delay is not consistent with a spherical distribution of clouds, but it can match a flattened configuration with a radius of $\sim$ 100 lt-days viewed at an inclination of $\sim 45°$.

While Krolik et al. (1991) find a geometrical configuration of clouds with self-consistent properties that can be described by photoionization modeling, the ambiguities of interpreting transfer functions are shown by the re-analysis of the NGC 5548 data by Done & Krolik (1993). By using the regularized

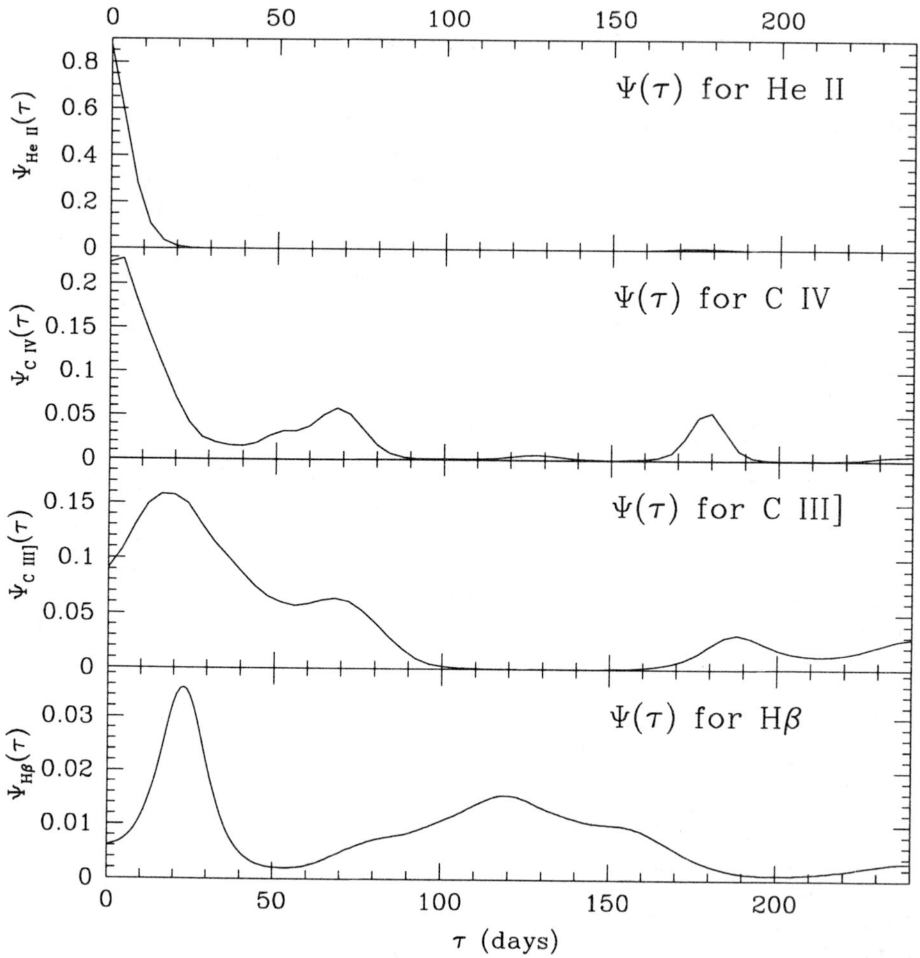

FIGURE I Transfer functions derived using the maximum entropy method are shown for He II λ1640, C IV λ1549, and C III] λ1909 from Krolik et al. (1991) and for Hβ from Horne, Welsh & Peterson (1991). The high ionization lines peak at shorter lags than those for the lower ionization lines, and they are compatible with a spherical distribution of emitting clouds. The lower ionization lines, in contrast, show little response at zero lag, a character that is more indicative of a flattened distribution.

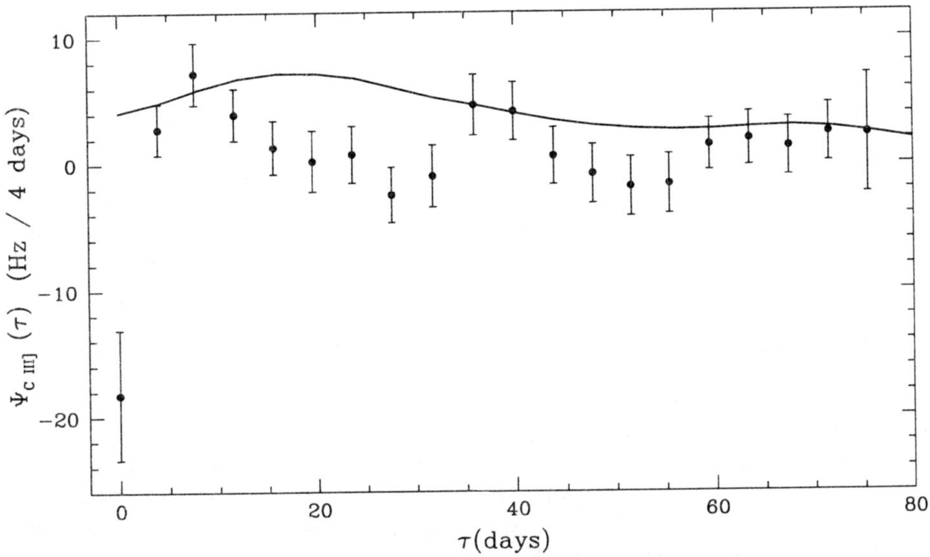

FIGURE II The transfer function for C III] λ1909 derived using the regularized linear inversion technique of Done & Krolik (1993) (solid dots with error bars) peaks at a shorter lag than that derived using the maximum entropy technique (solid line), and also is strongly negative at zero lag.

linear inversion technique to derive the transfer functions, they find significantly different shapes, as shown in Figure II for C III] λ1909. Relaxing the constraint in the maximum entropy technique that forces $\Psi(\tau)$ to be positive everywhere leads to a solution for the transfer function that is strongly negative at delays of zero. This could be consistent with a more spherical distribution for C III, with a highly ionized inner zone that is depleted as the continuum brightens. Both solutions for the transfer function give acceptable χ^2 fits to the data, but they are strikingly different, and at present there is no objective way to choose which is more likely to be correct.

Addressing the dynamics of the BLR from the IUE data is a more difficult problem. Clavel (1991) finds that the flux in the core of the C IV line ($|v| < 3000 km\ s^{-1}$) lags that in the wings by $\sim$ 2 days. Taken at face value, this is indicative of a radial velocity gradient that suggests virial dynamics. However, by analyzing the drop in flux after the last peak in the IUE campaign, Crenshaw & Blackwell (1990) find that the red wing of C IV falls more rapidly in response to the continuum decline than does the blue wing and argue for a net inflow of gas in the BLR. Unfortunately the low signal-to-noise ratio of the IUE data and the low sampling frequency relative to the short lags for C IV make it impossible to extract two-dimensional transfer functions from these data.

THE 1993 HST CAMPAIGN ON NGC 5548

To move beyond the initial results of the 1988-1989 campaign on NGC 5548 requires more frequent sampling and higher signal-to-noise spectra. Therefore a campaign to monitor NGC 5548 every day for 39 consecutive days using the Faint Object Spectrograph (FOS) on the Hubble Space Telescope (HST) was successfully proposed and carried out from 19 April to 27 May 1993. The prime goals of this program were to determine the smallest scales on which line emitting material exists, particularly for the highest ionization zone of the BLR using the He II $\lambda1640$ emission line, to obtain the two-dimensional transfer function for the C IV $\lambda1549$ line profile in order to study the dynamics of the BLR, and to set even tighter constraints on the delay between UV and optical continuum variations.

The 39 HST/FOS UV spectra of NGC 5548 were obtained over two to three orbits spaced at roughly 24 hour intervals. (One additional spectrum, for a total of 40, was obtained in February 1993 as a test of the target acquisition procedure.) Each spectrum represents a 1295 s exposure through the 4.3 arcsecond aperture using grating G190H (covering 1580-2330 Å), and a 1750 s exposure with grating G130H (covering 1150-1600 Å). The individual spectra have S/N better than 20 per resolution element at all wavelengths longward of 1200 Å. The weighted mean spectrum of all 40 exposures shown in Figure III is arguably the highest S/N UV spectrum of a Seyfert 1 galaxy ever obtained. Numerous emission lines typical of AGN are easily visible as well as several unidentified features which may be emission from Fe II multiplets. All common galactic absorption lines are apparent. In addition, slightly blue-shifted absorption intrinsic to NGC 5548 is visible in Lyα, N V $\lambda\lambda1238, 1242$, and C IV $\lambda\lambda1548, 1550$. Absorption by lower ionization states in NGC 5548 such as C II $\lambda1335$ is not apparent. While some minor calibration problems are still inherent in the preliminary results presented here, the data appear adequate to address each of the prime goals of the campaign.

CONCLUSIONS

Reverberation mapping is showing us a far more detailed picture of the broad-line region than could be obtained using simple photoionization models. The BLR is highly stratified, with high ionization zones closer to the ionizing continuum source than lower ionization gas. Density indicators such as C III] $\lambda1909$ are not applicable to the whole of the BLR, nor are ionization-parameter diagnostics such as the ratio of C III] $\lambda1909$ to C IV $\lambda1549$. A puzzle that is re-surfacing is the Lyα to Hβ ratio problem. While it has long been presumed that Kwan & Krolik (1981) laid this to rest by producing most of the Hβ through collisional excitation at high optical depths deep inside the broad-line clouds, the transfer functions for Lyα and Hβ in NGC 5548 clearly show that these lines are produced in physically distinct regions. This exacerbates the problem of their ratio since most of the Hβ is produced in a region producing little Lyα. Mechanisms for suppressing Lyα or enhancing Hβ in this zone are necessary, and perhaps embedded dust needs a resurgence in popularity.

Sophistication in the analysis of reverberation mapping data continues to

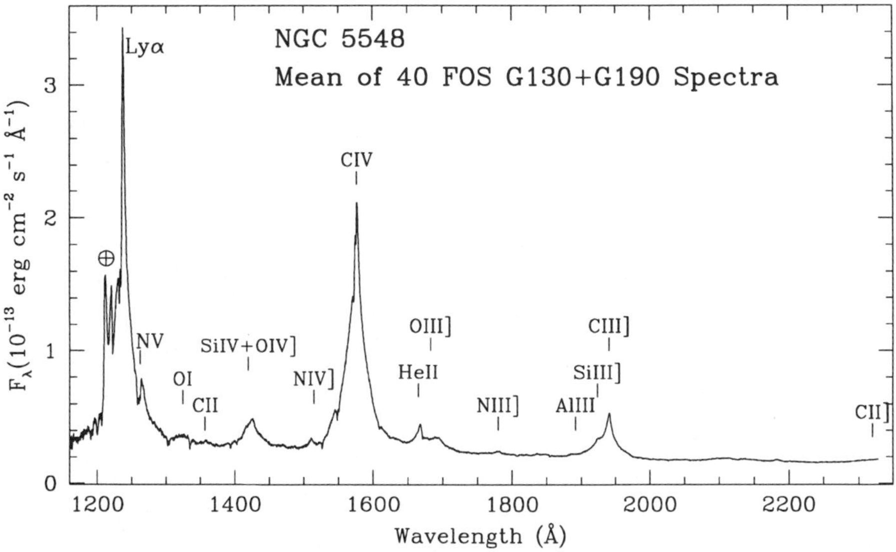

FIGURE III The weighted mean spectrum of NGC 5548 from the HST campaign is shown with prominent emission lines marked (from Korista et al. 1994).

increase. We've moved beyond the simple first-moment analysis of the peak in the cross-correlation function to deriving transfer functions for individual lines. While transfer functions are in principle more informative than the cross-correlation function, they have problems of their own. Transfer functions are not unique, and the solution one obtains can depend critically on the technique used and its operative constraints. To extract the physical and geometric properties of the gas distributions from them requires additional assumptions. In spite of these drawbacks, current reverberation mapping techniques are revealing unprecedented details of the structure of the BLR, and the near future holds promise for insights into its dynamics as well.

REFERENCES

Antonucci, R. R. J., & Cohen, R. D. 1983, *ApJ*, **271**, 564

Bahcall, J. N., Kozlovsky, B.-Z., and Salpeter, E. E. 1972, *ApJ*, **171**, 467

Blandford, R.D., & McKee, C.F. 1982, *ApJ*, **255**, 419

Capriotti, E. R., Foltz, C. B., & Peterson, B. M. 1982 *ApJ*, **261**, 35

Clavel, J. 1991, in *Variability of Active Galactic Nuclei*, ed. H. R. Miller & P. J. Wiita (Cambridge: Cambridge University Press), p. 301

Clavel, J., et al. 1991, *ApJ*, **366**, 64 (Paper I).

Crenshaw, D.M., & Blackwell, J.H., Jr. 1990, *ApJ*, **358**, L37

Dietrich, M., et al. 1993, *ApJ*, **408**, in press (Paper IV)

Done, C., & Krolik, J. H. 1993, in preparation

Espey, B. R., Carswell, R. F., Bailey, J. A., Smith, M. G., & Ward, M. J. 1989, *ApJ*, **342**, 666

Ferland, G.J., Peterson, B.M., Horne, K., Welsh, W.F., & Nahar, S.N. 1992, *ApJ*, **387**, 95

Gaskell, C.M., & Sparke, L.S. 1986, *ApJ*, **305**, 175

Horne, K., Welsh, W.F., & Peterson, B.M. 1991, *ApJ*, **367**, L5

Kallman, T. R., Wilkes, B. J., Krolik, J. H., & Green, R. 1993, *ApJ*, **403**, 45

Korista, K., et al. 1994, in preparation

Krolik, J.H., Horne, K., Kallman, T.R., Malkan, M.A., Edelson, R.A., & Kriss, G.A. 1991, *ApJ*, **371**, 541

Kwan, J., & Krolik, J. H. 1981, *ApJ*, **250**, 478

Maoz, D., et al. 1991, *ApJ*, **367**, 493

Maoz, D., et al. 1993, *ApJ*, **404**, 576

Pérez, E., Robinson, A., & de la Fuente, L. 1992, *MNRAS*, **256**, 103

Peterson, B.M. 1993, *PASP*, **105**, 247

Peterson, B.M., et al. 1985, *ApJ*, **292**, 164

Peterson, B.M., et al. 1991, *ApJ*, **368**, 119 (Paper II)

Peterson, B.M., et al. 1992, *ApJ*, **392**, 470 (Paper III)

Peterson, B.M., et al. 1993, *ApJ*, submitted (Paper VII)

Reichert, G. A., et al. 1993, *ApJ*, in press (Paper V)

Robinson, A., & Pérez, E. 1990, *MNRAS*, **244**, 138

Sparke, L.S. 1993, *ApJ*, **404**, 570

Stirpe, G.M., et al. 1993, *ApJ*, submitted (Paper VI)

Ulrich, M.-H., et al. 1984, *MNRAS*, **206**, 221

Wanders, I., et al. 1993, *A&A*, in press

Welsh, W. M., Horne, K. 1991, *ApJ*, **379**, 586

THE CENTRAL ENGINE AND ITS ENVIRONS

Contributed Papers

The First Stromlo Symposium: The Physics of Active Galaxies
ASP Conference Series, Vol. 54, 1994
G. V. Bicknell, M. A. Dopita, and P. J. Quinn (eds.)

OSSE OBSERVATIONS OF IC 4329A

A. CELOTTI, A.C. FABIAN, K. NANDRA, M.J. REES
Institute of Astronomy, Madingley Road, Cambridge, CB3 0HA, U.K.

J.E. GROVE, W.N. JOHNSON
Hulburt Center for Space Res., Naval Res. Lab., Washington, USA

INTRODUCTION

The hard X-ray spectra of Seyfert galaxies are of considerable interest for understanding the origin of the X-ray/γ−ray continuum. Nonthermal models for the formation of the hard continuum in compact sources predict that electron-positron pairs are generated leading to a γ−ray cutoff, a Compton downscattering feature and a strong annihilation line. Thermal models predict a quasi-exponential hard X-ray cutoff and a weaker annihilation line. Discriminating between these possibilities requires hard X-ray/soft γ−ray data of the brighter sources.

We present the first hard X-ray/soft γ-ray observations of the bright Seyfert 1 galaxy IC 4329A from the Oriented Scintillation Spectrometer Experiment (OSSE) on the Compton Gamma-Ray Observatory (GRO) (Fabian et al. 1993).

OBSERVATIONS AND SPECTRAL ANALYSIS

IC 4329A was the brightest Seyfert 1 galaxy observed with *Ginga* in the energy range 20 – 30 keV. The *Ginga* data are well modelled by the sum of an underlying power-law with a photon index of 2 and a reflection spectrum (Piro et al. 1990).

The source was well detected (at 13.5σ) in the 50 – 200 keV OSSE band. Significant variability was seen over the 3 weeks observing period (the flux dropped by a factor of $\sim$ 2–3 over a period of $\sim$ 10 days).

The 50 – 200 keV spectrum is steep. For an assumed power-law continuum, the photon index is determined to be 2.84 ± 0.36. The data are inconsistent with both the underlying power-law continuum observed 3.5 yr earlier from *Ginga* and the best-fitting (power-law) reflection model extrapolated into the OSSE band. The OSSE data clearly require that there is a break in the spectrum at an energy between 30 and 50 keV.

A break of approximately exponential form is expected in the radiation from a thermal electron distribution. We parametrize the OSSE data by modelling the underlying spectrum as $E^{-\Gamma} \exp(-E/E_c)$, where E is the photon energy. Fitting only the OSSE data and provided that $\Gamma = 2$ and $E_c > 40$ keV the model yields $E_c = 134^{+150}_{-50}$ keV. A good fit is also obtained if the normalization is fixed to be that of the *Ginga* data (see Figure I). If we jointly consider changes in the normalization and E_c, again a cutoff is required at $E_c < 400$ keV; if $E_c > 150$ keV, the source has to have been fainter than any previous X-ray

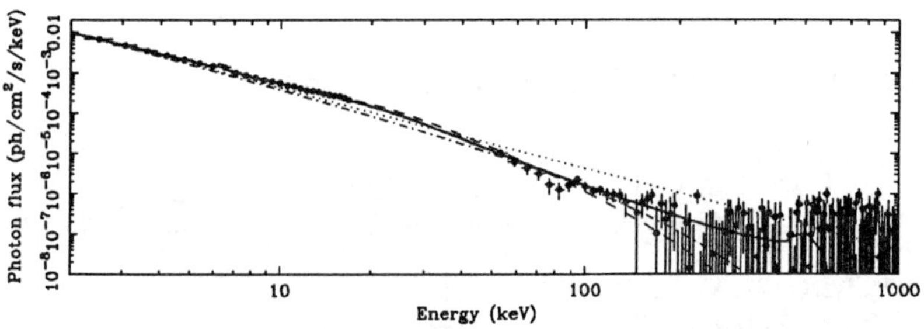

FIGURE I Joint *Ginga*/OSSE incident photon spectrum. Models shown are the (dotted) underlying power-law of photon index 2, (dot-dashed) $E^{-2}\exp(-E/E_c)$ spectrum which best fits the OSSE data and has the same normalization as the underlying *Ginga* spectrum, (solid) joint best-fitting $E^{-2}\exp(-E/E_c)$ spectrum with reflection added

observation. The fits have been repeated with similar results on the joint OSSE and *Ginga* data set with reflection components included.

The OSSE data are also well-fitted by the thermal models that represent the OSSE spectrum of Cygnus X-1 (Grabelsky et al. 1993). Seyfert galaxies and Galactic black hole candidates appear to show similar X-ray/soft $\gamma-$ray spectra.

We also found acceptable nonthermal pair model fits (including reflection) for the joint OSSE and *Ginga* spectra, by using the model supplied with XSPEC (due to A. Zdziarski, see Turner et al. 1993), modified to allow for power–law energy injection of electrons. For the values of the best fit model parameters see Fabian et al. (1993). We have not found any acceptable monoenergetic electron injection model for the joint data set, although the OSSE data alone can be fit.

We have obtained separate limits on a gaussian annihilation line at 511 keV. Any line carrying more than 10% of the total power has to have FWHM$>$ 350 keV (i.e. $\sigma > 150$ keV).

In conclusion, the OSSE data of IC 4329A indicate the presence of a soft $\gamma-$ ray cutoff, possibly indicative of emission from thermal processes. However acceptable nonthermal pair model fits can be found. Our results are consistent with the $\gamma-$ ray cutoff reported from observations of NGC 4151 (Jourdain et al. 1992, Maisack et al. 1993), suggesting that the cutoff may be a general property of all Seyfert galaxies.

REFERENCES

Fabian, A.C., Nandra, K., Celotti, A., Rees, M.J., Grove, J.E., Johnson, W.N. 1993, *ApJ*, in press

Grabelsky, D.A., et al. 1993, in *Proc. of First Compton Symposium*, in press

Jourdain, E., et al. 1992, *A&A*, **256**, L38

Maisack, M., et al. 1993, *ApJ*, **407**, L61

Piro, L., Yamauchi, M., Matsuoka, M. 1990, *ApJ*, **360**, L35

Turner, T.J. et al. 1993, *ApJ*, **407**, 556

The First Stromlo Symposium: The Physics of Active Galaxies
ASP Conference Series, Vol. 54, 1994
G. V. Bicknell, M. A. Dopita, and P. J. Quinn (eds.)

FAST COLLIMATION OF MHD WINDS

JIANKE LI
RCfTA, School of Physics, Sydney University, NSW 2006, AUSTRALIA

ABSTRACT MHD winds are collimated at a large distance. This collimation mechanism is believed to relate to bipolar flows or jets in star forming regions or active galactic nuclei. However, a fast collimation (*i.e.* collimation at a near distance) of MHD winds may exist if there is strong twist of the low latitude boundary magnetic field in a rotating object.

INTRODUCTION

Bipolar flows and jets represent two kinds of anisotropic mass and angular momentum loss from either stellar objects in star forming regions or active galactic nuclei. There are various models proposed to explain these phenomena. Among them, the MHD wind model succesfully demonstrates the asympyotic nature of wind collimation (Hevaerts & Norman 1989 and references therein). However, the formulation of MHD winds is based on the standard wind theory in which the wind passes the slow critical point, Alfvénic singular point and the fast critical point. The required toroidal magnetic field in the standard wind theory must then be very small, and this may not be true.

THE EFFECT OF A LARGE SPECIFIED TOROIDAL MAGNETIC FIELD

If we impose a large toroidal field, then the wind does not cross the fast critical point but the wind is still well behaved (Li 1992). Such a wind is called a non-standard wind. The non-standard wind solution is interesting in that the toroidal magnetic field and density decrease much more slowly than those in a standard wind. The velocity profile in an equatorial flux tube for the monopolar field structure is shown as curve SA in Figure I.

TWO DIMENSIONAL CONSIDERATION

Now we consider a two-dimensional wind structure and a strong toroidal magnetic field associated with the low latitude flux tube, so that the wind is standard in the higher latitude flux tube S and non-standard in the lower latitude flux tube N (see Figure II). There must exist a huge pressure difference (due to the gas pressure and toroidal magnetic pressure) between these two flux tubes.

The expansion of the low latitude flux tube does not create a new critical point, as shown in Fig. I, so does not effectively reduce the pressure difference between two flux tubes. We believe that this physics will lead to a substantial bending of the magnetic field torwards the rotation axis, leading to a fast wind collimation.

Further research is required to resolve this problem.

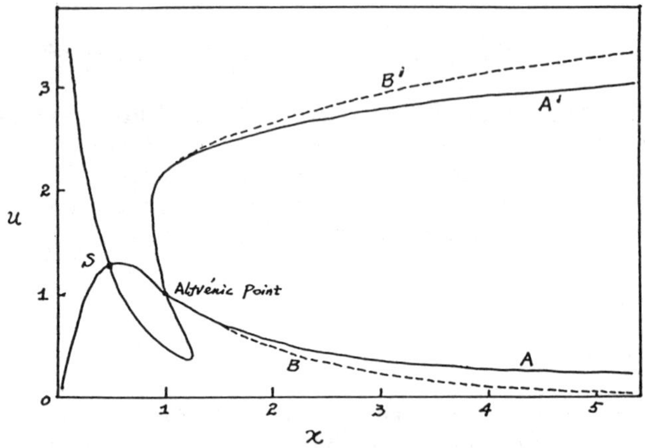

FIGURE I Dimensionless velocity profiles of a non-standard wind. Solutions SA' and SB' are not physical. Curve SA corresponds to a flux tube belonging to a monopolar field, and curve SB to a flux tube which diverges quicker than a monopolar field.

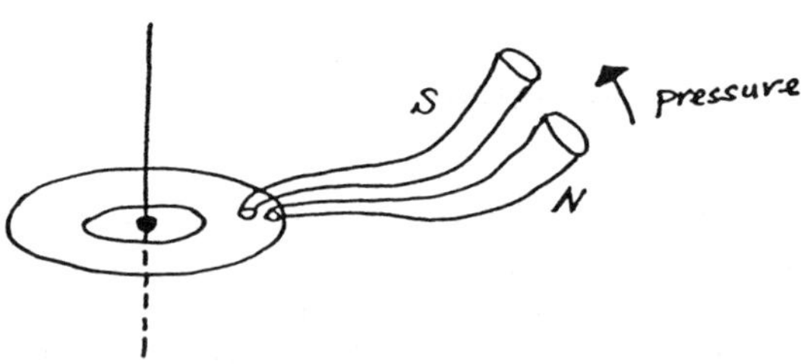

FIGURE II An illustration of collimation effect. The flux tube S is associated with a standard wind but the flux tube N with a non-standard wind. A large pressure difference is then expected, leading to a fast collimation.

REFERENCES

Hevaerts J. & Norman C. A. 1989, *ApJ*, **347**, 1055

Li, J. 1992, in *D Phil. Thesis*, University of Sussex, U.K.

The First Stromlo Symposium: The Physics of Active Galaxies
ASP Conference Series, Vol. 54, 1994
G. V. Bicknell, M. A. Dopita, and P. J. Quinn (eds.)

SHEAR INSTABILITIES IN MAGNETIZED ACCRETION DISKS

QINGHUAN LUO
Research Centre for Theoretical Astrophysics, School of Physics, University of Sydney, NSW 2006, Australia

ABSTRACT The global stability of magnetized accretion disks under an axisymmetric perturbation is discussed. It is shown that under certain conditions a shear instability can occur in the weakly magnetized, differentially rotating disk, which can be responsible for angular momentum transport. It is also shown that, similar to the result in the local approximation, the instability does not explicitly dependent on the azimuthal component of the magnetic field provided that the magnetic field is weak and not strongly depends on the radial distance to the center of the disk.

INTRODUCTION

The essential ingredients of most AGN models involve a compact object, presumably a supermassive black hole, and a differentially rotating accretion disk. The basic mechanism of disk accreting matter to the central object is via angular momentum transport outwards due to some kind of a turbulent viscosity. One possible mechanism for angular momentum transport involves a shear instability associated with a magnetic field (Balbus and Hawley 1991). Knobloch (1992) studied the shear instability using global analysis and obtained a result that seemingly contradicts to the result given in Balbus and Hawley (1991). It is shown here that their difference is due to the fact that local approximation is valid only when the angular velocity, magnetic field, etc. do not change appreciately within the length scale $1/k_r$, with k_r the radial wave vector. The global analysis is applicable to the more general case of strong differential rotation (the angular velocity strongly depends on the radial distance). It is shown that the result derived from the global analysis can reduce to the local result provided that the local approximation conditions can be met.

SHEAR INSTABILITY

To examine the global stability, consider a equilibrium state described by a stationary flow $\mathbf{v}^{(0)} = V(r)\mathbf{e}_\varphi$ and magnetic field $\mathbf{B}^{(0)} = B_\varphi(r,z)\mathbf{e}_\varphi + B_z(r,z)\mathbf{e}_z$ where cylindrical coordinates (φ, r, z) are used. All perturbed quantities are assumed to be proportional to $\exp i(k_z z - \omega t)$. From the ideal MHD equations, in the Boussinesq approximation one obtains a differential equation for $v_r(r)$, the velocity perturbation in the radial direction. One uses $\Omega = V(r)/r$, the angular velocity of a fluid element on a ring defined by a radial distance r, and $V_{Az} = B_z/(\mu_0\rho)^{1/2}$ the Alfvén speed associated with the magnetic field component along $\mathbf{e}_z$ (where ρ is the fluid density). For illustrative purposes, consider the boundary condition $v_r(r_a) = v_r(r_b) = 0$ for $r_a < r_b$ and consider $\omega = i\omega_I$ with real ω_I (cf. Knobloch 1992). From the differential equation for $v_r(r)$ (e.g. Acheson 1973), using $\tilde{\omega}_I^2 = \omega_I^2 + k_z^2 V_{Az}^2$ one has

$$\tilde{\omega}_I^4 = -(k_z^2/N) \int_{r_a}^{r_b} dr[r^2(d\Omega^2/dr)\tilde{\omega}_I^2 + 4\omega_I^2 \Omega V]|v_r|^2,$$

with $N = \int_{r_a}^{r_b} dr(r|\psi'|^2 + k_z^2 r|\psi|^2) > 0$, with $\psi \equiv iv_r(r)/\omega$. Both the Alfvén speeds (denoted by $V_{A\varphi}$ and V_{Az}, respectively) associated with B_φ and B_z are assumed to be weakly depends on r and z and the terms $V_{A\varphi}/r$ can be neglected [in Knobloch's paper (1992), this term is retained, which may be important for large B_φ].

In the weak field approximation, instead of the Rayleigh criterion (the stability condition when the angular momentum increases outward), for instability to occur one has the condition $d\Omega^2/dr < 0$, which implies that the instability may occur if the angular velocity decreases outward (from the rotation axis). This condition can easily be satisfied for accretion disks. This result is consistent with the local analysis by Balbus and Hawley (1991).

The shear instability discussed here is due to the presence of a vertical component of the magnetic field, which always tends to enforce corotation on any displaced fluid element. If a fluid element is displaced outward in the radial direction, as a consequence of the enforced corotation, the fluid element is subject to an excess centrifugal force which may not be balanced by restoring forces (magnetic field tension). In this case the shear instability can occur.

In the local approximation, using $v_r \propto \exp(ik_r r)$, one has the usual dispersion relation (Balbus and Hawley 1991). Similar to the geometric optical approximation in plasma theory, in which the plasma inhomogeneity is often treated as background in deriving the dispersion relation (Luo 1993), the dispersion equation is valid only when $|(d\xi/dr)/\xi| \ll |k_r|$ with ξ the local angular velocity, magnetic field, or density. Thus a sufficient condition for stability is similar to that derived from the global analysis, *i.e.* the angular velocity must decreases outward.

CONCLUSIONS

In conclusions, the global analysis presented here shows that the shear instability in magnetized accretion disks can occur as suggested by the local analysis given by Balbus and Hawley (1991). The result presented here is general, and can reduce to the result given in the local approximation provided that Ω, B_φ and B_z vary slowly with r and also provided that the magnetic field (B_φ, B_z) is weak.

REFERENCES

Acheson, D. J. 1973, *J. Fluid Mech.*, **61**, 609

Balbus, S. A. and Hawley, J. F. 1991, *ApJ*, **376**, 214

Knobloch, E. 1992, *MNRAS*, **255**, 25p

Luo, Q. 1993, Ph.D thesis, University of Sydney

Velikhov, E. P. 1959, *Soviet JETP*, **36**, 1398

THE X-RAY AND Γ-RAY BACKGROUND

Invited Paper

The First Stromlo Symposium: The Physics of Active Galaxies
ASP Conference Series, Vol. 54, 1994
G. V. Bicknell, M. A. Dopita, and P. J. Quinn (eds.)

WHAT ARE X-RAY OBSERVATIONS OF SKY BACKGROUND AND DISCRETE SOURCES TELLING US ABOUT THE PHYSICS OF AGN EVOLUTION?

ELIHU BOLDT
Laboratory for High Energy Astrophysics, NASA/Goddard Space Flight Center, Greenbelt, MD 20771

DARRYL LEITER[1]
Laboratory for High Energy Astrophysics, NASA/Goddard Space Flight Center, Greenbelt, MD 20771

ABSTRACT Our program to obtain a quantitative unified explanation of the cosmic X-ray background compatible with recent deep surveys of discrete sources and AGN broadband X-ray spectra has reached a challenging stage. Under the assumption that the AGN phenomenon is powered by accretion, we find that precise considerations of black hole mass provide severe constraints. In particular, it appears that compatibility can be achieved in terms of the currently estimated X-ray luminosity evolution of canonical AGN only if there exists a significant residual background of faint sources that are Eddington limited compact objects at large redshifts. It is found that such sources would represent the early-growth precursor phase ultimately responsible for the supermassive black hole galactic nuclei found in the present epoch, be they dormant or active.

INTRODUCTION

We examine currently available information about the extragalactic X-ray flux under the assumption that essentially all this radiation is generated via black hole accretion. In particular, the extent to which canonical AGN can account for it is evaluated within this context. By combining the deep *ROSAT* X-ray selected AGN sample (median $z = 1.5$) with the larger EMSS AGN sample (median $z = 0.3$) Boyle *et al.* (1993) have succeeded in providing us with candidate solutions for the AGN X-ray luminosity function and its evolution needed for addressing this issue quantitatively. Using this information, we are now able to directly compare the number of AGN that *should* be accessible to the *ROSAT* bandpass (< 2 keV) with the actual number of *all* sources obtained from the deepest surveys of resolved objects and surface brightness fluctuations observed for the residual background of faint unresolved ones (Hasinger *et al.* 1993). By considering 1) candidate solutions to the soft X-ray luminosity function of relatively unobscured AGN readily visible to *ROSAT*, 2) the more obscured

[1]NRC Senior Resident Research Associate

related AGN implied by the unified model (Antonucci and Miller 1985; Awaki 1991) and 3) recently obtained *Compton* GRO broadband Seyfert X-ray spectra (Maisack *et al.* 1993), we are able to infer the total present-epoch black hole mass density emerging from the AGN phenomenon in previous epochs. We can then compare this value with that implied by the broadband X-ray flux associated with the entire extragalactic sky, foreground plus background. From this black hole integral mass constraint, as provided by the CXB (cosmic X-ray background), and the mass spectrum observed for present-epoch AGN (Padovani, Burg and Edelson 1990) we can determine which one of the two basic solutions for AGN cosmological evolution (i.e., corresponding to $\Omega \simeq 1$ or $\Omega \ll 1$) is most tenable. The implications of this are then further investigated in the context of a residual background of faint sources corresponding to PAG (precursor active galaxies) indicative of a much earlier epoch (Leiter and Boldt 1982, 1992; Boldt and Leiter 1987) when all galactic nuclei presently with black holes were amply fueled and clearly active.

AGN NUMBER

We consider the best-fit power-law AGN X-ray luminosity function obtained by Boyle *et al.* (1993), from observations up to $z = 3$, for both $\Omega = 0$ (Case D) and $\Omega = 1$ (Case I). These correspond to pure luminosity evolution in that the underlying comoving density is invariant and that all changes with cosmic time can be characterized by the redshift dependence of the break luminosity L^* at which the power law function steepens. The luminosity evolution required for $\Omega = 1$ is found to be substantially different from that to be associated with $\Omega = 0$. For $\Omega = 0$, $L^* = 8.1 \times 10^{43}(1+z)^{2.8}$ ergs s^{-1} for all redshifts considered whereas the best fit with $\Omega = 1$ requires a constant $L^* = 1.2 \times 10^{45}$ ergs s^{-1} for $z > 2$ (at least out to $z = 3$), with an evolving $L^* = 6.2 \times 10^{43}(1+z)^{2.73}$ ergs s^{-1} only for $z < 2$. There is increasing evidence to suggest that the strong power-law evolution of the optical luminosity function also cuts off at $z \simeq 2$ (Boyle *et al.* 1991) and that AGN evolution is relatively weak over the redshift interval $2 < z < 3$ (Schmidt, Schneider and Gunn 1991). As discussed by Murphy, Cohn and Durisen (1991) within the context of galactic dynamics, this sort of plateau in early AGN evolution is the kind of behavior that can be expected following an Eddington limited initial growth phase (according to their model 2C); it provides a physically plausible framework for our subsequent considerations about the onset of the phenomenon of AGN.

The invariant comoving density of sources, corresponding to $L > 10^{42}$ ergs s^{-1} (0.3 - 3.5 keV) at $z = 0$, is approximately the same for both cases described above, *viz:* 1.9×10^{-5} Mpc^{-3} (for $H_o = 50$ km s^{-1} Mpc^{-1}, as assumed throughout this text), numerically equal to the present-epoch density of Seyfert 1 galaxies at $M_B < -17.5$ (Huchra and Burg 1992). The total number of such AGN (at $z < 4$) for $\Omega = 0$ is then $\sim 4 \times 10^3$ $degree^{-2}$ (including ~ 50 quasars $degree^{-2}$ at $L > L^*$), reached at a minimum flux level $S \simeq 4 \times 10^{-17}$ ergs s^{-1} cm^{-2} (0.3 - 3.5 keV). In sharp contrast, with $\Omega = 1$ the total number of such AGN accessible to *ROSAT* would be only $\sim 6 \times 10^2$ $degree^{-2}$ (including ~ 14 $degree^{-2}$ at $L > L^*$), reached at $S \simeq 2 \times 10^{-16}$ ergs s^{-1} cm^{-2}. However, from an analysis of surface brightness fluctuations in the residual background detected with *ROSAT* Hasinger *et al.* (1993) conclude that the total number of sources at $S > 2 \times 10^{-16}$ ergs s^{-1} cm^{-2} must exceed 2×10^3 degree^{-2}. Hence, if in fact $\Omega \simeq 1$ we are confronted with the situation that most of the faintest sources making up the unresolved CXB (cosmic X-ray background) would have to be other than canonical AGN. In such case, as we shall exhibit, they could well be PAG (precursor active galaxies). Although the number of AGN alone could be adequate if $\Omega \ll 1$ were actually more appropriate, we have found that the AGN X-ray luminosity evolution associated with this solution does not appear to be consistent with the scenario in which these sources are powered by accretion.

BLACK HOLE GALACTIC NUCLEI

The present-epoch mass density of supermassive black hole nuclei associated with Seyfert 1 galaxies having $M_B < -17$ is $\rho(\text{Seyfert 1}) \simeq 6.5 \times 10^2\ M_\odot$ Mpc^{-3} (Padovani, Burg and Edelson 1990). Since the ratio of the number density of Seyfert 2 galaxies to that of Seyfert 1 is ~ 2.3 (Huchra and Burg 1992), the unified scenario for such AGN implies that $\rho(\text{Seyfert}) \simeq 2.1 \times 10^3\ M_\odot$ Mpc^{-3}. Taking the number density of quasars as 4.2×10^{-7} Mpc^{-3}, that corresponding to $L > L^*$ [for $\Omega = 1$; Case I of Boyle *et al.* (1993)], and a mean nuclear black hole mass of $\sim 10^9\ M_\odot$ yields a somewhat smaller density, viz: $\rho(\text{quasars}) \simeq 4.2 \times 10^2\ M_\odot$ Mpc^{-3}. The mass density corresponding to all AGN is then $\rho(\text{AGN}) \simeq 2.5 \times 10^3\ M_\odot$ Mpc^{-3}, significantly less than the estimated total supermassive BH (black hole) mass density $\rho(\text{BH}) \simeq 1.4 \times 10^5\ M_\odot$ Mpc^{-3} associated with galactic nuclei that are relatively inactive (Kormendy and Richstone 1992). The fraction f of galactic nuclei with black holes that are currently active is therefore clearly small, viz: $f \equiv \rho(\text{AGN})/[\rho(\text{BH}) + \rho(\text{AGN})] \simeq 2\%$. We interpret this fraction f as the probability of such a nucleus being in the "on" state (Cannizzo 1992) and indicative of the duty cycle for such activity in the past as well. The population of black holes underlying AGN then has a relatively large number density, $[n(\text{AGN})]/f$, possibly comparable to that of "normal" galaxies.

As emphasized by Soltan (1982), the total present-epoch black hole mass density arising from accretion can be determined directly from the observed total sky flux of all the resultant radiation, independent of the parameter values (Ω and H_o) characterizing the underlying Friedmann cosmology. Since the omnidirectional energy flux $(4\pi I)$ of the CXB is observationally very well defined (Boldt 1987, 1992; Gruber 1992) it provides us with an especially good measure for this. In particular, the mass density associated with generating the CXB via accretion is given by

$$\rho(\text{CXB}) = (\epsilon^{-1} - 1)\ \eta\ (1 + \langle z \rangle)(4\pi I)/c^3, \text{ where}$$

$$\epsilon \equiv L(\text{bolometric})/[c^2(dM/dt)],$$

$$\eta \equiv L(\text{bolometric})/L(\text{all X-rays}), \text{ and}$$

$$\langle z \rangle \equiv [(\int z\, dI)/I]$$

where $< z >$ is the mean redshift of CXB photons with respect to observed energy flux.
The principal component of the CXB has a spectrum which may be integrated in closed form. This spectrum characteristic of the CXB (Boldt 1992), based on observations over the 3 - 50 keV band with HEAO-1 A2 (Marshall *et al.* 1980), is well described by the simple function

$$I_E \equiv dI/dE = 7.8E^{-0.29} \exp(-E/40 \text{ keV}) \text{ cm}^{-2}\text{s}^{-1}\text{sr}^{-1}.$$

Expressed as $E\,I_E$, it exhibits a well-defined pronounced peak at 28keV. The integrated omnidirectional flux associated with this spectrum is

$$4\pi I = 2.8 \times 10^{-6} \text{ ergs s}^{-1} \text{ cm}^{-2}.$$

Considering harder components of the background observed with HEAO-1 A4 as well (Gruber 1992) increases this integral flux by only $\sim 20\%$ (for energies up to ~ 1 MeV); the peak in $E\,I_E$ associated with the corresponding six-parameter three-component broadband spectral fit is within 3% of that obtained with the simple fit given above, both as regards magnitude and photon energy. Based on this principal single component that dominates the CXB we obtain that, for $\epsilon \simeq 0.1$,

$$\rho(\text{CXB}) = 1.4 \times 10^4\ \eta\ (1 + \langle z \rangle)\ M_\odot \text{ Mpc}^{-3}.$$

Considering the limits $\eta = 2 - 3$ (discussed later) and $\langle z \rangle = 1 - 4$ we obtain the bounds

$$\rho(\text{CXB}) = (0.6 - 2.1) \times 10^5\ M_\odot \text{ Mpc}^{-3}.$$

Since the mean black hole seed mass (i.e., that prior to any growth due to accretion) is expected to be no larger than a Jeans mass ($\sim 10^6 M_\odot$), much smaller than that of a canonical AGN supermassive black hole, the *total* mass density of present-epoch black holes should be dominated by ρ(CXB), the density of accreted matter inferred by the CXB. We note that this density measured with the CXB is definitely much higher than that of present-epoch AGN. It is important to recognize, however, that this is entirely consistent with the present-epoch black hole mass density of *dormant* galactic nuclei ρ(BH) estimated by Kormendy and Richstone (1992) and that it can *not* be substantially greater.

The luminosity density (volume emissivity) corresponding to the X-ray luminosity function for AGN [$\int L(dn/dL)dL$] can be used to evaluate the associated mass per AGN (ΔM) accreted from $z = 3$ to $z = 0$, yielding

$$\Delta M \propto (\epsilon^{-1} - 1)\ \eta/n,$$

where n is the comoving number density of the AGN involved. We have done this for the two cases (corresponding to $\Omega = 0$ and $\Omega = 1$) considered by Boyle *et al.* (1993) already cited. In doing this we have taken the observed X-ray luminosity in the band (0.3 - 3.5 keV) as 10% of the bolometric AGN luminosity, compatible with IR-UV estimates (Mushotzky, Done and Pounds 1993) but probably somewhat of an upper bound for sources with a "big bump" in the EUV. For Seyfert 1 galaxies (i.e., $L < L^*$) we take the spectrum used by Boyle *et al.* (i.e., $\alpha = 1.3$ at E < 1.5 keV, $\alpha = 1$ for E > 1.5 keV) for energies < 3.5 keV and the spectrum observed with *Ginga* (Pounds, Nandra and Stewart 1992) at higher energies, with an exponential roll-off introduced to match the peak in $E\,S_E$ observed for NGC4151 with OSSE (Maisack *et al.* 1993). From this we calculate that $\eta(\text{Seyfert 1}) = 2.3$ (i.e., a luminosity in broadband X-rays comparable to but somewhat less than that for all other bands combined), intermediate between the limits used for interpreting the CXB. The accreted mass density ρ_A associated with the AGN phenomenon ($z = 3$ to $z = 0$) is then obtained as follows:

$$\rho_A = [n(\text{Seyfert})]\Delta M_S + [n(\text{quasar})]\Delta M_q$$

where $n(\text{Seyfert}) = 3.3[n(\text{Seyfert 1})]$ and the unified model implies that the same value of ΔM_S applies to all types of Seyfert galaxies. The results of this analysis of the AGN X-ray luminosity function, under the assumption that $\epsilon \simeq 0.1$, are given in Table I.

TABLE I AGN Black Hole Mass Accretion ($z = 3$ to $z = 0$)

Ω	ΔM_S ($< L^*$)	ΔM_q ($> L^*$)	ρ_A
0	$1.23 \times 10^9\ M_\odot$	$4.10 \times 10^{10}\ M_\odot$	$8.8 \times 10^4\ M_\odot/$ Mpc^3
1	$6.65 \times 10^8\ M_\odot$	$1.64 \times 10^{10}\ M_\odot$	$4.9 \times 10^4\ M_\odot/$ Mpc^3

From Table I we see that the accreted mass density corresponding to $\Omega = 0$ is substantially larger than that for $\Omega = 1$ by a factor of ~ 1.8. It is interesting to note that the associated spectral densities (at 2 keV) of the entire AGN foreground for these two cases, as given by Boyle *et al.* (1993), exhibit the same trend, as expected, with a comparable factor. Most important, we point out that the mass accreted per AGN for $L < L^*$ (i.e., ΔM_S) is in both cases clearly larger than the mean black hole mass of present-epoch Seyfert 1 galactic nuclei estimated by Padovani, Burg and Edelson (1990), viz: $2.0(+1.9, -0.8) \times 10^7 M_\odot$. This then implies that the AGN effective "on-time" fraction for accretion ($z = 3$ to $z = 0$) must indeed be small, viz: $f < 3\%$ for the case with $\Omega = 0$ and $f < 6\%$ for $\Omega = 1$.

The two solutions to the AGN luminosity function and its evolution considered here also imply two solutions for the mean initial black hole mass (M_i) at $z = 3$. These solutions depend, of course, on the actual initial value of the Eddington ratio $\lambda \equiv L(\text{bolometric})/[1.3 \times 10^{38}(M/M_\odot)$ erg/s] at $z = 3$, still unspecified. These solutions for M_i are exhibited in Table II in terms of λ and η.

TABLE II Initial Black Hole Mean Mass M_i (at $z = 3$)

Ω	$(\lambda/\eta)M_i$ ($< L^*$)	$(\lambda/\eta)M_i$ ($> L^*$)
0	$1.1 \times 10^7 \; M_\odot$	$3.6 \times 10^8 \; M_\odot$
1	$4.6 \times 10^6 \; M_\odot$	$1.1 \times 10^8 \; M_\odot$

The final mean black hole mass M_f (at $z = 0$) may be obtained from the entries on Tables I and II, as follows:

$$M_f = M_i + (\Delta M)f.$$

The final *total* mass density ρ of the underlying black hole population is given by:

$$\rho = \rho_A + \{[n_S(M_S)_i + n_q(M_q)_i]/f\}.$$

Noting that $\lambda \leq 1$ and taking $\eta = 2.3$, we then find that, for $\Omega = 0$:

$$(M_S)_f \geq [2.5 \times 10^7 + 1.23 \times 10^9 f] \; M_\odot,$$

and

$$\rho \geq 8.8 \times 10^4 [1 + (0.020/f)] \; M_\odot \; \mathrm{Mpc}^{-3}.$$

To meet the constraint that $M_S < 3.9 \times 10^7 \; M_\odot$ then requires that $f < 0.011$, which then implies that $\rho > 2.5 \times 10^5 \; M_\odot \; \mathrm{Mpc}^{-3}$, larger than the upper bound obtained from the CXB energy flux. This means that any scheme for the CXB corresponding to $\Omega \ll 1$ is unacceptable if it involves a unified model of accretion-fueled AGN up to $z \geq 3$ for which the comoving luminosity density (0.3 - 3.5 keV) equals or exceeds that derived by Boyle *et al.* (1993), viz: $\langle n \langle L \rangle \rangle = 1.6 \times 10^{38} (1 + z)^{2.8}$ ergs s^{-1} Mpc^{-3}. Unified AGN schemes recently proposed for the CXB that provide some of the best overall fits to the broadband spectrum correspond to $\Omega = 0$ (Zdziarski, Zycki and Krolik 1993; Comastri, Setti, Zamorani and Hasinger 1993) and are actually unacceptable, according to this criterion on luminosity density, if we require that the sources indeed be powered by accretion. If the contribution of Seyfert 2 galaxies is excluded, Zdziarski, Zycki and Krolik (1993) find that using Seyfert 1 galaxies alone to fit the broadband CXB spectrum requires that the present-epoch luminosity density be unacceptably large, more than twice that derived by Boyle *et al.* (1993). Furthermore, since this large value was obtained for the case where $\Omega = 0$ it constitutes a ráther severe violation of the constraint on mass (Leiter and Boldt 1993).

For $\Omega = 1$:

$$(M_S)_f \geq [1.1 \times 10^7 + 6.65 \times 10^8 f] \; M_\odot,$$

and

$$\rho \geq 4.9 \times 10^4[1 + (0.016/f)]\ M_\odot\ \mathrm{Mpc}^{-3}$$ for the unified model.

In this instance where $\Omega = 1$ the constraint on Seyfert mass requires $f < 0.042$, which then implies $\rho > 0.7 \times 10^5\ M_\odot\ \mathrm{Mpc}^{-3}$, allowing for the distinct possibility of being within the bounds obtained from the CXB energy flux. Hence, scenarios for the CXB involving a unified model of accretion-fueled AGN do indeed appear to have the potential of being consistent with the solution for the luminosity function and its evolution obtained by Boyle *et al.* (1993) within the context of $\Omega = 1$.

SOURCES OF THE CXB

Madau, Ghisellini and Fabian (1993) have succeeded in applying a unified AGN model to account for the CXB spectrum within the context of an Einstein - de Sitter cosmology (i.e., $\Omega = 1$) but they need strong continuous evolution of the comoving luminosity density up to $z = 5$ to do it. This is clearly inconsistent with the *ROSAT* derived result that, for $\Omega = 1$, the AGN luminosity evolves strongly only up to $z \simeq 2$ (Boyle *et al.* 1993). We find that a unified AGN model similar to one considered by Zdziarski, Zycki and Krolik (1993) for $\Omega = 0$, but with the form of evolution needed for $\Omega = 1$ (here taken up to $z = 3$), can indeed be made to fit the broadband CXB spectrum within the context of an Einstein - de Sitter cosmology (Leiter and Boldt 1993). Since this fit is found to require Seyfert spectra which exhibit a peak in $E L_E$ at $\sim$ 60 keV the effective redshift corresponding to the observed CXB peak in $E I_E$ would then be at z(effective) $\simeq 1$. However, such a peak in Seyfert spectra may not be universal [e.g., although IC 4329A has a 130 keV e-folding energy characterizing its spectral roll-off (Fabian *et al.* 1993) the power-law involved here is too steep for yielding an actual peak in $E L_E$]. The Seyfert 2 spectra needed for this CXB fit correspond to self-absorption by $N_H \simeq 10^{23}\mathrm{cm}^{-2}$, exhibiting unit optical depth at $\sim$ 3.5 keV (Morrison and McCammon 1983), enough to render these sources essentially inaccessible to the bandpass (< 2 keV) of *ROSAT* (as assumed). We emphasize, though, that the Seyfert 1 normalization required for the unified model to adequately account for the CXB in this instance corresponds to an unacceptably large value for the present-epoch luminosity density, viz: $(n\langle L\rangle)_o = 3.2 \times 10^{38}$ ergs s^{-1} Mpc^{-3} (0.3 - 3.5 keV). This not only exceeds the value obtained by Boyle *et al.* (1993) for the situation where $\Omega = 1$ [i.e., $(n\langle L\rangle)_o = 1.7 \times 10^{38}$ ergs s^{-1} Mpc^{-3}] but is so large as to present us with a definite violation of the mass constraints previously described.

Using the actual AGN X-ray luminosity function and evolution derived by Boyle *et al.* (1993) for $\Omega = 1$ we are in fact able to construct an all-sky X-ray foreground arising from such sources out to at least $z = 4$ *without* violating any of the mass constraints coming from the assumption of accretion-powered emission. For the unified model we have addressed (Leiter and Boldt 1993) this is readily achieved with $f \simeq 2\%$. In addition to Seyfert (1 and 2) galaxies (at $L < L^*$) with a hard X-ray peak in $E L_E$ such as indicated for NGC 4151 (Maisack *et al.* 1993; Zdziarski, Lightman and Maciolek-Niedzwiecki 1993), we also consider the contribution of quasars (at $L > L^*$) having pure power-law

X-ray spectra characterized by the energy spectral index $\alpha = 0.8$ (Williams *et al.* 1992). The resulting composite spectrum, however, falls substantially short of accounting for the broadband CXB actually observed. In particular, there remains an unexplained residual background for which EI_E peaks at E = 26 keV and which amounts to ~ 50% of the total observed CXB at that energy. We find, however, that this residual spectrum is remarkably close to that expected for highly compact Eddington-limited sources of the type discussed by Zdziarski (1988) for PAG (Leiter and Boldt 1982, 1993), provided that they are located at $z \geq 4.4$. Such sources would be indicative of the growth phase responsible for producing the underlying black hole masses present at the onset of canonical AGN. We expect that the amply fueled compact PAG (precursor active galaxies), during this Eddington-limited phase, are essentially always "on", unlike AGN (Cannizzo 1992). If we assume that the invariant comoving density of these PAG at ≥ 4.4 is that of the underlying black hole population, $n(AGN)/f$, then we obtain sufficient flux to account for the residual CXB *without* impacting final mass considerations (at $z = 0$) in any way. Our calculations (Leiter and Boldt 1993) indicate that these objects could well also account for the many additional faint sources indicated from the analysis of deep *ROSAT* fields (Barcons *et al.* 1993; Georgantopoulos *et al.* *1993*; Hasinger *et al.* 1993).

SUMMARY

Although X-ray observations of AGN out to $z \approx 3$ indicate that they could readily account for the CXB (cosmic X-ray background) using the evolution obtained within the cosmological framework of $\Omega \ll 1$, the assumption of accretion-powered emission is then found to be incompatible with the mass function for present-epoch black hole galactic nuclei (both active *and* dormant). While the AGN luminosity function and evolution inferred for $\Omega = 1$ *are* compatible with the mass constraint, this solution falls short of accounting for the CXB, even when considering a unified model. We note, however, that the underlying supermassive black holes which already exist at the onset of the canonical AGN phenomenon of supply-limited accretion must have undergone a previous growth phase where the accretion would be expected to be radiation (Eddington) limited. In such case the residual CXB needed for $\Omega \approx 1$ is naturally explained by the characteristic X-ray emission from highly compact PAG (precursor active galaxy) sources associated with these numerous black holes, at redshifts just beyond the earliest AGN. These objects should be well represented among the many as-yet unresolved faint sources inferred from CXB surface brightness studies.

ACKNOWLEDGMENTS

One of us (EB) thanks Sarah, Bruce, Ben, Alex and Peter Lay for their encouragement in preparing the initial draft of this manuscript during a pleasant visit with them at Sydney and Hardy's Bay, following the symposium, before returning to the US. We thank Greg Madejski for suggestions that improved the final version.

REFERENCES

Antonucci, R. and Miller, J. 1985, *ApJ*, **297**, 621

Awaki, H. 1991, PhD thesis, Nagoya University

Barcons, X., Branduardi-Raymont, G., Warwick, R., Fabian, A., Mason, K., McHardy, I. & Rowan-Robinson, M. 1993, *MNRAS*, submitted

Boldt, E., & Leiter, D., 1987, *ApJ*, **322**, L1

Boldt, E. 1987, *Phys. Rept.*, **146**, 215

Boldt, E. 1992, in *The X-ray Background*, ed. X. Barcons & A. Fabian (Cambridge: Cambridge Univ. Press), p. 115

Boyle, B. Jones, L., Shanks, T., Marano, B., Zitelli, V. & Zamorani, G. 1991, in Crampton D., ASP Conference Series **21**, in *The Space Distribution of Quasars*, Astron. Soc. Pacif. San Francisco, p. 191

Boyle, B., Griffiths, R., Shanks, T., Stewart, G. & Georgantopoulos, I. 1993, *MNRAS*, **260**, 49

Cannizzo, J. 1992, *ApJ*, **385**, 94

Comastri, A., Setti, G., Zamorani, G. & Hasinger, G. 1993, this symposium (Paper B-2)

Fabian, A., Nandra, K., Celotti, A., Rees, M., Grove, J. & Johnson, W. 1993, this symposium (Paper A-1)

Georgantopoulos, I., Stewart, G., Shanks, T., Griffiths, R. & Boyle, B. 1993, *MNRAS*, in press

Gruber, D. 1992, in *The X-ray Background*, ed. X. Barcons & A. Fabian (Cambridge: Cambridge Univ. Press), p. 229

Hasinger, G., Burg, R. Giacconi, R., Hartner, G., Schmidt, M., Trumper, J. & Zamorani, G. 1993, *A&A*, in press (MPE Preprint 238)

Huchra, J. & Burg, R. 1992, *ApJ*, **393**, 90

Kormendy, J. & Richstone, D. 1992, *ApJ*, **393**, 559

Leiter, D. & Boldt, E. 1982, *ApJ*, **260**, 1

Leiter, D. and Boldt, E. 1992, in *Testing the AGN Paradigm*, ed. S. Holt, S. Neff & M. Urry (New York: AIP) AIP **254**, p 370

Leiter, D. & Boldt, E. 1993, in preparation

Madau, P., Ghisellini, G. & Fabian, A. 1993, *ApJ*, **410**, L7

Maisack, M., Johnson, W., Kinzer, R., Strickman, M., Kurfess, J., Cameron, R., Jung, G., Grabelsky, D., Purcell, W. & Ulmer, M. 1993, *ApJ*, **407**, L61

Marshall, F., Boldt, E., Holt, S., Miller, R., Mushotzky, R., Rose, R., Rothschild, R. & Serlemitsos, P. 1980, *ApJ*, **235**, 4

Morrison, R. & McCammon, D. 1983, *ApJ*, **270**, 119

Murphy, B., Cohn, H. & Durisen, R. 1991 *ApJ*, **370**, 60

Mushotzky, R., Done, C. & Pounds, K. 1993, *Ann Rev. Astron. Astrophys.*, in press

Padovani, P., Burg, R. & Edelson, R. 1990, *ApJ*, **353**, 438

Pounds, K., Nandra, K. & Stewart, G. 1992, in *X-ray Emission from AGN and the CXB* (eds W. Brinkmann and J. Trumper), MPE Report 235 p. 1

Schmidt, M., Schneider, D., & Gunn, J. 1991, in Crampton D., ASP Conference Series **21**, *The Space Distribution of Quasars*, Astron. Soc. Pacif. San Francisco, p. 109

Soltan, A. 1982, *MNRAS*, **200**, 115

Williams, O., Turner,M., Stewart, G., Saxton, R., Ohashi, T., Matkishima, K., Kii, T., Inoue, H., Makino, F. Hayashida, K. & Koyama, K. 1992, *ApJ*, **389**, 157

Zdziarski, A. 1988, *MNRAS*, **233**, 739

Zdziarski, A., Zycki, P. & Krolik, J. 1993, *ApJ*, in press

Zdziarski, A., Lightman, A. & Maciolek-Niedzwiecki, A. 1993, *ApJ*, submitted

THE X-RAY AND Γ-RAY BACKGROUND

Contributed Papers

The First Stromlo Symposium: The Physics of Active Galaxies
ASP Conference Series, Vol. 54, 1994
G. V. Bicknell, M. A. Dopita, and P. J. Quinn (eds.)

AGNS MAKE THE X–RAY BACKGROUND

ANDREA COMASTRI AND GÜNTHER HASINGER
Max Planck Institut für extraterrestrische Physik, 85740 Garching bei München, Federal Republic of Germany

GIANCARLO SETTI[1] AND GIOVANNI ZAMORANI[1,2]
[1] *Istituto di Radioastronomia, via Irnerio 46, 40126 Bologna, Italy*
[2] *Osservatorio Astronomico, via Zamboni 33, 40126 Bologna, Italy*

We present preliminary results on an AGN model for the X-ray background (XRB). The model takes into account the observed spectral properties of different classes of AGNs over a broad energy range, including the recent GINGA and OSSE observations in the energy range $\sim 2-20$ keV and $\sim 60-300$ keV respectively, and is based on the X-ray properties of AGN unified schemes (Setti & Woltjer 1989). The model is checked in relation to the XRB spectrum, up to ~ 100 keV and the number counts relationship obtained with ROSAT (Hasinger et al. 1993). The main ingredients of the model are the following:
i) The X-ray spectrum of Seyfert 1 galaxies has been described by the "reflection model" where about half of the flux of the primary spectrum, a power law with slope $\alpha_E = 0.9$, is reprocessed by relatively cold and optically thick matter (Pounds et al. 1990).
ii) For the Seyfert 2 objects we have assumed a Seyfert 1 spectrum modified by the absorption effects due to a molecular torus surrounding the central nucleus (Awaki & Koyama 1992). The amount of the absorption has been modeled with a distribution of column densities in the range $logN_H = 21.5 - 25.0$ (units of $atoms\ cm^{-2}$). Finally we have introduced a break to a steeper power law ($\alpha_E = 2.0$) in the typical Seyfert 1 spectrum as indicated by the recent OSSE observations (Cameron et al. 1993).
iii) For the high luminosity quasars we have assumed a single power law spectrum with slope $\alpha_E = 0.9$ without reflection component (Williams et al. 1992).

We have integrated the source spectra over redshift using the parameters obtained by Boyle et al. (1993, model A). The comoving volume emissivity in the 0.3-3.5 keV energy range changes with z for $z \leq z_{max}$ as $nL = nL_0(1+z)^{\beta}$, where $\beta = 2.75$, $z_{max} = 3$ in a Friedmann cosmology with $H_0 = 50$ $km\ s^{-1}\ Mpc^{-1}$ and $q_0 = 0$. A very good fit to the XRB (Fig. 1a) can be obtained with a flat distribution of the column densities of the absorbed objects in the range $logN_H = 21.5-25.0$, and with a number ratio between absorbed and unabsorbed objects of 2.5, in good agreement with the value (2.3 ± 0.7) found by Huchra and Burg (1992) for a complete sample of optically selected Seyfert galaxies. The same model produces a soft X-ray $Log\ N - Log\ S$ in agreement with the recent ROSAT measurements (Fig. 1b). A more detailed analysis including a discussion on the hard ($2-10$ keV) X-ray counts will be presented elsewhere.

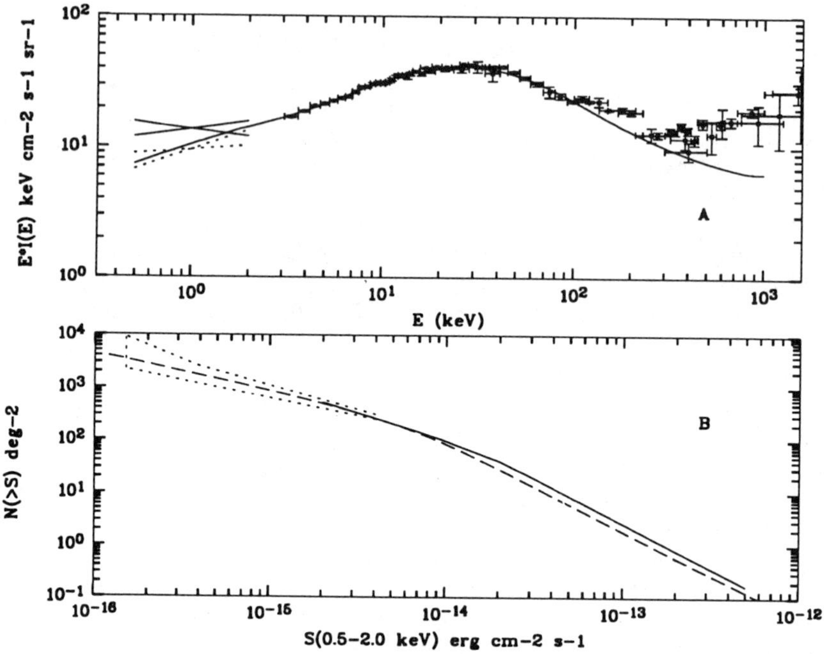

FIGURE I **a)** The XRB produced by our model compared with the data. The soft (0.5-2.0 keV) XRB spectrum is from ROSAT (Solid lines Hasinger et al. 1993; dashed lines Wang & McCray 1993), while the high energy data are a compilation of best results from various experiments by Gruber (1992). **b)** The observed soft X-ray counts (solid line) and fluctuation analysis (dotted line) compared with the model predicted counts (dashed line).

REFERENCES

Awaki, H. & Koyama K. 1992, in the *Proceedings of the COSPAR symposium "Recent results in X-ray and EUV astronomy"*, Washington, D.C.

Boyle, B. J. *et al.* 1993, *MNRAS*, **260**, 49.

Cameron, R. A. *et al.* 1993, in the *Proceedings of the Compton Symposium*, St. Louis, MO, ed. N. Gehrels.

Gruber, D. E. 1992, in the *Proceedings of "The X-ray background"* , ed. X. Barcons & A. C. Fabian (Cambridge: Cambridge Univ. Press), p. 44.

Hasinger, G. et al. 1993 *A&A*, **275**, 1.

Huchra, J. & Burg, R. 1992, *ApJ*, **393**, 90.

Pounds, K. A. *et al.* 1990, *Nature*, **344**, 132.

Setti, G. & Woltjer, L. 1989, *A&A*, **224**, L21.

Wang, D. Q. & McCray R. 1993, *ApJ*, **409**, L37.

Williams, O. R. *et al.* 1992, *ApJ*, **389**, 157.

The First Stromlo Symposium: The Physics of Active Galaxies
ASP Conference Series, Vol. 54, 1994
G. V. Bicknell, M. A. Dopita, and P. J. Quinn (eds.)

A.G.N. CENTRAL ENGINES AND THE DIFFUSE GAMMA RAY BACKGROUND

P.A. JOHNSON, R.J. PROTHEROE, A.P. SZABO
Dept. Physics and Math. Physics, Univ. of Adelaide, Australia, 5005

A. MASTICHIADIS
Lab. for H.E. Astrophysics, NASA/Goddard Space Flight Center, USA

T.S. STANEV
Bartol Research Institute, University of Delaware, USA

All the AGN detected so far by the Compton GRO are blazars, but here we consider the γ-ray emission from the more numerous 'ordinary' AGN (e.g. Seyferts and radio-quiet AGN) where the emission arises in the central region rather than a jet. We calculate the expected γ-ray output of individual AGN and use this together with the X-ray luminosity function of AGN to estimate the diffuse γ-ray background intensity.

The AGN model we adopt is that described by Protheroe and Kazanas (1983) and developed by Kazanas and Ellison (1986) in which acceleration takes place at a black hole accretion shock which is assumed to be supported by the pressure of relativistic particles. Full details of the present model are given by Szabo and Protheroe (1992a,b), Protheroe and Szabo (1992) and Johnson *et al.* (1993).

We assume that all γ-rays emerging from the AGN result from a pair–synchrotron cascade fed mainly by *pp* collisions and calculate the spectrum of γ-rays per accelerated proton as described by Johnson *et al.* (1993). To obtain the total γ-ray output of the AGN we multiply by the rate at which protons are injected into the AGN shock accelerator which we obtain by dividing the AGN continuum luminosity by the energy going into the electromagnetic cascade in the central region per injected proton.

We perform an integration over redshift and luminosity to obtain the γ-ray intensity from AGN. We use the parameters of Model A of Morisawa *et al.* (1990) for the luminosity function of AGN and its evolution and take account of cascading on the infrared/microwave background as described by Protheroe and Stanev (1993). Our result is shown in the Figure for the cases in which the diffusion coefficient at the shock takes 100 times the minimum value, and for a range of ratio of the shock radius to the Schwartzschild radius, R/r_s, chosen to bracket those expected. In the Figure we compare our result with the SAS-2 γ-ray background observations (Thompson and Fichtel 1982), and the expected backgrounds at high energy due to interactions of cosmic rays in the microwave background ($p\gamma$) and interstellar medium (pp) calculated by Halzen *et al.* (1990) and Berezinsky *et al.* (1993) respectively. We also show the contribution of blazars as estimated by Stecker et al. (1993, as corrected by Protheroe). The diffuse γ-ray backgrounds due to both blazars (Stecker *et al.* 1993) and central engines of 'ordinary' AGN (present work) are comparable to that observed by SAS-2. We conclude there is likely to be an important contribution from 'ordinary' AGN as well as blazars to the diffuse γ-ray background.

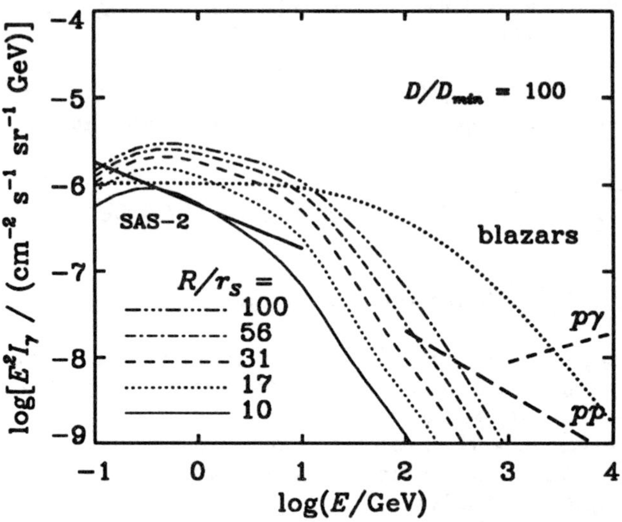

FIGURE I

REFERENCES

Berezinksky V.S. et al., 1993, *Astroparticle Physics*, **1**, 281-287

Halzen F. et al., 1990, *Physical Review D*, **41**, 342-346

Johnson, P.A. et al., 1993: *Proc. 23rd Int. Cosmic Ray Conf.*, **1**, 432-435

Kazanas, D., and Ellison, D.C., 1986, *ApJ*, **304**, 178

Morisawa, K. *et al.*, 1990, *A&A*, **236**, 299

Protheroe, R.J., and Kazanas, D., 1983, *ApJ*, **265**, 620

Protheroe, R.J., and Szabo, A.P., 1992, *Phys. Rev. Lett.*, **69**, 2885.

Protheroe, R.J., and Stanev, T., 1993, *MNRAS*, in press.

Stecker, F.W., De Jager, O.C., and Salamon, M.H., 1992, *ApJ*, **390**, L49

Szabo, A.P., and Protheroe, R.J., 1992b: in *Particle Acceleration in Cosmic Plasmas*, eds. G.P. Zank and T.K. Gaisser, A.I.P., New York, 301

Stecker, F.W., Salamon, M.H., and Malkan, M.A., 1993, *ApJ*, in press and erratum

Szabo, A.P., and Protheroe, R.J., 1992a: in *High Energy Neutrino Astrophysics*, edited by V.J. Stenger *et al.* (World Scientific,Singapore), p. 24

Thompson, D.J., and Fichtel, C.E., 1982, *A&A*, **109**, 354

UNIFIED SCHEMES

Invited Papers

The First Stromlo Symposium: The Physics of Active Galaxies
ASP Conference Series, Vol. 54, 1994
G. V. Bicknell, M. A. Dopita, and P. J. Quinn (eds.)

THE UNIFICATION OF ACTIVE GALAXIES: SEYFERTS AND BEYOND

JOSEPH S. MILLER
Lick Observatory, University of California, Santa Cruz, CA 95064

ABSTRACT The detection of obscured broad-line regions in several type 1 Seyfert galaxies has raised the possibilty that all Seyfert galaxies are of one type, and their classification depends on viewing direction. At present the evidence for this is inconclusive. Alternatively, there are indications that the obscured Seyfert 1's may be transitory, the obscuration being the result of recent interactions. The case for unification among the various types of QSOs is less clear, but it may be that broad absorption line QSOs are normal QSOs seen from a special direction and radio-loud quasars can be united with powerful radio galaxies.

INTRODUCTION

Recently there has been a great deal of interest in the unification of different types of active galaxies. The fundamental question is to what extent the apparent differences observed among the various types result from intrinsic differences as opposed to differences caused by viewing similar objects in different orientations. In fact, there are several versions of so-called "unified schemes." The simplest, and perhaps oldest scheme unites core-dominated and lobe-dominated radio sources. These are suggested to be essentially the same kind of object; the lobe-dominated radio sources appear as core-dominated sources if they are viewed along the line connecting the lobes. This simple scheme has been generalized to include optical characteristics. Seyfert 2 galaxies may be Seyfert 1 galaxies seen in a direction from which the broad line region is obscured. Similarly, narrow-line radio galaxies may be quasars with obscured broad- line regions. Broad absorption line QSOs (BAL QSOs) may be normal QSOs seen from a special direction. Furthermore, radio-optical correlations may be aspect dependent; BL Lac objects and optically violently variable (OVV) QSOs are viewed along the radio axis and hence are core dominated radio sources. There is considerable observational evidence in support of these unified schemes, though it is by no means proved that the various unifications listed above include all objects. Antonucci (1993) has written an excellent review of unified schemes, and I refer the reader to it for a comprehensive, if somewhat personal view of the subject. In this paper I will concentrate on the results of our group at Lick Observatory, briefly reviewing past results and describing some very recent ones. The major emphasis will be on Seyfert galaxies, though I will also discuss some new results on Cygnus A and BAL QSOs.

OPTICAL EVIDENCE FOR ASPECT-DEPENDENT CLASSIFICATION

The clearest optical evidence that active galactic nuclei may have an aspect-dependent appearance comes from studies which reveal the existence of "hidden" active nuclei. Numerous studies have shown that, with proper techniques, it is possible to demonstrate that some AGN would unquestionably be classified in another way if they were viewed from a different direction. One of the most dramatic examples of this comes from the studies of the type 2 Seyfert galaxy NGC 1068 by Antonucci, me, and others. Using spectropolarimetry, we (Miller and Antonucci 1983, Antonucci and Miller 1985) showed that the featureless continuum in this object is not seen directly, but as a result of scattering. Furthermore, the scattering also permits detection of a hidden broad line region (BLR). We proposed that the central continuum source and BLR are hidden from direct view by an obscuring torus, and electrons outside of this torus were the mirrors which allowed us to detect the obscured regions in polarized light. Subsequently, observations by Miller, Goodrich, and Mathews (1990, MGM) identified an off-nucleus dusty region which reflects the light of the nucleus. We showed that if the nucleus of NGC 1068 were observed from the direction of this region, it would appear as a type 1 Seyfert galaxy. Furthermore, the luminosity would also be that of a type 1. These results raised the possibility that Seyfert 1 and 2 galaxies are in fact the same type of object, and their classification depends on whether the central BLR is directly visible (type 1) or is obscured (type 2); I will return to this issue below. The identification of regions in AGN hidden from direct view is strong evidence for some kind of unification. There are at least four ways of detecting such hidden regions. The first is the one we used for NGC 1068: identify "mirrors" which reflect light from the region hidden from direct view. Spectropolarimetry is a great asset for this type of study. Using it, Miller and Goodrich (1990) showed that five Seyfert 2 galaxies in addition to NGC 1068 contain hidden BLR. Antonucci (1984) found that the narrow line radio galaxy 3C 234 shows broad lines in its polarized flux spectrum. The second technique is to observe at longer, infrared wavelengths which penetrate the obscuring fog. This has had mixed success, detecting hidden BLR in some cases, while not detecting them in others for which the first technique had already established the existence of a BLR; see Goodrich, et al. (1994) for an example of this kind of investigation. The remaining two ways are more indirect. One can infer the existence of a hidden active nucleus from the presence of "ionization cones" (see, for example, Pogge 1988). One can also deduce that such a region must exist from a study of the emission-line luminosities compared to the amount of radiation being received directly from the featureless continuum (as in Kinney, et al. 1991). The next section will discuss recent results on Seyfert 2's with hidden BLR. The detection of broad lines in these objects demonstrates that at least for some objects, the classification as type 1 or 2 is purely a question of viewing direction. In a later section I will discuss the question of whether in fact all Seyfert galaxies are of one type.

SEYFERT 2 GALAXIES WITH OBSCURED BROAD-LINE REGIONS

The study by Miller and Goodrich (1990) of Seyfert 2's with hidden type 1 regions showed that a number of these objects appeared to be different from NGC 1068 when their polarization spectrum is studied carefully. For example, nearly all show higher polarizations in their broad lines than in the continuum. Furthermore, some showed rotations in position angle across the broad-line profiles. For NGC 1068, the broad lines appear to have the identical polarization to the continuum. The results for the other objects appears to be in contradiction with the simple picture we adopted for NGC 1068: a single scattering located outside of the torus scatters both the central continuum source and the BLR located inside the torus. One might explain a lower polarization for the continuum source if it were more extended than the BLR, so that there is a larger range of scattering angles and hence more polarization-canceling for the continuum. However, the standard picture for AGN would indicate that the central continuum source should be more compact and centrally located than the BLR. Thus, if anything, we might expect the broad lines to be less polarized than the continuum. A different polarization position angle for the lines further confuses the situation. Our study was hampered by data of mixed quality. In order to investigate further the polarization characteristics of Seyfert 2 galaxies, Hien Tran carried out a careful spectropolarimetric study of them with our equipment on the Lick 3-m telescope as the focus of his Ph.D. thesis research (Tran 1993). Ten Seyfert 2 galaxies with obscured BLR were observed extensively in order to obtain data of uniformly high quality. Tran also obtained data for stars near the line of sight to the objects in order to estimate the contribution from galactic interstellar polarization and made corrections for interstellar reddening using information in the literature. He also obtained spectroscopic data for a number of normal galaxies, in order to provide template spectra for the subtraction of the stellar component in the Seyfert galaxies. Tran found that careful removal of the galactic interstellar polarization contribution removed the rotation in position angle across the broad lines: the lines and continua have identical position angles. By careful matching of the stellar spectra components with those of the template galaxies, he was able to obtain excellent removal of all features attributable to the stellar component. The derived stellar light fractions at 550 nm were on average higher than Goodrich and I derived, ranging from a low of 0.15 for Was 49b to 0.88 for Mrk 3. The continuum polarizations after starlight subtraction were wavelength-independent for all objects except NGC 7212, which showed a rise in polarization toward the blue. One of the most interesting results of Tran's study concerns the detectability of the broad lines in the spectra of these objects. The question has been raised (see Fernandes and Terlevich 1992 for a clear discussion) of why it is so difficult to detect the broad lines in normal light. In general, one might expect four components to be present in the spectra: the spectrum of the stars, the contribution from the narrow line region (NLR), and (by scattering) the BLR emission and the featureless continuum (FC). Since the starlight typically is not significantly polarized, and the narrow lines have low polarization, these two components are either eliminated or reduced considerably in the polarized flux spectrum; this makes the broad lines quite visible. One would then expect that, if one removed the stellar spectrum from the normal flux spectrum, the broad wings would appear at the same equivalent width with respect to the remaining FC as they do in the polarized flux spectrum. They do not. The broad wings are typically much harder to detect, that is, have much lower equivalent widths in the star-subtracted spectra than they do in the polarized flux spectra. There are two straightforward ways to explain this difference. The broad lines could indeed actually be more highly polarized than the FC, as appears to be the case, or there could be another source of continuous emission that is not polarized; this second continuum would dilute the polarization of the FC more than that of the lines. As I suggested above, it might be expected in the standard picture from geometrical considerations that FC would be more highly polarized than the continuum, so the first of these two possibilities appears more unlikely. Adopting the more likely hypothesis that a second, unpolarized continuum (FC2) is present in addition to the polarized FC (FC1), Tran was able to derive the flux levels for FC2 needed to make the equivalent widths of the broad lines the same in the polarized flux as in the starlight- and FC2-subtracted normal flux spectra. The results are quite striking. In the seven objects for which reliable estimates of FC2 were possible, FC2 contributed between 50 and 89% of the total continuum emission. With both the starlight and FC2 removed, the derived polarizations of the continua (of course now equal to that of the broad lines) are all above 15%; Mrk is 20%, Mrk 348 is 35%, and WAS 49b is 18%, to give a few examples. Not only does the hypothesis of a large contribution from an unpolarized FC2 explain the difficulty of detecting broad lines in normal spectra, but it also resolves a problem noted by Miller and Goodrich (1990). From geometrical modeling one would expect many highly polarized Seyfert 2 galaxies in the set of those with obscured BLR, while most show polarization of only a few percent. NGC 1068 was the exception at 16%, but in fact the data for it indicate little or no FC2 is present. When the others are corrected for FC2, it turns out that NGC 1068 has the one of the lowest polarizations. The hypothesis of the presence of an unpolarized FC2 is very attractive, as it resolves at least two previous difficulties with the standard picture for these objects. What is FC2? Tran was able to derive the energy distribution for FC2 in most of these objects and found it to be very similar to that of FC1. The starlight-subtracted spectra of objects for which FC2 dominates FC1 show no evidence for absorption lines; if FC2 comes from stars, they must be hot stars. However, the energy distribution of FC2 (and FC1) is much redder than that produced by hot stars, and we consider hot stars a very unlikely explanation. Alternatively, FC2 could be free-free emission, the emission coming from the same gas that is doing the electron scattering of the obscured FC1 and BLR. We found that a temperature of several times 10^5 for the electron- scattering gas in NGC 1068 gave a good fit to the observations. Free-free emission near this temperature produces a good match to the spectra derived for FC2. Furthermore, calculation of the fluxes in the Balmer lines produced by gas emitting the required levels of FC2 are well below the observed narrow line fluxes. Barvanis and his co-workers (see Barvanis 1993), have proposed that the optical- UV featureless continua of AGN are produced by free-free emission with an energy distribution similar to that observed for FC1 and 2. Thus it is possible and in fact quite consistent with our model for NGC 1068 (MGM) that FC1 is produced by free-free emission from an obscured region whose outer regions extend well above the obscuring torus and produce

FC2 and provide the mirrors which allow us to see the obscured regions. Spectropolarimetry of these objects in the far UV with HST may be able to add additional evidence for this picture.

WHAT OBSCURES THE BROAD LINE REGIONS?

The question of what provides the obscuration of the BLR in Seyfert galaxies like NGC 1068 remains open. While we commonly use terms like "obscuring torus," there is really no evidence that the obscuration is indeed caused by a torus or even disk- shaped object, for that matter. All that is required is an opaque region that covers both the BLR and central continuum source, but not the scattering region. However, the existence of ionization cones and the pattern of scattered light in NGC 1068 (MGM) indicate that radiation emerges from the central region in a cone-like distribution, which suggests a flattened or toroidal obscuring region. Also, it is widely held that the central source of energy in these objects is accretion in a disk onto a massive, compact object. However it is important to emphasize that, if the obscuring region is a torus, it is not the accretion disk itself. The height of even the thickest accretion disks is orders of magnitude smaller than the size of the BLR. Attempts to model appropriately thick obscuring disks have had some success (see Krolik and Begelman 1988 for example), but rather special conditions have had to be invoked. There is a relatively straightforward way to account for thick obscuring regions. Tran noted that six of the ten objects in his study showed evidence for interactions or had double nuclei. It could be the case that Seyfert 2 galaxies with hidden BLR are all cases where there has been a recent merger or strong interaction. This could have deposited large amounts of material in the central regions which provides the obscuration. In this case, the obscuration could be transitory. Perhaps NGC 1068 appeared as a Seyfert 1 in the past and will again in the future, when the central regions have settled down. Since the settling process might be expected to take 10^6 years or longer, NGC 1068 should appear as Seyfert 2 for a long time.

ARE ALL SEYFERT GALAXIES ONE TYPE?

The discovery that a number of Seyfert 2 galaxies would appear as type 1's if viewed from another direction raises the possibility that there is only one type of Seyfert galaxy, the classification of any particular object depending entirely on viewing direction. There are a number of arguments for and against this view, and the recent reviews by Lawrence (1991) and Antonucci (1993) summarize the situation well. Briefly, the arguments for complete unification of Seyfert galaxies are the following:

1. There is no question that at least some Seyfert 2's contain Seyfert type 1 regions which would be directly observable from appropriate directions. 2. The number of Seyfert 2 galaxies appears to be greater than Seyfert 1's, though this remains in dispute. If the opening angles of the ionization cones observed correspond to the angles over which the central regions can be observed directly, then one would expect the larger number of Seyfert 2 galaxies. This is of course more of a consistency argument than a direct one. 3. The narrow line regions

are the same in both types of objects, suggesting they are basically similar.

The arguments against are as follows:

1. The radio properties of the two types appear to be different. While these radio characteristics are aspect-independent, it is not entirely clear that selection effects haven't produced the apparent difference.

2. The properties of the off-nuclear gas are different (Pogge 1989). However, the number of objects for which this comparison is quite small, and the type 1 and type 2 objects compared have not been matched for luminosity, as Antonucci has stressed.

3. Seyfert 2 galaxies have a perpendicular relationship between continuum polarization and radio morphology, while type 1's have a parallel relationship. Our model for NGC 1068 (MGM) would maintain the perpendicular relationship for all angles, unless a new scattering region of different geometry dominates for more pole-on views.

While I believe that there are real differences between Seyfert 1 and 2 galaxies, the above arguments are inconclusive one way or the other. Recently Goodrich and I (ApJ, submitted) carried out a study of polarized Seyfert 1 galaxies with the goal of looking for evidence that some of them are Seyfert 2's like NGC 1068 seen pole-on. While the results were very interesting, with a variety (and confusing) collection of polarization behaviors being observed, we found no clear evidence for this reverse direction of unification. In any event, the fact that all high luminosity objects, the true QSOs, are of type 1 (assuming we would recognize them if they were type 2) suggests that the luminosity may play a role in the structure of the central objects. The observations are completely consistent with the view that, as the luminosity of AGN increase, the thickness of surrounding disks tends to decrease. The next section will discuss attempts to unify the various classes of QSOs and related objects.

QUASI-STELLAR OBJECTS

Because of their optical faintness, relatively limited evidence is available concerning unification of QSOs. The so-called optically violently variable (OVV) QSOs are members of the blazar class, which also includes BL Lac objects, and the accepted explanation for their dramatic flux and polarization variability is that they contain relativistic jets pointed more or less directly at us. Misaligned blazars would be classified as radio-loud, rather normal-appearing elliptical galaxies. What the distinction is between objects classified as BL Lac objects and OVV QSOs, if any, remains unclear; it could largely be an arbitrary division in a continuous range of luminosities. A substantial fraction of non-OVV QSOs show low polarizations, typically under 2% and non-variable. The polarization position angles tend to line up with extended radio structures, as is the case for Seyfert 1 galaxies, and what limited evidence we have suggests that the emission lines have much lower polarizations than the continuum. Furthermore, the polarization appears to be wavelength-independent over a wide range in wavelength. The simplest interpretation is that QSOs generally have thin surrounding disks, if any, and thus the central FC source and BLR are rarely hidden. In fact, observations of one object, OI 287, have been interpreted by Goodrich and Miller (1988) in terms of it being that rare QSO for which our view of the central continuum source and BLR are blocked. The polarizations of the FC and the broad lines are the same, high 8% and wavelength-independent; the polarization is parallel to the radio structure. We propose that the disk blocks the BLR, but not the NLR, which accounts for the unusually high equivalent widths of the narrow lines. The polarization alignment is simply explained if the polarization is produced by scattering by material along the surface of the disk, and the radio structure is aligned with the disk axis. Barthel (1989) has proposed that the FR2 radio galaxies are actually quasars seen from a direction from which the central quasar is obscured. Statistics of the 3C catalogue and the number of quasars as a function of redshift support this hypothesis if the cone of visibility of the central regions has an opening half-angle of roughly 45°. While the analogy with Seyfert galaxies makes this proposal attractive, direct evidence in support of it is rather limited. The narrow line radio galaxy 3C 234 was shown by Antonucci (1984) to have a hidden BLR, and it would definitely appear as a quasar, both in spectrum and luminosity, if it were viewed from the appropriate direction. Cygnus A has also been considered as an excellent candidate for actually being an obscured quasar, but searches to date for a hidden BLR (Miller and Tran 1993, Jackson and Tadhunter 1993) have failed to detect one. For the present, Barthel's proposal remains attractive in spite of the limited evidence for it, and it is worth devoting further observations to its test. It is tempting to interpret BAL QSOs as another example of normal QSOs seen from a special direction. Weymann et al. (1991) have found little or no difference between the emission-line spectra of BAL QSOs and normal QSOs and propose that the BAL QSOs represent those 10-14% of all QSOs where the line of sight passes through BAL- producing region. These objects are the only radio-quiet QSOs with relatively high polarizations (some are 3-5%). For PHL 5200 Goodrich and I (in preparation) have found that the emission lines are unpolarized, while the continuum both in and out of the absorption troughs has wavelength-independent polarization. A simple interpretation of BAL QSOs is that they are those QSOs we view along a line of sight that passes close to the surface of a surrounding disk. The central continuum source may be partially blocked; thus some of its light reaches us by scattering and contributes the polarized component. Much more observational work must be done in order for this picture to be tested carefully.

CONCLUDING REMARKS

The following general remarks can be made based on the material above:

1. There are some systematic relationships between the line spectra, polarizations, and radio structures of AGN and QSOs. Some time ago Miley and I (Miley and Miller 1979) found a correlation between line widths and profiles and the overall radio structure. Parallel alignments of radio structures and continuum polarization angles are common for objects with broad lines, while narrow-line objects show a perpendicular relationship.

2. The polarization data give considerable evidence for the existence of flattened distributions of matter surrounding the central objects. A simple interpretation of the data suggests that type 2 objects have relatively thick disks, and the scattering takes place along the disk axis direction. Type 1 objects could have thin disks, with the scattering taking place predominantly in the disk surface direction.

3. Recent results for Seyfert 2 galaxies with hidden BLR give strong evidence for a second, very extended source of continuum emission. This emission could be free-free radiation from the electron-scattering region itself.

4. The relationship between obscuring disks and accretion disks is not clear. It could be that extended, warped disks have the appropriate geometry to hide central BLR. Alternatively, thick obscuring disks could be the result of recent interactions or mergers of galaxies and hence be transitory. In this picture, type 1 and 2 Seyfert galaxies would not be unified in general, as type 2's with obscured BLR would in fact be type 1's going through a transitory phase; their normal state would be type 1.

5. It appears that some Seyfert 2 galaxies have central objects similar to those in some narrow-line radio galaxies. This would imply that spiral and elliptical galaxies can have the same kind of active central object.

6. To date, evidence for unification of QSOs with other classes is quite limited; though it remains quite possible that radio-loud quasars are closely related to radio galaxies. BAL QSOs could well be normal QSOs seen from a special direction.

7. The central continuum sources of some objects may be intrinsically anisotropic (MGM).

As a final remark, unification of the different types of active galaxies is a very worthwhile goal toward which there has been some real success. It still remains an open question, though, to what extent the diversity of active galaxy types results from aspect effects, evolutionary effects, or intrinsic differences. It does seem likely, however, that the same basic kind of structure is common to the vast majority of them.

It is a pleasure to acknowledge the extensive collaborations I have had on this research with R. Antonucci, R. Goodrich, L. Kay, H. Tran, and W. Mathews. This research was supported in part by NSF grant AST 88-18925.

REFERENCES

Antonucci, R. 1984, *ApJ*, **278**, 499

Antonucci, R. 1993, *ARA&A*, **31**, 473

Antonucci, R., and Miller, J. S. 1985, *ApJ*, **297**, 621

Barthel, P. 1989, *ApJ*, **336**, 606

Barvanis, R. 1993, *ApJ*, **412**, 513

Fernandes, R. C., and Terlevich, R. 1992, in *Relationships Between Active Galactic Nuclei and Starburst Galaxies*, A. V. Filippenko, ASP: San Francisco, 241

Goodrich, R. W., and Miller, J. S. 1988, *ApJ*, **331**, 332.

Goodrich, R. W., Veilleux, S., and Hill, G. S. 1994, *ApJ*, in press

Jackson, N., and Tadhunter, C. N. 1993, *A&A*, **272**, 105

Kinney, A., Antonucci, R.,Ward, M., Wilson, A., and Whittle, M. 1991, *ApJ*, **377**, 100

Krolik, J., and Begelman, M. 1988, *ApJ*, **329**, 702

Lawrence, A. 1991, *MNRAS*, **252**, 586

Miley, G., and Miller, J. S. 1979, *ApJ*, **228**, L55

Miller, J. S., and Antonucci, R. 1993, *ApJ*, **271**, L7

Miller, J. S., and Goodrich, R. W. 1990, *ApJ*, **355**, 456

Miller, J. S., and Tran, H. D. 1993, *BAAS*, **25**, 919

Miller, J. S., Goodrich, R. W., and Mathews, W. G. 1990, *ApJ*, **378**, 47 (MGM)

Pogge, R. 1988, *ApJ*, **328**, 519

Pogge, R. 1989, *ApJ*, **345**, 730

Tran, H. D. 1993, Ph.D. thesis, University of California, Santa Cruz

Weymann, R. J., Morris, S. L., Foltz, C. B., and Hewett, P. C. 1991, *ApJ*, **373**, 23

The First Stromlo Symposium: The Physics of Active Galaxies
ASP Conference Series, Vol. 54, 1994
G. V. Bicknell, M. A. Dopita, and P. J. Quinn (eds.)

DISK-DRIVEN HYDROMAGNETIC WINDS AS A KEY INGREDIENT OF AGN UNIFICATION SCHEMES

ARIEH KÖNIGL
Department of Astronomy & Astrophysics, University of Chicago, 5640 South Ellis Avenue, Chicago, IL 60637, U.S.A.

ABSTRACT Centrifugally driven winds from the surfaces of magnetized accretion disks are an attractive mechanism for removing the angular momentum of the accreted matter and of producing bipolar outflows and jets in compact astronomical objects. They have unique observational manifestations stemming from their highly stratified structure and from their exposure to the radiation field of the central object. In the context of AGNs, such winds may underlie the classification of broad-line and narrow-line sources (e.g., the Seyfert 1/Seyfert 2 dichotomy) as well as the apparent dearth of luminous Seyfert 2 galaxies. More generally, they could strongly influence the spectral characteristics of Seyfert galaxies, QSOs, and BL Lac objects.

INTRODUCTION

It is now widely believed that, in many important instances, active galactic nuclei (AGNs) with seemingly disparate properties are, in fact, members of the same class of objects and are perceived to be different only because their appearance depends on their orientation relative to the observer. In this view, the generic AGN consists of a massive black hole and a circumnuclear accretion disk, and the liberated rotational and gravitational energy of this system powers the strong observed thermal and nonthermal emission as well as the oppositely directed, supersonic jets that sometimes emanate from these objects. The orientation-dependent variations in the apparent characteristics of AGNs have been attributed to either one of two effects: obscuration (at large angles to the symmetry axis) by a geometrically thick molecular torus or by an inclined disk, and Doppler beaming (along the symmetry axis) of the nonthermal emission from a relativistic jet component (e.g., Blandford & Rees 1978).

Orientation effects of this type have been proposed as the basis for the general classification of AGNs, including Seyfert 1 (Sy 1) and Seyfert 2 (Sy 2) galaxies, high-luminosity QSOs and ultraluminous IR galaxies, and broad-line radio galaxies (and radio-loud QSOs) and narrow-line radio galaxies (e.g., Antonucci 1993). In this picture, the AGN is surrounded by an obscuring 'torus' that prevents a direct view of the broad-line region (BLR) and strongly absorbs the soft X-ray emission from the center. Spectropolarimetric measurements, high-resolution images of 'ionization cones,' and statistical analyses of galaxy samples have indicated that Seyfert galaxies are characterized by mean torus opening half-angles $\sim 30° - 40°$. The opening angles have been inferred to increase with luminosity: in particular, there are no reported Sy 2 counterparts to Sy 1 galaxies with high X-ray luminosities. Furthermore, the 'torus' appears to have a fuzzy, time-variable edge: this is manifested in the optical properties of Seyfert 1.8/1.9 galaxies and in the range of X-ray absorbing columns inferred in Sy 2's.

As discussed by Königl & Kartje (1993), previous interpretations of the 'torus' in terms of an ensemble of molecular clouds (Krolik & Begelman 1988) or of a warped galactic disk (Phinney 1989) are problematical either in their underlying physical mechanism or in their ability to account for the observations. Instead, motivated by previous results in the study of young stellar objects (e.g., Safier 1993), these authors have proposed to interpret the 'torus' (as well as other, related attributes of AGN unification schemes) in terms of a centrifugally driven wind from the surface of a magnetized accretion disk in the galactic nucleus. This contribution highlights the key elements of this model.

CENTRIFUGALLY DRIVEN WIND MODEL

To simplify the treatment of the outflow dynamics, we have adopted the self-similar (in the spherical radial coordinate R), cold, centrifugally driven wind model put forward by Blandford & Payne (1982) [BP]. We have assumed a continuous outflow, and in this respect our picture complements the Emmering et al. (1992) interpretation of broad emission-line clouds in terms of discrete, centrifugally driven clumps. The basic physical principle of the wind is captured by the 'bead on a rotating wire' analogy: in a Keplerian disk, matter is flung out centrifugally if the magnetic field lines to which it is attached are inclined at an angle $\leq 60°$ to the surface. Such a wind has the capacity to transport a significant fraction of the angular momentum of the accreted matter and could naturally evolve into a bipolar outflow or a pair of jets (e.g., Blandford & Königl 1993). Owing to the rapid acceleration of the gas, the density distribution in the wind is highly stratified in the vertical direction. The velocity distribution is also stratified, with $V \propto R^{-1/2}$ and increasing with decreasing polar angle θ (see Fig. I). We have generalized the BP model by considering an arbitrary power-law dependence of the density and magnetic field on the radius: $\rho \propto R^{-b}$, $B \propto R^{-(b+1)/2}$ ($b = 1.5$ in BP).

The detection of centrally concentrated molecular line emission in many AGNs suggests that the nuclear disks are dense and dusty. For typical wind parameters, the dust would be readily uplifted from the disk surface by gas–grain collisions and should remain embedded in the wind along streamlines originating beyond the dust sublimation radius $\sim 0.1 L_{44}^{1/2}$ pc. If the bolometric luminosity $L = 10^{44} L_{44}$ ergs s^{-1} exceeds the effective (dust opacity-based) Eddington luminosity $L_{\mathrm{Edd,eff}} \approx 3 \times 10^{42} M_7$ ergs s^{-1} (where $M = 10^7 M_7 M_\odot$ is the central mass), then radiation-pressure effects would tend to 'open up' the dust-loaded streamlines. In a further generalization of BP, we have included a simple representation of this effect in our model by introducing the gravity-reduction factor $(1-\Gamma)$, where $\Gamma \approx e^{-\tau} L/L_{\mathrm{Edd,eff}}$ and τ is the optical depth to the origin. We verified that the asymptotic opening half-angle θ_∞ of the streamlines increases with Γ but that the magnetic field could nevertheless be strong enough to prevent the gas from being blown away even for $\Gamma \gg 1$.

APPLICATIONS

Optical Obscuration

The disk-driven wind model provides an appealing physical explanation of the origin of the 'obscuring tori' in AGNs. In particular, the optical shadowing of the BLR can be attributed to dust extinction in the outer regions of the wind. The visual extinction is negligible for $\theta < \theta_\infty$ and then increases gradually until $A_V = 1$ at $\theta = \theta_s$ (see Fig. II). If evolutionary effects are relatively unimportant, one can identify the angular ranges $\theta < \theta_\infty$, $\theta_\infty < \theta < \theta_s$, and $\theta_s < \theta$ with Sy 1–1.5, Sy 1.8–1.9, and Sy 2 galaxies, respectively. For example, by analyzing two different samples, Osterbrock & Shaw (1988) and Osterbrock & Martel (1993) inferred $\theta_\infty \approx 29°$, $\theta_s \approx 39°$ and $\theta_\infty \approx 38°$, $\theta_s \approx 58°$, respectively, although it is possible that these values (particularly θ_s) are overestimated due to light-scattering effects in the dust-free regions of the wind. The model also provides a natural explanation of the apparent dearth of luminous Sy 2 galaxies, which can be attributed to the flattening of the dust-loaded streamlines in high-Γ sources.

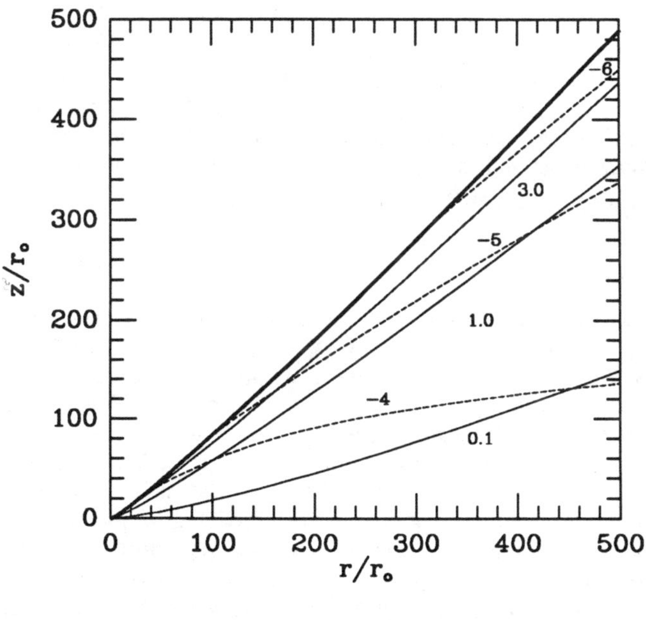

FIGURE I A representative self-similar wind solution, corresponding to dimensionless BP parameters $\kappa = 0.06$ (mass-to-magnetic flux ratio) and $\lambda = 43$ (total angular momentum), $b = 1$, and $\Gamma(\theta) \propto (0.5\pi - \theta)^2$ (with $\Gamma = 10$ at the Alfvén point). This solution is characterized by $\theta_\infty = 44°$ and a mass outflow rate $\dot{M}_{out} \approx 1.0\, M_\odot\ \mathrm{yr}^{-1}$ between $r_0 = 10\, r_G$ and 1 pc (assuming $M_7 = 1$ and a disk magnetic field of 10^4 G at r_0). The *solid* and *dashed* curves are contours of constant poloidal speed (in units of the Keplerian speed at r_0) and $log[\rho/\rho(r_0)]$, respectively, while the *heavy* line marks the innermost streamline of the wind.

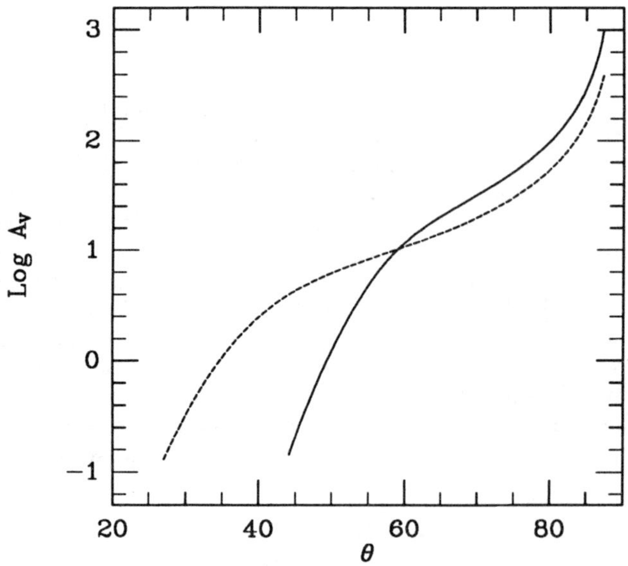

FIGURE II Visual extinction as a function of the polar angle for the wind model of Fig. I (*solid* curve) as well as for another model (*dashed* curve) characterized by $\kappa = 0.025$, $\lambda = 78$, $\theta_\infty = 27°$, and $\dot{M}_{\rm out} \approx 0.4\, M_\odot\ {\rm yr}^{-1}$.

X-Ray Absorption and Emission

In the original molecular-torus scenario, the cold gas that gave rise to the optical extinction was also responsible for the strong X-ray absorption indicated in Sy 2's. However, there is now compelling evidence for the presence of a 'warm' (partially ionized) absorbing component in Seyfert galaxies (e.g., Turner et al. 1992). There is also evidence for the presence of a 'hot' (highly ionized) gas that may share the same volume as the warm component (e.g., Marshall et al. 1993). These components have a straightforward interpretation in our model: they correspond to the inner (atomic) region of the wind that is photoionized and heated by the central radiation field. Using an adaptation of T. Kallman's XSTAR photoionization code, we have explicitly demonstrated that it is possible to account for the inferred physical parameters of these components in our model. In this picture, electron scattering by the ionized gas right above the disk surface may give rise to the large widths of the broad emission lines (e.g., Emmering et al. 1992), whereas scattering at a height $\gtrsim 10^{20}$ cm (after the bulk of the gas has cooled from $\sim 10^6$ K to $\sim 10^5$ K) could be responsible for the polarized BLR emsission component measured in Sy 2's (e.g., Antonucci & Miller 1985). Owing to the strong stratification of the outflow, one expects that objects observed at intermediate angles to the axis would exhibit only partial soft X-ray attenuation, in analogy with the partial obscuration inferred in Sy 1.8–1.9 galaxies (see Fig. III). For observation angles $\lesssim \theta_\infty$, it may be possible to measure significant X-ray absorption but negligible visual extinction: this could account for the properties of a source like NGC 4151 (Evans et al. 1993).

Infrared Continuum of Seyfert Galaxies and QSOs

The dusty regions of the wind scatter and reprocess the continuum radiation from the central object. As we have verified with the help of a radiative transfer code that implements the 'accelerated lambda operator' perturbation technique, comparatively tenuous $b \approx 1$ outflows can account for the flat (in νF_ν) $\sim 3 - 30$ μm spectra that are often measured in Sy 1's and radio-quiet QSOs. However, for typical parameters, the winds are dense enough to be moderately opaque to their own thermal emission, which could explain why Sy 2's (which are viewed at a large angle to the symmetry axis) have a steeper spectrum in this energy range (cf. Pier & Krolik 1992). The extended mid-infrared emission detected in objects like NGC 1068 (e.g., Cameron et al. 1993) is naturally associated in this picture with outflowing dust that is irradiated by continuum photons propagating through the dust-free central funnel. Using a Monte Carlo scheme, we plan to verify that the dust embedded in the magnetized wind could also account (through scattering or absorption/emission by aligned grains) for the measured infrared polarization properties of Seyfert galaxies and QSOs.

X-Ray Absorption Features in BL Lac Objects

BL Lac objects generally have featureless spectra that are commonly attributed to a nonthermal emission component from a relativistic jet. However, there is now growing evidence for an apparent $\sim$ 0.6 keV absorption feature in the spectra of several such objects (e.g., Madejski et al. 1991). The best-studied object so far is PKS 2155-304, where the feature has been interpreted as an O VIII Lyα resonant absorption trough arising in a $\sim 0.1\,c$ outflow (Canizares & Kruper 1984). We have verified that a clumped, disk-driven wind can reproduce the basic characteristics of this feature when the steep ionizing continuum that is measured in objects like PKS 2155-304 (energy spectral index $\sim$ 1.6) is substituted for the relatively flat continuum of Seyferts and QSOs. The detailed shape of the predicted absorption trough could in principle be used to distinguish this model from a spherically symmetric wind interpretation (Krolik et al. 1985).

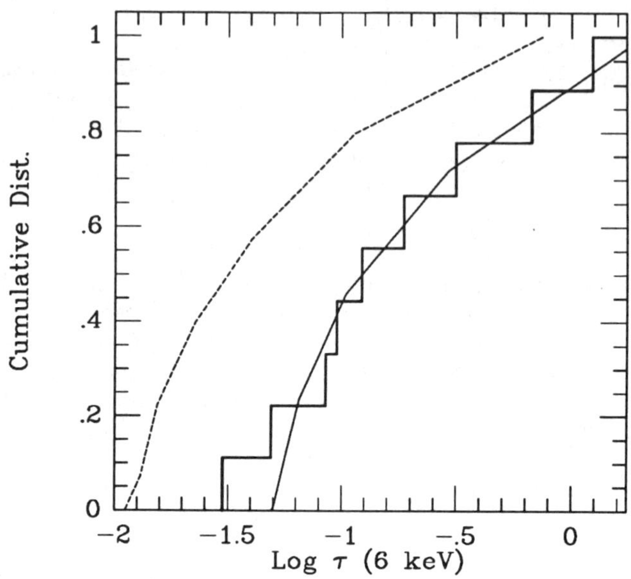

FIGURE III Cumulative 6-keV optical-depth distribution for the $\theta_\infty = 44°$ (*solid* curve) and $\theta_\infty = 27°$ (*dashed* curve) wind models of Fig. II, calculated for a big-blue-bump–dominated continuum spectrum with $L_{44} = 1$. The absorption within the dust sublimation radius has been calculated using a wind model with $\Gamma = 0.1$ at the Alfvén point. The *heavy* curve is the observed distribution for the (incomplete) sample of UV-detected Sy 2's compiled by Mulchaey et al. (1992).

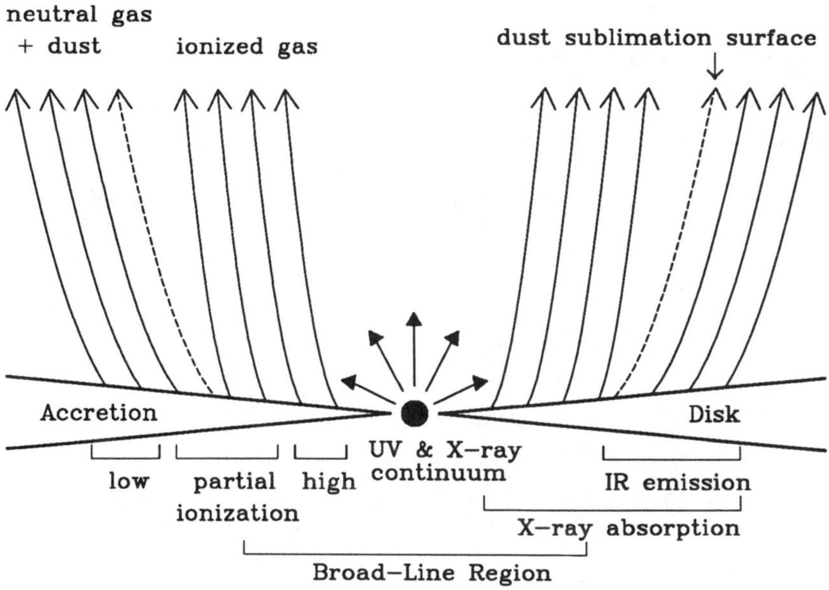

FIGURE IV Schematic representation of how a disk-driven hydromagnetic wind interacts with the AGN continuum emission. The innermost regions are heated and ionized by the powerful radiation field, with the temperature and degree of ionization decreasing both with distance and with the polar angle, whereas the outer regions (beyond the dust sublimation radius) are cooler and contain dust. The radiation-pressure force on the dust causes the outer streamlines to have a larger opening angle.

CONCLUSION

Hydromagnetic disk-driven winds may represent the basic physical mechanism that produces the apparent dichotomy between broad-line and narrow-line AGNs and between low-luminosity and high-luminosity sources. Furthermore, these winds could profoundly affect the infrared through X-ray characteristics of Seyfert galaxies and QSOs as well as some of the observed properties of BL Lac objects. They may thus need to be incorporated as a key ingredient into all AGN unification schemes.

ACKNOWLEDGMENTS

I am grateful to John Kartje, my collaborator on this project, for his invaluable input and for his help in producing this contribution. I also thank the organizers of this Symposium for their encouragement and generosity. This work was supported in part by NASA grants NAGW-1636 and NAG 5-2265.

REFERENCES

Antonucci, R. 1993, *ARA&A*, in press

Antonucci, R. R. J., & Miller, J. S. 1985, *ApJ*, **297**, 621

Blandford, R. D., & Payne, D. G. 1982, *MNRAS*, **199**, 883

Blandford, R. D., & Königl, A. 1993, *S&T*,**85**, No. 3, 40

Blandford, R. D., & Rees, M. J. 1978, in *Pittsburgh Conf. on BL Lac Objects*, A. M. Wolfe, Pittsburgh: Univ. Pittsburgh Press, 328

Cameron, M., et al. 1993, *ApJ*, in press

Canizares, C. R., & Kruper, J. 1984, *ApJ*, **278**, L99

Emmering, R. T., Blandford, R. D., & Shlosman, I. 1992, *ApJ*, **385**, 460

Evans, I. N., Tsvetanov, Z., Kriss, G. A., Ford, H. C., Caganoff, S., & Koratkar, A. P. 1993, *ApJ*, in press

Kartje, J. F., & Königl, A. 1993, *ApJ*, submitted

Krolik, J. H., & Begelman, M. C. 1988, *ApJ*, **329**, 702

Krolik, J. H., Kallman, T. R., Fabian, A. C., & Rees, M. J. 1985, *ApJ*, **295**, 104

Madejski, G. M., Mushotzsky, R. F., Weaver, K. A., Arnaud, K. A., & Urry, C. M. 1991, *ApJ*, **370**, 198

Marshall, F. E., et al. 1993, *ApJ*, **405**, 168

Mulchaey, J. S., Mushotzky, R. F., & Weaver, K. A. 1992, *ApJ*, **390**, L69

Osterbrock, D. E., & Martel, A. 1993, *ApJ*, **414**, 552

Osterbrock, D. E., & Shaw, R. A. 1988, *ApJ*, **327**, 89

Phinney, E. S. 1989, in *Theory of Accretion Disks*, F. Meyer, W. J. Duschel, J. Frank, & E. Meyer-Hofmeister, Dordrecht: Kluwer, 457

Pier, E., & Krolik, J. H. 1992, *ApJ*, **401**, 99

Safier, P. N. 1993, *ApJ*, **408**, 115

Turner, T. J., Done, C., Mushotzky, R., Madejski, G., & Kunieda, H. 1992, *ApJ*, **391**, 102

The First Stromlo Symposium: The Physics of Active Galaxies
ASP Conference Series, Vol. 54, 1994
G. V. Bicknell, M. A. Dopita, and P. J. Quinn (eds.)

COMPACT RADIO NUCLEI IN SEYFERT GALAXIES

A.L.ROY
School of Physics, University of Sydney 2006, Australia, and Australia Telescope National Facility, CSIRO, PO Box 76, Epping NSW 2121, Australia

R.P.NORRIS, M.J.KESTEVEN, E.R.TROUP, AND J.E.REYNOLDS
Australia Telescope National Facility, CSIRO, PO Box 76, Epping NSW 2121, Australia

ABSTRACT

We have observed a sample of 159 Seyfert galaxies with a 275-km baseline radio interferometer to search for compact, high brightness temperature radio emission from the active nucleus. We obtained the surprising result that compact radio cores are much more common in Seyfert 2 galaxies than in Seyfert 1 galaxies, which at first seems to be inconsistent with orientation unification schemes. We propose a model, involving optical depth effects in the narrow-line region, which can reconcile our result with the standard unified scheme.

INTRODUCTION

A number of schemes have been devised to account for the difference between the two types of Seyfert galaxies. Of the four schemes reviewed by Lawrence (1987), orientation unification has enjoyed particular success. This scheme invokes a dense torus of dust and molecular gas in Seyfert galaxies surrounding the nucleus. This causes the galaxy to appear as a Sy2 if our view of the broad-line region (BLR) is blocked by the torus, and as a Sy1 only when we view from within a cone aligned with the polar axis of the dust torus, so that we can see down to the BLR. In both cases the nucleus is the same type of object, and the apparent differences are due simply to the effects of our viewing angle.

In this paper, we use radio observations to test this orientation model. Dust surrounding the nucleus should be optically thin at radio wavelengths, and so orientation should have no effect on the radio appearance.

We have selected 159 Seyfert galaxies and observed them with the 275-km baseline Parkes-Tidbinbilla Interferometer (PTI) at 2.3 GHz (Norris et al. 1988). The PTI is sensitive to structures with brightness temperatures greater than 10^5 K and sizes less than 0.1 arcsec, which corresponds to 20-200 pc ($H_0 = 75$ km s^{-1} Mpc^{-1}) over the redshift range $0.01 < z < 0.1$, typical of the sample. Thus the PTI can detect radio emission from compact non-thermal sources associated with the AGN, and is blind to extended HII regions associated with star-formation activity, which have typical brightness temperatures of 10^4 K.

Detectable radio supernovae are rare in starburst regions (Norris et al. 1990).

Preliminary results have been published by Norris et al. (1992). Here we have increased the sample sizes, and used a refined statistical treatment.

SAMPLE

As our aim is to test whether orientation effects cause the differences between Sy1s and Sy2s, we need samples which are unbiased by orientation. Selection in the far infra-red (FIR) offers the least bias of any wavelength since dust is probably transparent to FIR emission. Our two primary samples are therefore FIR selected, one drawn from Norris et al. (1990) (34 galaxies), and the other from de Grijp et al. (1987) (102 galaxies).

For comparison, we also selected three samples on the basis of optical or mid-infra-red (mid-IR) properties. These were drawn from the CfA Seyfert sample (Edelson 1987) (16 galaxies), from the optically selected sample of Norris et al. (1990) (34 galaxies), and from the mid-IR-selected Seyfert sample of Spinoglio and Malkan (1989) (29 galaxies).

To avoid confusion from borderline classifications, we used only those galaxies with a reliable classification of Sy1 or Sy2, rejecting intermediate classes.

We found no significant difference between the total radio luminosities of Sy1s and Sy2s (measured at 2.4 GHz using the Australia Telescope as part of a related project, Roy et al. 1993), or between the FIR luminosities, or between the redshift distributions in the FIR selected sample. The optically selected sample showed a hint that Sy2s were closer than Sy1s on average. However, the difference was not formally significant.

The observations with the PTI consisted typically of a 20-min coherent integration at one position angle on each source, which produced a 5-σ sensitivity limit of 3 mJy. To simplify the statistical analysis we only considered whether or not the galaxies were detected above this limit.

RESULTS

All five Seyfert samples were observed with the PTI and the detection rates in the combined sample are illustrated by Figure I(a). Surprisingly, the detection rate of Sy2s (46%) was much greater than that for Sy1s (26%). The difference was significant at the 1.4% level (two-tailed difference-of-two-proportions test with Yates continuity correction).

To examine this curious difference further, we looked for the effect in the samples which contributed to Figure I. We divided the data from Figure I into the FIR-selected and optically selected samples, and again compared the detection rates. The results are illustrated in Figure I (b).

In both the FIR and the optical+mid-IR-selected samples almost half of the Sy2s were detected, as opposed to only one quarter of the Sy1s. This was significant at the 2.0% level for the FIR-selected sample, but was only at 14.7% for the optically selected sample due to the small sample size. Had the FIR-selected sample been as small as the optically selected sample the significance would have been 17%. So, the FIR-selected sample, and possibly the optical+mid-IR-

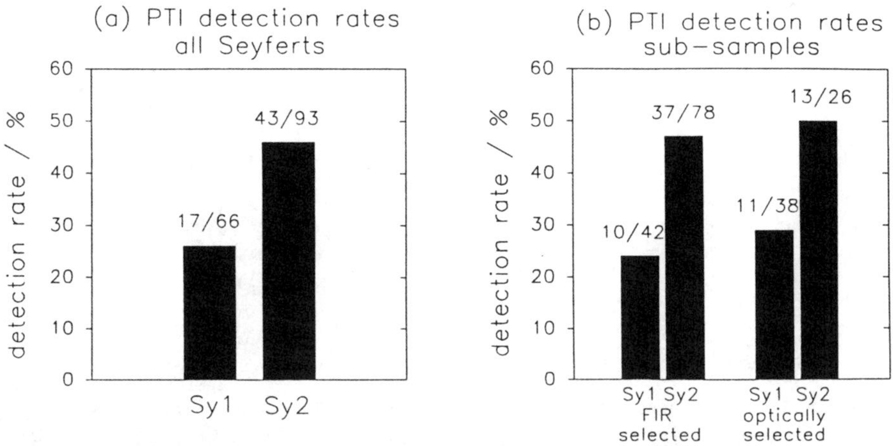

FIGURE I PTI detection rates for (a) all 159 Seyfert galaxies, separated into types 1 and 2. The ratios indicate the number detected/total in sample. Also shown are the PTI detection rates separately for (b) FIR-selected Seyferts and for the combined optical and mid-IR-selected Seyfert sample.

selected sample, displays a difference between the Sy1 and Sy2 detection rates.

We further strengthened the 'difference-of-two-proportions' test by considering the samples individually as far as possible, by using one-tailed tests where justified, and only then combining the results for all samples. We found that the detection rates differed with 0.36% significance, which reflects our increased confidence from having seen the effect in independent samples. The observed difference between Seyfert types is extremely unlikely to occur by chance.

DISCUSSION

The simplest explanation for the different detection rates is that the two Seyfert types are intrinsically different. However, this interpretation is inconsistent with the orientation model. Here we offer two variants of a model, first proposed by Norris et al. (1992), which attempt to reconcile our result with the standard model by invoking free-free absorption in the narrow-line region (NLR) clouds at 2.3 GHz.

In the first model, radio cores within the BLR are obscured by the NLR. In Sy1 galaxies we view the radio core through the full depth of the NLR since our view passes nearly down the NLR axis. If the optical depths are above unity, as discussed below, then the NLR clouds will block our view of the core and of each other (Fig. II). In Sy2 galaxies we view the compact core through very little of the NLR because our line of sight is nearly perpendicular to the NLR axis. Since there are fewer NLR clouds which could block the view of either the core or other clouds, it follows naturally that the PTI should detect nuclear emission more often in Sy2s than in Sy1s.

The second model also invokes absorption by NLR clouds, but applies only

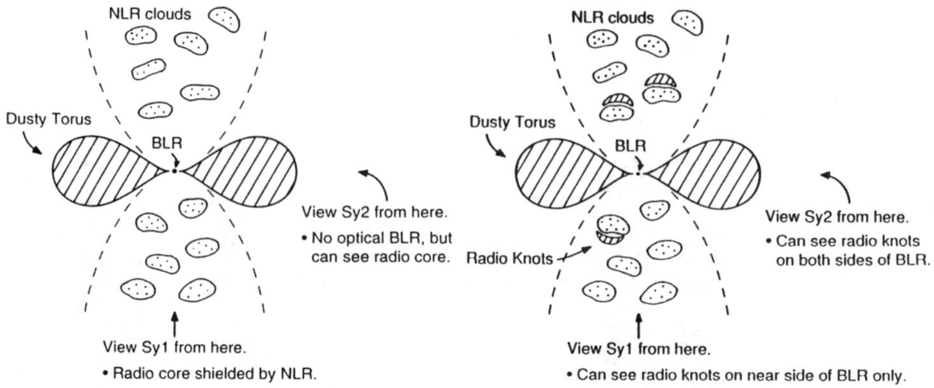

FIGURE II The proposed models to explain why we see radio cores in Sy2 galaxies but not in Sy1 galaxies. The NLR clouds are optically thick at 2.3 GHz, and so in Sy1s the core cannot be seen, and only radio emission from NLR clouds on one side of the nucleus is visible. However, in Sy2s, the radio emission from both regions is visible.

if the PTI cores are located near the NLR clouds rather than in the BLR, and is also illustrated by Figure II. Compact radio emission is closely associated with individual NLR clouds, and there may be an offset between radio and NLR emission (e.g. Whittle et al. 1988). When observing Sy1s, radio emission from clouds beyond the nucleus would be blocked from view by the cloud itself, and only the components on the near side would be visible in the radio. Sy2s, on the other hand, are viewed perpendicular to the axis of the NLR and so our view of the radio components would not be obscured by the NLR clouds, and the PTI detection rate would be lower for Sy1s than for Sy2s.

Both variants of this model predict that the NLR clouds will become optically thin at higher frequency, so the model can be tested by checking whether the Sy1 and Sy2 detection rates become equal at a sufficiently high frequency.

We will now attempt to anticipate some potential criticisms. We have some justification for assuming the clouds to be optically thick. In the NLR clouds the electron temperature is 10^4 to 5.10^4 K, the electron density is $10^{9.5\pm1}$ m^{-3}, and the cloud sizes are 6-12 pc (Osterbrock 1987; Evans et al. 1991; Lawrence 1987). These conditions imply that the optical depth per cloud at 2.3 GHz is 0.01 to 10^3, and so clouds may be optically thick.

Seyfert cores generally have steep radio spectra, whereas absorbed spectra are flat, which seems to argue against our models. We counter-argue that our model predicts steep spectra since the absorbed sources will be much weaker than the directly visible steep spectrum sources.

Photoionization models of the NLR region diagnose small (0.01) covering factors which may be too small for absorption in the NLR to be significant. However Netzer and Laor (1993) show that dust in the NLR can strongly suppress line emission, which causes these models to underestimate the filling and covering factors. A better estimate of the covering factor is 0.2 to 0.5. This should be ample to permit absorption by NLR clouds to be significant.

CONCLUSION

We have observed well-defined samples of Seyfert galaxies with a long-baseline interferometer. We found that compact radio structures are much more common in Seyfert 2 galaxies than in Seyfert 1 galaxies. We have proposed a model which reconciles our observations with the orientation unification model, by invoking large radio optical depth in the NLR clouds which surround the nuclear radio sources. This scenario predicts that further observations at a sufficiently high frequency should find no difference between the Sy1 and Sy2 detection rates, since the NLR clouds should then become optically thin.

ACKNOWLEDGMENTS

We thank David Allen, who has been involved in this project since its inception, and the staff and Director of the Tidbinbilla Tracking Station for their assistance. This research used the NASA/IPAC Extragalactic Database.

REFERENCES

de Grijp, M.H.K., Miley, G.K., and Lub, J. 1987, *A&AS*, **70**, 95

Edelson, R.A. 1987, *ApJ*, **313**, 651

Evans, I.N., Ford, H.C., Kinney, A.L., Antonucci, R.R.J., Armus, L., and Caganoff, S. 1991, *ApJ*, **369**, L27

Lawrence, A. 1987, *PASP*, **99**, 309

Netzer, H., and Laor, A. 1993, *ApJ*, **404**, L51

Norris, R.P., Kesteven, M.J., Wellington, K.J., and Batty, M.J. 1988, *ApJS*, **67**, 85

Norris, R.P., Allen, D.A., Sramek, R.A., Kesteven, M.J., and Troup, E.R. 1990, *ApJ*, **359**, 291

Norris, R.P., Roy, A.L., Allen, D.A., Kesteven, M.J., Troup, E.R., Reynolds, J.E. 1992, in *Relationships between active galactic nuclei and starburst galaxies*, ed. Alexei V. Filippenko, ASP conference series vol. 31, 71

Osterbrock, D.E. 1987, in *Spectroscopy of Astrophysical Plasmas*, ed. Dalgarno, A., and Layzer, D., 59

Roy, A.L., Norris, R.P., Kesteven, M.J., Reynolds, J.E., and Troup, E.R. 1993, in preparation

Spinoglio, L., and Malkan, M.A. 1989, *ApJ*, **342**, 83

Ulvestad, J.S., and Wilson, A.S. 1989, *ApJ*, **343**, 659

Whittle, M., Pedlar, A., Meurs, E.J.A., Unger, S.W., Axon, D.J., and Ward, M.J. 1988, *ApJ*, **326**, 125

The First Stromlo Symposium: The Physics of Active Galaxies
ASP Conference Series, Vol. 54, 1994
G. V. Bicknell, M. A. Dopita, and P. J. Quinn (eds.)

UNIFIED SCHEMES OF FR2 RADIO GALAXIES AND QUASARS

PETER D. BARTHEL
Kapteyn Astronomical Institute, P.O. Box 800, NL–9700 AV Groningen, The Netherlands

ABSTRACT Arguments pro and contra unification of radio-loud quasars and powerful radio galaxies through the effects of relativistic beaming and anisotropic obscuration are reviewed.

INTRODUCTION

Unification of Seyfert galaxies – radio-quiet objects – was discussed earlier by Dr. Joe Miller. The radio-loud FR2 denomination became part of the Unification Church in 1988/89, followed soon thereafter by the FR1 denomination. Indeed, unification is kind of a religion: one has to believe in things one cannot see. The SOC has asked me to preach today in the FR2 denomination (Dr. Meg Urry will follow in the FR1 denomination), and it is a pleasure to see that the church is nearly full. I will not ask you for your money, but do hope that you will see the (hidden) light at the end of the sermon!

As is probably known, radio galaxies with luminosities above the Fanaroff & Riley (1974) break luminosity of $10^{25.2}$ W/Hz (Class FR2) have radio morphologies such as displayed by Cygnus A: double lobed, edge brightened. We will call such powerful radio galaxies PRGs hereafter. Radio-loud QSOs, which we'll call QSRs in the following, have either the Cygnus A morphology (be it that the radio core and a one-sided jet are typically more prominent), or a core-halo morphology with dominant core emission. In the optical the QSRs are of course characterized by an unresolved bright continuum source and luminous broad and narrow emission lines; PRGs on the other hand have faint optical continua and generally only narrow emission lines. Although these objects are seemingly different, there are now strong reasons to believe that they are actually one and the same, the orientation of their radio axis with respect to the line of sight being the discriminator between the classes. In this review I will address the various arguments, but first go into a bit of history.

HOW FR2 UNIFICATION DEVELOPED

From 1978 onwards astronomers have suspected that AGN radiation – in any part of the electromagnetic spectrum – could be aspect angle dependent. Effects such as relativistic beaming, shadowing, or emission from a non-spherical optically thick body were envisaged. It is of paramount importance to understand these effects, for the following reasons:

• they introduce selection biases for classes of objects;
• they make us to infer incorrect luminosities, since we naively multiply the measured flux densities by $4\pi D_L^2$;
• they lead us to overestimate the AGN diversity.

The arguments for relativistic beaming in core-dominated QSRs are well known and will not be repeated here. Relativistic beaming was invoked by Blandford & Rees (1978) and Blandford & Königl (1979) to model blazars, and subsequently by Orr & Browne (1982) to unify core-dominated (CD) and lobe-dominated (LD) QSRs. Whereas *extrinsic optical* effects in the form of shadowing were invoked to unify Seyfert galaxies of Types 1 and 2 (Antonucci & Miller 1985), *intrinsic radio* effects in the form of relativistic beaming played the dominant role in our understanding of the interrelation of different types of QSRs and PRGs. Indeed, improved dynamic range radio mapping at arcsec and milliarcsec scale provided the drive behind unified schemes for radio-loud objects. Radio haloes around blazars, *being beamed objects*, were examined in the mid-eighties by Wardle et al. (1984) and Antonucci & Ulvestad (1985). On the basis of high dynamic range radio maps, these authors were forced to conclude that unbeamed counterparts must exist among non-blazar 3C/4C-objects. In his excellent review Antonucci (1993) describes this forced conclusion as his 'religious conversion.' Phinney (1985) examined the relative occurrence of LD and CD QSRs on a relativistic beaming hypothesis. As an explanation for the fact that too many CD QSRs existed for all extended LD QSRs he suggested as one possible explanation that "unbeamed QSRSs are not classified as such" but dismissed that explanation. This last idea however was taken more or less seriously by Bridle & Perley (1984), Scheuer (1987), Barthel (1987), and Peacock (1987), who mention the possibility that QSRs near the sky plane might be classified as radio galaxies. The philosophical question from the beginning of the 1980s "there MUST be unbeamed blazars – what will they look like?" led to another burning question at the end of the 1980s "there MUST be unbeamed QSRs – what will they look like?" This last question was particularly stimulated by the discovery of superluminal motion (SLM) in cores of very large double-lobed QSRs (e.g., Barthel et al. 1986, Barthel et al. 1989), their pronounced jet/counterjet brightness asymmetry (e.g., Wardle & Potash 1984) and their pronounced lobe/counterlobe depolarization asymmetry (e.g., Laing 1988, Garrington et al. 1988). Recall that these properties can all be ascribed to beaming and preferred orientation.

Borrowing upon anisotropic dust obscuration ideas (Antonucci & Miller 1985), Barthel (1989) then demonstrated that in the $0.5 < z < 1.0$ redshift range core-dominated QSRs, lobe-dominated QSRs, and PRGs may well belong to one and the same population, and that this unified scheme is not contradicted by far-IR, optical, or X-ray data. 3CRR (Laing et al. 1983) statistics imply LD QSRs to be favourably oriented PRGs, within a $\sim 45°$ cone, and CD QSRs similarly within a $\sim 15°$ cone. I will conclude this 'historical' section by recalling the main (1989) arguments behind unification of FR2 QSRs and PRGs:

• apparently preferred orientation for ALL QSRs;
• QSR and PRG cosmic evolution broadly comparable – read the last sentence in the abstract of the classical Laing et al. (1983) paper!
• narrow emission line luminosities comparable (recall Spinrad 1986!)
• distributions of relative numbers/sizes supportive;
• optical polarization properties supportive.

It was, and still is my view, that these arguments make FR2 unification unavoidable. Let us consider the 1993 arguments.

1993 STATUS/PROSPECTS OF FR2 UNIFICATION

Proof of the complete, simple FR2 Unification Scheme requires detection of the obscured BLR and continuum (e.g., in scattered light) in whichever PRG. This is not yet possible. Moreover, not *every* PRG will harbour an obscured QSO, as will be discussed below. So far only consistency checks have been carried out, some of which are supportive, and some are not. It should be stressed that proper consistency checks need completely identified, randomly oriented samples, over a wide range of redshift/luminosity. I will deal with these consistency checks, which are mostly phenomenological, but supported by theory of course.

Argument PRO1

FR2 unification explains the frequent occurrence of superluminal motion in QSRs, including some very large ones (Barthel et al. 1989, Vermeulen et al. 1993). Such (VLBI) observations strongly suggest that QSRs in the sky plane do not exist. In fact, for 3CR QSRs all relationships involving apparent velocity, radio core strength and variability, projected source linear size and curvature are in the sense as predicted by orientation and relativistic beaming (Hough and Readhead 1987, 1989, incl. updates). Kapahi & Murphy (1990) examined the relative core strength in 3CR PRGS and QSRs and found continuity, i.e., consistency with the simple scheme. Superluminal motion and other beaming indicators were investigated for large samples by Ghisellini et al. (1993): consistency with the simple schemes (FR1 and FR2) was found.

Argument PRO2

FR2 unification solves the overall jet brightness asymmetry in QSRs as being due to orientation effects. Many symmetric double lobed QSRs, including the largest known (4C 74.26: Riley & Warner 1990, Pearson et al. 1992) display one-sided jets, from milliarcsec to arcsec scale. Jet prominence for PRGs was investigated by Fernini et al. (1993), who found general agreement with the simple scheme. It is interesting to note that the sole jet detection in the Fernini et al. sample, 3C 22, has the highest core fraction (1% at 5 GHz; compare Cyg A with 0.3%). This fits with radio core strength distributions in QSRs (see PRO1).

Intermezzo: Why Counterjets Are Really There

There are currently three beautiful observations telling us that counterjets must be present. From long slit spectroscopy Axon et al. (1989) discovered that the

[OIII] emission line doublet in 3C 120 displays the same velocity splitting at the jet side and the – invisible – counterjet side, with ΔV = 300km/sec. Optical continuum (synchrotron) emission at the counterjet side in M 87 was discovered by Stiavelli et al. (1992), implying particle lifetimes of order 10^3 yr, hence recent particle acceleration by the invisible counterjet. Thirdly, Carilli et al. (1993) discovered the VLBI scale counterjet in Cyg A. Whereas the jet : counterjet ratio in Cyg A is about 2:1 on arcsec scales, this ratio is about 10:1 on milliarcsec scale. Together with the measured VLBI speed of $\sim 0.4c$ (for $H_0 = 100$) and assuming jet deceleration on the kpc scale, these data are consistent with orientation of Cyg A at about 35° from the sky plane (Vestergaard & Barthel 1993).

End intermezzo

Argument PRO3

FR2 unification solves the overall depolarization asymmetry as being due to orientation effects. QSRs generally display a considerable Faraday dispersion asymmetry: the dispersion towards the unjetted side is substantially greater than towards the jetted side (e.g., Garrington & Conway 1991). Jets in PRGs are usually absent, so PRGs cannot be plotted in the QSR Faraday dispersion plot. However, the Faraday dispersion ratio values for PRGs appear to be closer to unity, which is in agreement with simple unification predictions (Garrington & Conway 1991). McCarthy et al. (1991) discovered that lobe distance asymmetries in high z PRGs are correlated with asymmetries observed in the extended emission line regions (EELRs). This suggests that an asymmetric (ISM/ICM) environment is present, which may be responsible for the depolarization asymmetry as well. However, although mild environmental effects cannot be ruled out, the majority of sources are more strongly depolarized at the counterjet side, regardless of the EELR location. See also Scheuer et al. (1993) for a discussion of correlations with radio lobe spectral index. QSR 1318+113 (4C 11.45) provides a good example: the depolarization was mapped by Garrington et al. (1991), the one-sided jet was mapped by Lonsdale & Barthel (1987), and the asymmetric Lyα EELR (extended towards the jet side) can be found in Heckman et al. (1991). This nevertheless leaves the asymmetric EELRs to be explained: destruction of the 'backside' nebula through resonant scattering in the host galaxy is the probable explanation.

Argument PRO4

FR2 unification is broadly consistent with the observed QSR/PRG number densities, linear size distributions and luminosity functions. At intermediate and high redshifts the observed fractions of (3CRR) PRGs and QSRs are in rough agreement with their linear size distributions (Barthel 1989, 1993), using cone angles in the range 35°– 55° to divide the classes. In addition, the observed radio luminosity functions (for powerful steep and flat spectrum sources) can be modeled with the effects of relativistic beaming and orientation/shadowing taken into account (Padovani & Urry 1992). Using an average Lorentz factor $\gamma = 11$, the average CD QSR appears at an angle $\theta = 11°$, the average LD QSR at $\theta = 26°$, and PRGs occur outside a 38° cone. Hence these two different approaches to unify FR2 QSRs and PRGs are entirely consistent.

Intermezzo: Caveat

Singal (1993a,b) pointed out that life is not as simple as Barthel (1989) suggested. He and also Kapahi (1990) showed that the simple n(QSR)/n(PRG) $\sim 1/2.5$ model only works at intermediate redshift: at high redshift there are too many (3CRR) QSRs, at low redshifts there are too many PRGs. Moreover, the size distributions as implied by the number densities do not fit the observed distributions, and the cosmic linear size evolutions for the classes do not occur in parallel. These are valid criticisms, but they should be viewed with some caution. Completeness of the samples is a first requirement, and history has taught us that the most difficult identifications are the high redshift radio galaxies – which tend to have fairly large linear sizes (e.g., Röttgering 1993). This weakens Singal's (1993b) evolutionary objections. The 3CRR catalogue suffers from low number statistics, and in addition its low z range suffers from contamination. To illustrate the former, 3CRR (Laing et al. 1983) at $z > 1$ has 17/32 which is 53% QSRs. This implies a dividing cone angle of 62° (Singal 1993a). Going one Jansky deeper, to 3CR (Spinrad et al. 1985 plus updates), we find 21/48 which is 44% QSRs, implying a 56° cone angle (Barthel 1993). Given these low number statistical uncertainties, it is no surprise that the QSR fraction in the small 408 MHz 1-Jy sample is discrepant (Singal 1993a). Moreover, the selection criteria for the 1-Jy and 3CRR samples are quite different. At the low redshift end (z a few tenths) contamination with dull FR2 PRGs is important. Although often cited, Hine & Longair (1979) is not always well read! The FR2 classification does NOT guarantee the Hine & Longair Class A classification indicative of the presence of strong high excitation emission lines. This is particularly valid for PRGs one or two orders of magnitude more luminous than the FR-break luminosity (i.e., low z PRGs), and less so for the most luminous PRGs (i.e., intermediate and high redshift PRGs.) Inspection of Figure 2 in Hine & Longair (1979) shows that the fraction of Class A objects ranges from about 50% to 70% for PRGs above the FR-break. In addition, optical studies by Heckman et al. (1986) and Smith & Heckman (1989) demonstrated that only about half of the low redshift Class A sources have morphological peculiarities resembling QSR hosts. In summary, there is a fairly substantial population of dull FR2 PRGs, which – although they may have (extended, low excitation) line emission (Baum et al. 1993) – are not likely to harbour an obscured QSO. This is entirely consistent with spectroscopic results by Laing et al. (1993), and explains the low z discrepancies (Singal 1993a and Kapahi 1990). **End intermezzo**

Argument PRO5

Combining radio and optical data, the [OIII]λ5007 equivalent width in QSRs appears anti-correlated with $\log R$ (relative radio core strength), in the sense predicted by orientation and beaming (Jackson & Browne 1989). Baker et al. (1993) measured an even stronger effect for the [OII]λ3727 line. These measurements imply modest optical flux boosting (Browne & Wright 1985, Kapahi & Shastri 1987).

Argument PRO6

Direct evidence for radiation anisotropy is provided by the extended emission line regions (EELRs), commonly observed in particularly high z PRGs (e.g., Fosbury

1992, McCarthy 1993). The ionized gas is observed broadly along the radio axes. Two scenario's can be envisaged: a nuclear anisotropic ionizing radiation field which is not directly visible, or local ionization by young stars formed under the influence of radio jets. The EELRs in some cases extend beyond the radio sources, thereby favouring an obscured ionizing continuum over jet induced star formation. Highly ionized species are observed in some EELRs, requiring a hard ionizing spectrum which is not seen directly (Robinson et al. 1987). Shock auto-ionization cannot be excluded (Bicknell 1993), but the origin of the shocks remains to be explained.

Argument PRO7

Besides EELR alignment mentioned above, alignment of optical continuum emission with the radio source is also frequently observed in high redshift PRGs. Although jet induced star formation plays a role, the dominant mechanism explaining the extended optical structures (in the eyes of this reviewer) is scattering of an obscured AGN (QSO) continuum. I refer to McCarthy (1993) for a full discussion of this issue. Most important recent findings include significant amounts of rest frame UV/optical polarization, with the electric vector perpendicular to the radio axis, or better the extended continuum itself (Januzzi & Elston 1991, Tadhunter et al. 1992, di Serego Alighieri et al. 1993). A beautiful 'local' example is provided by Cyg A (Tadhunter et al. 1990). The UV-continuum appears better aligned with the EELR than with the radio axis (Rigler et al. 1992, Cimatti et al. 1993). This tells us that dust and/or electrons associated with the EELRs are the scattering agents. The strength of the extended UV-component is probably related to the jet power (Dunlop & Peacock 1993, see also PRO9). Goodrich & Cohen (1992) observed a scattered QSO continuum in the broad-lined PRG 3C 109, in agreement with X-ray data for this object (Allen & Fabian 1992).

Argument PRO8

Not only the morphologies but also the EELR luminosities provide support for the simple FR2 unification scheme. A powerful test of any unification scheme is of course an intercomparison of the different classes in their luminosity of some type of *isotropically emitted* radiation. Jackson & Browne (1990) for instance measured the [OIII]λ5007 EELR luminosities of samples of QSRs and PRGs. Assuming that this emission line is being emitted isotropically, the measured luminosity differences were in disagreement with simple unification. Hes et al. (1993a) made a similar intercomparison, using the [OII]λ3727 emission line. Since no luminosity differences were found and it can be argued that the [OII] line is better suited, this result is supportive of the simple scheme. See also the figures in Hes et al. (1993b).

Heckman et al. (1991) measured the morphologies and luminosities of extended Lyα haloes around high redshift QSRs. Comparison with PRG EELRs (see PRO6) shows no differences in luminosities. On average, QSRs appear to display smaller Lyα haloes. These observations are consistent with the simple unification scheme.

Argument PRO9

An interesting comparison was made by Rawlings & Saunders (1991). These authors determined the radio jet power for a sample of intermediate redshift QSRs and PRGs, and plotted the resulting values against the EELR (h.l. [OIII]) luminosities. Over four orders of magnitude the EELR luminosity was about one per cent of the jet power; QSRs and PRGs show roughly the same continuous distribution.

Continuity was also observed in the near-IR K-band. Whatever its physical origin may be (e.g., McCarthy 1993), QSRs appear to lie on the $K(z)$-diagram as established for PRGs (Lehnert et al. 1992).

Argument PRO10

A powerful consistency check is obviously intercomparison of hosts and environments. Whereas the jury is out on the former (see below), the latter provides support for simple FR2 unification. Smith & Heckman (1990) found no differences in the local environments for redshift $z < 0.3$. Evolution in local environment has been observed at intermediate redshifts: Yee & Green (1987) found $z \sim 0.5$ QSRs to reside in moderately rich clusters. The same was discovered for PRGs: Hill & Lilly (1991) compared the galaxy overdensity within 0.5 Mpc for QSRs and PRGs, and found no differences. Bahcall & Chokshi (1992) confirmed this.

Argument PRO11

An elegant proof of anisotropic obscuration in QSRs is provided by the Lyman-edge test. The virtual absence of QSR-intrinsic Lyman continuum edges (Antonucci et al. 1991, Koratkar et al. 1992, Tytler & Davis 1993), together with the strong evidence that broad line region clouds occupy a flattened configuration perpendicular to the radio axis (Wills & Browne 1986) implies that we apparently never observe optical QSR continua *through* their associated BLR. This is consistent with the model of opaque tori surrounding the BLR in the equatorial plane, the standard ingredient of FR2 unification.

Argument PRO12

A substantial cosmological population of obscured AGN together with a population of unobscured AGN may make up the cosmic X-ray background (Setti & Woltjer 1989, Madau et al. 1993). Both populations should be transparent for energies in excess of 10 keV, but the obscured (narrow-line) subpopulation should block the lower energy X-rays through photo-electric absorption. The observed X-ray background spectrum could be fitted, assuming AGN space density ratios as proposed by the unification scheme.

Argument PRO13

The nearby PRG Cyg A provides several supporting facts for FR2 unification:
- the light of the galaxy in the central regions is diluted by extended featureless continuum radiation (Pierce & Stockton 1986);
- the bipolar reflection nebula (Tadhunter et al. 1990) argues for a hidden origin of this continuum source;

• near-IR images confirm the presence of this nuclear continuum source (Djorgovski et al. 1991);
• there is a strongly reddened nuclear narrow line region spot-on the radio core, arguing for obscuring dust at sub-kpc scale (Vestergaard & Barthel 1993);
• the mass of this nuclear dust component is $10^{6.5}$ M$\odot$ (Djorgovski et al. 1991); molecular gas is probably also present (Mazzarella et al. 1993);
• the inferred nuclear extinction A_V (broad lines) exceeds 25^m; idem A_V (nuclear continuum) $\sim 50^m$ (Djorgovski et al. 1991, Ward et al. 1991); this extinction is consistent with the observed low energy X-ray cutoff;
• as mentioned above, the jet : counterjet brightness ratios at arcsec and milliarcsec scale are consistent with the radio axis of Cyg A being at $\sim 55°$ to the line of sight.

Despite these supportive arguments, scattered broad emission lines have not yet been seen in Cyg A (see below). We will now turn to the arguments which provide evidence against simple FR2 unification.

Argument CON1

Far-IR radiation in AGN is probably emitted isotropically (Sanders et al. 1989). PRGs were found to be luminous far-IR emitters (Golombek et al. 1988), as were QSRs (Neugebauer et al. 1986). Detailed comparison of the 25, 60 and 100μm emission revealed that QSRs as a class are a factor three to four more luminous than PRGs (Heckman et al. 1992), providing a powerful counterargument. These differences may be explained however to aspect dependence of optically thick dust emission (Pier & Krolik 1993), and/or inclusion of some nonthermal Blazar component in the QSR class.

Argument CON2

As for the underlying host galaxies, which should obviously be similar for QSRs and PRGs, the situation is not clear. Observations by Smith et al. (1986), Hutchings (1987), Hutchings et al. (1989), Smith & Heckman (1990), Véron-Cetty & Woltjer (1990) indicate that the average host M_V-values may differ by 0.7mag. These differences between QSR and PRG hosts have been reported by some authors, but denied by others. Color differences have low significance. Abraham et al. (1992) presented simulations arguing that host absolute magnitudes can be determined at such accuracy that the reported differences are significant. However, as convincingly shown by Heckman et al. (1986), QSR and PRG hosts certainly do not have simple exponential disk or $r^{1/4}$-profiles (used as input in the Abraham et al. simulations). I am therefore not convinced that the reported magnitude differences are significant.

Argument CON3

Despite attempts, few convincing proofs of scattered broad lines in narrow line PRGs have been given (e.g., Antonucci 1993). One of the best candidates to look for a scattered BLR is obviously the nearby PRG Cygnus A. Despite a sensitive search, Jackson & Tadhunter (1993) have not found a broad Hα component in its polarized spectrum; although there is little doubt about the presence of a hidden continuum source, the presence of an obscured BLR in Cyg A remains to be proven. Allowing for some dilution due to non-scattering processes however lowers the significance for its absence.

CONCLUSIONS

It is my view that – apart from some problems – the model unifying FR2 quasars and radio galaxies unification is healthy: QSRs in the sky plane must masquerade as PRGs. More precisely:

Q: is every QSR beamed (= favourably oriented) ?
A: yes
Q: does every PRG contain a hidden QSO ?
A: no, but most if not all at high radio power do, and a good fraction at intermediate power do.

In order to get the FR2 unification picture complete, we still need:
- continued/expanded VLBI and radio-imaging projects;
- larger samples of completely identified radio sources, covering a greater redshift/luminosity range; in addition, high quality optical spectra;
- spectropolarimetry at 8m-class telescopes;
- more imaging polarimetry at UV-wavelenghts;
- hard X-ray as well as deep far-IR data;

We should also take seriously the possibility of unification of radio-quiet QSOs with luminous far-IR galaxies (e.g., Hines 1991). It is an embarrassing discovery to find out that after 30 years we still do not have complete, high-quality radio and optical data for the extragalactic 3CR population (Laing et al. 1993): this hinders exploration of unification theories to their full extent. However, let me finish positively by recalling that as far as the radio-loud situation is concerned, we are fortunate to have two marvelous test cases in our southern and northern back yards: FR1 unifiers should concentrate on Cen A (DEC −40°), FR2 unifiers should focus on Cyg A (DEC +40°)!

ACKNOWLEDGMENTS

I acknowledge financial support by the organizers of this excellent Symposium, and careful reading of the manuscript by Ronald Hes and Bob Sanders.

REFERENCES

Abraham, R.G., et al., 1992, *ApJ*, **401**, 474

Allen, S.W., Fabian, A.C., 1992, *MNRAS*, **258**, 29P

Antonucci, R., 1993, *ARA&A*, **31**, 473

Antonucci, R.R.J., Ulvestad, J.S., 1985, *ApJ*, **294**, 158

Antonucci, R.R.J., Miller, J.S., 1985, *ApJ*, **297**, 621

Antonucci, R.R.J., et al., 1991, *ApJ*, **342**, 64

Axon, D.J., et al., 1989, *Nature*, **341**, 631

Bahcall, N.A., Chokshi, A., 1992, *ApJ*, **385**, 33

Baker, J.C., et al., 1993, this volume

Barthel, P.D., 1989, *ApJ*, **336**, 606

Barthel, P.D., 1993, in prep.

Barthel, P.D., et al., 1986, *ApJ*, **310**, L7

Barthel, P.D., et al., 1989, *ApJ*, **336**, 601

Baum, S.A., et al., 1993, this volume

Bicknell, G.V., 1993, this volume

Blandford, R.D., Königl, A., 1979, *ApJ*, **232**, 34

Blandford, R.D., Rees, M., in *Pittsburgh Conf. on BLLac Objects*, ed. A.M. Wolfe, p. 328

Bridle, A.H., Perley, R.A., 1984, *ARA&A*, **22**, 319

Browne, I.W.A., Wright, A.E., 1985, *MNRAS*, **213**, 97

Carilli, C., et al., 1993, priv. comm.

Cimatti, A., et al., 1993, *MNRAS*, **264**, 421

di Serego Alighieri, S., et al., 1993, *ApJ*, **404**, 584

Djorgovski, S., et al., 1991, *ApJ*, **372**, L67

Dunlop, J.S., Peacock, J.A., 1993, *MNRAS*, **263**, 936

Fanaroff, B.L., Riley, J.M., 1984, *MNRAS*, **167**, 31P

Fernini, I., et al., 1993, *AJ*, **105**, 1690

Fosbury, R.A.E., 1992, in *33th Herstmonceux Conference: The Nature of Compact Objects in AGN*, in press.

Garrington, S.T., Conway, R.G., 1991, *MNRAS*, **250**, 198

Garrington, S.T., et al., 1988, *Nature*, **331**, 147

Garrington, S.T., et al., 1991, *MNRAS*, **250**, 171

Ghisellini, G., et al., 1993, *ApJ*, **407**, 65

Golombek, D., et al., 1988, *AJ*, **95**, 26

Goodrich, R.W., Cohen, M.H., 1992, *ApJ*, **391**, 623

Heckman, T.M., et al., 1986, *ApJ*, **311**, 526

Heckman, T.M., et al., 1991, *ApJ*, **390**, 100

Heckman, T.M., et al., 1992, *ApJ*, **391**, 39

Hes, R., et al., 1993a, *Nature*, **326**, 362

Hes, R., et al., 1993b, this volume

Hill, G.J., Lilly, S.J., 1991, *ApJ*, **367**, 1

Hine, R.G., Longair, M.S., 1979, *MNRAS*, **188**, 111

Hines, D.C., 1991, *ApJ*, **374**, L9

Hough, D.H., Readhead, A.C.S., 1987, *ApJ*, **321**, L11

Hough, D.H., Readhead, A.C.S., 1989, *AJ*, **98**, 1208

Hutchings, J.B., 1987, *ApJ*, **320**, 122

Hutchings, J.B., et al., 1989, *ApJ*, **342**, 660

Jackson, N., Browne, I.W.A., 1990, *Nature*, **343**, 43

Jackson, N., et al., 1989, *Nature*, **338**, 485

Jackson, N., Tadhunter, C.N., 1993, *A&A*, **272**, 105

Januzzi, B.T., Elston, R., 1991, *ApJ*, **366**, L69

Kapahi, V.K., 1990, in *Parsec-scale radio jets*, ed. J.A. Zensus & T.J. Pearson, p. 304

Kapahi, V.K., Murphy, D.W., 1990, in *Parsec-scale radio jets*, ed. J.A. Zensus & T.J. Pearson, p. 313

Kapahi, V.K., Shastri, P., 1987, *MNRAS*, **224**, 17P

Koratkar, A.P., et al., 1991, *ApJ*, **400**, 435

Laing, R.A., 1988, *Nature*, **331**, 149

Laing, R.A., et al., 1983, *MNRAS*, **204**, 151

Laing, R.A., et al., 1993, this volume

Lehnert, M.D., et al., 1992, *ApJ*, **393**, 68

Lonsdale, C.J., Barthel, P.D., 1987, *AJ*, **94**, 1487

Madau, P., et al., 1993, *ApJ*, **410**, L7

Mazzarella, J.M., et al., 1993, *ApJ*, **409**, 170

McCarthy, P.J., 1993, *ARA&A*, **31**, 000

McCarthy , P.J., et al., 1991, *ApJ*, **371**, 478

Neugebauer, G., et al., 1986 *ApJ*, **308**, 815

Orr, M.J.L., Browne, I.W.A., 1982, *MNRAS*, **200**, 1067

Padovani, P. and Urry, C.M., 1992, *ApJ*, **387**, 449

Peacock, J.A., 1987, in *Astrophysical Jets and their Engines*, ed. W. Kundt, p. 185

Pearson, T.J., et al., 1992, *MNRAS*, **259**, 13P

Phinney, E.S., 1985, in *Astrophysics of Active Galaxies and Quasi-Stellar Objects*, ed. J.S. Miller, p. 453

Pier, E.A., Krolik, J.H., 1993, *ApJ*, **401**, 99

Pierce, M.J., Stockton, A., 1986, *ApJ*, **305**, 204

Rawlings, S., Saunders, R., 1991, *Nature*, **349**, 138

Rigler, M.A.., et al., 1992, *ApJ*, **385**, 61

Riley, J.M., Warner, P.J., 1990, *MNRAS*, **246**, 1P

Robinson, A., et al., 1987, *MNRAS*, **227**, 97

Röttgering, H., 1993, Ph. D. Thesis (Leiden Univ.)

Sanders, D.B., et al. 1989, *ApJ*, **347**, 29

Scheuer, P.A.G., 1987, in *Superluminal Radio Sources*, ed. J.A. Zensus & T.J. Pearson, p. 104

Scheuer, P.A.G., et al., 1993, this volume

Setti, G., Woltjer, L., 1989, *A&A*, **224**, L21

Singal, A.K., 1993a, *MNRAS*, **262**, L27

Singal, A.K., 1993b, *MNRAS*, **263**, 139

Smith, E.P., et al., 1986, *ApJ*, **306**, 64

Smith, E.P., Heckman, T.M., 1989, *ApJ*, **341**, 658

Smith, E.P., Heckman, T.M., 1990, *ApJ*, **348**, 38

Spinrad, H., 1986, *PASP*, **98**, 269

Spinrad, H., et al., 1985, *PASP*, **97**, 932

Tadhunter, C.N., et al., 1990, *MNRAS*, **246**, 163

Tadhunter, C.N., et al., 1992, *MNRAS*, **256**, 53P

Stiavelli, M., et al., 1992, *Nature*, **355**, 802

Tytler, D.R., Davis, C., 1993, priv. comm.

Vermeulen, R.C., et al., 1993, *ApJ*, in press

Véron-Cetty, M.-P., Woltjer, L., 1990, *A&A*, **236**, 69

Vestergaard, M., Barthel, P.D., 1993, *AJ*, **105**, 456

Ward, M.J., et al. 1991, *ApJ*, **382**, 115

Wardle, J.F.C., et al. 1984, *ApJ*, **279**, 93

Wardle, J.F.C., Potash, R.I., in *Physics of Energy Transport in Extragalactic Radio Sources*, ed. A.H. Bridle & J.A. Eilek, p. 30

Wills, B.J., Browne, I.W.A., 1986, *ApJ*, **302**, 56

Yee, H.K.C., Green, R.F., 1987, *ApJ*, **319**, 28

The First Stromlo Symposium: The Physics of Active Galaxies
ASP Conference Series, Vol. 54, 1994
G. V. Bicknell, M. A. Dopita, and P. J. Quinn (eds.)

JET SIDE VERSUS SPECTRAL INDEX IN QUASARS

A.H. BRIDLE
NRAO, Edgemont Rd., Charlottesville, Va., USA

R.A. LAING
Royal Greenwich Observatory, Madingley Rd., Cambridge, UK

P.A.G. SCHEUER AND S. TURNER
MRAO, Cavendish Laboratory, Madingley Rd., Cambridge CB3 0HE, UK

ABSTRACT Out of eight bright quasars for which we have detailed observations so far, seven show a flatter spectrum on the jet side and one clearly shows a steeper spectrum on the jet side. What is new in these observations is that they have enough angular resolution to show that this spectral asymmetry is not confined to the hot-spots. Detailed examination of the changes in spectral asymmetry with surface brightness shows that both orientation-dependent and environmental factors play well-defined parts, though the relevant mechanisms are not obvious.

INTRODUCTION

There are two strong correlations involving differences between the two sides of an extragalactic radio source:-

1. The lobe on the side with the jet (or with the more prominent jet) is the less depolarized (Garrington et al. 1988, Laing 1988),
2. The lobe which is less depolarized has the flatter radio spectrum (Liu and Pooley, 1991a).

The Laing-Garrington effect has a natural explanation as an effect of orientation: according to relativistic beaming theories, the observed jet is that on the near side, and radiation from the near-side lobe passes through less of the depolarizing plasma around the source. The Liu-Pooley effect, on the other hand, is hard to interpret as anything but intrinsic to the source; it is generally agreed that the lobe material moves slowly (i.e. speed $\ll c$), and therefore the observed radio spectrum of the lobe is unlikely to be affected significantly by the orientation of the source. Furthermore, it appears that steeper radio spectrum and greater depolarization both occur in the shorter lobe(Liu and Pooley, 1991b), which is often associated with emission-line regions (Liu and Pooley, 1991b; McCarthy et al. 1991).

Since both correlations are very strong, it is a small step to put them together and conclude

3. The lobe containing the observed jet has the flatter radio spectrum.

That step would have momentous implications, though, for it implies that the jets are intrinsically one-sided and all conventional ideas of relativistic beaming are nonsense. However, that step may be unjustified, for one notices that the Laing-Garrington sample consists chiefly of quasars (naturally enough, since quasars have prominent jets) while the Liu-Pooley and Pedelty et al. (1989) samples consists chiefly of radio galaxies. Specifically, Pedelty et al. showed that, in sources without prominent jets, the longer lobe is the less depolarized. Therefore we can suggest two alternative interpretations:-

3a. Jets are intrinsically one-sided, and correlate with flatter radio lobe spectra, etc.; relativistic beaming theories are wrong, or

3b. Two effects act independently: (i) the Laing-Garrington correlation due to orientation, which dominates in quasars because their radio axes are at large angle to the plane of the sky, and (ii) the Liu-Pooley correlation between depolarization, lobe spectral index, emission line gas and lobe length, which dominates in radio galaxies because their radio axes are fairly close to the plane of the sky so that orientation effects are weak.

An obvious way to decide between hypotheses 3a and 3b is to look for a direct correlation between jet side and lobe radio spectrum, i.e. to make a direct observational test of statement 3 above.

PUBLISHED EVIDENCE

Garrington, Conway and Leahy (1991) examined 40 quasars and 7 radio galaxies. They confirmed the Laing-Garrington correlation between depolarization and jet side in this larger sample. They also found a weaker correlation between jet side and flatter radio spectrum, but could not be sure that their result was not due to a larger contribution from the (perhaps mildly relativistic) hot-spot on the jet side: '...the asymmetry in spectral index may simply be caused by the trend, noted above, for the jet-side hot-spot to be the brighter of the two'. Images with more beam-widths across the source were needed to resolve the question decisively.

OBSERVATIONS AND THEIR INTERPRETATION

The first requirement is a sample of radio sources with extended lobes as well as unambiguous one-sided jets, preferably sources for which images made with good UV coverage already exist at one or more frequencies. Bridle et al.(1993). have made 5 GHz images of a sample of 12 bright quasars, with excellent resolution and dynamic range; from these we selected the 9 which fulfil our other conditions. For some we made new maps at 1.4 GHz with the A configuration of the VLA, for others adequate archival data were available and for one (3C208) MERLIN observations are essential to provide sufficient resolution at 1.4 GHz; we therefore report on 8 sources at this stage. This is a small sample, but adequate to shed light on correlations which are almost one-to-one.

Details of the observations will be reported elsewhere when the data are complete. The first result of comparing the 1.4 GHz and 5 GHz maps (convolved to the same resolution) was clear: in 7 of the 8 sources the integrated lobe spectrum was flatter on the jet side. (The one exception, 3C263, runs counter to both rules; it clearly has a steeper lobe spectrum and greater depolarization on the jet side.) This finding confirmed that of Garrington Conway and Leahy (1991) and appeared inconsistent with relativistic beaming theories.

Before accepting such a disturbing conclusion, we need to analyse the observations in more detail, making use of the angular resolution of our data to separate the lobes from the hot-spots. That exercise is fraught with problems from the outset, since there is no single radio spectral index for a lobe; particularly in large sources, the spectrum steepens progressively as one proceeds from the hot-spot to more and more diffuse emission nearer the quasar. Obvious ways to proceed are (i) plot spectral index versus distance from the quasar (the usual procedure for studying 'spectral ageing'), or (ii) plot spectral index versus surface brightness. Initially we have chosen the latter route. We dissect each side of the source into a number of surface brightness bins, having first blanked out the nucleus and the jet. We do not blank out the hot-spots, for in our results they are distinguishable from the rest of the source by their high surface brightness, and the edge of the hot-spot would have to be fixed in an unsatisfactory, arbitrary way.

The results are shown in Fig.I There is no one simple pattern. In 3C175, for example, the radio spectrum is consistently flatter on the jet side (at given surface brightness), in 3C263 (only) it is consistently steeper on the jet side. In some other sources the spectrum is flatter on the jet side at high surface brightness but the spectra on the two sides converge or even cross over at progressively lower surface brightness.

The clue to unravelling these complicated patterns came from looking again at the image of 3C263, which is conspicuously lop-sided, having a much shorter lobe on the jet side, and we recall that the shorter lobe is associated with steeper spectrum in radio galaxies (Pedelty et al., 1989; Liu and Pooley, 1991b). This suggests that environmental factors determine the spectral index asymmetry in the largest-scale lobe structures. A quick examination of the other images showed immediately that

1. those with the jet in the longer lobe consistently have the flatter spectrum on the jet side, while
2. those with the jet in a significantly shorter lobe show crossing-over behaviour in Fig.I, i.e. flatter spectrum on the jet side at high surface brightness, steeper spectrum on the jet side at low surface brightness.

Equivalent information is presented more quantitatively and perhaps more clearly in Fig.II, where the spectral indices for the three highest and the three lowest surface brightness bins are plotted against the ratio of lobe lengths.

It shows clearly that the brightest parts of the source have a flatter spectrum on the jet side (except for 3C263, which has the most extreme lobe-length ratio), while in the dimmer parts of the source the flatter spectrum is in the longer lobe in every case (though it has to be said that this is purely accidental in 3C204, which has almost equal lobe lengths).

The images at 1.4 and 5 GHz also give information on the depolarization asymmetry. The Laing-Garrington correlation is broadly confirmed; six of our sample obey it and two (3C263 and 3C204) do not.

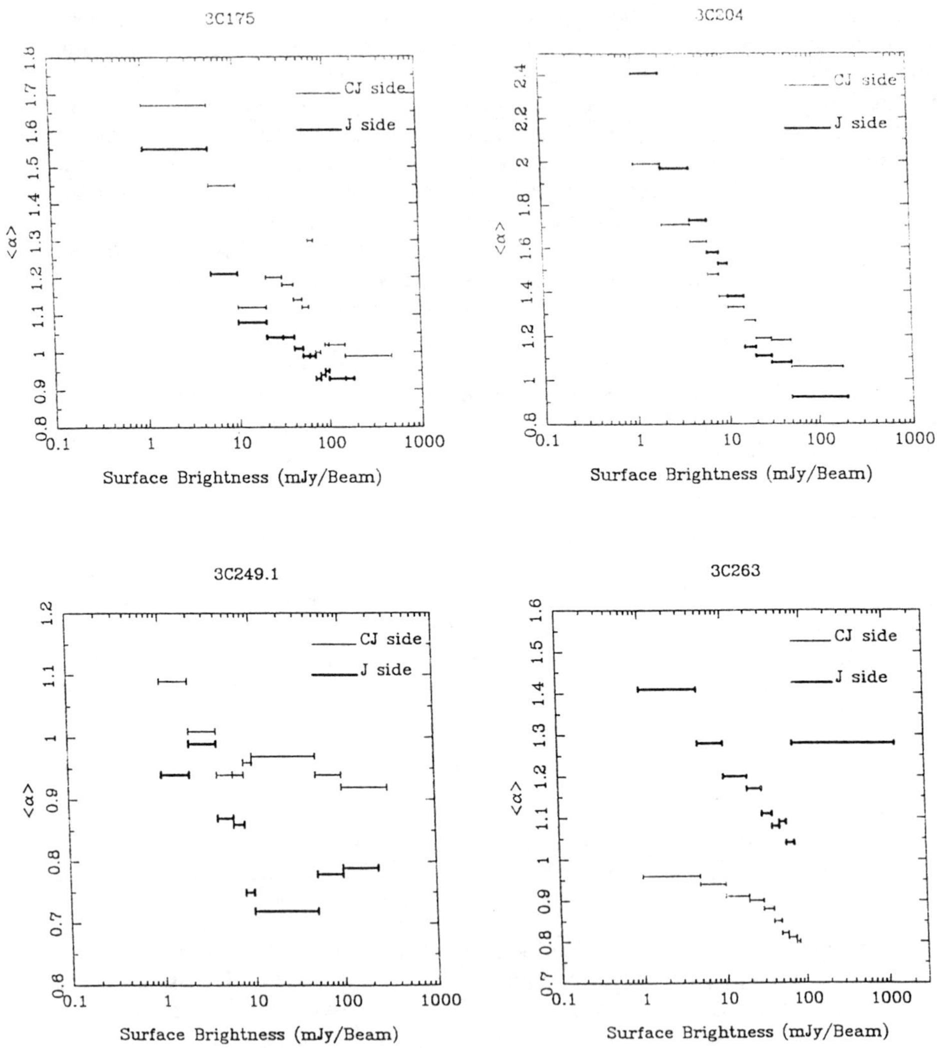

FIGURE I Spectral index $\alpha_{1.4}^{5}$ plotted against surface brightness.

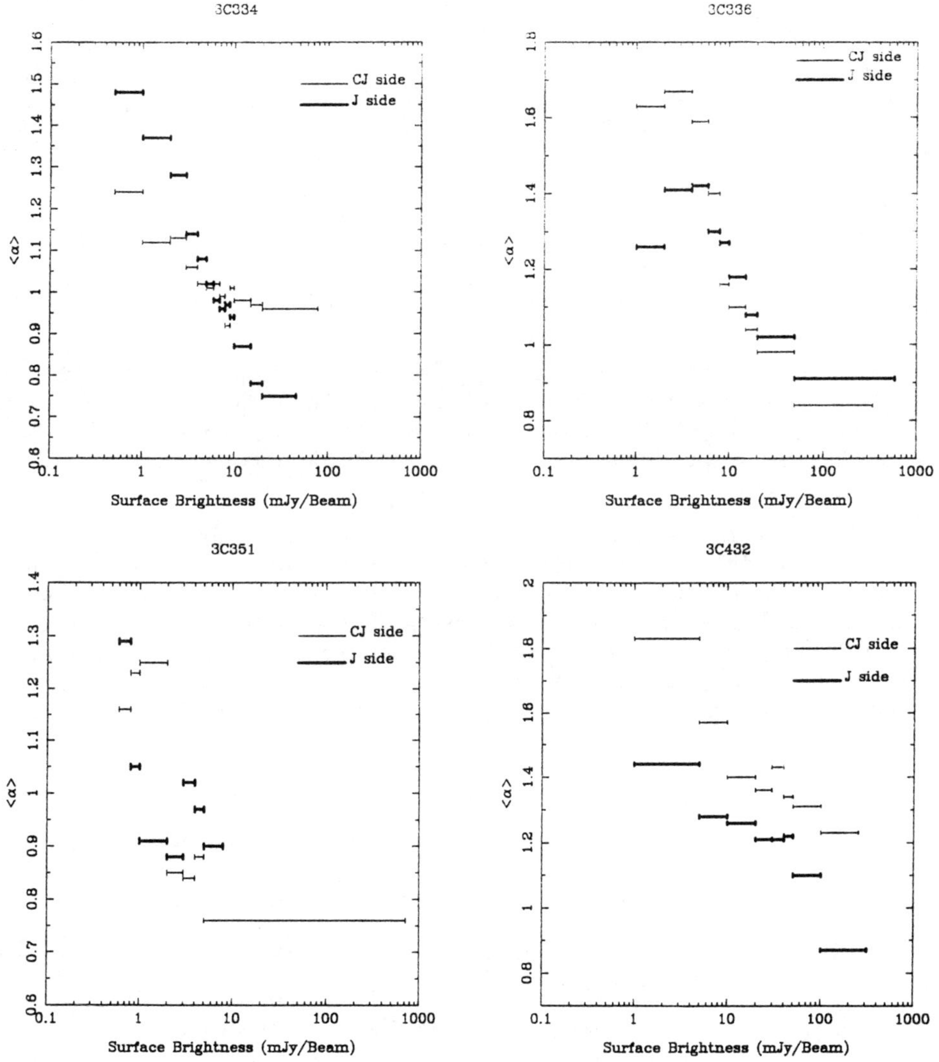

FIGURE I (continued)

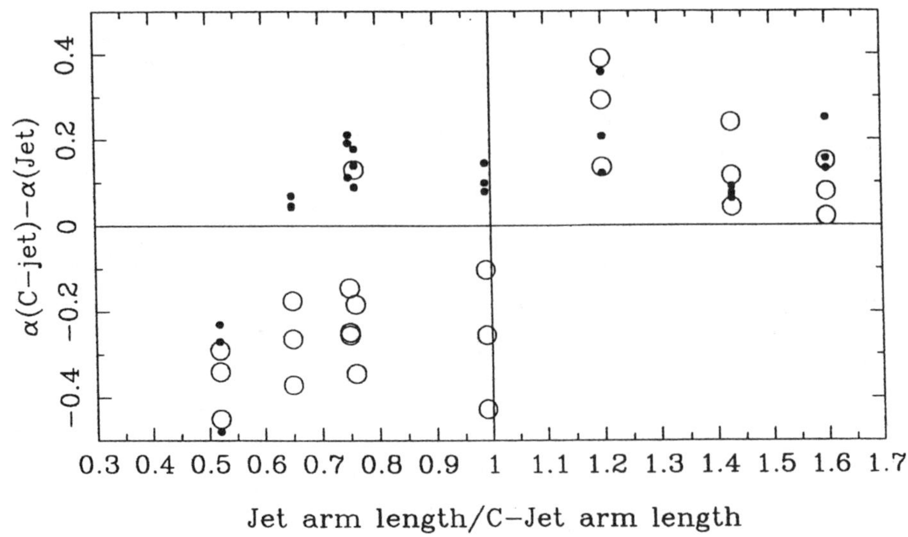

FIGURE II The spectral index $\alpha^{5}_{1.4}$ for the three highest surface brightness (•)and the three lowest surface brightness (○) intervals in each source, plotted against the ratio of lobe lengths on the jet and counter-jet sides. Here 'lobe length' refers to the distance from the nucleus to the furthest 3σ contour.

Our sample is a small one, and we ought to supplement it in any way we can. In the light of the regularities found above, can we make better use of published information? We have not yet returned to the data in Garrington, Conway and Leahy (1991), but we have looked again at the the data of Liu and Pooley (1991a). Three sources in their sample have observed jets, and it so happens that they all have the jet in the longer lobe; therefore it is no surprise that the flatter spectrum is associated with the jet side in those three sources.

CONCLUSIONS

1. Lobe-length asymmetry controls the spectral asymmetry of diffuse emission.
   Jet side controls the spectral asymmetry of bright compact emission.

2. Relativistic beaming theories survive this observational test, by the skin of their teeth (see 3. below).

3. The spectral asymmetries plotted in Fig. I indicate that the influence of the jet on spectral index α is not confined to the hot-spots. If confirmed by more detailed analysis (and if relativistic beaming theories are broadly correct) this implies that the flow beyond the hot-spots must still have speeds that are a substantial fraction of the speed of light, so that orientation can influence the measured value of α.

4. We do not know the mechanism by which

   (a) orientation affects α in and around hot-spots,

   (b) lobe length affects α in low brightness regions.

Garrington, Conway and Leahy (1991) suggested tentatively that the correlation they found between flatter lobe spectrum and jet side might be caused by contamination of the spectrum with a brighter, Doppler-boosted, hot-spot on the jet side, and Tribble (1992) has advocated this explanation. Our results do not support this idea: on the jet side, the hot-spots themselves have a flatter spectrum, and the hot-spots on the jet side do not appear to be systematically brighter than hot-spots on the counterjet side. On the other hand, we cannot at present exclude the possibility that there are structures smaller than our resolution limit which have flat spectra and are contaminating the hot-spots as they appear on our images.

ACKNOWLEDGMENTS

We are indebted to the astronomers who produced the 5 GHz images of our sample for their collaboration. The VLA is operated by Associated Universities Inc. under a cooperative agreement with the National Science Foundation. ST acknowledges support by the Carnegie Trust for Scottish Universities.

REFERENCES

Bridle, A.H., Hough, D.H., Lonsdale, C.J., Burns, J.O. and Laing, R.A. 1993. To be submitted.

Garrington, S.T., Leahy, J.P., Conway, R.G. and Laing, R.A. 1988. *Nature*, **331**, 147.

Garrington, S.T., Conway, R.G. and Leahy, J.P. 1991. *MNRAS*, **250**, 171.

Laing, R.A. 1988. *Nature*, **331**, 149.

Liu, R. and Pooley, G.G. 1991a. *MNRAS*, **249**, 343.

Liu, R. and Pooley, G.G. 1991b. *MNRAS*,**253**, 669.

McCarthy, P.J., van Breughel, W. and Kapahi, V.K. 1991. *ApJ*, **371**, 478.

Pedelty, J.A., Rudnick, L., McCarthy, P.J. and Spinrad, H. 1989. *AJ*, **97**, 647.

Tribble, P. 1992. *MNRAS*, **256**, 281.

The First Stromlo Symposium: The Physics of Active Galaxies
ASP Conference Series, Vol. 54, 1994
G. V. Bicknell, M. A. Dopita, and P. J. Quinn (eds.)

EVIDENCE FOR A BEAMED OPTICAL CONTINUUM IN RADIO QUASARS

J. C. BAKER AND R. W. HUNSTEAD
Department of Astrophysics, University of Sydney, NSW 2006, Australia

V. K. KAPAHI AND C. R. SUBRAHMANYA
N.C.R.A., T.I.F.R., Pune, India

ABSTRACT
We have defined a new *complete* sample of southern quasars chosen to be relatively free from selection biases and use this to study the aspect dependence of the optical continuum emission in radio quasars. As a test of enhancement of the optical continuum, we show that the narrow [O III] and [O II] emission line equivalent widths decrease systematically with the degree of radio core dominance. The effect is much stronger in [O II] than [O III] and is in excellent agreement with a simple relativistic beaming model.

INTRODUCTION

If we assume that the radio core emission from radio-loud quasars is relativistically beamed towards us, do we see evidence for an accompanying beamed optical component?

Enhanced optical emission has been suggested to account for the disparate observed properties of differently selected samples of quasars (Kapahi and Shastri, 1987; Browne and Wright, 1985). Studies of selection effects require that core-dominated quasars are optically brighter than lobe-dominated quasars by at least one magnitude. Furthermore, an aspect dependent optical continuum would imply that virtually *all* flux limited quasar samples are subject to an orientation bias.

THE MOLONGLO QUASAR SAMPLE

The Molonglo Quasar Sample (MQS) has been defined to minimise the effects of orientation bias, present in most other samples, by initially selecting radio sources > 0.95 Jy from the 408 MHz Molonglo Reference Catalogue (MRC; Large *et al.* 1981) in the declination band $-30° < \delta < -20°$. Sources selected at a low radio frequency are dominated by their unbeamed extended emission and so should have their radio jet axes oriented randomly in the sky.

Deep optical identifications have been obtained from the UK Schmidt IIIaJ-plates and are considered to be complete down to the plate limit ($B_J = 22.5$). The optical magnitude distribution peaks well above the plate limit, which precludes any bias in our sample introduced by an optical flux limit. Follow up radio imaging has been carried out for the quasar candidates, which cover the RA range $09^h - 14^h$ and $20^h - 06^h$, with the Molonglo Observatory Synthesis Telescope at 843 MHz and also the VLA at 5 GHz. More recently, core flux densities have been obtained with the Parkes-Tidbinbilla Interferometer at 2.3 GHz. Low resolution optical spectroscopy for the MQS, covering the wavelength range 3000–10000Å, has been obtained at the Anglo-Australian Telescope and is now virtually complete. At present, the number of quasars in the MQS is 69 but we anticipate a small increase following recent CCD imaging of blank fields at Las Campanas as a completeness check. However, we do not expect that our preliminary results will change significantly as a result. MQS quasars cover the redshift range $z = 0.3 - 2.9$.

Thus we have carefully defined a complete and unbiased quasar sample (Baker *et al.* 1993) with the primary aim of studying orientation dependent effects in radio quasars within the framework of the 'unified schemes'. The MQS is ideal for this purpose as it is low frequency radio selected, with a flux density limit five times lower than the 3CR sample, and has effectively no imposed optical flux limit.

THE EQUIVALENT WIDTH TEST

Any systematic enhancement of the optical continuum with radio core power should manifest itself as an anticorrelation of isotropic line equivalent width, W_λ, with radio core dominance, R (the ratio of core to extended radio flux, in the quasar rest frame). Such an effect has been reported by several authors, most recently by Jackson and Browne (1989) who use the narrow [O III]λ5007 emission line as an isotropically emitted standard.

Jackson and Browne find a decrease in W_λ of up to one order of magnitude for a range in R spanning about 4 orders of magnitude. However, they note in a later paper (Jackson and Browne 1990) that the [O III] line luminosity distributions for matched radio samples of quasars and galaxies are different, such that the quasars have more luminous [O III] emission. In the context of the 'unified scheme' of Barthel (1989), if radio galaxies are indeed the unbeamed parent population of quasars, then these differences in narrow line luminosity imply either that the two populations are intrinsically different or that [O III] emission must be emitted anisotropically.

The quasar-radio galaxy unification picture has gained a new lease of life from the recent finding of Hes *et al.* (1993) that radio galaxies and quasars have indistinguishable luminosity distributions in [O II]λ3727. Thus [O II] appears to be emitted more isotropically than [O III]. This might be expected as [O II] has a lower ionisation potential and critical density than [O III] and would presumably be emitted from a larger volume (see Hes *et al.*, these proceedings).

In light of these results, what are our preliminary findings using the MQS? As in previous studies, we find a moderate decrease of W_λ [O III] with R (see Figure I). However, we find a much steeper anticorrelation of the more isotropic [O II] equivalent width with R (see Figure II). The steeper slope of this latter relation is consistent with a simple beaming model (Browne and Murphy 1987) although we note that the scatter is large, particularly for the low-R sources. The

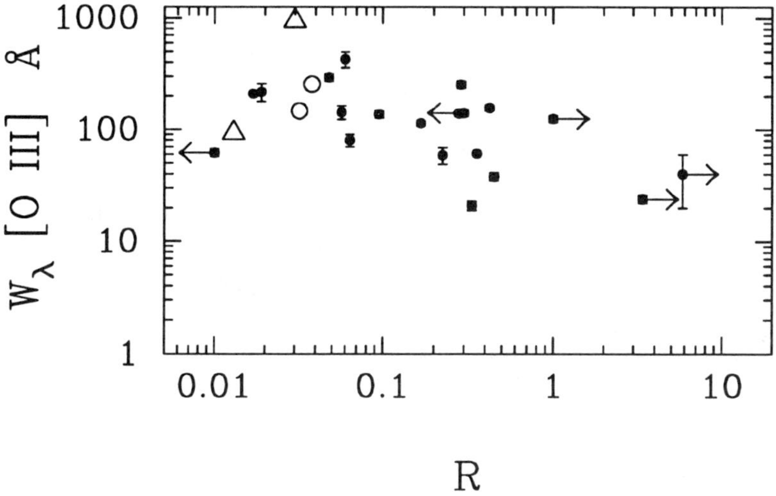

FIGURE I Preliminary plot of W_λ versus R using [O III] $\lambda\lambda 4959, 5007$ for the 24 quasars with $z < 0.9$, showing a moderate anticorrelation. Arrows indicate upper limits and quasars that may be re-classified as radio galaxies are shown as triangles.

beaming model shown in Figure II is for a single component beamed continuum with spectral index, $\alpha = 0.7$ and Lorentz factor, $\gamma = 5$ (as in Browne and Murphy 1987). It has been postulated (Wills *et al.* 1992) that a two component (flat spectrum thermal plus steep spectrum beamed) model continuum may apply, with a consequent steepening in gradient at high R. However, our preliminary data do not appear to show evidence of this.

As confirmation that our anticorrelations are primarily due to enhancement of the optical continuum and not a spurious result due to variation of the narrow line luminosities, we note that no correlation is found between narrow line luminosity and R. Narrow line luminosity has, however, been shown to scale with extended radio luminosity (eg. Saunders *et al.* 1989) but we assume both are related to the intrinsic source power and are aspect independent. Therefore we consider the anticorrelation of line equivalent width with R to be a direct indicator of relative continuum strength versus core power.

The weaker than expected anticorrelation of [O III] equivalent width with R has previously been used as an argument against optical beaming in quasars (Kellermann and Owen 1988). However, our new findings considering the [O II] line as an isotropic benchmark are consistent with predictions of an enhancement in the optical continuum of 2–3 magnitudes for core dominated quasars (Browne and Wright 1985, Kapahi and Shastri 1987).

Scatter in the $W_\lambda - R$ plots (about one order of magnitude at a given R, $R < 1$) could be due to (i) variability of the radio core/optical continuum with respect to the narrow lines on different timescales, (ii) dilution of the beamed optical synchrotron component by thermal emission, perhaps the Big Blue Bump, (iii) obscuration of the lines and/or continuum by dust or (iv) measurement errors. We believe the last to be negligible — individual line flux measurement

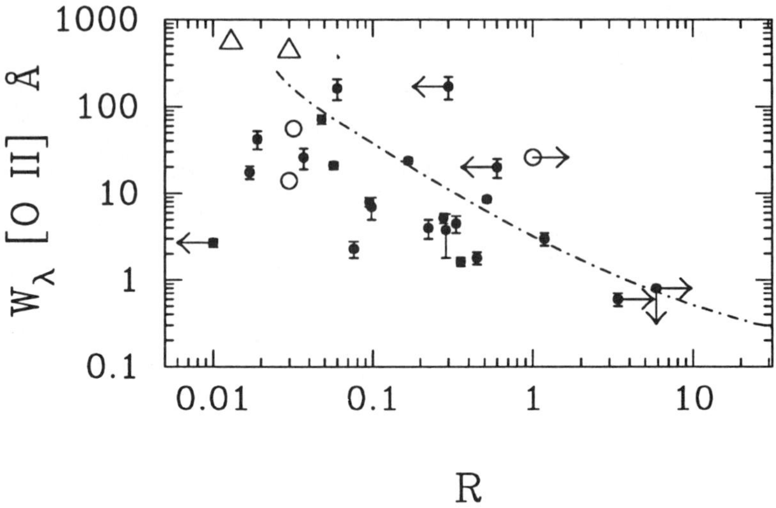

FIGURE II Preliminary W_λ versus R plot for the MQS using [O II] λ3727, for 28 quasars with $z < 1.6$. A strong anticorrelation is observed with a slope which is consistent with a simple Doppler beaming model ($\gamma = 5$ shown). Arrows indicate upper limits, quasars which may be re-classified as radio galaxies are plotted as triangles and open circles denote data drawn from the literature.

errors are only of the order of $\sim 20\%$, and photometric errors are minimised by considering line equivalent widths rather than straight continuum fluxes. Contaminating emission from the small 3000Å bump has been taken into account for measurements of the [O II] equivalent widths — we have used the underlying continuum flux where it could be estimated and any difference has been included within the errors. Possible causes of scatter within the W_λ [O III] – R plot have been discussed by Wills *et al.* (1992), particularly in relation to a strong unbeamed optical component.

BROAD LINE EQUIVALENT WIDTHS

In contrast to the results for the narrow lines, no significant anticorrelation is observed for $W_\lambda - R$ plots using the high ionisation broad lines, C IV or C III]. Perhaps a slight anticorrelation exists for Mg II. In each case the scatter is greater than that observed for [O II] and the overall lack of a significant correlation supports the idea that these broad lines are emitted anisotropically.

An interesting result is seen in the case of Hβ, where W_λ appears to be constant over the range of $R = 0.01 - 10$, with a remarkably small dispersion. This supports the previously reported close proportionality between optical continuum luminosity and Hβ line luminosity (Yee 1980, Miller *et al.* 1992) which has been taken as indicative of recombination of Hβ following photoionisation. However, such a proportionality is hard to reconcile if the optical continuum is presumed to be relativistically boosted, as indicated above by the narrow lines in the same spectral region as Hβ.

CONCLUSIONS

By demonstrating a strong inverse dependence of [O II] line equivalent width on radio core dominance, we have effectively quashed the objection to beaming models based on the small observed range of narrow line equivalent widths. Our observed anticorrelation is steeper than that reported previously by Jackson and Browne (1989) for [O III]. Such a steep slope at high R can only be modelled adequately in terms of Doppler boosting of the optical continuum. Aspect dependent geometric models, such as emission from a hot disk, are inadequate.

Further analysis of the MQS optical data, in particular the apparently contrary $W_\lambda - R$ relation for Hβ, is required to establish whether beaming is the only viable model for enhancement of the optical continuum.

ACKNOWLEDGMENTS

JCB is supported by a postgraduate scholarship from the Research Centre for Theoretical Astrophysics and RWH would like to acknowledge financial support from the ARC.

REFERENCES

Baker, J. C., Hunstead, R. W., Kapahi, V. K. and Subrahmanya, C. R., 1993, in *The Nature of Compact Objects in AGN*, eds. A. Robinson, R. J. Terlevich, C. U. P., (in press)

Barthel, P. D., 1989, *ApJ*, **336**, 606

Browne, I. W. A. and Murphy, D. W., 1987, *MNRAS*, **226**, 601

Browne, I. W. A. and Wright, A. E., 1985, *MNRAS*, **213**, 97

Hes, R., Barthel, P. D. and Fosbury, R. A. E., 1993, *Nature*, **362**, 326

Hine, R. G. and Sheuer, P. A. G., 1980, *MNRAS*, **193**, 285

Jackson, N., Browne, I. W. A., Murphy, D. W. and Saikia, D. J., 1989, *Nature*, **338**, 485

Jackson, N. and Browne, I. W. A., 1990, *Nature*, **343**, 43

Kapahi, V. K. and Shastri, P., 1987, *MNRAS*, **224**, 17p

Kellermann, K. I. and Owen, F. N., 1988, in *Galactic and Extragalactic Radio Astronomy*, eds. K. I. Kellermann and G. L. Verschuur, 563

Large, M. I., Mills, B. Y., Little, A. G., Crawford, D. F., and Sutton, J. M., 1981, *MNRAS*, **194**, 693

Miller, P., Rawlings, S., Saunders, R. and Eales, S., 1992, *MNRAS*, **254**, 93

Saunders, R., Baldwin, J. E., Rawlings, S., Warner, P. J. and Miller, L., 1989, *MNRAS*, **238**, 777

Wills, B. J., Wills, D., Breger, M., Antonucci, R. R. J. and Barvainis, R., 1992, *ApJ*, **398**, 454

Yee, H. K. C., 1980, *ApJ*, **241**, 894

The First Stromlo Symposium: The Physics of Active Galaxies
ASP Conference Series, Vol. 54, 1994
G. V. Bicknell, M. A. Dopita, and P. J. Quinn (eds.)

SPECTROPHOTOMETRY OF A COMPLETE SAMPLE OF 3CR RADIO SOURCES: IMPLICATIONS FOR UNIFIED MODELS

R.A. LAING, C.R. JENKINS, J.V. WALL & S.W. UNGER
Royal Greenwich Observatory, Madingley Road, Cambridge CB3 0HE, U.K.

ABSTRACT We present spectrophotometric data obtained with the Faint-Object Spectrograph on the William Herschel Telescope for a complete subsample of 3CR radio sources with $z < 0.88$. The wavelength coverage extended to 1μm and this proved to be crucial in the detection of broad Hα emission in a number of objects previously classified as narrow-line radio galaxies. Our results suggest that the simple unified scheme proposed by Barthel (1989) requires significant modification. We can divide the narrow-line spectra of FRII radio sources into two fairly well-defined classes: low and high excitation. Within the high-excitation group, the fractions of objects with and without detectable broad lines are roughly independent of redshift, the statistics at $z < 0.4$ having been modified significantly by the new broad-line detections. The high-excitation objects could be related in the way suggested by Barthel, with the broad lines visible over a cone of half-angle 50°. We suspect that the low-excitation group forms an isotropic sample: no broad lines are detected, but a few objects show strong radio cores and other indications of beaming. Thus, a picture more reminiscent of the BL Lac/FRI unified scheme seems appropriate for them.

INTRODUCTION

Unified models

The idea that orientation has a major influence on the appearance of Active Galaxies has aroused considerable recent interest (Antonucci 1993 and references therein). In this paper, we are concerned with the suggestion that powerful radio galaxies and quasars are the same type of object seen from different directions (Scheuer 1987; Barthel 1989). In the simplest version of this model, the central continuum source and broad-line region are surrounded by an opaque torus whose axis coincides with that of the radio source. If the object is viewed from a direction within some angle $\theta_c \approx 50°$ of this axis, then the BLR can be seen, and the object is classified as a broad-line radio galaxy or quasar. From other directions, it appears to be a narrow-line radio galaxy. The broad-line objects are expected to have stronger radio cores and one-sided radio jets as a result of Doppler favouritism.

Whilst this simple hypothesis works reasonably well for a sample of 3CR radio sources with $0.5 < z < 1.0$ (Barthel 1989), it was already clear to Browne (1987) that there were significant problems in extending it to lower redshifts, even within the same radio sample. Modifications to and difficulties with the idea were discussed by Browne & Jackson (1992), Lawrence (1991) and Kapahi (1990). Most of these discussions were based on heterogeneous optical spectra, some of considerable antiquity, the criteria for recognising broad-line radio galaxies being especially dubious. We therefore decided to obtain spectrophotometry for a well-defined sample of radio sources, with the aim of testing Barthel's hypothesis.

The sample

The sources were selected from the 3CR radio sample defined by Laing, Riley & Longair (1983), which has complete redshift information except for one object. In addition to the radio selection criteria, we imposed the conditions:

- $0^h < \mathrm{RA} < 13^h$ and
- $z < 0.88$ or
- $V < 20$

The redshift limit was chosen in order to observe all objects for which [OIII]λ5007 was visible. The objects with $z > 0.88$ (all quasars) are not discussed here.

OBSERVATIONS AND DATA REDUCTION

The observations were made with the FOS-2 spectrograph on the William Herschel Telescope. 400 spectra of 88 objects were obtained over a 3-week period, using a slit of 2×20 arcsec^2. In this paper, we discuss the first-order spectra (in most cases, the objects were too weak to give much flux in second order). Optimal extraction was used unless the object appeared to have spatially-resolved lines, in which case a uniformly-weighted extraction was performed. All of the data were tied to short exposures taken under photometric conditions.

The key features of this study are that the sample is complete, the data are homogeneous and accurately calibrated and that the wavelength range extends to 1μm.

THE SPECTRA

Figure I shows spectra of three FRII radio sources with $0.2 < z < 0.3$. One of these, 3C 67, shows broad wings on its Hα and Hβ lines, but was previously classified as a narrow-line radio galaxy! This illustrates the value of observations with good sensitivity in the red, since broad Hα is much stronger than Hβ in this object. The second object, 3C 171, has a spectrum characteristic of high-excitation narrow-line radio galaxies and Seyfert 2's. The final galaxy, 3C 132, is quite different in character. We have no reason to doubt the identification, which was made on the basis of sub-arcsecond positional coincidence with a central radio component, but the spectrum shows weak, narrow Hα, with no signs of high-excitation lines such as [OIII]λ5007.

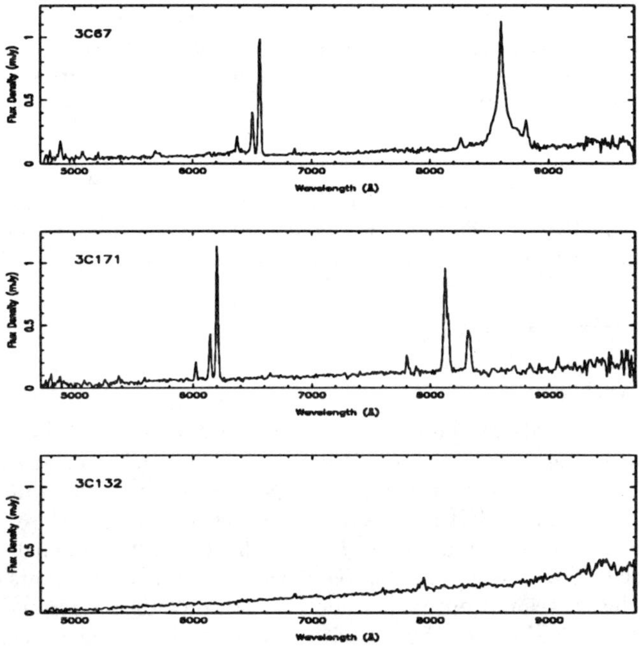

FIGURE I Examples of spectra: (a) 3C 67 (broad-line, z = 0.3102); (b) 3C 171 (narrow-line, z = 0.2384); (c) 3C 132 (low-excitation, z = 0.214).

DISCUSSION

Broad-line radio galaxies and quasars

We have detected four new broad-line radio galaxies with $z < 0.4$. Broad wings are clearly visible at Hα, but are difficult to see at Hβ (almost certainly explaining why they were missed in earlier surveys). Two of the new detections are compact steep spectrum sources (3C 67 and 268.3); the other two are classical doubles with one-sided jets (3C 33.1 and 219).

Galaxies without broad lines: differences in excitation

It is clear from Figure I that some FRII radio sources have low-excitation spectra, by which we mean that lines such as [OIII] are very weak by comparison with the hydrogen lines or undetectable. These objects have been known for a considerable time (Hine & Longair 1979), but their existence has tended to be ignored in discussions of unified schemes. They are important, since they are nearly as common as broad-line objects in our sample.

We have divided the objects without detectable broad Hα at $z < 0.45$ into two classes:

- Narrow-line (high-excitation). These have [OIII]/Hα (narrow) > 0.2 and equivalent widths of [OIII] > 3Å.

- Low-excitation, with relatively weaker [OIII].

This quantifies the distinction between "narrow-line" and "weak-emission or absorption-line" spectra often made in the literature. A similar classification can be made at higher redshifts using the Hβ line. All of the broad-line objects clearly have strong [OIII] lines.

FRI sources

We confirm the result that FRI radio sources tend to have low-excitation spectra (Hine & Longair 1979). Exceptions do exist, but there are none in our sample. We do not consider the FRI galaxies further in this paper.

TESTS OF UNIFIED MODELS

Redshift Distribution

Our division of the optical spectra into three classes: broad-line, narrow-line, and low-excitation and our discovery of new broad-line galaxies at low redshifts makes a significant change to the redshift-distribution test for unified models. We have considered only FRII radio sources, omitting 3C 48 and 3C 138 (which do not have double outer structure) and 3C 16 (where the identification is in doubt). It is clear that the low-excitation objects cannot be members of either of the other two classes seen at different orientations, since their narrow-line spectra are totally different. Many of the broad and narrow-line objects have spatially extended [OIII] regions and it is inconceivable that enough of the emission could be obscured to change the spectral classification (*cf.* Hes *et al.* 1993). We therefore pose the question: "are the redshift distributions of the narrow-line and broad-line objects identical for FRII radio sources in our sample?" The answer is that they are, except for a marginally-significant deficit of broad-line objects at the lowest redshifts, where our ability to differentiate between the classes is open to question. Barthel (1989)'s analysis was only done for $0.5 < z < 1.0$ and it was a cause of some worry that a naive extrapolation to lower redshifts in the same sample led to contradictions (*e.g.* Lawrence 1991). Our results are consistent with the idea that the fractions of broad and narrow-line objects are independent of redshift, and that the half-angle of the putative obscuring cone is roughly 50° (we find 16 broad and 29 narrow-line objects).

Narrow-line luminosities

In the simplest version of the unified model, the luminosity from narrow forbidden lines should be independent of orientation (but see Browne & Jackson 1992 and Hes *et al.* 1993). The distributions of [OIII] luminosity for the broad and narrow-line objects overlap, although there is still a significant tendency for the broad-line objects to have higher [OIII] power at given redshift. This is much less than the effect found by Jackson & Browne (1990) because of our omission of the low-excitation objects, and it may be that a slightly different excitation criterion or some orientation dependence of [OIII] luminosity would eliminate the difference. We do not regard this result as a serious problem for the unified model.

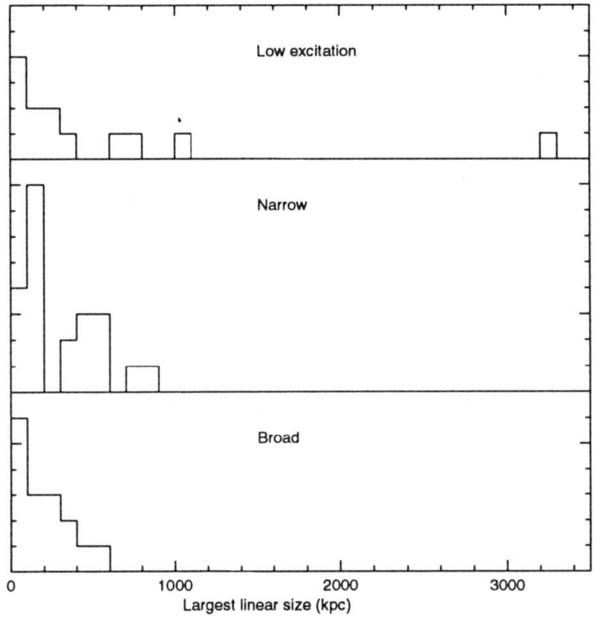

FIGURE II Histograms of linear size for the three spectral classes discussed in the text.

Radio-source sizes

Figure II shows the linear size distributions of the three classes. If the narrow and broad-line objects together form an isotropic sample, then we would expect that the narrow-line objects would be larger, the ratio of mean projection factors being 1.74. We measure the mean ratio to be 1.4, in the correct sense, but with weak evidence for an increase with redshift. The test is marginally consistent with the predictions of the simple unified model (*cf.* Kapahi 1990, who found that the broad-line objects were larger at low redshift: the difference is again due to our use of a complete sample together with the omission of the low-excitation objects). The low-excitation objects have a wide range of linear sizes: the two largest objects in the sample have spectra of this type.

Incidence of radio jets

Bridle & Perley (1984) showed that jets are significantly more common in FRII quasars than in galaxies of comparable luminosity. This result may be made more precise and extended to a larger range of power and redshift using the present data. A comparison between the broad-line and high-excitation narrow line classes shows that the former have a much higher incidence of radio jets (all one-sided) than do the latter, regardless of redshift. 11/16 broad-line objects show one-sided jets (by Bridle & Perley's criteria) compared with 1/29 of the

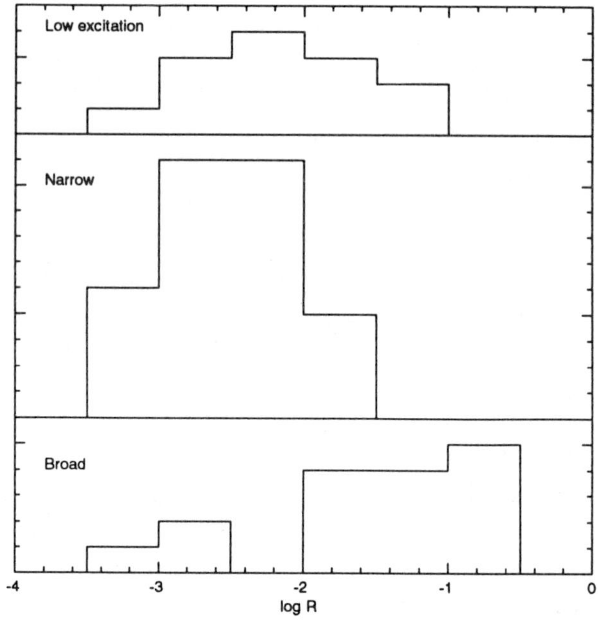

FIGURE III Histograms of radio core fraction for the three spectral classes discussed in the text.

narrow-line objects (a further 5/29 of the latter show knots of emission along the radio axis which may be bright parts of jets, but fail to meet the strict criteria). 3/13 of the low-excitation objects have jets: one of these (3C 200) is morphologically similar to the broad-line objects; 4C 12.03 has two-sided jets of the type found in FRI sources and the last is the giant radio galaxy 3C 236. The condition for a jet to be detected is poorly defined, but it is clear that broad-line objects are much more likely to to show one-sided jets than members of either of the other classes. This is as expected in the unified models if radio jets are two-sided and relativistic, since Doppler beaming makes the approaching jet brighter as the angle to the line of sight decreases.

Radio cores

Figure III shows the distributions of R (the ratio of core to extended luminosity at a fixed emitted frequency of 5 GHz) for our three spectral classes. The broad and narrow-line objects have very different R distributions, the broad-line objects having relatively brighter cores at all redshifts. There are three interesting outliers: 3C 67 and 268.3 (compact steep-spectrum sources) and 3C 196 (a quasar), all of which have anomalously weak cores. The difference between the two classes is in the sense expected from Doppler favouritism in the unified model, confirming the result of Barthel (1989) for a larger redshift range. The median values of R for the two classes differ by a factor $\approx$10. The unified model predicts that the median angles to the line of sight are 35° and 71°. If we adopt a simple beaming model with a spectral index of 0 and a Lorentz factor of 5 for the flow, we predict that the ratio of medians should be ≈12, in good enough agreement with observations, especially given the crudity of the assumptions.

The low-excitation objects have a wide range of R, consistent with the sum of the distributions for the narrow and broad-line objects. This could be taken as evidence that they have similar intrinsic core properties to the high-excitation objects and form an isotropic population, but this explanation is clearly oversimplified because of their dispersion in radio morphology and linear size.

What are the low-excitation FRII galaxies?

We have emphasised that not all galaxies associated with FRII radio sources have high-excitation narrow-line spectra. The low-excitation galaxies are heterogeneous in their properties. A few of them (mostly with $z < 0.2$) have radio morphologies intermediate between the FRI and FRII classes. The remainder are bona fide FRII's with secure identifications established by precise coincidence between the radio core and optical positions. The two largest objects in the sample, 3C 236 and DA 240, have spectra of this type, but the higher-redshift examples are mostly quite small. We suspect that some of the latter are bright central cluster galaxies similar to 3C 295.

It seems likely that these objects form an isotropic sample, albeit a hereogeneous one, and it is quite possible that they form part of the parent population of BL Lac objects. It was supposed until recently that this population consisted entirely of FRI radio galaxies, but recent evidence (*e.g.* Murphy, Browne & Perley 1993) suggests that some BL Lac objects have extended radio emission with powers and structures more characteristic of FRII's. The parent population could, therefore, consist of a mixture of FRI's and a few FRII's with low-excitation spectra.

CONCLUSIONS

The present study generally supports the unified scheme for high-luminosity radio sources. The most important new results are that we have detected a number of previously unknown broad-line radio galaxies and have drawn attention to a significant population of FRII radio galaxies with low-excitation spectra. Once the latter are excluded, the simple hypothesis that the fraction of broad-line objects is independent of redshift is consistent with our data. Whilst slight discrepancies remain in the comparison of [OIII] powers and linear sizes, they are much smaller than in previous studies. The radio core and jet properties are in good agreement with expectations. The low-excitation class contains a mixed bag of radio sources. We expect it to be an isotropic sample and speculate that it could form part of the parent population of BL Lac objects.

REFERENCES

Antonucci, R. 1993, *ARA&A*, **31**, 473.

Barthel, P.D. 1989, *ApJ*, **336**, 60.

Bridle, A.H. & Perley, R.A. 1984, *ARA&A*, **22**, 319.

Browne, I.W.A. 1987, in *Superluminal Radio Sources*, ed. Zensus, J.A. & Pearson, T.J., (Cambridge: CUP), 110.

Browne, I.W.A. & Jackson, N. 1992, in *Physics of Active Galactic Nuclei*, ed. Duschl, W.J. & Wagner, S.J., (Berlin: Springer), 618.

Hes, R., Barthel, P.D. & Fosbury, R.A.E. 1993, *Nature*, **326**, 362.

Hine, R.G. & Longair, M.S. 1979, *MNRAS*, **188**, 111.

Jackson, N. & Browne, I.W.A. 1990, *Nature*, **343**, 43.

Kapahi, V.K. 1990, in *Parsec-scale Radio Jets*, ed. Zensus, J.A. & Pearson, T.J., (Cambridge: CUP), 304.

Laing, R.A., Riley, J.M. & Longair, M.S. 1983, *MNRAS*, **204**, 151.

Lawrence, A. 1991, *MNRAS*, **252**, 586.

Murphy, D.W., Browne, I.W.A. & Perley, R.A. 1993, *MNRAS*, in press.

Scheuer, P.A.G. 1987, 1987, in *Superluminal Radio Sources*, ed. Zensus, J.A. & Pearson, T.J., (Cambridge: CUP), 104.

The First Stromlo Symposium: The Physics of Active Galaxies
ASP Conference Series, Vol. 54, 1994
G. V. Bicknell, M. A. Dopita, and P. J. Quinn (eds.)

A COMPARISON OF THE INFRARED PROPERTIES OF RADIO-LOUD ACTIVE GALAXIES

CHRISTOPHER P. O'DEA, TIMOTHY M. HECKMAN, AND STEFI A. BAUM
STScI, 3700 San Martin Dr., Baltimore, MD 21218

EIJA LAURIKAINEN
Turku Univ. Obs., Tuorla, SF-21500, Piikkiö, Finland

ABSTRACT We report on a study of the IRAS flux densities of different classes of radio loud AGN. The samples were matched in redshift and radio power of optically thin emission. We have compared the mid and far infrared flux densities from IRAS in order to constrain the bolometric luminosity of the central engines. The IR flux densities should be less subject to obscuration and orientation effects than the optical because (1) the optical depth of obscuring material will be much smaller in the IR than in the optical and (2) the IR is thought to be thermal emission in most AGN (except Blazars which are not discussed here). We find the following. (1) GHz peaked spectrum (GPS) and Compact Steep Spectrum (CSS) radio galaxies tend to be similar to non-GPS-CSS radio galaxies in the mid and far IR. GPS-CSS quasars are also similar to (but may be a factor of two weaker than) non-GPS-CSS quasars. (2) FRII radio galaxies tend to be stronger than FRI radio galaxies in the FIR by a factor of about 3. (3) Broad line radio galaxies tend to be stronger than narrow line radio galaxies by a factor of 4-5 in the mid-IR, but are similar in the FIR.

INTRODUCTION

One of the most important topics today is the relationship between the different classes of AGNs. It is now widely recognized that orientation and obscuration may be important in determining the appearance of AGNs (e.g., Antonucci 1993). Thus, to investigate the similarities and differences between different types of AGNs we should concentrate at least initially on the wavelength regimes where the radiation is most likely to be emitted isotropically and the least likely to be affected by obscuration. Here we consider the radio emission from extended optically thin structures and the mid-to-far infrared (MFIR) emission. The MFIR emission (except in Blazars which are not considered here) is thought to be dominated by the thermal emission from dust which has been heated by the optical and UV light from the central source.

Thus, in order to compare the MFIR properties of different types of radio-loud AGNs we have done the following. We have generated samples which are matched in redshift and radio power. We have used the SUPERSCANPI algorithm at IPAC to determine the mean, weighted mean, and median flux densities for the ensembles of objects at the four wavelengths observed by IRAS. In these proceedings we briefly discuss our results on the relative flux densities of the samples. Further details and additional results are given by Heckman *et al.* (1993). Our results on the properties of these ensembles of objects extends and complements earlier work on individual objects, e.g., Golombek, Miley & Neugebauer (1988) and Impey & Gregorini (1993).

THE SAMPLES

A summary of the global properties of the samples is given in Table I.

TABLE I Mean Values of Sample Parameters.

Sample [a] (1)	Number [b] (2)	log P_{5GHz} [c] Watts/Hz (3)	redshift [d] (4) (5)	S_{5GHz} [e] Jy
RG	51	27.0	0.753	0.875
GPS-CSS RG	33	27.1	0.766	1.035
QSR	28	27.3	0.945	0.964
GPS-CSS QSR	22	27.5	0.934	2.289
FRI	42	24.6	0.090	0.826
FRII	29	24.6	0.087	0.576
NLRG	45	25.7	0.162	1.496
BLRG	9	25.8	0.165	1.799

[a]Sample.

[b]Number of objects in Sample.

[c]Mean power in Watts per Hz at 5 GHz in the rest frame of the source. Assumes $H_o = 75$ km/s/Mpc and $q_o = 0.1$.

[d]Mean heliocentric optical redshift.

[e]Mean flux density at 5 GHz.

The Gigahertz-Peaked Spectrum and Compact Steep Spectrum Sources

GPS and CCS sources are powerful but physically small sources with turnovers in their radio spectra (e.g., O'Dea, Baum, & Stanghellini 1991; Fanti *et al.* 1990a,b). The GPS sources have sub-kpc radio sources and turnovers at frequencies of-order a GHz; while the CSS sources have 1-10 kpc scale sizes and turnovers at frequencies ~ one-to-several hundred MHz. They comprise approximately 5-10% (GPS) and ~ 20% (CSS) of radio-selected AGN samples, and are thought to be isotropically-emitting radio sources whose small size reflects their confinement by exceptionally dense circumnuclear gas (e.g., Fanti *et al.* 1990a; O'Dea, Baum, & Stanghellini 1991). It is possible that the dense ambient medium would absorb and reprocess an unusually high fraction of the optical and UV output of the AGN in GPS-CSSs, making them abnormally strong MFIR sources. However, it has also been suggested (e.g., Gopal-Krishna and Wiita 1991; De Young 1993) that jets which propogate through a medium of high density will have convert jet mechanical energy into radio synchrotron emission with unusually high efficiency. In this case a GPS-CSS source of a given radio power will be an intrinsically less luminous AGN than would another equally radio-powerful non-GPS-CSS AGN. This argument might then suggest that the GPS-CSS sources would be deficient in their MFIR emission. The availibility of the IRAS database and new large samples of GPS-CSSs, allows us to test both these ideas.

Our GPS-CSS galaxies and quasars came mainly from the samples defined by Fanti *et al.* (1990a) and Stanghellini *et al.* (1990). The comparison samples of non-GPS-CSS radio galaxies and quasars that matched the corresponding GPS-CSS samples in mean redshift and radio power were constructed from the 3CR catalog (after excluding all the GPS and CSS sources, and also those few sources with complex or inverted radio spectra - e.g. 3CR273, 3CR345, and 3CR454.3).

The Fanaroff-Riley Class I and Class II Radio Galaxies

There are clear differences in the environments and global properties of FRIs and FRIIs (e.g., Baum, Heckman & van Breugel 1992 and references therein, Zirbel *et al.* and Morganti *et al.* in these proceedings; however, see also Owen, these proceedings).

The lack of strong optical manifestations of nuclear activity in FRIs (cf. Hine & Longair 1979) has led to the idea that FRIs may have a qualitatively different sort of central engine than other types of AGNs - one that directs most of its output into the jet and produces relatively little optical and UV radiation. For most well-studied AGN classes, the MFIR is one of the most energetically-important spectral regions (e.g., Sanders *et al.* 1989). It is therefore of interest to compare the MFIR properties of FRI and FRII radio galaxies within the range of radio luminosity where the two classes overlap (between about 10^{41} and 10^{43} ergs s^{-1}). The FRI and FRII samples were compiled from the following references: Laing, Riley, & Longair (1983); Owen & Laing (1989); Fanti *et al.* (1987); and Caganoff (1989). Radio sources with intermediate morphologies, including members of the 'fat double' class were excluded from all samples. We constructed samples that provided a good match in redshifts and radio power between the two classes, while simultaneously including as many radio galaxies as possible.

The Broad-Line Radio Galaxies and Narrow-Line Radio Galaxies

The BLRGs and NLRGs are hypothesized to be drawn from the same parent population, but the BLRGs/RLQs are viewed more nearly parallel to the polar axis of a central dusty torus (which in turn aligns with the radio jet axis), while the NLRGs are viewed more nearly in the equatorial plane of the torus (e.g., Barthel 1989).

Heckman, Chambers, & Postman (1992, hereafter HCP) used IRAS MFIR data for high- redshift ($z > 0.3$), high-luminosity samples of RLQs and NLRGs drawn from the 3CR sample. HCP concluded that the MFIR emission (at typically 6 to 60 μ in the AGN rest-frame) is about four times stronger on-average in the RLQs than in NLRGs with the same mean redshift and low-frequency radio flux. HCP concluded that at least one of the following two statements must be true. (1) The RLQs are intrinsically stronger sources of MFIR emission than the NLRGs. If true, then the simple unification scheme must be incorrect or incomplete. (2) The MFIR radiation from RLQs and NLRGs is not emitted isotropically. If true, then the popular model (Sanders *et al.* 1989; Pier & Krolik 1992,1993) in which the MFIR is thermal emission from dust requires that the radiation source is both highly nonspherical and is also optically thick at wavelengths below about 60 μ.

Here we present a similar comparison, but focus on a low-redshift sample (3CR radio galaxies with z = 0.05 to 0.30). The advantages are the following. (1) We observe the sample objects at longer rest wavelengths. This is important, because the central IR source will become increasingly optically thin (and will hence radiate more nearly isotropically) at longer IR wavelengths (Pier & Krolik). (2) Our sample is of lower radio and infrared luminosity than the HCP sample. Thus, we can check for a luminosity dependence in the wavelength behavior of the flux density ratio broad/narrow line objects.

RESULTS AND IMPLICATIONS

Our results for the samples are presented in the form of ratios of the IRAS flux densities which are normalized by the mean radio flux densities of the samples (Table II).

TABLE II MFIR flux density ratios normalized by mean radio flux densities.

Sample[a] (1)	Type [b] (2)	12μ (3)	25μ (4)	60μ (5)	100μ (6)
GPS-CSSRG / RG	Mean	...	...	1.41 ± 0.64	< 2.37
GPS-CSSRG / RG	Median	...	< 0.79	< 0.91	< 1.69
GPS-CSSQSR / QSR	Mean	...	0.31 ± 0.21	0.75 ± 0.21	0.21 ± 0.16
GPS-CSSQSR / QSR	Median	< 0.28	0.44 ± 0.41	0.32 ± 0.17	0.34 ± 0.29
FRII /FRI	Mean	< 1.34	> 2.86	5.12 ± 1.23	2.82 ± 0.62
FRII /FRI	Median	1.32 ± 0.42	> 1.53	4.85 ± 1.18	> 3.88
BLRG /NLRG	Mean	5.50 ± 2.24	4.12 ± 1.05	1.22 ± 0.20	0.67 ± 0.32
BLRG /NLRG	Median	3.74 ± 1.51	4.16 ± 1.38	2.07 ± 0.56	1.28 ± 0.54
HCP low-z QSR/RG	Mean	...	> 2.80	3.95 ± 1.57	6.03 ± 2.81
HCP low-z QSR/RG	Median	...	2.59 ± 1.22	2.51 ± 0.90	1.88 ± 0.96
HCP total QSR/RG	Mean	> 1.18	2.65 ± 1.16	3.16 ± 1.04	2.81 ± 1.69
HCP total QSR/RG	Median	> 1.60	1.83 ± 0.69	1.51 ± 0.48	1.25 ± 0.43

[a]Sample.
[b]Type of statistic used – straight mean and median

GPS-CSS galaxies and quasars

The GPS-CSS *quasars* and comparison samples were detected at 25, 60, and 100 μ. Within the uncertainties, we can conclude that the GPS-CSS quasars and non-GPS-CSS quasars have mean and median MFIR fluxes that are similar, though the GPS-CSS quasars may be about a factor of two weaker than the comparison sample. In the case of the galaxies, there are only marginal detections of the GPS-CSS sample at 60 μ and of the comparison sample at 60 and 100 μ. We can only say that the mean flux densities at 60 & 100 μ for the GPS-CSS and comparison galaxies are similar to within a factor of about two or three.

Thus, the data are consistent with the MFIR properties of GPS-CSS galaxies and quasars being *comparable* to comparison samples of extended sources. This result is consistent with all of the following. (1) Jet efficiency is not very different in the GPS-CSS sources and the comparison samples, i.e., our results are consistent with any difference in jet efficiency being no more than a factor of about two. (2) GPS-CSS sources do not reprocess significantly more optical-uv radiation than 'normal' extended galaxies and quasars. Thus, either (a) GPS-CSS central engines and nuclear environments are similar to those in the comparison samples; or (b) their central engines are in denser environments, but there is only a modest increase in covering factor which is not enough to increase the reprocessed radiation significantly. (Since 'normal' galaxies and quasars already have moderate covering factors, it may not be possible to gain a large factor). (3) At least some of the GPS-CSS sources could be young "3CR-type" radio sources.

FRI and FRII galaxies

The FRIIs are detected at better than 5σ at 25, 60, and 100 μ. In contrast, the FRI sample is only marginally detected ($\sim 3\sigma$) at 60 and 100 μ. Using the detections at 60 and 100 μ we conclude that FRII/FRI $= 3.8 \pm 0.8$. Thus,. we find yet one more difference between FRIs and FRIIs, i.e., The FRIIs are more powerful in the MFIR.

BLRG and NLRG galaxies

The BLRG/NLRG flux density increases monotonically with decreasing wavelength. While the ratio is about five at 12 μ, it decreases to near unity at 100 μ. HCP conducted a similar analysis of samples of 3CR radio galaxies and quasars at $z > 0.3$, and found that the quasars were consistently brighter by a factor of two to four in all the IRAS bands. Taking into account the uncertainties of the individual ratios the simplest interpretation is that the HCP results are consistent with the new results, and that the flux ratio is indeed an inverse function of MFIR rest-wavelength. However, it is also possible that there is an intrinsic luminosity (or redshift) dependence of the MFIR properties of radio-loud AGNs.

Our results suggest that the MIR in the low z, low luminosity BLRGs and NLRGs is *anisotropic*. The optically thick torus models of Pier & Krolik (cf. Konigl these proceedings) are qualitatively consistent with this. Comparing our results with the higher z, higher luminosity samples of HCP suggests that there may be a luminosity effect. In the higher luminosity objects, the anisotropy may persist out to longer wavelength. This would require the following. (1) The existence of an additional source of FIR flux density in the high luminosity quasars. This would be inconsistent with the simple unifying scheme. (2) The FIR is also anisotropic in the high luminosity objects. This is not currently accounted for in the Pier & Krolik models and may require a change in the geometry of their torus in the higher luminosity objects. Alternately, the high luminosity objects may have a non-thermal (blazar) component to the MFIR.

ACKNOWLEDGMENTS

We thank Ski Antonucci for helpful discussions. T. Heckman received support from the NASA ADP (NAG5-1368). Chris O'Dea received support from the STScI visitor program for a visit by E. Laurikainen. IPAC is funded by NASA as part of the IRAS extended mission program under contract to JPL.

REFERENCES

Antonucci, R. 1993, *ARA&A*, in press

Barthel, P. D. 1989, *ApJ*, 336, 606

Baum, S. A., Heckman, T. M., & van Breugel, W. 1992, *ApJ*, 389, 208

Caganoff, S. 1989, PhD dissertation, The Australian National University

De Young, D. S. 1993, *ApJ*, 402, 95.

Fanti, C., Fanti, R., de Ruiter, H. R., & Parma, P. 1987, *A&AS*, 69, 57

Fanti, R., Fanti, C., Schilizzi, R. T., Spencer, R. E., Rendong, N., Parma, P., van Breugel, W. J. M., & Venturi, T. 1990a, *A&A*, 231, 333

Fanti, C., Fanti, R., O'Dea, C. P., & Schilizzi, R. T. (editors), 1990b, Proceedings of the Dwingeloo Workshop *Compact Steep Spectrum and GHz Peaked Spectrum Radio Sources*, (Bologna: Istituto di Radioastronomia)

Golombek, D., Miley, G. K., & Neugebauer, G. 1988, *AJ*, 95, 26

Gopal-Krishna & Wiita, P. J. 1991, *ApJ*, 373, 325

Heckman, T. M., Chambers, K. C., & Postman, M., 1992, *ApJ*, 391, 39

Hine, R. G., & Longair, M. S. 1979, *MNRAS*, 188, 111

Impey, C., & Gregorini, L., 1993, *AJ*, 105, 853

Laing, R. A., Riley, J. M., & Longair, M. S., 1983, *MNRAS*, 204, 151

O'Dea, C.P., Baum, S.A., & Stanghellini, C. 1991, *ApJ*, 380, 66

Owen, F. N. & Laing, R. A. 1989, *MNRAS*, 238, 357

Pier, E. A. & Krolik, J. H., 1992, *ApJ*, 401, 99

Pier, E. A. & Krolik, J. H., 1993, *ApJ*, in press

Sanders, D. B., Phinney, E. S., Neugebauer, G., Soifer, B. T., & Matthews, K., 1989, *ApJ*, **347**, 29

Stanghellini, C., O'Dea, C. P., Baum, S. A., & Fanti, R. 1990, in proceedings of the Dwingeloo Workshop *Compact Steep Spectrum and GHz Peaked Spectrum Radio Sources*, eds, C. Fanti, R. Fanti, C. P. O'Dea, & R. T. Schilizzi, (Bologna:Istituto di Radioastronomia), 17

The First Stromlo Symposium: The Physics of Active Galaxies
ASP Conference Series, Vol. 54, 1994
G. V. Bicknell, M. A. Dopita, and P. J. Quinn (eds.)

UNIFICATION OF BL LAC OBJECTS AND FR 1 RADIO GALAXIES

C. MEGAN URRY
Space Telescope Science Institute, 3700 San Martin Drive, Baltimore, MD 21218, USA

PAOLO PADOVANI
Department of Physics, II University of Rome, Via della Ricerca Scientifica 1, I-00133 Rome, Italy

ABSTRACT We discuss the hypothesis that BL Lac objects are Fanaroff-Riley type 1 radio galaxies whose emission is dominated by a relativistic jet roughly aligned with our line of sight. A critical test is a comparison of the relative numbers of FR 1 radio galaxies and BL Lac objects, which can be done by comparing observed and predicted luminosity functions. The present data support this unification scheme and our analysis yields estimates for the bulk Lorentz factor and the angle to the line of sight of the jet emission at different wavelengths. Increasing collimation of the jet radiation with increasing wavelength is indicated both by our analysis and by a direct examination of the X-ray number counts of radio- and X-ray-selected BL Lac objects. In contrast to the success of the beaming hypothesis, there are problems with the competing picture of BL Lac objects as microlensed optically violently variable (OVV) quasars. Systematic differences between BL Lac objects and OVV quasars argue against microlensing or any other direct connection between the two AGN classes. Finally, there is no evidence that BL Lac objects need be further separated into a low-redshift, low-luminosity, narrow-lined class and a high-redshift, high-luminosity, broad-lined, OVV-type class.

INTRODUCTION

There is considerable evidence that relativistic beaming is important in at least some AGN, notably the blazars (BL Lac objects and optically violently variable quasars). Classic arguments include superluminal velocities, jet asymmetries and systematic depolarization asymmetries, high brightness temperatures (the "Compton catastrophe"), and correlations of various beaming-affected quantities with the ratio of core to extended radio power (the parameter R). Interpretation of any of these in terms of relativistic beaming is satisfying but not compelling, as each can probably be explained in other ways as well. For example, superluminal velocities do not require the emitting plasma to be moving relativistically nearly along the line of sight; instead, a moving pattern and/or the movement of a light beam along a screen could produce the same effect. The recent detections of high energy gamma-rays from a large number of blazars with the Compton

Gamma-Ray Observatory EGRET experiment, however, provides an essentially model-independent argument for relativistic beaming. Specifically, the gamma-ray variability puts a limit on the size scale and the observed gamma-ray flux gives an estimate of the implied luminosity. Together these must satisfy the requirement that the optical depth to pair production is small so that the gamma rays can escape, which in turn implies the gamma rays are relativistically beamed ($\gamma \gtrsim 3$). The gamma-ray detections make relativistic beaming an extremely compelling explanation for all of the observed characteristics of blazars.

If one accepts the importance of relativistic beaming in at least some AGN, then inevitably the appearance of those AGN is strongly aspect dependent, with observed flux going as the power p (where $p \sim 2-3$; Lind & Blandford 1985) of the kinematic Doppler factor, $\delta = (\gamma(1-\beta\cos\theta))^{-1}$, where θ is the angle to the line of sight. Furthermore, there must exist a large number of identical but misaligned objects in which there need be no obvious evidence of relativistic beaming. For likely Lorentz factors, the ratio of beamed to misaligned objects is $\lesssim 10\%$ This ratio is sufficiently small that the space distribution of the parent population (all functionally equivalent objects, regardless of orientation) is effectively the same as the space distribution of the misaligned objects (parents minus beamed objects).

The idea that apparently different classes of AGN differ only in their orientation is the most thoroughly explored unification scheme. With the numerous gamma-ray detections of blazars, this scheme becomes considerably more compelling. *If relativistic beaming is important in AGN, then some form of unification of AGN has to hold.*

Given a class of AGN believed to be dominated by relativistic beaming, such as blazars, we can devise and test a unified scheme as follows. First, we identify the parent population on the basis of isotropic properties, such as host galaxy type, narrow-line luminosities, AGN environment, extended radio emission, far-IR or X-ray emission, etc. Given well-defined samples of the parent population and of the beamed objects, one can then make some consistency checks, such as comparing the average isotropic properties between the two samples, or comparing the numbers and luminosities of the two. Relativistic beaming makes clear predictions about the ratio of beamed to parent objects as a function of luminosity (Urry & Shafer 1984, Urry & Padovani 1991).

A "well-defined sample" almost invariably means a flux-limited sample, and in most cases, a flux limit introduces a selection bias, which in the case of relativistic beaming can be particularly strong. Most people working on unified schemes try to compare parent and beamed populations using "unbiased" samples, meaning samples selected by unbeamed, unobscured flux, typically low-frequency radio flux, or possibly far-infrared or hard-X-ray flux. For several reasons, we have chosen a different approach. First, any or all of these wavelengths can be significantly affected by beaming (e.g., beamed hot spots in giant radio lobes or significant contribution from beamed core flux in the low-frequency radio regime; beaming and/or partially optically thick obscuration in the far-infrared; partial obscuration in the X-ray, depending on column density). Second, even if the emission is wholly isotropic, the sample selection imposes restrictions on content. For example, in the 3CR catalog, there are very few flat-spectrum radio sources (which are the blazars). Therefore, using the 3CR sample to compare unbiased samples excludes blazars, making it impossible to test blazar unification

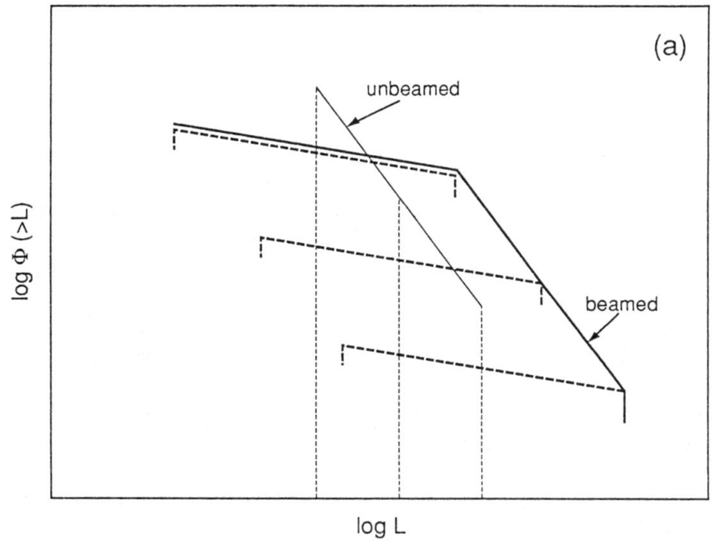

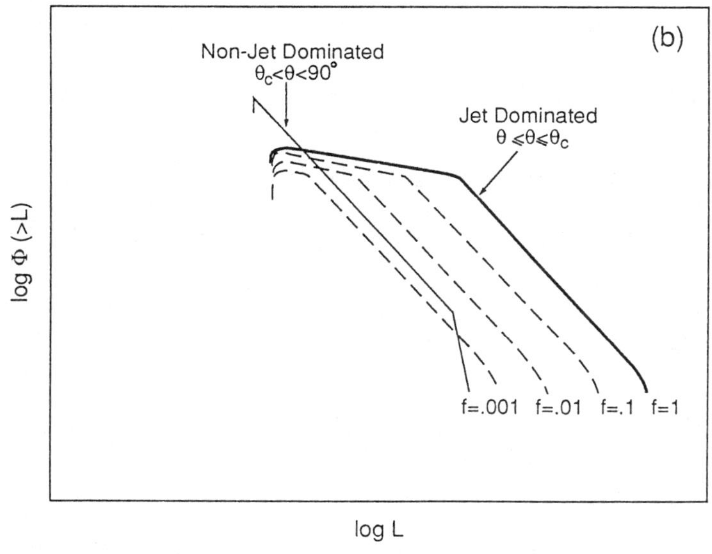

FIGURE I The effect of relativistic beaming on observed luminosity functions. *(a)* Simple case where all emitted flux comes from a randomly oriented relativistic jet. For a delta-function intrinsic luminosity (thin dashed line), the observed luminosity function after beaming (thick dashed line) is a flat power law of differential slope $1 \lesssim (p+1)/p \lesssim 1.5$ (Eqn. 1). A power-law intrinsic luminosity function (thin solid line), after integration of Eqn. (2), gives rise to an observed luminosity function that is a double power law (thick solid line), with flat slope $(p+1)/p$ at low luminosities and the same slope as the intrinsic power law at high luminosities. *(b)* Case where the intrinsic luminosity in the jet is a fraction $f = 0.001$, 0.01, 0.1, or 1 of the unbeamed luminosity. As in *(a)*, the beamed luminosity function has slope $(p+1)/p$ below the break and the parent luminosity function slope above the break. In both *(a)* and *(b)*, the high luminosity objects are dominated by those closest to the line of sight ($\theta \sim 0°$).

schemes. Another example is far-infrared samples, where isotropic re-radiation by dust is assumed, and thus intrinsic dust content determines sample content. Also, as anyone who has looked at the IRAS sample knows, the number of AGN detected in the faint point source catalog is much smaller than the number of galaxies, so pulling an AGN sample out of such a survey is painful. Finally, in the hard X-ray band, existing all-sky surveys include fewer than 100 AGN (e.g., the Piccinotti sample of $\sim$ 30 AGN); no future hard X-ray surveys are planned; and even if they were, the spatial resolution is limited, making sample identification difficult.

These considerations led us to conclude that to apply the key test of unified schemes — namely, to compare the numbers of beamed and parent objects — we must account for the biases and selection effects inherent in most well-defined samples. Our method uses a model-dependent description of effect of beaming on luminosity functions (Urry & Shafer 1984; Urry & Padovani 1991); it is not strongly sensitive to the details of the beaming model and is fairly robust. We have applied our technique to the unification of radio-loud quasars and luminous radio galaxies (Padovani & Urry 1992), as well as to BL Lac objects and low-luminosity radio galaxies (Padovani & Urry 1990, 1991; Urry, Padovani & Stickel 1991). The latter unified scheme is discussed in detail below.

A UNIFIED SCHEME FOR BL LAC OBJECTS AND FR 1 RADIO GALAXIES

Connection between BL Lacs and FR 1s

It has been known for some time that the unbeamed properties of BL Lac objects are similar to those of Fanaroff-Riley type 1 (FR 1) radio galaxies (Wardle et al. 1984, Antonucci & Ulvestad 1985, Browne 1989, Ulrich 1989). Specifically, the extended radio powers in BL Lac objects are in the range $10^{23} - 10^{26}$ W/Hz, the host galaxies are typically luminous ellipticals, and emission lines (when observed) are usually narrow lines with luminosities typical of those in radio galaxies (see Padovani 1992a). FR 1 galaxies are often found in clusters of galaxies, as are many BL Lac objects (Falomo et al. 1993; Wurtz et al. 1993).

It was also realized early on that if BL Lac objects represented a special orientation, the number density of the parents would have to be significantly higher than the number density of BL Lacs, and that low-luminosity radio galaxies were a promising candidate population on that basis (Schwartz & Ku 1983; Browne 1983; Perez-Fournon & Biermann 1984); however, the first comparisons of number densities did not correct adequately for the relativistic beaming. Incorporating the effects of beaming, we compared the luminosity functions of BL Lac objects and FR 1 galaxies in a recent series of papers addressing radio-selected and X-ray-selected samples, as well as the scant optical information available (Padovani & Urry 1990, 1991; Urry, Padovani & Stickel 1991). The results from these studies are described below.

The Effect of Relativistic Beaming on Luminosity Functions

Our comparison of number densities is based on understanding the effect of relativistic beaming on the observed luminosity functions (Urry & Shafer 1984).

Consider an ensemble of emitters all having the same intrinsic luminosity ($\mathcal{L}$) and all moving with the same relativistic bulk speed (β) at random angles (θ) to the line of sight. The probability of having a particular Doppler factor is $P(\delta) = d(\cos\theta)\ d\theta/d\delta = (\beta\gamma\delta^2)^{-1}$. The probability of observing luminosity L given intrinsic (emitted) luminosity $\mathcal{L}$ is then

$$P(L|\mathcal{L}) = P(\delta)\,\frac{d\delta}{dL} = \frac{1}{\beta\gamma\delta}\,\mathcal{L}^{1/p}\,L^{-(p+1)/p}\ . \qquad (1)$$

For fixed $\mathcal{L}$, as in the present example, the distribution of observed luminosities is a flat power law (with index in the range 1-1.5) extending from $L \sim (2\gamma)^{-p}\mathcal{L}$ to $L \sim (2\gamma)^{p}\mathcal{L}$. (See figure Ia.) The low-luminosity cutoff corresponds to an object moving directly away from us ($\theta = 180°$) and the high-luminosity cutoff to perfect alignment ($\theta = 0°$). The normalization of this flat power law decreases with increasing beaming (higher β, γ, or p) because the beaming cone angle gets smaller.

For a distribution of intrinsic luminosities (i.e., a luminosity function), the observed luminosity function is just the integral of the intrinsic luminosity distribution with the conditional probability function in Eqn. (1):

$$\Phi_{obs}(L) = \int d\mathcal{L}\ P(L|\mathcal{L})\ \Phi_{intr}(\mathcal{L})\ . \qquad (2)$$

This integral is basically the envelope of the beamed luminosity functions for fixed intrinsic luminosities, shown as a thick solid line in Figure Ia, where the intrinsic luminosity function is shown as a thin solid line.

In practice one expects an unbeamed component to be present as well. If we use the simple parametrization that the intrinsic luminosity in the jet is a fixed fraction of the unbeamed luminosity, $\mathcal{L}_{jet} = f\mathcal{L}_{unbeamed}$, then

$$L = \mathcal{L}_{unbeamed} + L_{jet} = \mathcal{L}_{unbeamed}(1 + f\delta^p)\ . \qquad (3)$$

If we define the critical angle, θ_c, to be where the beamed and unbeamed luminosities are comparable (i.e., $f\delta^p = 1$), then for $\theta < \theta_c$ the luminosity will be dominated by beamed emission and we can identify these sources as blazars. The observed parent and beamed luminosity functions for the case of a single-power-law intrinsic luminosity function are shown in Figure Ib, for four values of f.

The key point is that the luminosity function of the beamed population has a characteristic broken power law form, flat at the low luminosity end and steep at the high luminosity end. This remains approximately the case even when the intrinsic luminosity function has a more complicated form, without sharp cutoffs (Urry & Padovani 1991). This means that the comparison of the number densities of beamed and parent populations is a strong function of luminosity, and can be evaluated only by measuring the luminosity functions for each.

Population Statistics for the FR 1/BL Lac Unified Scheme

To test whether the observed number densities are commensurate with the hypothesis that BL Lacs are the aligned versions of FR 1s, we derive the FR 1 luminosity function, beam it according to the prescription outlined above, and compare the predicted and observed BL Lac luminosity functions. The free parameters are γ, the bulk Lorentz factor of the jet, and f, the fractional luminosity of the jet. We can further constrain these two parameters either from R, the ratio of beamed to unbeamed flux (core to extended flux), or from the ratio of maximum BL Lac luminosity to maximum FR 1 luminosity (the approach we used for the X-ray comparison, since X-ray spatial resolution is insufficient to measure R).

Population Statistics for X-Ray Samples

Several years ago, sizable well-defined samples of BL Lac objects became available. The first such samples were from X-ray surveys, including the HEAO1-A2 all-sky survey (only 5 BL Lac objects; Piccinotti et al. 1982), the Einstein Medium Sensitivity Survey (30 BL Lacs; Maccacaro et al. 1989), and the EXOSAT High Galactic Latitude Survey (11 BL Lacs; Giommi et al. 1989). The last two of these, which went down to fairly low X-ray fluxes, $\sim 2 \times 10^{-13}$ ergs cm^{-2} s^{-1} in the 0.3-3.5 keV band, revealed a decided flattening of the counts at low flux, in marked contrast to the steep X-ray logN-logS curves for other kinds of AGN. We compared predictions of the FR 1 unified scheme with the X-ray counts from these three samples and found good agreement for a bulk Lorentz factor $\gamma_x \sim 3$ (Fig. II; Padovani & Urry 1990). The ratio of the total number of BL Lac objects to FR 1 galaxies is ~ 0.14 — according to the best-fit luminosity function, more than 90% of all BL Lac objects are at lower X-ray luminosities than those currently observed — and the critical angle separating BL Lacs from FR 1s is $\theta_c \sim 30°$.

Our comparison of X-ray samples was done in terms of the counts, which represent the combination of luminosity function plus cosmology plus evolution, rather than luminosity functions, because determining the latter from the Einstein Medium Sensitivity Survey (EMSS) is complicated. Since then the EMSS-derived X-ray luminosity function of BL Lac objects has been published (Morris et al. 1991) and our results do not change much if we fit that luminosity function directly. The principal uncertainties in this calculation are due to uncertain and/or incomplete redshift information for the BL Lac objects, lack of X-ray data for FR 1 radio galaxies, and the limited statistics of the small BL Lac and FR 1 samples. Some of these issues will be addressed by new X-ray samples now being optically identified and this will enable us to greatly refine this calculation.

Population Statistics for Radio Samples

The 1 Jy sample of BL Lac objects, a radio-selected sample of size comparable to the EMSS X-ray-selected sample, has also become available recently (Stickel et al. 1991). These BL Lac objects were selected from a catalog of 518 radio sources with 5 GHz fluxes ≥ 1 Jy on the basis of their flat radio spectrum ($\alpha_r \leq 0.5, f_\nu \propto \nu^{-\alpha}$) and weak or absent emission lines (rest-frame equivalent width ≤ 5 Å). Of the 34 1 Jy BL Lacs in the sample, 25 have redshift information from emission lines, 4 have lower limits to their redshifts from absorption lines, while in 5 objects neither emission or absorption lines have been detected but a lower limit of $z > 0.2$ can be estimated from their stellar appearance.

The radio data are better suited than the X-ray data for testing the beaming

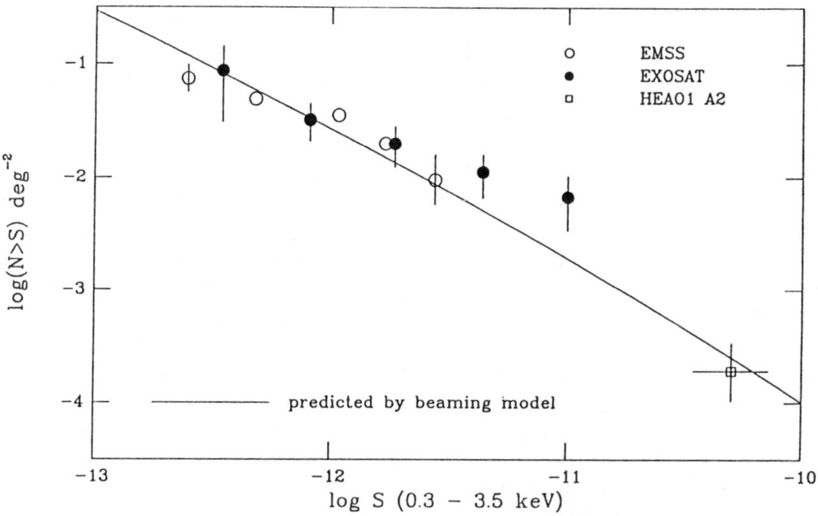

FIGURE II Integral X-ray counts of BL Lacertae objects. The fit derived from the predicted beamed luminosity function (solid line) is compared to data from three surveys: EMSS (*open circles*), EXOSAT, $\alpha = 1.5$ (*filled circles*), HEAO1-A2 (*open squares*). Details in Padovani & Urry (1990).

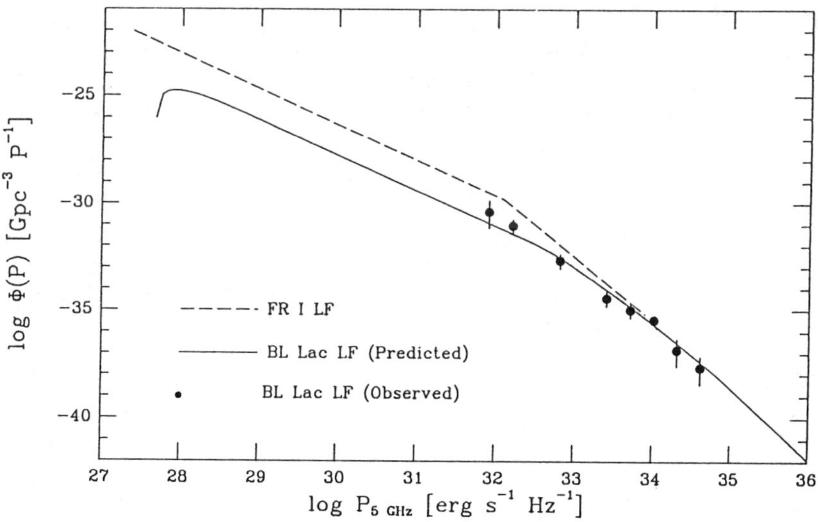

FIGURE III The differential radio luminosity function of BL Lacertae objects. Prediction of the beaming model (*solid line*) is compared to observational estimates (*filled circles*) from Stickel et al. (1991). The observed differential luminosity function of FR 1 galaxies is also shown (*dashed line*). Details in Urry, Padovani & Stickel (1991).

hypothesis for BL Lac objects. For one thing, the radio sample of FR 1 galaxies is larger than the X-ray sample, and while the BL Lac samples are comparable in size, the radio sample has a higher percentage of known redshifts. Also, the radio flux limit is fixed, so the derivation of the radio luminosity function of BL Lac objects is straightforward and the comparison with data is direct. Finally, the higher spatial resolution available with radio interferometry allows estimates of the ratio between beamed and unbeamed radio flux (Antonucci and Ulvestad 1985), adding an important constraint to the combination of the parameters γ and f.

Perhaps because of this additional information, it is impossible to fit the observed radio luminosity function of BL Lacs by beaming the radio luminosity function of FR 1s with a single Lorentz factor (Urry, Padovani & Stickel 1991). Instead, we require a distribution of Lorentz factors in the range $5 \lesssim \gamma_r \lesssim 30$. While the form of the distribution is not really constrained, it is weighted toward the lower values; e.g., for a power law of the form $N(\gamma_r) \propto {\gamma_r}^G$, the best-fit index, $G \sim -4$. The predicted (solid line) and observed (solid circles) radio luminosity functions for the 1 Jy sample are compared in Fig. III. The mean Lorentz factor (approximately independent of the exact form of the distribution), $\langle\gamma_r\rangle \sim 8$, corresponding to a ratio between BL Lacs and FR 1s $\sim 1:70$, about an order of magnitude smaller than in the X-ray case. The total number of BL Lacs is most sensitive to the lowest value of γ_r, while the peak ratio of beamed to unbeamed flux is sensitive to the highest value. Our best-fit model predicts that radio-selected BL Lacs are aligned within $\theta_c \sim 10°$ of the line of sight.

Additional information on bulk Lorentz factors is available from measurements of superluminal motion. All eleven BL Lac objects that have been observed more than once with VLBI techniques show superluminal motion. For the BL Lacs the inferred Lorentz factors are higher than ~ 4, and are probably distributed to at least $\gamma_r \sim 15$. These results agree roughly with the distribution in γ_r required to fit the luminosity function and core-halo ratios of radio-selected BL Lacs (Urry, Padovani & Stickel 1991).

Relative Population Statistics for X-Ray and Radio Samples

We note that a distribution of Lorentz factors might well be required for X-ray-selected samples if the available data were better. In any case, the mean Lorentz factor derived for the radio-selected sample is considerably higher than for the X-ray-selected sample. This can be seen directly, independently of the beaming model assumed, from the X-ray counts of X-ray-selected and radio-selected BL Lac objects. The number density of X-ray-selected BL Lac objects on the sky is 10 times the number density of radio-selected BL Lac objects for comparable X-ray fluxes (Urry, Padovani & Stickel 1991.) This too is consistent with the picture of RBLs being more beamed, and therefore more rare.

A possible explanation for the apparent wavelength dependence of the beaming solid angle is that the Lorentz factor increases along the jet. This also explains the observed broad-band spectral characteristics of X-ray- and radio-selected BL Lac objects (Ghisellini et al. 1985; Maraschi et al. 1986; Ghisellini & Maraschi 1989). Alternatively, a fan beam with fixed Lorentz factor but increasing collimation from X-ray to radio can also explain the data (Celotti et al. 1993).

Optical data for BL Lac objects are still so sparse that no strong constraints can be derived, although a consistent picture, with $\gamma \sim 10$ (very roughly), is allowed (Padovani & Urry 1990).

IMPLICATIONS OF THIS UNIFIED SCHEME FOR BLAZARS

Comparison of X-Ray-Selected and Radio-Selected BL Lac Objects

The properties of X-ray-selected and radio-selected BL Lac objects are systematically different (Ledden & O'Dell 1985, Stocke et al. 1985, Maraschi et al. 1996, Padovani 1992a). On average, X-ray-selected BL Lacs have lower polarization, less variability, and higher starlight fractions (e.g. Stocke et al. 1985, Morris et al. 1991, Jannuzi et al. 1993), while radio-selected BL Lacs are more polarized, more variable, and more luminous. Radio-selected BL Lacs have roughly the same X-ray luminosities as X-ray-selected BL Lacs (on average, a factor of four higher), but 10-100 times (or even more) the radio luminosities (Maraschi et al. 1986). Comparison of X-ray and radio-selected BL Lac samples to an FR 1 sample suggests that in fact the radio emission in radio-selected BL Lac objects is "more beamed", i.e., boosted by a higher Lorentz factor, than in the X-ray-selected BL Lac objects Padovani (1992a).

Since the X-ray luminosities of the two BL Lac "types" are comparable, X-ray surveys should be relatively unbiased (or if anything, should find more radio-bright BL Lacs since they are slightly more luminous). In fact, radio-bright BL Lacs are very rarely detected in X-ray surveys, implying that they are intrinsically rare objects. Consider a jet where the X-ray beam is broader than the radio beam, for whatever reason. The probability of intersecting the X-ray beam is considerably higher, so the number density of such objects (typical, radio-faint, X-ray-selected BL Lacs) would be higher than the number density of objects for which our line of sight falls in the radio beam (radio-bright BL Lacs). That is, the observed difference in number densities of radio-selected and X-ray-selected BL Lac objects can be explained easily if $\gamma_x < \gamma_r$.

Comments on Gravitational Microlensing

Determination of the redshifts of BL Lac objects is not easy as they are by definition line-less objects. Often redshifts are measured from absorption features in the galaxy surrounding the point-like BL Lac nucleus. Ostriker and Vietri (1985, 1990) suggested that the apparent host galaxy in these cases is instead an intervening, lensing galaxy, and that BL Lac objects are actually background optically violently variable (OVV) quasars with their continuum emission greatly amplified relative to the line emission through microlensing by stars in the foreground galaxy; hence the low equivalent line widths. They estimated that the magnification of the compact, microlensed continuum source relative to the line emission has to be about a factor of 10. In this picture, redshifts measured from the putative host galaxies would give an erroneously local distribution of the BL Lac objects.

We now know that the redshift distribution of BL Lac objects is not anomalous, but can be easily explained by a beaming model. Moreover, there are various problems with the microlensing picture: (1) the optical amplification factors derived from numerical simulations under the lensing hypothesis are much smaller than the required factor of 10 to swamp the emission lines of quasars (Padovani 1992b); (2) the observed optical number counts of BL Lacs (Padovani and Urry 1991; Hawkins et al. 1991) probably do not flatten at $B \sim 16-18$ as predicted; (3) the host galaxies of BL Lacs appear always centered on the BL Lac object itself (e.g., Abraham et al. 1991), while the microlensing model predicts some to be off-center; and (4) the differences in VLBI polarization structure, that is on milliarcsecond scales, should not be present if a BL Lac were simply a microlensed OVV quasar (Gabuzda et al. 1992).

In summary, microlensing may be important in some BL Lacs — for example, the high-redshift ones with known intervening absorption systems (e.g., Stickel et al. 1991) — but it is unlikely to be a widespread phenomenon.

The Relation of OVV Quasars to BL Lac Objects

Radio selected BL Lacs are defined as flat spectrum radio sources with a rest-frame equivalent width of the strongest line of less than 5 Å. Similar criteria were used for the selection of the EMSS BL Lac sample (Stocke et al. 1991), although the equivalent width limit was based on observed rather than rest-frame equivalent width (the sample is sufficiently low in redshift that the distinction is minor). Although there are undoubtedly some borderline objects in which the equivalent widths of emission lines exceed 5Å when the continuum is in a low state, this criterion seems to separate the two classes quite well (see discussion in Stickel et al. 1991). Some have suggested there might be two populations of BL Lacs: those associated with nearby ellipticals, showing spectral lines which can be considered to indicate the presence of a galaxy, typically at $z \lesssim 0.3$, and those with weak but broad emission lines, more distant ($z \gtrsim 0.3$), to be associated with quasars (Burbidge and Hewitt 1989).

The existence of two types of BL Lac objects — a local, low-luminosity, narrow-lined, "true" BL Lac population and a more distant, higher-luminosity, broad-lined quasar-like population — does not seem to be required by the data, however. There are systematic differences in the properties of BL Lac objects and OVV quasars, quite apart from the optical emission line properties. For example, the average soft X-ray spectral index for BL Lacs is $\alpha_x \sim 1.0$, while OVVs have $\alpha_x \sim 0.5$ (Worrall and Wilkes 1990). Even BL Lacs with faint but broad emission lines have X-ray spectra more similar to the other BL Lacs than to the OVVs. Similarly, BL Lacs have a characteristic VLBI polarization structure quite different from that of quasars, with magnetic fields perpendicular rather than parallel to the jets (Gabuzda et al. 1992). Furthermore, the milliarcsecond polarization structure of low- and high-redshift BL Lacs, divided as Burbidge and Hewitt would prescribe, seem to be indistinguishable. Using the 1 Jy BL Lac sample, Padovani (1992b) has shown that the distributions of extended radio power for the low- and high-luminosity (low- and high-redshift) BL Lacs are not significantly different, while the same distributions for OVV quasars and high-redshift BL Lac objects differ at the 99.9% level. Another radio property, the ratio between the compact and the extended emission, is not significantly different for the low- and high-redshift BL Lacs. The [O III] line luminosities of OVVs and BL Lacs also differ by two orders of magnitude (Padovani 1992b). Even detection rates with GRO appear to differ, with OVVs more numerous in the γ-ray band (which is in qualitative agreement with the radio number counts). Finally, while some BL Lacs appear to be associated with FR 2 parents rather than FR 1s (Kollgaard et al. 1992), based on their large extended radio powers and morphologies, the evidence is still preliminary. In practical terms, it is not a problem for the scheme presented above because a few objects at the high luminosity end will not make much of a difference and the morphology of FR 1s is in any case pretty varied (Parma, these Proceedings)

SUMMARY

The best physical explanation for BL Lac objects is that they represent a special orientation of FR 1 radio galaxies with relativistically beamed jets. Relativistic beaming is definitely important in some blazars, and is very likely important in BL Lac objects. It then follows that, due to the strong anisotropy induced by the relativistic bulk motion, some form of unification has to be important.

The evidence for a connection between FR 1 radio galaxies and BL Lac objects is strong, both in terms of isotropic properties and in terms of luminosity functions of X-ray- and radio-selected BL Lac objects. One of last difficulties of the FR 1-BL Lac unification scheme — the presence of relativistic bulk motion in the cores of FR 1 galaxies — appears to be gaining acceptance on both observational and theoretical grounds (Parma, Venturi, Laing, Bicknell; these Proceedings). One problem still open is the relatively small Lorentz factors inferred for the few FR 1s with VLBI maps compared with the Lorentz factors inferred both from superluminal motion and from the fit to the luminosity functions for BL Lacs. More data on FR 1s are certainly needed to improve the statistics.

Different bulk Lorentz factors are required for radio- and X-ray-emitting regions within single BL Lac objects. This could be explained by increasing collimation of the radiation or by increasing collimation of the jet at longer wavelengths. At present both models seem to be viable.

ACKNOWLEDGMENTS

We thank the organizers of the First Mount Stromlo Symposium for a very stimulating and productive meeting.

REFERENCES

Abraham, R. G., McHardy, I. M., and Crawford, C. S. 1991, *MNRAS*, **252**, 482

Antonucci, R. J. R., and Ulvestad, J. S. 1985, *ApJ*, **294**, 158

Burbidge, G., and Hewitt, A. 1989, in *BL Lac Objects*, ed. L. Maraschi, T. Maccacaro, & M.-H. Ulrich (New York: Springer-Verlag), p. 412

Browne, I. W. A. 1983, *MNRAS*, **204**, 23p

Browne, I. W. A. 1989, in *BL Lac Objects*, ed. L. Maraschi, T. Maccacaro, & M.-H. Ulrich (New York: Springer-Verlag), p. 401

Celotti, A. *et al.* 1993, *ApJ*, **416**, 118

Falomo, R., Pesce, J. E., and Treves, A. 1993, *ApJ*, **411**, L63

Gabuzda, D. C. *et al.* 1992, *ApJ*, **388**, 40

Ghisellini, G., Maraschi, L., and Treves, A. 1985, *A&A*, **146**, 204

Ghisellini, G., and Maraschi, L. 1989, *ApJ*, **340**, 181

Giommi, P., et al. 1989, in *BL Lac Objects*, ed. L. Maraschi, T. Maccacaro, & M.-H. Ulrich (New York: Springer-Verlag), p. 231.

Hawkins, M. R. S., Véron, P., Hunstead, R. W., and Burgess, A. M. 1991, *A&A*, **248**, 421

Jannuzi, B. T., Smith, P. S., and Elston, R. 1993, *ApJ*, in press

Kollgaard, R. I., Wardle, J. F. C., Roberts, D. H., and Gabuzda, D. C. 1992, *AJ*, **104**, 1687

Ledden, J. E., and O'Dell, S. L. 1985, *ApJ*, **298**, 630

Lind, K. R., and Blandford, R. D. 1985, *ApJ*, **295**, 358

Maccacaro, T. *et al.* 1989, in *BL Lac Objects*, ed. L. Maraschi, T. Maccacaro, & M.-H. Ulrich (New York: Springer-Verlag), p. 222.

Maraschi, L., Ghisellini, G., Tanzi, E. G., and Treves, A. 1986, *ApJ*, **310**, 325

Morris, S.L., Stocke, J.T., Gioia, I.M., Schild, R. E., Wolter, A., Maccacaro, T., Della Ceca, R., 1991, *ApJ*, **380**, 49

Ostriker, J. P., and Vietri, M. 1985, *Nature*, **318**, 446

Ostriker, J. P., and Vietri, M. 1990, *Nature*, **344**, 45

Padovani, P. 1992a, *A&A*, **256**, 399

Padovani, P. 1992b, *MNRAS*, **257**, 404

Padovani, P., and Urry, C. M. 1990, *ApJ*, **356**, 75

Padovani, P., and Urry, C. M. 1991, *ApJ*, **368**, 373

Padovani, P., and Urry, C. M. 1992, *ApJ*, **387**, 449

Perez-Fournon, I., and Biermann, P. 1984, *A&A*, **130**, L13

Piccinotti, G. *et al.* 1982, *ApJ*, **253**, 485

Schwartz, D. A., and Ku, W. H.-M. 1983, *ApJ*, **266**, 459

Stickel, M. *et al.* 1991, *ApJ*, **374**, 431

Stocke, J. T. *et al.* 1985, *ApJ*, **298**, 619

Ulrich, M.-H., 1989, in *BL Lac Objects*, ed. L. Maraschi, T. Maccacaro, & M.-H. Ulrich (New York: Springer-Verlag), p. 45

Urry, C. M., and Padovani, P., 1991, *ApJ*, **371**, 60

Urry, C. M., Padovani, P., and Stickel, M. 1991, *ApJ*, **382**, 501

Urry, C. M., and Shafer, R. A. 1984, *ApJ*, **280**, 569

Wardle, J. F. C., Moore, R. L., and Angel, J. R. P. 1984, *ApJ*, **279**, 93

Worrall, D. M., and Wilkes, B. J. 1990, *ApJ*, **360**, 396

Wurtz, R., Ellingson, E., Stocke, J. T., and Yee, H. K. C. 1993, *AJ*, **106**, 869

The First Stromlo Symposium: The Physics of Active Galaxies
ASP Conference Series, Vol. 54, 1994
G. V. Bicknell, M. A. Dopita, and P. J. Quinn (eds.)

DECELERATING RELATIVISTIC JETS IN FRI RADIO SOURCES

R.A. LAING
Royal Greenwich Observatory, Madingley Road, Cambridge CB3 0HE, U.K.

ABSTRACT A model for the velocity and magnetic-field structure in FRI radio jets is presented. It is assumed that the jets are intrisically two-sided and symmetrical, and that they decelerate from relativistic speeds on parsec scales to non-relativistic speeds at large distances from the nucleus. The model jets consist of a transverse-field core surrounded by a slower, parallel-field shear layer. The predicted brightness and polarization structures are in good qualitative agreement with those observed.

THE MODEL

Unified schemes for FRI radio galaxies and BL Lac objects, observations of asymmetries in total intensity and depolarization in twin-jet sources and models of the energetics of decelerating flows (Urry, Parma *et al.*, Bicknell; these proceedings) all suggest that weak radio jets may have relativistic speeds on kiloparsec scales. The present paper considers the appearance of decelerating jets, in an attempt to understand the well-known relation between sidedness and field structure: two-sided jets have transverse apparent magnetic fields, whilst one-sided jets have longitudinal fields (Bridle & Perley 1984).

Two antiparallel, but otherwise identical jets propagate away from a galactic nucleus along a direction which makes an angle θ to the line of sight. The jet cores have velocity β_cc and contain a magnetic field which has no longitudinal component but is otherwise random (Laing 1980). The surrounding shear layer has a range of velocities β_sc ($\beta_s < \beta_c$). Its field is entirely longitudinal. In order to match observations, the shear layer is given a fraction f $\approx$ 0.1 of the total intensity of the core for a static jet in the plane of the sky. Two effects modify the appearance of the jet if its speed is a significant fraction of c. Aberration causes it to be observed at an angle differing from θ in its rest frame, so the emissivities of both components and the degree of polarization of the core are modified. Doppler boosting also changes the relative flux density of core and shear layer.

OBSERVATIONAL CONSEQUENCES

If FRI jets have field and velocity structures of this type and decelerate away from the nucleus, the observable consequences are as follows:

- Jets in the plane of the sky are precisely symmetrical. They have weak (Doppler-dimmed) $\mathbf{B}_{\parallel}$ bases on both sides of the nucleus, as shown in Figures I and II.

- Jets tilted by a moderate angle to the line of sight ($30° < \theta < 90°$) are one-sided close to the nucleus and undergo a field flip from longitudinal to transverse as they slow down, since the shear layer is slower than the core and suffers less beaming away from the line of sight (Figures III and IV). The approaching jet has a maximum in degree of (perpendicular) polarization where $\beta_c \approx \cos\theta$.

- As was pointed out originally by Komissarov (1988), the emission from both jet components is heavily suppressed close to the nucleus unless $\theta < 30°$, giving a possible explanation for the observed "gaps" (Bridle & Perley 1984). The rest-frame surface-brightness falls more rapidly with distance from the nucleus than appears from observation.

- Jets very close to the line of sight (within the beaming cone of the core) are one-sided, but have $\mathbf{B}_{\perp}$ (Figures V and VI). These are BL Lac objects in the unified models, consistent with their observed transverse polarization.

- $\mathbf{B}_{\perp}$ jets are always centre-brightened but $\mathbf{B}_{\parallel}$ jets are limb-brightened, since the emission from the centre must be suppressed for the longitudinal component to dominate. This is observed in at least some cases (M87, for example).

- Depolarization asymmetry is expected, as the lobe with the brighter jet base is seen through less magnetoionic material in the galaxy halo.

- The degree of asymmetry is expected to increase with total radio power (as the kinetic luminosity of the jet increases and it decelerates less rapidly) and with core prominence (which should decrease with the angle to the line of sight: Parma *et al.*, these proceedings).

SUMMARY

The hypothesis that jets in FRI sources decelerate from relativistic speeds on kiloparsec scales is consistent with many of their observed properties and especially with the correlation between sidedness and field direction. The present study emphasises the importance of velocity gradients along and across relativistic jets, which radically modify their appearance.

REFERENCES

Bridle, A.H. & Perley, R.A., 1984. *ARA&A*, **22**, 319.

Komissarov, S., 1988, *Ap&SS*, **150**, 59.

Laing, R.A., 1980. *MNRAS*, **193**, 439.

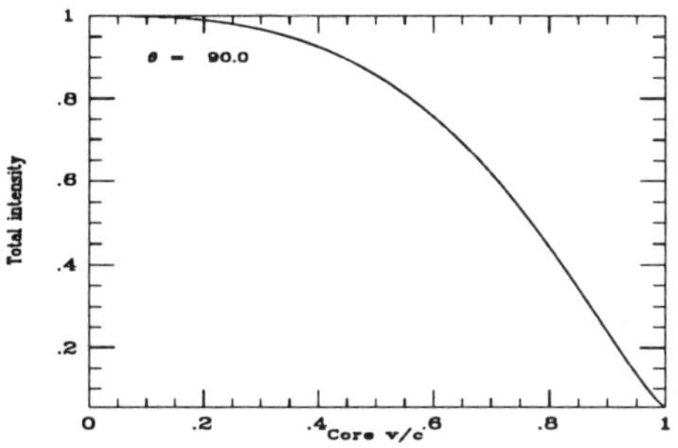

FIGURE I A plot of total intensity against core velocity for a jet at 90° to the line of sight. The total intensity has been integrated across the jet and the shear layer has 0.1 of the flux of the core in the rest frame.

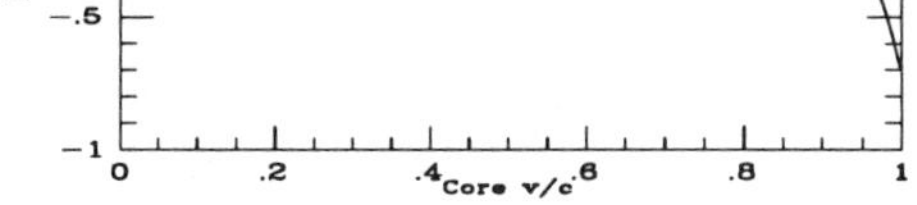

FIGURE II A plot of the degree of polarization against core velocity for a model jet at 90° to the line of sight, as in Figure I. Positive and negative degrees of polarization correspond to transverse and longitudinal magnetic fields, respectively.

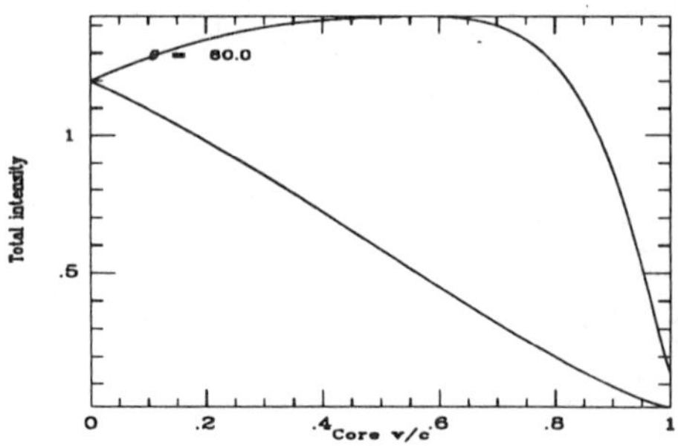

FIGURE III A plot of total intensity against core velocity for a pair of model jets at 60° to the line of sight. The approaching jet is represented by the upper curve.

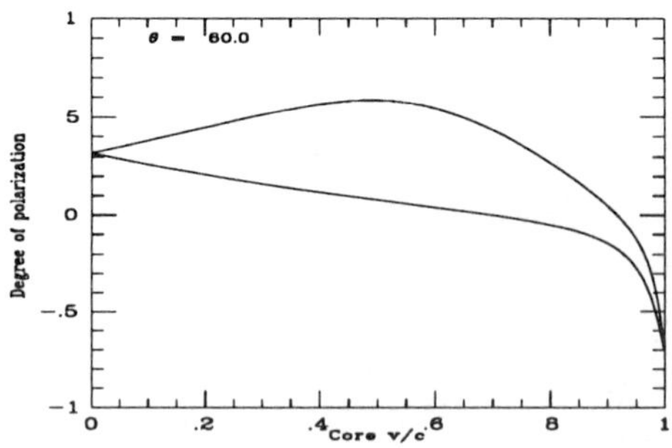

FIGURE IV A plot of degree of polarization against core velocity for model jets at 60° to the line of sight. The approaching jet is represented by the upper curve.

The First Stromlo Symposium: The Physics of Active Galaxies
ASP Conference Series, Vol. 54, 1994
G. V. Bicknell, M. A. Dopita, and P. J. Quinn (eds.)

BEAMED RADIATION IN RADIO GALAXIES

GEOFFREY V. BICKNELL
Mt. Stromlo & Siding Spring Observatories, Australian National University

ABSTRACT This paper focuses on two aspects of the Physics of Radio Galaxies in which it has been proposed that relativistic beaming of radiation is necessary to account for the observations. Physical arguments are presented which show that it is quite plausible that the jets in FRI radio galaxies are relativistic on the parsec scale and decelerate to subrelativistic flow on the kiloparsec scale. On the other hand it is shown that beaming of the optical continuum from the core of Centaurus A is not necessary to account for the excitation of the emission line filaments in the ISM of this galaxy. Models involving autoionizing shocks produced by the interaction between a low Mach number jet and dense clouds along its path are capable of explaining the morphology of the filaments and the simultaneous high excitation and low excitation lines in their spectra.

INTRODUCTION

The difference between FRI and FRII radio jets is often expressed in terms of turbulent, transonic and subrelativistic, for the former and supersonic and possibly relativistic for the latter. The kpc-scale emission from FRI's is thought to be largely unbeamed; beaming is widely thought to be implicated in the one-sided jet structures of class II radio sources and quasars. However, this standard dichotomy is not without some outstanding issues. For example, (1) The kpc-scale bases of FRI jets are frequently one-sided; (2) The small number of FRI jets that have been observed with VLBI on the pc-scale show a one-sided jet pointing to the brighter kpc-scale jet; (3) BL Lac objects, in which beaming and other relativistic effect are obviously important are believed to originate from a parent population of FRI radio galaxies, implying that the jets in the latter are relativistic on the parsec scale. (See, for example Ulrich, 1989.) (4) If FRI jets are relativistic on the parsec scale, why are the motions of knots in at least some pc-scale FRI jets, *sub*luminal (Reid *et al.*, 1989; Venturi *et al.*, 1993). An obvious solution to some of these problems is for FRI jets to decelerate from a relativistic to a subrelativistic velocity somewhere between the parsec and kiloparsec scales. Since deceleration is implicated in the physics of the kiloparsec scale structure anyway (e.g. Bicknell, 1986) this does not seem too radical a suggestion. Nevertheless, it remains to be shown that this is dynamically consistent.

Optical observations of emission line regions in radio galaxies present another series of challenges, for the resolution of which a number of workers have also sought an explanation involving the physics of beamed radiation. For example, there is a well known deficit between the ionizing flux detected from the core of a radio galaxy and the flux required to account for the extended Balmer line luminosity. Moreover, there is a serious problem in that the temperature of Extended Emission Line Regions (inferred from the [OIII]$\lambda4363/\lambda5007$ ratio) is higher than predicted by standard power-law photoionization models. In some cases the "beaming" may be induced by obscuration in a thick torus surrounding the active core; in others the beaming must be relativistic to be consistent with the lack of reradiated IR flux from a putative obscuring torus. In this paper I summarize a different approach to the emission line data on one important source Centaurus A. This summary is based on collaborative work with Ralph Sutherland, Mike Dopita and David Singleton. Our approach involves the physics of autoionizing shocks and the theory has ramifications for line-emitting gas in numerous other active galaxy contexts.

RELATIVISTIC ENTRAINING FLOW

Since there is no known expression for turbulent viscosity in the case of non-relativistic supersonic flow, let alone relativistic flow, an approach based on jet conservation laws (Bicknell, 1993) is useful. Such laws are obtained by integrating the relativistic hydrodynamic equations throughout a control volume incorporating the jet and the surrounding ISM. These laws relate jet parameters at an arbitrary cross-section (area A_2) to core-jet parameters (cross-sectional area A_1). Neglecting the initial rest-mass flux of the pc-scale jet, particle number, momentum and energy conservation imply that

$$\gamma_2 \rho_2 v_2 A_2 \approx -\int_{S_E} \rho \mathbf{v} \cdot \mathbf{n} \, dS \tag{1}$$

$$\left(\gamma_2^2 \frac{w}{c^2} v_2^2 + \Delta p_2\right) A_2 - \left(\gamma_1^2 \frac{w}{c^2} v_1^2 + \Delta p_1\right) A_1 \approx -\int_{r_1}^{r_2} dr \left[\frac{dp_{\rm ext}}{dr} \int_A \left(1 - \frac{\rho^*}{\rho_{\rm ext}}\right) \frac{z}{r} \, dS\right] \tag{2}$$

and

$$\left[(\gamma_2^2 - \gamma_2)\rho_2 c^2 + 4\gamma_2^2 p_2\right] v_2 A_2 - 4\gamma_1^2 p_1 v_1 A_1 \approx -\int_{S_E} \rho h \mathbf{v}_{\rm ent} \cdot \mathbf{n} \, dS \tag{3}$$

Here, S_E is the "entrainment surface" located in the ISM.

The jet momentum is affected by the external pressure gradient. However, if the jet is initially free or of high Mach number, momentum is approximately conserved. The jet energy is affected by the entrainment of internal energy which, for the high enthalpy jets under consideration, is negligible. With these simplifications, the above equations become:

$$\gamma_2^2 \beta_2^2 (1 + \mathcal{R}) \approx \frac{1 + 4\gamma_1^2\beta_1^2}{4} \left(\frac{p_1 A_1}{p_2 A_2}\right) \tag{4}$$

$$[(\gamma_2 - 1)\mathcal{R} + \gamma_2] \gamma_2 \beta_2 \approx \gamma_1^2 \beta_1 \left(\frac{p_1 A_1}{p_2 A_2}\right) \tag{5}$$

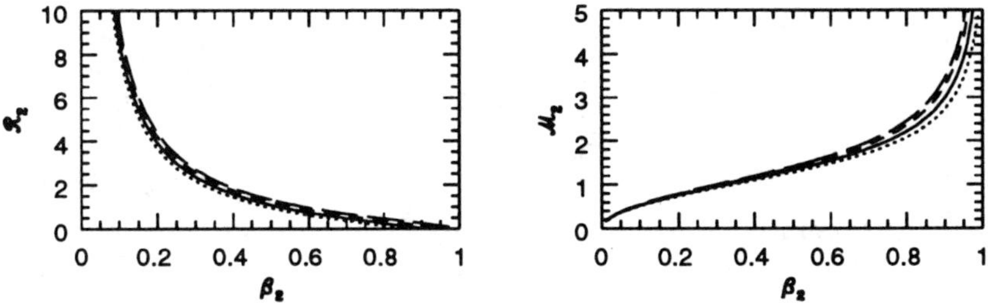

FIGURE I The ratio R of rest-mass energy density to enthalpy (left panel) and Mach number as a function of jet velocity (right panel) in an initially ultrarelativistic entraining jet. Solid curve: $\gamma_1 = 5$; dotted curve: $\gamma_1 = 2.5$; long-dashed curve: $\gamma_1 = 1.25$; short-dashed curve: $\gamma_1 = 1.1$

where $\mathcal{R} = \rho c^2/4p$, the ratio of jet rest-mass energy to enthalpy, and $\beta = v/c$. (The jet Mach number is given by $\mathcal{M}^2 = (2 + 3\mathcal{R})\gamma^2\beta^2$). Given jet pressures and cross-sections on both parsec and kiloparsec scales then one may estimate both $\mathcal{R}$ and β as a function of the initial jet Lorentz factor. Moreover, $\mathcal{R}$ can be determined as a function of jet velocity and initial β from the above equations. The parameter $\mathcal{R}$ and Mach number are plotted in figure I for a number of initial Lorentz factors. These solutions are extremely interesting since they show that when an initially ultra-relativistic jet slows to a Mach number say between 1 and 2, then its velocity is necessarily between 0.3c and 0.7c. Since the spreading rates of FRI jets are most naturally associated with transonic turbulence then this defines a natural velocity $\sim 0.3 - 0.7c$ for the kpc-scale bases of FRI sources. Such velocities are sufficient to explain the surface brightness ratios at the bases of Class I jets (see the papers by Parma and Laing in these proceedings). A velocity $\sim 0.3c$ corresponding to a Mach number of 1 has a simple physical interpretation: The ratio of energy to momentum fluxes in an ultrarelativistic jet is c. When the jet has slowed to a mildly relativistic velocity the same (conserved) ratio is $3v(1 + \mathcal{M}^2/6)\mathcal{M}^{-2}$. Equating the two gives, for $\mathcal{M} = 1$, $v/c = 2/7 \approx 0.29$.

DETAILED CONSIDERATION OF NGC 315

Figure II shows the calculated Mach number of the NGC 315 jet (Willis *et al.*, 1981; Venturi *et al.*, 1993) at $\Theta = 17''$ and $27''$ as a function of the initial Lorentz factor for $p/p_{\rm min} = 10$ and 100 and for jet inclinations of 90° and 30°. The implication of these plots is that a kpc-scale transonic Mach number and a moderate initial Lorentz factor in the 90° jet are consistent if the ratio of initial jet pressure to the minimum value is not much more than about 10. If the jet pressure is much higher than this then initially, the jet is at most mildly relativistic. The allowable range of pc-scale Lorentz factors increases if the jet is inclined at 30° to the line of sight (although for $p/p_{\rm min} \approx 100$ there is not much difference in the two curves because the jet does not decelerate for such a high

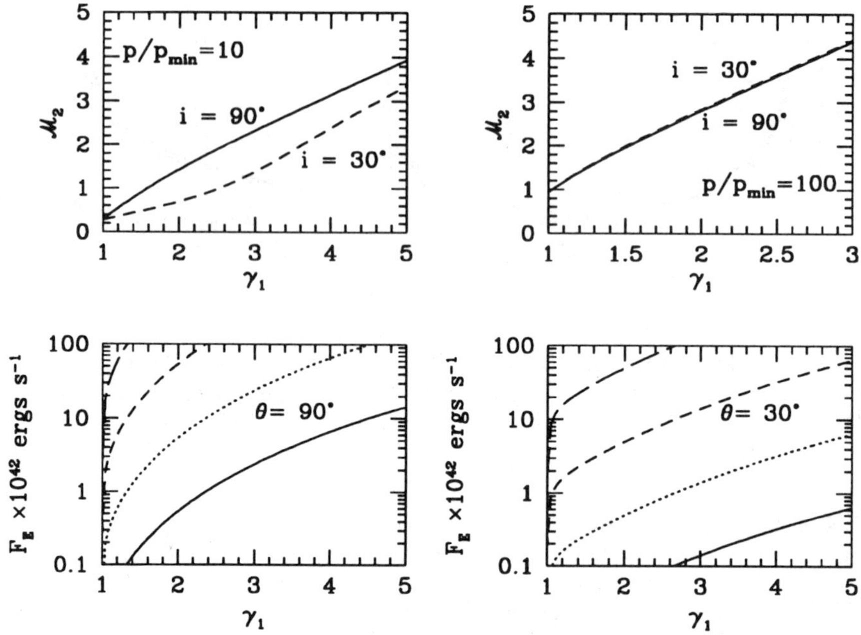

FIGURE II *Upper panels:* Mach Number at $\Theta = 17''$ as a function of pc-scale Lorentz factor and jet inclination. *Lower panels:* Jet Energy Flux for values of $p/p_{\rm min} = 0.1, 1, 10$ and 100(increasing from lower to upper curves).

initial pressure). The higher Lorentz factors at lower pressures are realted to the lower inferred minimum pressure, necessitating a higher velocity for given energy and momentum fluxes. The energy flux is also of interest since it is conserved and in principle, can be estimated from minimum energy and age estimates of the lobes. A roundabout estimate for NGC 315 is $\sim 10^{43}$ergs s^{-1}. This is more easily achieved for a range of $p/p_{\rm min}$ if the jet is inclined at 30° to the line of sight (see figure II).

Similar considerations also apply to NGC 6251 (see Bicknell, 1993). In that case the energy flux is probably better estimated from reliable spectral index variations.

Modeling of the large scale jet in NGC 315 (Bicknell, these proceedings) predicts a velocity decline between $\Theta = 17''$ and $\Theta = 100''$ by a factor of a few. If the jet velocity is approximately 0.6 c at $\Theta = 17$, this decline suffices to account for the equality of surface brightnesses at $\Theta \approx 100''$.

KNOTS AND SIDEDNESS

The small number of FRI sources that have been imaged with VLBI exhibit *sub*luminal motion (e.g. M87, Reid *et al.*(1989); NGC 315, Venturi *et al.*(1993))

giving one cause to question the validity of relativistic velocities on small scales in FRI sources. The physical case is neatly summarized in figure III where the brightness ratio of two oppositely directed jets is plotted against the apparent β. Obviously, a high brightness ratio and a low transverse velocity ($R > 50$ and $\beta < 0.5$ for the pc-scale jet in NGC 315) imply a prohibitively small angle to the line of sight for a source with already a large projected size on the sky. A similar angle is inferred for M87. Lind and Blandford (1985) emphasized that knots are most naturally interpreted as shocks. Given this, and the models of the flux variations of BL-Lac in which the knots are reverse shocks caused by variations in the flow velocity (Hughes, Aller and Aller, 1989a,b) it is natural to consider the implications of reverse shocks in FRI jets. Even a weak shock can move rapidly backwards against the jet flow (with a relative velocity of $c/\sqrt{3}$) leading to a low velocity in the observer's frame. This situation is summarized in the right panels of figure III where the observed pre- and post-shock fluid velocities are plotted for different observed shock velocities and shock angles as a function of the shock pressure ratio. Clearly, both velocities are considerably larger than the observed shock velocity especially for oblique shocks where the parallel component of the fluid velocity is unaffected. The limiting values for the post-shock velocity correspond to the cases of weak and strong shocks respectively.

THE EMISSION LINE FILAMENTS IN CENTAURUS A

As indicated in the introduction the extranuclear emission lines of radio galaxies pose interesting problems relating to their excitation and temperature. Although the emission line filaments in Centaurus A are of low luminosity ($\sim 10^{38}$ergs s^{-1}) compared to emission line regions in powerful FRII radio galaxies, the proximity of Centaurus A means that detailed study is extremely rewarding. Morganti *et al.*(1991) have obtained excellent [OIII] images together with high and low dispersion spectra of the "inner" and "outer" filaments of Cen A showing the presence of both high and low excitation lines and large turbulent velocities ($\sim 200 - 400\,\mathrm{km\,s^{-1}}$). Both the inner and outer filaments show the apparent effects of disruption by the radio plasma in Centaurus A.

Morganti *et al.*have argued that the filaments are excited by a narrow beam of radiation from the core of the galaxy, showing that if this is the case, then the beaming cannot be due to obscuration by a dusty torus and must be intrinsic. My colleagues (Ralph Sutherland, Mike Dopita, David Singleton) and I, in a series of papers (Bicknell, 1991, Sutherland, Bicknell and Dopita, 1993 (SBD), Bicknell, Dopita and Singleton 1993) have developed a model for these filaments which invokes interaction between a radio jet and dense clouds along its path. Low dynamic range images (Burns 1993) of the *Middle Lobe* region of Centaurus A show that the *Outer Filaments* are coincident with the southern region of the middle lobe but that there is no obvious radio emission coincident with the *Inner Filaments.* However, the inner filaments lie along the extrapolated path of the inner jet. Clearly higher dynamic range AT or VLA images are required in view of the fact that our model predicts that radio plasma is still flowing in the region of the inner filaments.

Our model involves: (1) Disruption of the molecular cloud by the Kelvin-Helmholtz instability acting at the jet-cloud interface (2) Production of super-

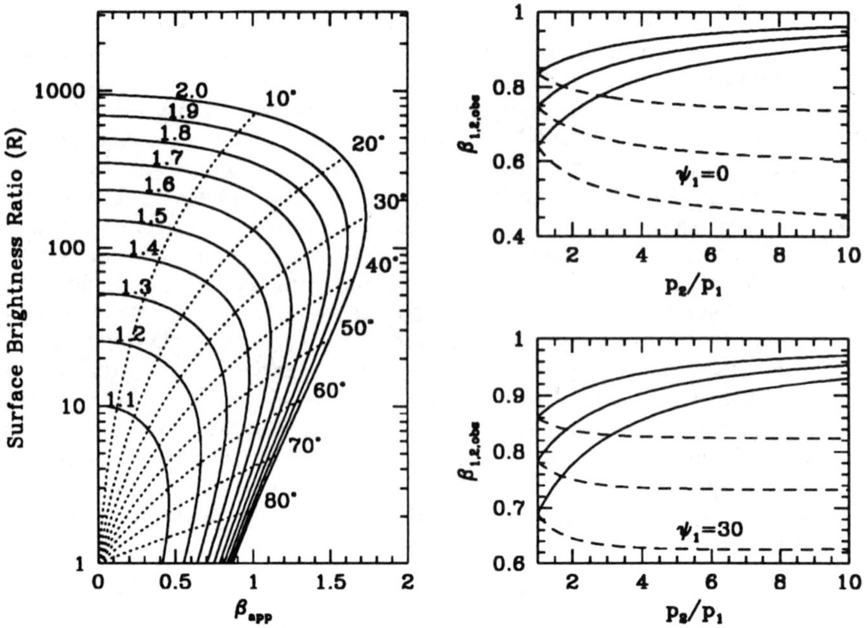

FIGURE III *Left Panel:* The surface brightness ratio of two equal, oppositely directed jets as a function of the apparent jet velocity for a range of Lorentz factors between 1 and 2. *Right Panels:* The pre- and post-shock velocities corresponding to shock velocities of 0.1, 0.3 and 0.5c as a function of shock pressure ratio both for a normal and an oblique shock (ψ_1 is the angle between the shock normal and the pre-shock velocity in the shock frame.)

sonic turbulent velocities in the molecular material (3) Dissipation of the turbulence via radiative *autoionizing* shocks and (4) Simultaneous production of high and low excitation lines through density and/or geometric effects. The idea of autoionizing shocks was first introduced by Binette, Dopita and Tuohy (1985). Let us now consider the key elements of this model.

The Kelvin-Helmholtz Instability

There is a well known limit on the Mach number for the K-H instability to occur, namely, $\mathcal{M}\, cos\phi < \left(1 + (c_{s,2}/c_{s,1})^{2/3}\right)^{3/2}$ where ϕ is the angle between the perturbation and the flow, c_s is the sound speed and subscripts 1 and 2 refer to the jet and cloud respectively (Miles 1958, Fejer and Miles 1963). Strictly, this inequality refers to the case of equal specific heat ratios; nevertheless it should be generally indicative. In the case in point, the jet Mach number clearly should not be much larger than unity in order for the instability to be effective, consistent with the idea that the jets in the outer parts of FRI sources are transonic.

Simulations of Cloud Ablation

In view of the potential relevance of the Kelvin-Helmholtz instability, simulations for flows with Mach numbers $\lesssim 1$ and a high density ratio are useful to ascertain whether the physical reason for some of the features of the filaments can be identified. Simulations with a density ratio $\sim 10^{-4}$ are difficult to perform because of the lower growth rates and the Courant criterion. However, simulations for a density ratio of 100 are relatively straightforward. Figure IV shows the evolution of the density in such a simulation for $\mathcal{M} = 0.5$. The jet is the upper gray region and the cloud is the lower dark region. The jet initially makes an angle of 5° with the cloud and this is the reason for the intermediate colored wedge of material at the left. This simulation exhibits many of the observed features of the filaments, in particular the wave-like region of knot A, B and C with its long drawn out filament, the filamentary structure of the surface waves and a socket-like feature at the end of the long filament in frame 7 which appears and disappears as the filament waves in and out of regions of differing velocity. Further downstream, arcs of gas which have been totally ablated from the main cloud. Another welcome feature of these simulations are the sound waves and weak shocks which are driven into the dense material by the turbulence at the interface. Some of these arise from the collision of filaments on the interface. These filaments have velocities of order 10 – 20% of the free stream velocity. The density ratio in this simulation is large, but not large enough to show the development of highly supersonic turbulence since the sound speed in the dense region is not low enough. However, given that the Kelvin-Helmholtz instability is capable of producing velocities in the dense material equal to a substantial fraction of the free-stream velocity, then as the density ratio increases, the turbulent velocities should become more supersonic. This point is currently being checked with higher density ratio simulations.

The Energy Budget

The jet energy flux incident on a filament of cross-sectional area A is:

$$\begin{aligned} F_E &= 1.0 \times 10^{37} \, \frac{\gamma_{\rm jet}^{3/2}}{\gamma_{\rm jet}-1} \, \eta^{-1/2} \left(\frac{n_{\rm ism}}{10^{-3}}\right) \left(\frac{T_{\rm ism}}{10^7}\right)^{3/2} \left(\frac{A}{10^4 {\rm pc}^2}\right) \\ & \qquad \mathcal{M}_{\rm jet} \left(1 + \frac{\gamma_{\rm jet}-1}{2} \mathcal{M}_{\rm jet}^2\right) \, {\rm ergs\,s^{-1}} \qquad (6) \end{aligned}$$

For $A \sim 4 \times 10^4$ pc, $\eta \approx 0.1$ and $\mathcal{M} \sim 1$ the incident power is $\approx 2 \times 10^{39}\,{\rm ergs\,s^{-1}}$ adequate to account for the $H\beta$ emission-line luminosity $\sim 10^{37}\,{\rm ergs\,s^{-1}}$ assuming that most of the incident power is dissipated in radiative shocks and allowing for a factor ~ 100 between the total emission line luminosity and the $H\beta$ luminosity.

Order of magnitude estimates for the Balmer Line Luminosity

Given that the turbulent supersonic velocity field produced by the Kelvin-Helmholtz instability is dissipated through the occurrence of strong shocks, it is possible to estimate the Balmer line luminosity from the data on the turbulent velocity field. The geometry and different ionization zones of cloud-cloud collisions is described in detail in Sutherland, Bicknell and Dopita (1993).

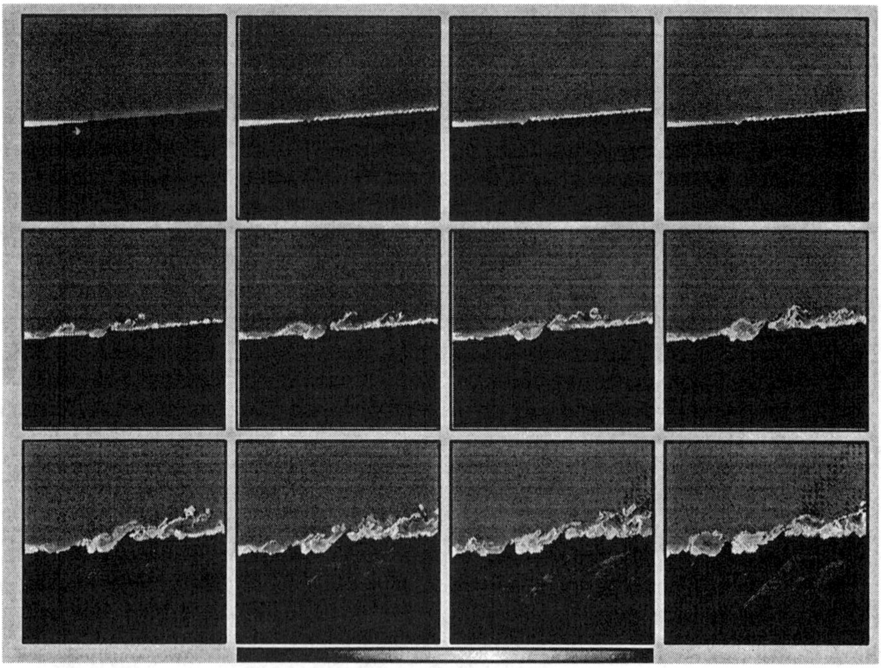

FIGURE IV The evolution of the density (log scale) in a 2D simulation of the Kelvin-Helmoltz instability. Note the production of filaments and ablated clouds at the interface.

When two equal density clouds collide with a velocity $\Delta v \approx 200\,\mathrm{km\,s^{-1}}$ the temperature produced behind each of the strong shocks propagated into the clouds is:

$$T_{\rm sh} \approx \frac{3}{64}\frac{\mu\, m_p}{k}(\Delta v)^2 = 1.3\times 10^5 \left(\frac{\Delta v}{200\,\mathrm{km\,s^{-1}}}\right)^2 \qquad (7)$$

and this can be high enough to produce an ionizing flux of UV and soft X-ray radiation. The cooling time behind the shocks is short so that all the internal energy is radiated into the upstream and downstream plasma. Thus, the $H\beta$ luminosity (for optically thick photoionization) is given by:

$$L_{H_\beta} = 1.5\times 10^{36}\,\mathrm{ergs\,s^{-1}}\, f\, n_0 \left(\frac{\Delta v}{200\mathrm{km\,s^{-1}}}\right)^3 \left(\frac{A_{\rm sh}}{10^4\mathrm{pc}^2}\right) \qquad (8)$$

where $f = \left[\int_{\nu_0}^{\infty}(\nu_0/\nu)\Lambda_\nu\, d\nu\right]\left[\int_0^{\infty}\Lambda_\nu\, d\nu\right]^{-1}$ parametrizes the ratio of the flux of ionizing photons to post-shock luminosity and is approximately 0.8 for non-equilibrium cooling as described in SBD.

When cross-sections of elements other than Hydrogen are taken into account, model calculations show that it the exponent of Δv is closer to 2.4 (see Dopita, these proceedings). Nevertheless, equation 8 is quite accurate in the velocity range of interest and shows that the luminosity of the knots is comfort-

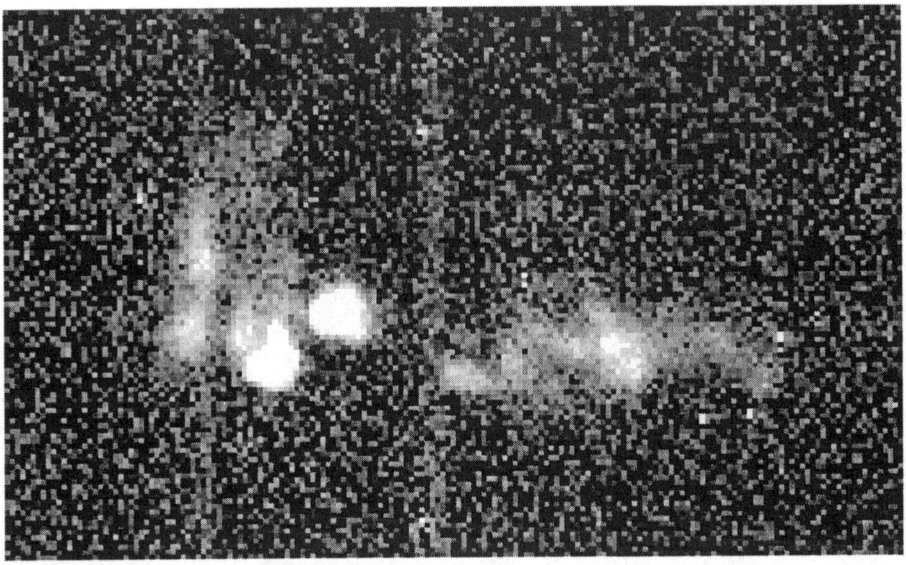

FIGURE V A spatial-velocity image of the inner filaments based on [OIII]λ5007. The horizontal axis is the spatial distance along the slit. The vertical axis is velocity. The direction of the nucleus is to the left.(From Bicknell, Dopita and Singleton, 1993)

ably accounted for by the combination of the velocity field observations and this simple physical picture.

The modeling carried out in SBD also showed that both high excitation and low excitation lines could be produced if different regions of the interface have different density. These models provided a good fit to the observed spectrum including the temprature sensitive line $[OIII]\lambda 4363$ which is produced in the high temperature region behind a shock.

RECENT AAT OBSERVATIONS OF CENTAURUS A

Figure V shows an AAT spectrum in the vicinity of [OIII]λ5007 with approximately 30 km s^{-1} resolution. The slit passes through the middle of the inner filaments and includes knots A, B and C. This spectrum shows some dramatic features including large regional differences in the velocity field, evidence for entrainment in the form of features emanating from the brightest knots and extending in the velocity direction by about 700 km s^{-1} and evidence for a surface-wave in the right hand part of the image. confirming the physical picture presented above.

CONCLUSIONS

That beaming is important in the determining the radio characteristics of Radio Galaxies seems to be unavoidable. The overwhelming impression that one gains from working on the VLA and VLBI data of FRI radio galaxies is that the notions of relativistic jet asymmetries, jet deceleration and FRI-BL-Lac unification are all consistent. The case for beamed optical radiation is not so clear cut. Nothwithstanding the popularity of beamed ionizing radiation, I have presented arguments that the emission line filaments in the nearest Active Galaxy, Centaurus A, are excited by their interaction with an, as yet undetected, radio jet. Perhaps one of the most appealing aspects of this model is its high degree of internal consistency: The morphology, internal velocities and spectra of the filaments are all explained in terms of the interaction between a $\sim 1500 - 50000\,\mathrm{km\,s^{-1}}$ radio jet and dense molecular clouds along its path. The model makes a clear prediction for the flux of the ultraviolet line CIVλ1549 (see SBD). Clearly, this model, if substantially correct, has ramifications to other emission line regions (see Koekemoer, Bicknell and Dopita, these proceedings). The bottom line in all of these calculations is the unification of the optical line emission and the local kinematics.

REFERENCES

Bicknell, G. V. 1986, *ApJ*, **305**, 109

Bicknell, G. V. 1991, *PASA*, **9**, 93

Bicknell, G. V., Dopita, M. A., and Singleton, D. 1993, in preparation

Binette, L., Dopita, M. A., and Tuohy, I. R. 1985, *ApJ*, **297**, 476

Fejer, J. A. and Miles, J. W. 1963, *J. Fluid Mech.*, **15**, 335

Hughes, P. A., Aller, H. D., and Aller, M. F. 1989a, *ApJ*, **341**, 54

Hughes, P. A., Aller, H. D., and Aller, M. F. 1989b, *ApJ*, **341**, 68

Lind, K. R. and Blandford, R. D. 1985, *ApJ*, **295**, 358

Miles, J. W. 1958, *J. Fluid Mech.*, **4**, 538

Reid, M. J., Biretta, J. A., Junor, W., Muxlow, T. W. B., and Spencer, R. E., 1989, *ApJ*, **336**, 112

Sutherland, R. S., Bicknell, G. V., and Dopita, M. A. 1993, *ApJ*, **414**, 510

Ulrich, M.-H. 1989, in L. Maraschi, T. Maccacaro, and M.-H. Ulrich (eds.), *BL Lac Objects*, Vol. 334 of *Lecture Notes in Physics*, 45, Springer-Verlag, Berlin

Venturi, T., Ferretti, L., Giovannini, G., Comoretto, G., and Wehrle, A. E., 1993, *ApJ*, in press

Willis, A. G., Strom, R. G., Bridle, A. H., and Fomalont, E. B. 1981, *A&A*, **95**, 250

The First Stromlo Symposium: The Physics of Active Galaxies
ASP Conference Series, Vol. 54, 1994
G. V. Bicknell, M. A. Dopita, and P. J. Quinn (eds.)

BRIGHTNESS ASYMMETRIES AND VELOCITIES OF JETS IN LOW LUMINOSITY RADIO GALAXIES

P. PARMA, H. R. DE RUITER[1] AND R. FANTI[2]
Istituto di Radioastronomia del CNR, Bologna, Italy

R. LAING
Royal Greenwich Observatory, Cambridge, U. K.

ABSTRACT

Although low luminosity B2 radio sources are very symmetric on large scales, the brightness ratio between the two sides tends to become asymmetric close to the core. The asymmetry is stronger in sources with higher total luminosities.

Adopting Doppler beaming as the cause of the asymmetry, our data are consistent with the idea that jets in the stronger FR I sources decelerate from $\beta \sim 0.6$ to $\beta \sim 0.1$ in the first ~ 5 kpc. In the weaker FR I sources the deceleration apparently occurs in the first 1-2 kpc. Thus Doppler favoritism may occur also in low luminosity sources.

The correlation between jet velocity (determined from the energy budget) at 10 kpc from the core and total power lends strong support to this interpretation.

The scheme discussed here is consistent with the Laing-Garrington effect and with the notion that weak FR I jets have $\beta << 0.1$ at large scales.

INTRODUCTION

Fanaroff-Riley (FR) type I radio sources are defined as sources without hot-spots, and with the peak in brightness close to the core (Fanaroff & Riley 1974). Such sources have a radio power below $\sim 10^{25}$ W Hz^{-1} at 1.4 GHz.

The present discussion is based on FR I sources selected from the B2 catalogue. A discussion of the two B2 samples of low luminosity radio galaxies, which contain in total 105 radio galaxies, can be found in Fanti et al. (1987).

The B2 sources have been studied in detail at 20 cm using the VLA in different configurations, and the results were summarized in De Ruiter et al. (1990). The present report uses this latter article as a starting point.

Note that in the following we often use the distinction low and high luminosity as meaning low and high luminosity end of the FR I luminosity range.

---

[1]Osservatorio Astronomico di Bologna, Bologna, Italy

[2]Dipartimento di Fisica, Università di Bologna, Bologna, Italy

JETS IN FR I SOURCES

Jets are found in about 60 % of the B2 sources. In most cases the jets are two-sided, but asymmetries are noticeable in the more powerful FR I sources, and among them even a few one-sided jets are present. The difference in brightness is always more conspicuous the closer one gets to the core; in fact a symmetrical twin jet on large scales may resemble a one-sided jet on smaller scales.

Application of Bicknell's model (Bicknell 1984) to a number of B2 sources shows that the opening angle and brightness distributions are consistent with the expected behaviour of low Mach number ($M \sim 2$) jets. The bulk velocities of such jets should be in the range $10^3 - 10^4$ km s^{-1} (Bicknell et al. 1990). This suggests that low luminosity sources are different from their high power (FR II) cousins, in that their bulk velocities are much lower: relativistic beaming effects should play no role. We will show, however, that low luminosity sources are more complicated than they may seem at first sight.

POLARIZATION

Recently we obtained 20 cm polarization data for most B2 objects. Assuming that, in a statistical sense, the polarization at 20 cm can be used as an indicator of *depolarization* (see Parma et al. 1993), it was found that low luminosity radio galaxies show the same kind of depolarization asymmetry as high power radio quasars (Garrington et al. 1988), although the degree of asymmetry is much less conspicuous. Laing (1988) has argued that the depolarization asymmetry is caused by relativistic beaming, which produces a main and a counter-jet from an originally symmetric twin jet.

What about the FR I jets? If they are moving so slowly that Doppler effects are unimportant how can they show the small but discernible depolarization asymmetry? Perhaps an answer may be given that saves the general idea of FR I sources as being non-relativistic on large scales (with velocities of order of at most a few times 10^4 km s^{-1} for the highest luminosities among FR I sources), and at the same time preserves Laing's scenario. Laing (1992) has formulated a new hypothesis for jets in low luminosity sources; it states that FR I jets are (mildly) relativistic on small (~pc) scales, but decelerate quickly and become non-relativistic on large (~kpc) scales. This hypothesis has led us to investigate more closely the brightness asymmetries present in the B2 sources.

ASYMMETRIES

We have analyzed 20 cm and in some cases 6 cm VLA radio images of most B2 sources, with resolution ranging from ~1 to ~12 arcsec. We made slices perpendicular to the jet axis in order to determine the jet diameter and brightness. This was done at various *linear* distances from the core, starting (if possible) at 1 kpc.

Dividing the sources into two classes of high and low total power, where high and low refer of course to the power range of the B2 sources, we have plotted distance from the core against the median brightness ratio R_I of each power

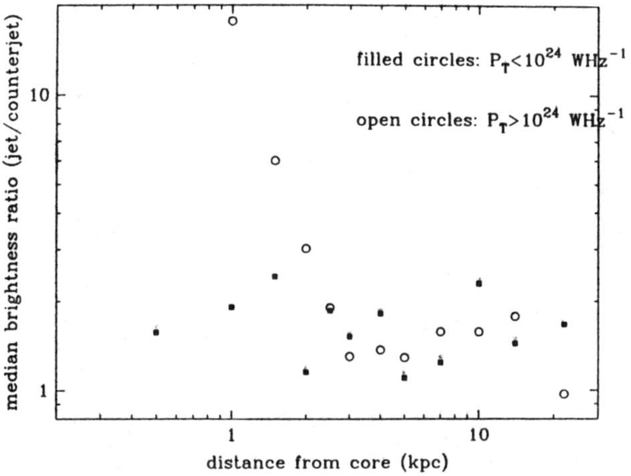

FIGURE I Median brightness asymmetry as a function of distance from core. Open circles: strong FR I sources. Filled: weak FR I sources.

class (see Figure I): high power tends to produce an increasing asymmetry as we come closer to the core, while for low power sources this is not true.

A similar plot can be constructed using the core power. However, it may be that the core power itself is beamed. We therefore use the *normalized* core power P_{cN}. Since there exists a relation between core and total power (see De Ruiter et al. 1990 and references therein), we divide core power by the median core power associated with the total power of the source. Therefore strongly beamed (asymmetric) sources are expected to have $P_{cN} > 1$ and weakly beamed sources (symmetric) $P_{cN} < 1$. Figure II shows that this is indeed the case. Our data are therefore consistent with the idea that both core power and brightness asymmetry of the jets are due to beaming.

The distribution of P_{cN} of the B2 radio galaxies suggests velocities in the range $0.6 < \beta < 0.9$: the jet velocities close to the core should thus be mildly relativistic. Does this agree with the velocities we deduce from the jet asymmetries? The answer is yes: if we plot the brightness asymmetry R_I against P_{cN} for the individual sources at different distances from the core, we find a correlation which is most significant at close distances but disappears beyond ~ 5 kpc.

At 1.5 to 2 kpc the distribution of R_I is consistent with $\beta \sim 0.6$, while from 3 to 5 kpc the required β decreases from ~ 0.5 to ~ 0.1. Dividing the sources into high luminosity ($P_t > 10^{23.5}$ W Hz^{-1}) and low luminosity objects there is a hint that the effect is more clear for the stronger sources, and absent in the weak sources (perhaps because the high velocities occur in this case much closer to the core).

If the picture given here is correct, jet velocities beyond, say, 5 kpc from the core should be $\leq 0.1c$, at least in the low luminosity sources considered here. There is another way to get a rough estimate of the jet velocity, using energy budget arguments (Bicknell 1986). It is possible to calculate $v_J \times \kappa$ (ignoring

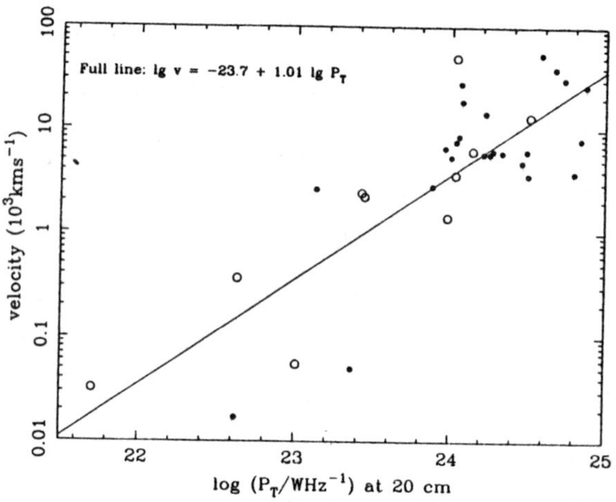

FIGURE II Median brightness asymmetry as a function of distance from core. Open circles: $P_{cN} > 1$. Filled: $P_{cN} < 1$.

a term involving the jet Machnumber) directly from the jet diameter and the lobe (or jet) luminosity. We take 10 kpc as a reference point, or, for shorter jets, the end of the jet, in order to stay well outside the region where Doppler boosting might be important. In Figure III we show v_J(10 kpc) in units of 10^3 km s^{-1} as a function of total source power; note that for four sources with jets the lobe emission is weak or absent ("naked jets"), so that the jet velocity is determined from the jet luminosity, while for all other objects a κ of 0.05 was assumed (which should be correct within a factor 2). As is clear from Figure III a strong correlation exists between P_T and v_J(10 kpc); a best fit gives:

$$\log v_J = -23.7 + 1.01 \pm 0.13 \log P_T. \quad (1)$$

Fits using only sources with data at 10 kpc or at the end point < 10 kpc are practically identical. Extrapolating this relation to higher source luminosities the jet velocity, which is << 0.1 (in terms of $\beta = v_J/c$) for $P_T < 10^{24}$ WHz^{-1}, quickly approaches unity beyond $P_T = 10^{25}$ WHz^{-1}. Thus at 10 kpc Doppler boosting should start to appear modestly only at the very highest end of the FR I, but will already be predominant at the low luminosity end of FR II sources. Our data therefore confirm the idea that the division between the FR I and II classes is basically due to the jet velocity, while it also explains quite simply why the transition is so sharp.

Our results may be summarized as follows: all galaxies considered here, even at the lowest radio luminosities, form twin radio jets moving with high velocities ($\beta \sim 0.7$) close to the core, that decelerate to $\beta \sim 0.1$ in the first ~ 5 kpc. However, sources of low total power decelerate more quickly (perhaps already in the first kpc). Although all sources decelerate such that no Doppler boosting is present anymore at 10 kpc, the jet velocities derived at this point, as derived from the energy budget, are proportional to total power: apparently stronger sources

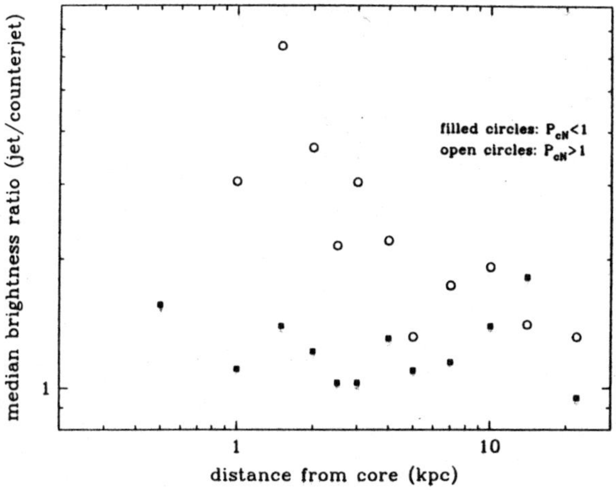

FIGURE III Jet velocity as a function of total source power. Filled circles: at 10 kpc. Open circles: end point of jet (< 10 kpc).

still have a higher residual velocity. Extrapolating this correlation to higher radio luminosities, relativistic jet velocities should occur beyond 10^{25} WHz^{-1}.

Since Doppler effects should be most important close to the core, it is obviously of great interest to study FR I sources on VLBI scales. One such programme has been undertaken (see Venturi et al, this conference).

REFERENCES

Bicknell, G.V. 1984, *ApJ*, **286**, 68

Bicknell, G.V. 1986, *ApJ*, **300**, 591

Bicknell, G.V., de Ruiter, H.R., Fanti, R., Morganti, R. and Parma, P. 1990, *ApJ*, **354**, 98

de Ruiter, H.R., Parma, P., Fanti, C., Fanti, R. 1990, *A&A*, **227**, 351

Fanaroff, B.L. and Riley, J.M. 1974, *MNRAS*, **167**, 31p

Fanti, C., Fanti, R., de Ruiter, H.R. and Parma, P. 1987, *A&AS*, **69**, 57

Garrington, S.T., Leahy, J.P., Conway, R.G., Laing, R.A. 1988, *Nature*, **331**, 147

Laing, R.A. 1988, *Nature*, **331**, 149

Laing, R.A. 1992, in *Astrophysical Jets*, ST Science Institute Symposium, Baltimore, May 1992, in press.

Parma, P., Morganti, R., Capetti, A., Fanti, R., de Ruiter, H.R. 1993, *A&A*, **267**, 31

The First Stromlo Symposium: The Physics of Active Galaxies
ASP Conference Series, Vol. 54, 1994
G. V. Bicknell, M. A. Dopita, and P. J. Quinn (eds.)

VLBI OBSERVATIONS OF FR-I RADIO GALAXIES

T. VENTURI
Istituto di Radioastronomia di Bologna

G. GIOVANNINI, L. FERETTI
Dipartimento di Astronomia and Istituto di Radioastronomia - Bologna

W.D. COTTON
National Radio Astronomy Observatory

L. LARA
CSIC, Instituto de Astrofisica de Andalucia

J. MARCAIDE
Universidad de Valencia

A.E. WEHRLE
Infrared Processing and Analysis Center - California

ABSTRACT

Information on the VLBI morphology of various classes of radio galaxies is available nowadays. We focussed our attention on the VLBI morphology and properties of FR-I radio galaxies. To this aim we used all the FR-Is with mapped VLBI structure belonging to a complete sample of low-intermediate power radio galaxies we have been studying in these years, and the maps available in the literature for objects of this class. The first preliminary results emerging from this analysis indicate that the most common VLBI morphology of FR-I radio galaxies is one-sided, with a core and a jet aligned with the strongest kpc-scale jet. The lower limits on the intrinsic plasma velocities along the parsec-scale jets, derived under the assumption that the observed asymmetry is caused by Doppler boosting, are typically in the range $\gtrsim$ 0.65c - 0.85c. This leads to values of the Lorentz factor γ in the range $\sim$ 1.3 - 2.

INTRODUCTION

The study of Fanaroff-Riley Type I (FR-I) radio galaxies on the VLBI scale is important for various reasons. On the arcsecond (kiloparsec) scale, low luminosity FR-I radio galaxies are very different from the more powerful FR-II radio galaxies. They show two-sided jets, whose brightness is usually asymmetric close to the core of the radio emission, the degrees of symmetry increasing further out along the jet. Jets in FR-II radio galaxies, on the other hand, are one-sided on the VLBI scale and tend to be very asymmetric even at large distances from the core of the radio source. This is explained by assuming that the plasma velocity in FR-II radio galaxies is highly relativistic from the very beginning to the very end of the radio jets, while in FR-Is the kpc-scale jets are at most subrelativistic (see Parma et al., these proceedings). It is very interesting to compare the parsec-scale (VLBI) morphology and properties of these two classes of radio galaxies in order to understand what makes them so different on the very large scale. Furthermore, FR-I radio galaxies are thought to be the parent population of BL-Lacs at large angles to the line of sight. If this is the case, then we expect the Lorentz factors for these two classes of objects to be very similar on the parsec scale.

In order to address these questions, we collected all the VLBI data available thus far for FR-I radio sources, for a total of 9 sources. The largest fraction of FR-I radio galaxies we used for our investigation, 6 out of 9, comes from our multifrequency VLBI observations of a complete sample of low-intermediate luminosity radio galaxies (see Giovannini et al., these proceedings, and references therein). They are NGC315 (see also Venturi et al. 1993), 0755+37, 0836+29, 1144+35, 3C338 (see also Feretti et al., 1993), 3C465. We will refer to these sources as Group A. We also included in the discussion M87 (Reid et al., 1989), NGC6251 (Jones et al., 1986), and Centaurus A (Meier et al., 1993), and these will constitute Group B.

RESULTS

The very first result is the occurence of one-sidedness (core + jet) on the parsec scale among the FR-I radio galaxies under investigation. Only two radio galaxies, 3C338 and 1144+35, have a different morphology. The first is symmetric on the VLBI scale, with two very similar jets: the ratio R between the jet and counterjet brightness is ~ 1.1 at 1 pc from the core (see Feretti et al., 1993). 1144+35 shows a more complex morphology, with two dominant components, and at this stage it is impossible to say whether it is a double or a triple source, since the location of the core is still unknown. All the other sources in both groups A and B are characterised by a strong core and a one-sided jet. In each case the VLBI jet is aligned with the strongest kpc-scale jet. Examples of the morphologies found on the parsec scale are reported in Fig. 1.

The brightness ratios R we derived suggest various degrees of asymmetry on the VLBI scale for the FR-Is in our sample. M87 is the most asymmetric source, with $R \gtrsim 290$ at 1.4 pc (Reid et al., 1989); in NGC315 the brightness ratio is $R \gtrsim 50$ at 10 pc from the core (Venturi et al. 1993). The values obtained for these two sources put very strong contraints on the intrinsic plasma speed and on the angle the radio source axis forms to the line of sight, as we will discuss later. On the other hand for 0836+29 the lower limit to the brightness ratio is $R \gtrsim 15$ at 6.5 pc from the core, a value which puts less severe constraints to the values of β_{min} and θ_{max}. For the other FR-Is in the sample the lower limits on R fall within this interval. In Table 1 we report the values of the jet to counterjet brightness ratio R for the radio galaxies in group A and B and the distance d from the core (given in pc) at which the ratio was computed. Given the asymmetry of each source (exception made for 3C338 only), we assumed the noise level of the map as upper limit for the counterjet brightness.

Using the lower limits of R derived from the observations and assuming that the parsec-scale asymmetry is due to Doppler boosting on two intrinsically symmetric jets, we can derive the constraints on the product $\beta cos\theta$, where β represents the ratio between the intrinsic jet speed v_{intr} and the velocity of light c ($\beta = \frac{v_{intr}}{c}$), and θ is the angle formed by the direction of the radio emission and the line of sight. For the sources available thus far, one should keep in mind that the one-sidedness gives us only lower limits to the product $\beta cos\theta$, which result into upper limits for the angle to the line of sight θ and into lower limits for the intrinsic jet velocity. The standard formula we used to constrain the product $\beta cos\theta$ is: $\beta cos\theta = (R^{\xi} - 1)/(R^{\xi} + 1)$ (Pearson and Zensus, 1987). Under the assumption of isotropy $\xi = \frac{1}{2+\alpha}$, α being the value of the spectral index along the jet ($S \propto \nu^{-\alpha}$).

The limit on the product $\beta cos\theta$ (and therefore the limits on β and θ) spans over a wide range of values. With the values of R given in Tab. 1 and assuming $\alpha \sim 0.5$ for all sources, we find $\beta cos\theta \leq 0.019$ for the symmetric 3C338, then we find values going from $\gtrsim$0.49 and $\gtrsim$0.50 for 0836+29 and 3C465 respectively to $\gtrsim$0.65 and $\gtrsim$0.81 for NGC315 and M87 respectively. With these values the upper limits to the angles to the line of sight are in the range 36°-61°. For 3C338 we derive $\theta \gtrsim 88°$. In Fig. 2a and 2b we plot the values derived for $\beta cos\theta$ and for θ_{max} respectively. The corresponding lower limits for β_{min} are in the range $\gtrsim$ 0.65 - 0.85. This leads to $\gamma \sim 1.3 - 2$.

TABLE I - Brightness Ratio for FR-Is

Name		d	R
0055+30	NGC 315	10	$\gtrsim$ 50
0755+37	NGC2484	1.1	$\gtrsim$ 20
0836+29	4C29.30	6.5	$\gtrsim$ 15
1228+12	M87	1.4	$\gtrsim$290
1626+39	3C338	1	$\gtrsim$ 1.1
2335+26	3C465	1.3	$\gtrsim$ 16
	Cen A	0.5	>> 100
	NGC6251	2.7	$\gtrsim$100

DISCUSSION

The results emerging from this analysis on a still small sample of FR-I radio galaxies suggest that the radio jets in this class of radio galaxies start out with relativistic speeds. Such result is expected by the Laing model (Laing, these proceedings) and by the changes in the plasma speed along the kpc-scale jets for the same class of objects (Parma et al., these proceedings). Furthermore, on average, the maximum viewing angles derived for these objects are consistent with the idea that FR-I radio galaxies are the parent population of BL-Lacs seen at large viewing angles, the rough limit between these two classes being $\theta \sim 30°$ (Ghisellini et al., 1993a).

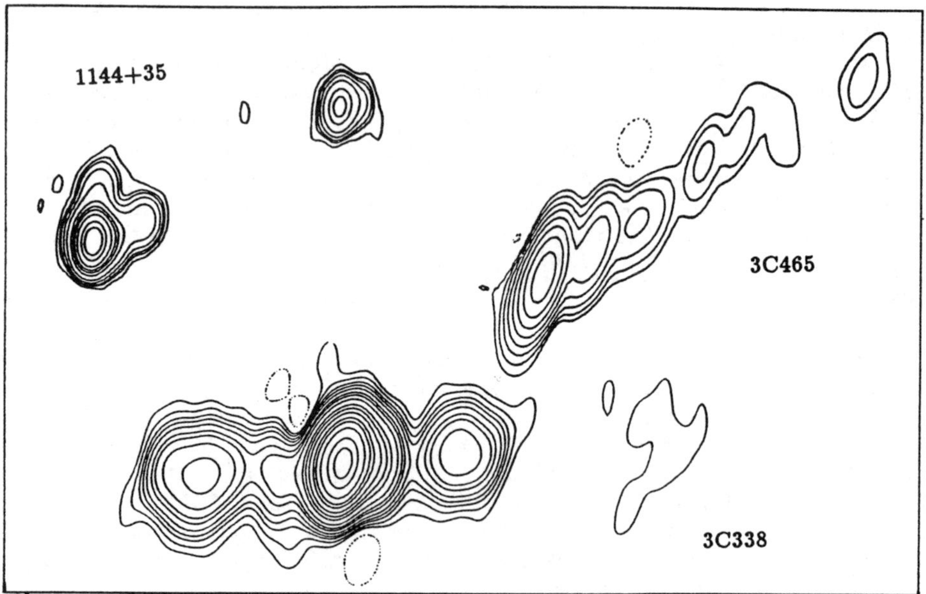

Fig. 1 - VLBI maps of three sources in our sample: 1144+35 - 5 GHz, restoring beam 2.7×1.5 mas, p.a. −4°, the peak in the map is 0.314 Jy/beam. 3C465 - 5 GHz, restoring beam 2.8×2.1 mas, p.a. −10°, the peak in the map is 132 mJy/beam. 3C338 - 1.6 GHz, restoring beam is 8×4 mas, p.a. −17°, the peak flux is 85 mJy/beam.

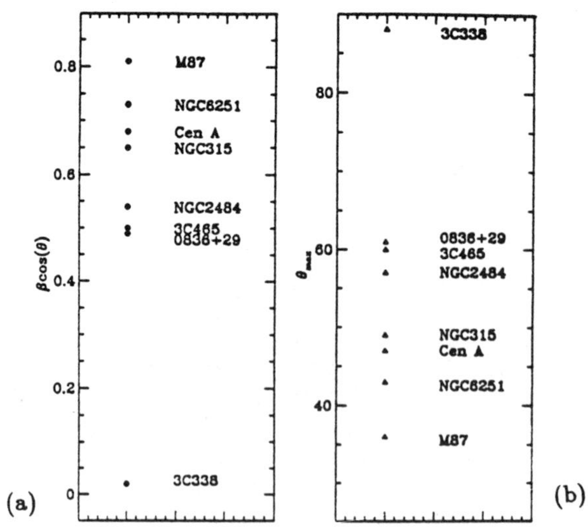

Fig. 2 - Distribution of (a) the product $\beta cos\theta$ and (b) the maximum allowed viewing angle $\theta_{\max}$ for the FR-I sources in groups A and B .

There is however a matter of concern. The frame we have just described presents some problems when applied to the best studied sources on the VLBI scale. For M87 and NGC315 the value of $\beta cos\theta$ and the derived limit on θ are very extreme for FR-I radio galaxy.

In M87 the asymmetry is really high, and the range of plausible values for β and θ is very small. Only for $\beta \rightarrow 1$ we obtain a viewing angle barely in agreement with the unification model. Furthermore, the subluminal motion detected along the parsec-scale jet in this source ($\beta_{app} \sim 0.3$, Biretta, 1993) is much lower than expected if $\beta \rightarrow 1$.

A similar problem occurs in NGC315. The source has a very large angular size, which we must take into account when we select the range of plausible values for θ. Only if we take $\beta \rightarrow 1$ we obtain an acceptable value for θ (Venturi et al., 1993). With such high value of β, we would expect to see proper motion in the knots along the VLBI jet. Comparison between old 6 cm data (Preuss, 1983) and our more recent data published at the same frequency indicates that the morphology of this source remained pretty stable in almost ten years, even allowing for the different quality of the two maps. On the basis of this comparison, we can place an upper limit on the intrinsic plasma speed $\beta < 0.5$ along the jet of NGC315, and this is in contrast with the high parsec-scale brightness asymmetry.

The same kind of inconsistency occurs in Centaurus A. Subluminal motion with $\beta_{app} = 0.26$ (Meier et al., 1993) is detected along the VLBI jet of this source, but this number is lower than expected on the basis of the VLBI asymmetry.

A possibile solution to this inconsistency is the assumption that the pattern speed is not representative of the fluid speed. If the bright knots in the VLBI maps of FR-I radio galaxies are standing shocks rather than moving blobs, no motion is expected. In the case of M87, Biretta (1993) suggested that $\beta_{intr} \sim 2\beta_{pattern}$. However, on the basis of other arguments Ghisellini et al. (1993b) concluded that for low luminosity radio galaxies $\beta_{intr} \sim \beta_{pattern}$.

Another way of overcoming the problem was suggested by Bicknell (1993, these proceedings). He pointed out that if the blobs are longitudinal shocks, then the pattern speed is much lower than expected on the basis of simple Doppler asymmetry.

REFERENCES

Biretta, J.A. 1993, in Astrophysical Jets, in press

Feretti, L. et al. 1993, *ApJ* **408**, 446

Ghisellini, G. et al., 1993a, *ApJ* **407**, 65

Ghisellini, G. et. al. 1993b, in testing the AGN Paradigm, in press

Jones, D.L. et al. 1986, *ApJ* **305**,684

Meier, D.L. et al. 1993, in Subarcsecond Radio Astronomy, in press

Pearson, T.J., Zensus A. 1987, in Superluminal Radio Sources, Eds. A. Zensus & T.J. Pearson, Cambridge University Press, 1

Preuss, E. 1983, in Astrophysical Jets, Eds. A. Ferrari & A.G. Pacholczyk, Dodrecht Reidel, 1

Reid, J.M. et al. 1989, *ApJ* **336**,112

Venturi, T. et al., 1993, *ApJ* **408**, 81

The First Stromlo Symposium: The Physics of Active Galaxies
ASP Conference Series, Vol. 54, 1994
G. V. Bicknell, M. A. Dopita, and P. J. Quinn (eds.)

TOWARDS A TRULY UNIFIED SCHEME FOR AGN

STEVE RAWLINGS
Astrophysics, Department of Physics, Keble Road, Oxford, OX1 3RH, UK

ABSTRACT

Truly 'Unified Schemes' require more variables than viewing angle θ if they are to encompass all AGN. Adopting the massive black hole (MBH) hypothesis, I argue that the additional variables accretion rate $\dot{M}$ and black hole mass M_{BH} are necessary, and may in broad terms be sufficient. For radio-loud AGN there is a large range of $\dot{M}$, with M_{BH} fairly constant at $\sim 10^{9.5}\ M_{\odot}$; for radio-quiet AGN, the range of $\dot{M}$ scales with M_{BH} so as to remain close to the Eddington rate $\dot{M}_{Edd}$. I present a plot which separates radio-loud and radio-quiet AGN according to the efficiency with which accretion power is channeled into relativistic jets. By noting new evidence that some radio-quiets are in elliptical galaxies, I suggest that radio-louds evolve from radio-quiets once $\dot{M}$ falls sufficiently below $\dot{M}_{Edd}$ to trigger an efficient-jet-producing mechanism.

UNIFIED SCHEMES: WHY θ IS NOT ENOUGH

There is now little doubt that AGN properties depend on viewing angle θ, but theories of AGN demand that any truly 'Unified Scheme' be based on more fundamental parameters (e.g. Blandford 1990). To constrain these parameters observationally, we have been measuring AGN output in ways which are as independent of θ as possible (e.g. Rawlings 1992). Two of the most useful measures, radio luminosity L_{rad} and [OIII] narrow-line luminosity $L_{[OIII]}$, are compared in Figure I.

Radio-selected AGN

The correlation between L_{rad} and $L_{[OIII]}$ for radio-selected AGN is seen in the upper part of Figure I. We have interpreted this correlation, its slope (≈ 0.85), and the origin of its scatter, in terms of a physical link between AGN photoionizing luminosity L_{phot} and bulk kinetic power Q in the jets (Rawlings and Saunders 1991); implications of this result are discussed by Rawlings (1992) and Saunders (1992). Radio-loud quasars and radiogalaxies of similar $L_{[OIII]}$ are unified using a dusty torus to provide a θ-dependent obscuration. One needs also a large range of $L_{[OIII]}$, and hence L_{phot}, which adopting the massive black hole (MBH) hypothesis, requires a large range of accretion rate $\dot{M}$.

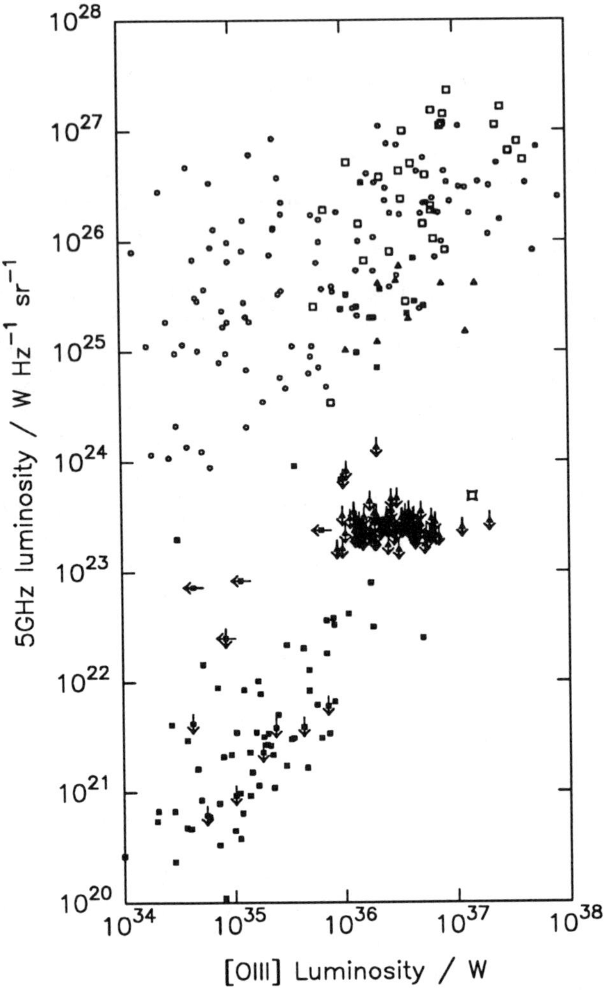

FIGURE I The radio luminosity, [OIII] luminosity plane for AGN; see Miller, Rawlings and Saunders (1993b) for full references and a more detailed discussion. Objects from complete radio-selected samples are plotted as open symbols, with squares representing broad-line AGN, and circles, narrow-line AGN. Objects from optically-selected samples are plotted as filled symbols: quasars from a complete $z < 0.5$ sample (Miller, Rawlings and Saunders 1993a) are marked by squares; triangles represent a complete sample of higher-z quasars (Miller, Peacock and Mead 1990). E1821+643, discussed in the text, is represented by the squashed-square.

Optically-selected AGN

Some AGN selected by UV-excess are precisely the same objects as those in radio-selected samples, and thus they too fall on the upper radio-loud correlation. Most, however, obey the different L_{rad}–$L_{[OIII]}$ correlation seen in the lower part of Figure I. We have also interpreted this correlation in terms of a link between L_{phot} and Q (Miller *et al.* 1993a), the key difference being that the 'radio-quiet' jets, although relativistic, have values of Q/L_{phot} which are $\sim 10^3$ lower than similar-L_{phot} radio-louds. The gap between the two correlations is a consequence of this; so too is the existence of a few objects in the gap since, when θ is small, L_{rad} becomes dominated by Doppler-boosted emission from the relativistic jet. A range of $\dot{M}$ is again required to explain the range of $L_{[OIII]}$.

Other AGN, and a definition of radio-quiet

By plotting any AGN on Figure I one can determine which of the two correlations it is likely to belong to, and thus whether it is radio-loud or radio-quiet. Some warnings are, however, in order: (i) both L_{rad} and $L_{[OIII]}$ can be polluted by non-AGN processes, e.g. those associated with a starburst (see Miller *et al.* 1993a), moving points up and to the right; (ii) Doppler-boosting of the emission from small-θ jets can move objects upwards from either correlation (e.g. quasars in the gap in Figure I, or the BL-Lac objects shown in Figure 3 of Rawlings 1992); (iii) there exist objects with anomalously weak [OIII] (Miller *et al.* 1993a), due e.g. to low narrow-line covering factor or excessive reddening, which can appear to the left of either correlation.

As an example of the usefulness of Figure I, consider the case of E1821+643. We see that this object lies very clearly on the radio-quiet correlation. However, because this quasar has extreme $L_{[OIII]}$ and is relatively nearby ($z \approx 0.3$), it has enough radio flux to make it a 'radio-quiet' member of a relatively bright radio-selected sample (Lacy, Rawlings & Warner 1992).

UNIFIED SCHEMES: WHY θ AND $\dot{M}$ ARE NOT ENOUGH

The properties of the broad emission lines in UV-selected quasars suggests that black hole mass M_{BH} is necessarily a third variable in our Unified Scheme. The most straightforward explanation of the correlation between broad-line FWHM and $L_{[OIII]}$ is that the broad-line region is gravitationally bound, and that $\dot{M}$ scales with M_{BH} so as to remain close to the Eddington rate $\dot{M}_{Edd}$ (see Figure 7 of Miller *et al.* 1992). Radio-loud quasars have values of $\dot{M}/\dot{M}_{Edd}$ which are within the range spanned by the radio quiets, but which are systematically a little low, with $M_{BH} \sim 10^{9.5} M_{\odot}$ (Miller *et al.* 1993b). There are too few low-redshift broad-line radio-louds to get similar estimates of M_{BH} at lower $L_{[OIII]}$, but the small spread in their stellar luminosities is compelling evidence that we are dealing with similar systems. Remembering too the huge stored energies in the lobes of some low-$L_{[OIII]}$ radiogalaxies (Rawlings and Saunders 1991), it seems probable that there is no scaling between $\dot{M}$ and M_{BH}, and that $M_{BH} \sim 10^{9.5} M_{\odot}$ for all radio louds.

UNIFIED SCHEMES: WHY θ, $\dot{M}$ AND M_{BH} MAY BE ENOUGH

How do we unify radio-quiets and radio-louds? First, recall that the gap in Figure I results not from a lack of relativistic jets in radio-quiets, but from values of Q/L_{phot} which are $\sim 10^3$-times lower than the near-unity values achieved by efficient radio-loud jets.[1] This suggests two distinct modes of accretion which channel power into relativistic jets with very different efficiencies.

Second, let us question the 'fact' that AGN in elliptical galaxies always produce radio-loud objects (e.g. Scheuer 1992). We now know of at least one radio-quiet, E1821+643, which since it is almost certainly the cD galaxy of a very rich cluster (Lacy, Rawlings and Hill 1992), is almost certainly a massive elliptical. Moreover, by comparing $z \sim 0.2$ radio-quiets and radio-louds of similar L_{phot}, Dunlop *et al.* (1993) find similar galaxy luminosities for each class, and find that both follow the same near-infrared Hubble relation as radiogalaxies. Giant elliptical galaxies with super-massive black holes (SMBH; e.g. those with $\sim 10^{9.5}\ M_{\odot}$) can seemingly foster radio-quiets as well as radio-louds.

Can we explain these properties using $\dot{M}$ and M_{BH} or, as Blandford (1990) suggests, do we also require a parameter to measure MBH spin? I will conclude by outlining a scenario which requires that MBH are spinning, *but does not use a spin parameter to unify radio-quiets and radio-louds.*

Suppose that the efficient-jet-producing mechanism shuts down at some critical value of $\dot{M}$ near $\dot{M}_{Edd}$, e.g. if the inner disc must form an ion-supported torus to extract spin energy from the MBH (Rees *et al.* 1982). Conditions in the present-day Universe are not conducive to feeding SMBHs at close to $\dot{M}_{Edd}$ (for possible explanations see Ellingson, Green and Yee 1991; and Small and Blandford 1992). If SMBHs are found only in giant ellipticals, we can see why all radio-loud AGN accrete at sub-Eddington rates (Rawlings and Saunders 1991; Miller *et al.* 1993b), and why all low-z radio-louds are in ellipticals.

At earlier times in the Universe, we will find SMBHs accreting close to the Eddington limit, and thus we expect ellipticals with high $\dot{M}$ but inoperative efficient-jet-producing mechanisms. I suggest that the newly-discovered radio-quiet ellipticals are the first known examples of such objects. The unification I propose therefore is that radio-louds are radio-quiets in which $\dot{M}$ has fallen sufficiently to allow the operation of an efficient-jet-producing mechanism.[2]

How do we test this hypothesis? Since the switch from radio-quiet to radio-loud involves only a change in the 'inner disc', one expects little change in L_{phot}, and only subtle changes in associated properties. Apart from differences which are *caused* by a powerful jet, radio-quiet and radio-loud AGN are remarkably similar, and in cases of minor difference, e.g. in the optical FeII lines, the likely

---

[1] Using estimates of Q, Celotti and Fabian (1993) argue in favour of electron-proton jets for radio-loud AGN. By changing to electron-positron jets, while keeping other parameters constant, Q is lowered by a factor $\sim 10^3$, just as we require for the radio-quiets.

[2] Is this the ultimate fate of all radio-quiet MBH? It is possible that many spiral-based AGN have too low M_{BH}, and too plentiful a supply of fuel for $\dot{M}$ to have dropped sufficiently below $\dot{M}_{Edd}$. The nearest starved-Seyfert candidate is the Galactic Centre: Falcke, Mannheim and Biermann (1993) claim this has $Q/L_{phot} \sim 1$, in line with values achieved by the efficient-jet-producing mechanism of radio-loud AGN.

explanation is a changed geometry on small scales (Miller *et al.* 1993a).

A more fruitful approach might be to consider the implications of radio-quiets evolving into radio-louds. For example, one expects similar cluster environments for both classes: Ellingson, Yee and Green (1991) have begun to address this question, but not as yet with samples which are sufficiently well matched in L_{phot} and redshift. Other implications arise if a radio-loud phase is followed immediately by a radio-quiet phase, predicting, for example, a population of young radiosources (the GHz peaked-spectrum sources?) with a narrow range in de-reddened $L_{[OIII]}$.

ACKNOWLEDGMENTS

I thank the organisers for arranging financial support, and the UK SERC for an Advanced Fellowship. I acknowledge my collaborators Philip Miller and Richard Saunders, and thank Mark Lacy and Stephen Serjeant for useful discussions.

REFERENCES

Blandford, R.D., 1990, in *Active Galactic Nuclei*, R.D. Blandford, H. Netzer and L. Woltjer, Springer-Verlag, 161

Celotti, A. and Fabian, A.C., 1993, *MNRAS*, **264**, 228

Dunlop, J.S., Taylor, G.L., Hughes, D.H. and Robson, E.I., 1993, *MNRAS*, in press

Ellingson, E., Green, R.F. and Yee, H.K.C, 1991, *ApJ*, **378**, 476

Ellingson, E., Yee, H.K.C. and Green, R.F., 1991, *ApJ*, **371**, 49

Falcke, H., Mannheim, K. and Biermann, P.L., 1993, *A&A*, in press

Lacy, M., Rawlings, S. and Hill, G.J., 1992, *MNRAS*, **258**, 828

Lacy, M., Rawlings, S. and Warner, P.J., 1992, *MNRAS*, **256**, 404

Miller, L., Peacock, J.A. and Mead, A.R.G., 1990, *MNRAS*, **244**, 207

Miller, P., Rawlings, S. and Saunders, R., 1993a, *MNRAS*, **263**, 425

Miller, P., Rawlings, S. and Saunders, R., 1993b, *MNRAS*, in preparation

Miller, P., Rawlings, S., Saunders, R. and Eales, S., 1992, *MNRAS*, **254**, 93

Rawlings, S., 1992, in *Extragalactic Radio Sources – From Beams to Jets*, eds. J. Roland, H. Sol and G. Pelletier, C. U. P., 332

Rawlings, S., and Saunders, R., 1991, *Nature*, **349**, 138

Rees, M.J., Begelman, M.C., Blandford, R.D. and Phinney, E.S., 1982, *Nature*, **295**, 17

Saunders, R., 1992, in *Extragalactic Radio Sources – From Beams to Jets*, eds. J. Roland, H. Sol and G. Pelletier, C. U. P., 344

Scheuer, P.A.G., 1992, in *Extragalactic Radio Sources – From Beams to Jets*, eds. J. Roland, H. Sol and G. Pelletier, C. U. P., 368

Small, T.A. and Blandford, R.D., 1992, *MNRAS*, **259**, 725

UNIFIED SCHEMES

Contributed Papers

The First Stromlo Symposium: The Physics of Active Galaxies
ASP Conference Series, Vol. 54, 1994
G. V. Bicknell, M. A. Dopita, and P. J. Quinn (eds.)

MODELS OF DUSTY DISCS IN ACTIVE GALACTIC NUCLEI

A. EFSTATHIOU AND J.H. HOUGH
Division of Physical Sciences, University of Hertfordshire, Hatfield, Herts AL10 9AB, UK

M. ROWAN-ROBINSON
School of Mathematical Sciences, Queen Mary & Westfield College, Mile End Road, London E1 4NS, UK

INTRODUCTION

One of the challenges faced by unified theories for Active Galactic Nuclei (AGN) is to match the observed infrared emission to that predicted for the optically and geometrically thick dusty discs believed to obscure the central source of radiation. We have used the method of Efstathiou & Rowan-Robinson (1990), generalised to take into account a distribution of grain species and sizes (Efstathiou & Rowan-Robinson 1993), to compute models that assume different density distributions. The method treats multiple scattering from grains.

RESULTS AND DISCUSSION

The assumed disc geometry is illustrated in Figure I. The dust density is assumed to follow a power-law in radius ($n \propto r^{-\beta}$) and the energy distribution of the central source a power-law in frequency ($F_\nu \propto \nu^{-\alpha}$). The total equatorial ultraviolet (1000Å) optical depth τ_{uv} is another model parameter. The grain mixture of Rowan-Robinson (1992) is used for the calculations. We have computed models with $\beta = 0, 1$ and 2 and τ_{uv} in the range 100 – 2000. The main conclusions from this study are as follows:

(1) All of the models predict significant anisotropy in the infrared emission. The anisotropy is minimised with decreasing τ_{uv} and increasing H/r_2.

(2) The predicted spectra of models with $\beta = 1, 2$ are narrower, peak at shorter wavelengths and also show 'shallower' $10\mu m$ absorption features compared with uniform density models.

(3) Models with $\beta = 1, 2$ show 'featureless' mid-infrared spectra over a wider range of parameter space than the uniform density models.

We have also compared predicted spectra with the observed spectrum of NGC1068. There is no direct estimate of the value of the extinction to its centre. It is usually assumed that the absence of hard X-rays implies an A_V of the order of 1000 (e.g. Pier & Krolik 1993.) There is equally good evidence, however, for an A_V in the range 20-30 (Bailey et al. 1988, Bridger et al. 1993.) In Figure II we present a very good fit to the spectrum of NGC1068 which assumes a line of sight A_V of 28. Clearly the extinction can not be constrained by the infrared

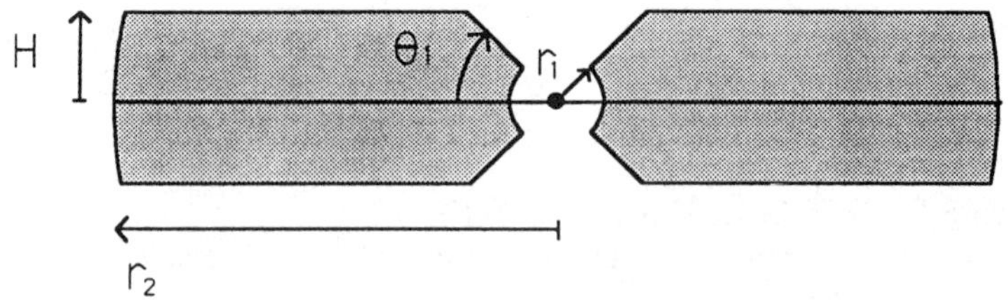

FIGURE I Schematic of the assumed disc geometry.

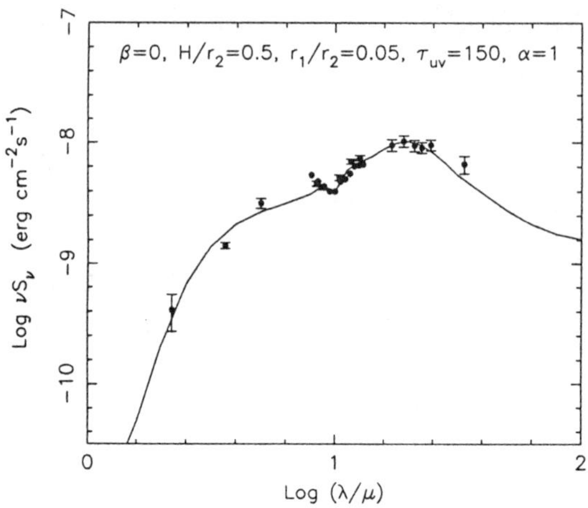

FIGURE II Fit to the spectrum of NGC1068 with a model assuming $\Theta_1 = 60^o$. The A_V along the line of sight is 28 mag. and the predicted viewing angle is 34° from the equator. Data from Rieke & Low (1975) and Roche et al. (1991).

spectrum alone. An independent approach such as spectropolarimetry is needed in order to make further progress.

REFERENCES

Bailey, J., Axon, D., Hough, J., Ward, M., McLean, I., and Heathcote, S. 1988, *MNRAS*, **234**, 899

Bridger, A., Wright, G.S., and Geballe, T.R. 1993, preprint

Efstathiou, A., and Rowan-Robinson 1990, *MNRAS*, **245**, 275

Efstathiou, A., and Rowan-Robinson 1993, *MNRAS*, in press

Pier, E., and Krolik, J., 1993, *ApJ*, submitted

Rieke, G.H., and Low, F.J. 1975, *ApJ*, **199**, L13

Roche, P.F., Aitken, D.K., Smith, and Ward, M.J., 1991, *MNRAS*, **248**, 606

Rowan-Robinson, M. 1992, *MNRAS*, **258**, 787

The First Stromlo Symposium: The Physics of Active Galaxies
ASP Conference Series, Vol. 54, 1994
G. V. Bicknell, M. A. Dopita, and P. J. Quinn (eds.)

VLBI OBSERVATIONS OF A COMPLETE SAMPLE OF RADIO GALAXIES

G. GIOVANNINI, L. FERETTI
Dipartimento di Astronomia and Istituto di Radioastronomia - Bologna

T. VENTURI
Istituto di Radioastronomia di Bologna

W.D. COTTON
National Radio Astronomy Observatory

L. LARA
CSIC, Instituto de Astrofisica de Andalucia

J. MARCAIDE
Universidad de Valencia

A.E. WEHRLE
Infrared Processing and Analysis Center - California

ABSTRACT We undertook a project of observation of a complete sample of radio galaxies with the VLBI technique, in order to test: a) current models on jet dynamics and b) the radio source unified schemes. A preliminary analysis based on the data obtained by us for 10 galaxies of our sample indicates that the parsec scale structure in low power radio galaxies (FR-I) and high power radio galaxies (FR-II) is essentially the same.

RESULTS

The knowledge of the structure of radio galaxies on the parsec scale is very important to test current models on jet dynamics and get new pieces of information to confirm the radio source unified schemes. To these aims, we undertook a project of observation of a **complete** sample of radio galaxies with the VLBI technique, selecting from the B2 and 3CR galaxy samples the objects with a core flux density larger than 100 mJy at 6 cm at the arcsecond resolution. This sample was presented in Giovannini et al. (1990, *ApJ* **358**, 159). It consists of 27 radio galaxies, 15 of which belong to the FR-I class, 5 to the FR-II class and 7 are unresolved or slightly resolved on the arcsecond scale.

The VLBI detailed mapping and analysis of the radio galaxies of the sample is in progress. A detailed study on the radio galaxy NGC315 (Venturi et al, 1993, *ApJ* **408**, 81) and 3C338 (Feretti et al., 1993, *ApJ* **408**, 446), has been published.

A preliminary analysis based on these data (Table I) indicates that an asymmetric VLBI morphology, i.e. core and one-sided jet, is the most frequent radio structure found among the sources mapped thus far. It is shared by 7 out of the 10 mapped radio galaxies. Among these 7 core-jet sources we find 4 FR-I, 2 FR-II and 1 CSS source. An important result is that the parsec scale structure in low and high power radio galaxies is essentially the same. Moreover in all cases the parsec scale jet is oriented on the same side of the kpc scale main jet. This correlation implies either that jets are intrinsically asymmetric or that parsec and kpc scale jets are both relativistic.

TABLE I Observed Radio Galaxies

Name		Kpc Morphology	VLBI Morphology
0055+30	NGC 315	FR-I	One-sided
0116+31	4C31.04	C	Compact Double
0410+11	3C109	FR-II	One-sided
0755+37	NGC2484	FR-I	One-sided
0836+29	4C29.30	FR-I	One-sided
1144+35		FR-I	Complex
1626+39	3C338	FR-I	Two-sided
1641+17	3C346	CSS	One-sided
1833+32	3C382	FR-II	One-sided
2335+26	3C465	FR-I	One-sided

The jet to counter-jet brightness ratio is limited by the noise level in our maps, it ranges from $\gtrsim$10 for 3C109 to $\gtrsim$ 50 for NGC315. The interpretation of these asymmetric structures as due to Doppler effect implies that all these sources are not in the plane of the sky but should have angles $<<60°$with respect to the line of sight (assuming $\beta \lesssim 1$). In fact with angles larger than 60°a jet to counter-jet brightness ratio smaller than 10 is expected even assuming $\beta \sim 1$. A counter-jet with a brightness 10 times lower than the main-jet should be visible in our VLBI maps. Therefore sources with angles in the range 90°- 60°with respect to the line of sight are expected to be seen as sources with two-sided parsec scale jets. In our sample at present only 3C338 shows a two-sided jet and therefore it is expected to be close to the plane of the sky.

A two-sided structure could be present also in 4C31.04 and 1144+35. While for the source 1144+35 VLBI data at different frequencies are necessary to understand its radio structure, in the source 4C31.04 (unresolved at arcsecond resolution) the map at 3.6 cm suggests the presence of a core emission in between of two nearly symmetric steep spectrum lobes. These complex structures with two-sided lobes, but no jet-like structure, are not fully understood yet; they resemble the double kpc scale structures (core + lobes), but on a scale smaller than 1 kpc. Similar structures have been found in the Caltech-Jodrell Bank VLBI survey and have been identified with high redshift galaxies or quasars (Conway et al., 1992, *ApJ* **396**, 62; Readhead in "Frontiers of Space and Ground-based Astronomy" by Wamsteker et al., 1993 in press). An important result of our survey is the existence of such a compact symmetric object at low redshift.

The First Stromlo Symposium: The Physics of Active Galaxies
ASP Conference Series, Vol. 54, 1994
G. V. Bicknell, M. A. Dopita, and P. J. Quinn (eds.)

A COMPARISON OF THE SPECTROPOLARIMETRICALLY IDENTIFIED HIDDEN BLR SEYFERT 2S WITH OTHER SEYFERT 2 GALAXIES

L. KAY
Barnard College and Columbia University

Spectropolarimetric observations of Seyfert 2 galaxies have provided compelling evidence that some are in fact Seyfert 1s, viewed at an angle such that the BLR is hidden from direct view. (Antonucci and Miller, 1985, Miller and Goodrich, 1990, and Tran, Miller, and Kay, 1992). However the generalizability of this model to *all* Sy 2 galaxies is still in question. Data from a spectropolarimetric study of the featureless continuum of 49 Seyfert 2 galaxies conducted at Lick Obs., and from published IR and radio data, was used to look for differences between these 9 hidden BLR Seyfert 2s (NGC 1068, Mrk 3, Mrk 348, Mrk 463E, NGC 7674, Mrk 477, Mrk 1210, NGC 7212, and Was 49) and the others.

The observations, sample selection, and method of estimating the fraction of the light in the Seyfert spectrum due to starlight from the host galaxy are described in Kay, 1990 and 1994a. The galaxy fraction F_g estimated for each Seyfert 2 was plotted versus its average polarization. The average F_g for the 9 hidden BLR objects is considerably less: 0.30, compared to 0.75 for the others. It is possible that 'real' Seyfert 2s are the ones with a higher starlight fraction; but more likely the spectropolarimetric method of looking for hidden BLRs by observing scattered light in polarized flux works best when the starlight fraction is low and the polarization signal is less complicated by dilution. There are a few Seyfert 2s with lower F_g but no hidden BLRs observed (yet), but this data did not go out to Hα where the broad line in polarized flux is most readily identified. Also there are many geometrical reasons why the spectropolarimetric method may not work in a given case (e.g. host galaxy dust lanes can block the view to the scattering region, the polarization can be contaminated by interstellar dust, or the scattering geometry may be different so that the electron scattering region which served as a mirror in some objects may not be present in others).

Two recent papers have suggested that these spectropolarimetrically detected hidden BLR Seyfert 2s differ from others in that they have considerably higher radio power (Moran *et. al.* 1992), and flatter IR spectra (Hutchings and Neff, 1991). A 20 cm distribution function for 40 Seyfert 2s was produced using data from the literature. The 9 hidden BLR objects are at the higher end of the 20 cm distribution function for these Sy 2s, but they are not alone there (Kay, 1994b). The objects first studied polarimetrically were the Markarian Seyfert 2s, originally selected for UV excess and high optical luminosity, and it is more difficult to conduct spectropolarimetry on the fainter ones. It is possible that the 'real' Seyfert 2s are the less luminous objects, or spectropolarimetry just reveals hidden BLRs best in the more luminous objects. (Several extremely luminous IRAS galaxies with Seyfert 2 spectra have also been shown by spectropolarimetry or IR spectroscopy to have hidden BLRs (e.g. Hough *et. al.* 1991, Kay, Antonucci, and Coleman, 1992). It is problematic to compare luminosity functions until at the least, Hα spectropolarimetry is available for a well-defined complete sample of Seyfert 2s.

An IR color luminosity diagram as in Hutchings and Neff (1991) was created using published IRAS data for 40 of these Seyfert 2s. The IR spectra of the hidden BLR Seyferts do tend to be flatter, but they do not comprise a distinct group (Kay, 1994b). Perhaps the flatness arises from the geometry of our viewing angle to the system. Since the observed featureless continuum polarizations are so low, Miller and Goodrich (1990) suggested that those detected in reflected light by spectropolarimetry are the ones we see at relatively high inclinations, observing just below the rim of the scattering cone. In their models of dusty tori around AGN, Pier and Krolik (1992) calculated that the infrared color of dust tori would be warmer for the face-on viewing angles.

Comparisons of the spectropolarimetrically identified hidden BLR Seyfert 2s with other Sy 2s are difficult: these objects do not form a complete sample, spectropolarimetry works best when the object is luminous and the dilution of the polarization by starlight is lower, and *not* seeing a BLR in the polarized flux can be due to the geometry of the system and the difficulty of the method. Additional strong candidates for Sy 2s that are hidden Sy 1s come from off-nuclear spectroscopy indicating a broad Hα component (Shields and Filippenko, 1988), from broad H lines identified in IR spectra (Blanco, Ward and Wright, 1990, Goodrich, Veilleux and Hill, 1992), and from ionization cones, often bipolar, seen in many Sy 2s, (e.g. Storchi-Bergman, *et. al.* 1992). The inclusion of these candidates in any comparison might diminish some of the selection effects arising from identifying hidden BLR Seyfert 2s solely by spectropolarimetry.

REFERENCES

Antonucci, R.R.J., and Miller, J.S. 1985, *ApJ*, **297**, 621.

Blanco, P.R., Ward, M. J., and Wright, G.S. 1990, *MNRAS*, **242**, 4p.

Goodrich, R.W., Veilleux, S. and Hill, G. 1992, *BAAS*, **24**, 1209.

Hough, J., Brindle, C., Wills, B., Wills, D., and Bailey, J. 1991, *ApJ*, **372**, 478.

Hutchings, J.B., and Neff, S.C. 1991, *AJ*, **101**, 434.

Kay, L.E. 1994a, 1994b, in prep.

Kay, L.E. 1990, Lick Observatory PhD thesis.

Kay, L.E. Antonucci, R.R.J., and Coleman, P. 1992, *BAAS*, **24**, 1173.

Miller, J.S. and Goodrich, R.W. 1990, *ApJ*, **355**, 456.

Moran, E., Halpern, J., Bothun, G., and Becker, R. 1992, *AJ*, **104**, 990.

Pier, E. and Krolik, J. 1992. *ApJ*, **339**, L23.

Shields, J.C. and Filippenko, A.V. 1988, *ApJ*, **332**, L55.

Storchi-Bergmann, T., Mulchaey, J.S., and Wilson, A.S. 1992, *ApJ*, **395**, L73.

Tran, H., Miller, J.S., and Kay, L. 1992, *ApJ*, **397**, 452.

The First Stromlo Symposium: The Physics of Active Galaxies
ASP Conference Series, Vol. 54, 1994
G. V. Bicknell, M. A. Dopita, and P. J. Quinn (eds.)

TESTING UNIFIED SCHEMES WITH A COMPLETE SAMPLE OF RADIO SOURCES

R. MORGANTI[1]
Istituto di Radioastronomia, via Irnerio 46, 40126 Bologna (Italy)

R.A.E. FOSBURY[2]
ST-ECF, Garching (Germany)

T.A. OOSTERLOO
Australian Telescope National Facility, CSIRO, Epping (Australia)

C.N. TADHUNTER
Department of Physics, University of Sheffield, (UK)

INTRODUCTION

Optical and radio data are both important and complementary in the attempt to understand the mechanisms at work in Active Galactic Nuclei (AGN). Many interesting results have been obtained in such studies (Morganti et al.1992 and references therein), especially concerning the existence of correlations between radio and optical properties and the use of *unified schemes* as tool to describe the observed characteristics of radio sources. In spite of this, a study of a complete sample of radio sources using an homogeneous set of both optical and radio data has never been presented. We undertook such a study for a complete subset of the Wall & Peacock (1985) 2.7 GHz sample, selected using the following criteria: declinations $\delta < 10°$ and redshifts $z < 0.7$. Our subsample is complete down to a flux density of 2 Jy. The resulting datasets (set of lines for each object, i.e. [OII]λ3727, [OIII]λ5007 and Hβ, and spectral index of the the underlying continuum) and radio (core and total flux and morphology) are presented in Tadhunter et al. (1993) and Morganti et al. (1993).

THE SIMULATION

These data confirm the existence of a correlation between core and total radio power (Feretti et al., 1984) and between [OIII] luminosity and total radio power (Figs.1a,b). Although already known, the latter looks, from our data, to be quite interesting since it shows two populations (see Fig 1b). These populations do not divide cleanly into FRI and FRII although the higher powered group contain only FRII and compact sources. We have applied a Montecarlo simulation to

[1]Visiting Astronomer, Australian Telescope National Facility, CSIRO, Epping (Australia)

[2]Affiliated to the Astrophysics Division, Space Science Department, European Space Agency

these data to ascertain if these correlations can be explained by unified schemes for AGN. From a luminosity function (de Ruiter et al. 1990) and assuming that the sources are uniformly distributed out to z_{max}=0.7, we reproduce the distribution of the extended radio power and redshifts. The power of the cores was estimated using the approach of Padovani & Urry, 1992. The observed core flux is: $F^{obs}_{core}=\delta^p\ F^{intr}_{core} = f\ \delta^p\ F_{ext}$, where δ ($\delta = [\gamma(1-\beta\ cos\Theta)]^{-1}$) and f the ratio between core flux and unbeamed radio emission (F_{ext}). Moreover a distribution of γ is assumed (again following Padovani & Urry 1992). Among all the objects drawn from the LF, only those with total radio flux (extended plus observed core flux) higher than 2Jy belong to the simulated sample.

The crosses in Fig.1a,b represent the objects in the simulated sample and they appear to be in very good agreement with the observed sample (filled circles). We can reproduce the slope of the logP_{core} vs logP_{tot} correlation by assuming that the observed core power is boosted or dimmed depending on the combination of opening angle of the jet ($1/\gamma$) and the angle formed by the jet with the line of sight. The observed data are described by *two different distributions of* γ *for low-powered (mainly FRI) and high-powered (mainly FRII) radio sources*: For the former γ should be around 1.5-2 while for the latter greater then 5. The log$L_{[OIII]}$ vs logP_{tot} correlation is derived assuming that $L_{[OIII]}$ is a function of the intrinsic core power and γ. The proportionality with γ is likely to tell us something about the capability of the "monster" to accelerate the radio plasma and to produce UV photons. To support this we have found that a correlation exists between the optical spectral index and the [OIII] luminosity.

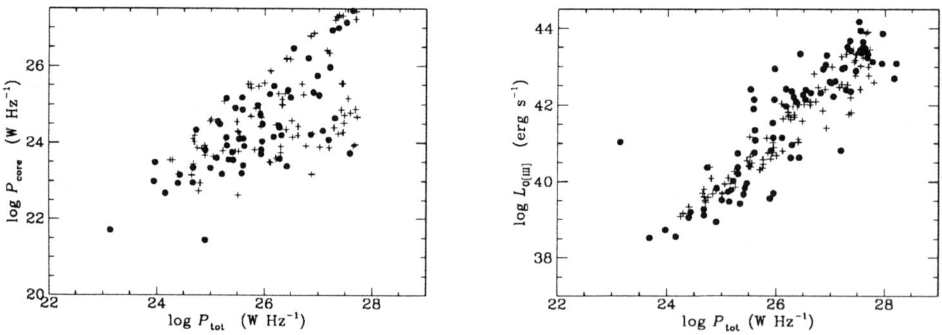

REFERENCES

Feretti, L. et al., 1984, *A&A*, **139**, 55

Morganti R., Ulrich M.-H. & Tadhunter C.N. 1992, *MNRAS*, **254**,546

Morganti R., Killeen N. & Tadhunter C.N. 1993, *MNRAS*, in press

de Ruiter, H. et al., 1990, *A&A*, **227**, 351

Padovani P.& Urry C.M. 1992, *ApJ*, **387**, 449

Tadhunter et al. 1993 *MNRAS*, in press

Wall & Peacock 1985, *MNRAS*, **216**,173

The First Stromlo Symposium: The Physics of Active Galaxies
ASP Conference Series, Vol. 54, 1994
G. V. Bicknell, M. A. Dopita, and P. J. Quinn (eds.)

FURTHER EVIDENCE AGAINST THE ORIENTATION BASED UNIFIED SCHEME FOR QUASARS AND FR II RADIO GALAXIES

ASHOK K. SINGAL
Netherlands Foundation for Research in Astronomy, Postbus 2, 7990 AA Dwingeloo, The Netherlands. E-mail: singal@nfra.nl

INTRODUCTION

According to the pure orientation based unified scheme, powerful radio galaxies and quasars both belong to the same parent population, but with quasars aligned at a smaller angle ($\psi < \psi_c \sim 45°$) with respect to the observer's line of sight (Barthel 1989). Under this hypothesis, the observed differences in the radio sizes of galaxies and quasars are merely due to the difference in their geometric projection effects. Strong evidence against this simple unified scheme for radio galaxies and quasars, from the observed large variations in the ratio of their number densities as well as that of their radio sizes across the luminosity-redshift (P–z) plane, has been published recently (Singal 1993a,b).

Here we present some additional evidence against the unified scheme models, where a popular notion appears to be that in the low frequency (meter–wavelengths) selected samples, the observed number density as well as the median value of angular sizes ($\theta_{\rm med}$) of quasars are about half as large as those of radio galaxies. This impression seems to have been formed based purely on the identification content of the 3CR sample (Laing et al. 1983) in a limited redshift range ($0.5 < z < 1$), and is indeed the reason for the oft-quoted value of $\psi_c \sim 45°$. But there are trends present in other low-frequency samples that appear to contradict this notion. We have necessarily kept our discussion to the samples formed at low frequencies where sources get selected based largely on their lobe emission. In fact in our discussion we have tried to exclude flat spectrum sources ($\alpha \leq 0.5$, with $S \propto \nu^{-\alpha}$) which mostly are core–dominant cases and where the relativistic beaming might introduce serious selection effects.

THE SAMPLE

Our results are based on a homogeneous radio sub-sample of B3 sources for which the relevant radio and optical information is given by Vigotti et al. (1989). We have retained all sources with $S_{408} > 0.63$ Jy in the sample of Vigotti et al. where they have shown explicitly from the magnitude distribution of the quasars (actually blue stellar objects) identified on the Palomar Sky Survey Prints (PSSPs), that for this radio flux density limit in their sample, there may only be a few quasars among the empty fields. This trend, already known in the literature for some time, allows us to estimate the number of quasars in a sufficiently radio-bright sample from its identification contents on PSSPs even if all sources in the sample have not yet been identified on deeper optical plates.

RESULTS AND DISCUSSION

We find that there are only about 56 quasars among a total of 401 sources (excluding all flat spectrum cases) in this sample (fraction of quasars, $f_q \sim 0.14$). This is quite different from the fraction seen in the total 3CR sample ($f_q \sim 0.25$, see Singal 1993a). In fact f_q appears to decrease steadily with decreasing flux density level. By subdividing our sample into different flux density limits we find that,
for $S_{408} \geq 1.6$ Jy ($S_{\rm med} \approx 2.5$ Jy), $f_q \sim 0.17$,
for $1.6 > S_{408} \geq 0.8$ Jy ($S_{\rm med} \approx 1$Jy), $f_q \sim 0.12$,
and for $0.8 > S_{408} > 0.63$ Jy ($S_{\rm med} \approx 0.7$Jy), $f_q \sim 0.11$.
Further credence to these values is lent by the 1Jy sample (Allington-Smith 1982) that has reached an almost complete optical identification level and where for $S_{\rm med} \sim 1.2$ Jy, $f_q \sim 0.14$. Such a variation of f_q with flux density is not expected in the popular unified scheme models (with $\psi_c \sim 45°$!), under the assumptions of which we expect the space and luminosity distributions (and as a consequence the flux density distributions) of quasars and galaxies to be similar. In fact, from the f_q values at $\lesssim 1$ Jy levels, one might infer that $\psi_c \lesssim 30°$. But then $\theta_{\rm med}$ for galaxies should be larger than that for quasars by a factor exceeding 2.5.

As was already pointed out in Singal (1988, Fig. 3), at the highest flux density levels ($S_{408} \gtrsim 10$ Jy) the angular sizes of quasars may indeed be smaller than those of the galaxies (by a factor of 2 or more), but at $\lesssim 1$ Jy levels there appears to be no significant difference between their measured sizes. ¿From our present B3 sample we find that $\theta_{\rm med}$ for galaxies is only 1.3 ± 0.2 times as large as that for quasars, implying $\psi_c \sim 72°$. It should be noted that a value of $\lesssim 1.5$ for the ratio of $\theta_{\rm med}$ between galaxies and quasars implies $\psi_c \gtrsim 60°$, and that in turn requires that in the selected sample the *number* of quasars seen should be *equal to* or *even larger* than that of galaxies. As discussed above, the observed number ratio always appears to be the other way around. Thus it is clear that even if we were to allow the value of ψ_c to be a free parameter, the observed number ratio and the size ratio cannot both simultaneously be explained in the orientation based unified scheme models for radio galaxies and quasars.

Although complete optical identifications and redshift determinations may ultimately be necessary for a final confirmation of these results, the trends that are already visible seem inconsistent with the unified scheme models.

REFERENCES

Allington-Smith, J. R. 1982, *MNRAS*, **199**, 611

Barthel, P. D. 1989, *ApJ*, **336**, 606

Laing, R. A., Riley, J. M., and Longair, M. S. 1983, *MNRAS*, **204**, 151

Singal, A. K. 1988, *MNRAS*, **233**, 87

Singal, A. K. 1993a, *MNRAS*, **262**, L27

Singal, A. K. 1993b, *MNRAS*, **263**, 139

Vigotti, M., Grueff, G., Perley, R., Clark, B. G., and Bridle, A. H. 1989, *AJ*, **98**, 419

The First Stromlo Symposium: The Physics of Active Galaxies
ASP Conference Series, Vol. 54, 1994
G. V. Bicknell, M. A. Dopita, and P. J. Quinn (eds.)

THE EVOLUTIONARY UNIFIED SCHEME: RADIO SOURCES IN THE $\theta - Z$ PLANE

F. VAGNETTI AND R. SPERA
Dipartimento di Fisica, Università di Roma "Tor Vergata", via della Ricerca Scientifica, I-00133 Roma, Italy

In the *evolutionary unified scheme* (Vagnetti et al. 1991, hereafter VGC), a slower evolution in the optical band is found for flat-spectrum quasars than for steep-spectrum quasars and this is attributed to the presence of a slowly evolving relativistically beamed component in addition to an isotropic component:

$$L = L_i + L_j \delta^{2+\alpha_{opt}}, \tag{1}$$

$\delta = [\Gamma(1 - \beta\cos\theta)]^{-1}$ being the Doppler factor. In turn, the slow evolution of the beam is attributed to a cosmological increase of Γ, rather than to an evolution of L_j/L_i. In the present work we consider the evolution of radio-loud quasars in the optical band as a function of the viewing angle, compare it with the average evolution of flat- and steep-spectrum quasars, and determine the cosmic evolution (increase) of Γ for three classes of models.

To study the distribution of quasars and radiogalaxies in the θ-z plane, we compare the quasar luminosity (eq. 1) with the luminosity of the host galaxies, arguing that some quasars should in fact be observed as radiogalaxies, depending on the redshift and viewing angle. This argument applies to models with constant Γ as well, in fact it is based only on two ingredients: (i) the presence of an anisotropic component, and (ii) the evolutionary nature of the quasar luminosity. We add this class of radiogalaxies to those which are possibly determined by obscuration at $\theta \gtrsim 45°$, after Barthel (1989).

It is also needed an appropriate viewing angle distribution. We generalize the formula introduced by Cohen (1989) to account for the probability that low-power sources exceed the observational flux limit, if their beamed component lies near the line of site. Including both isotropic and anisotropic components, we obtain:

$$p(\theta) \propto (1 + f\delta^{2+\alpha_r})^p \sin\theta, \tag{2}$$

p being the slope of the integral counts in the radio band.

In our evolutionary models, we assume pure luminosity evolution with exponential form $L = L_o \exp kT$ for each of the components, as it was assumed in VGC for the average evolutions of steep- and flat-spectrum quasars.

We present three classes of models.

A) Single $\Gamma(z)$. The evolutionary ratio of the two quasar classes can be written as:

$$\frac{L_s(z)}{L_f(z)} = ce^{(k_s - k_f)T} = \frac{a + \langle\delta^{2+\alpha_{opt}}\rangle_s}{a + \langle\delta^{2+\alpha_{opt}}\rangle_f}, \tag{3}$$

assuming that L_i and L_j have the same intrinsic evolution, c and a being appropriate luminosity ratios at $z = 0$. The terms in brackets are averaged (using eq. 2) within the contiguous intervals $0 - \theta^*$, $\theta^* - \theta_{max}$, where $\theta_{max} \simeq 45°$ (Barthel 1989), while θ^* can be evaluated counting flat- and steep-spectrum quasars in the same samples used by VGC, so that the integral $\int_0^{\theta^*} p(\theta)d\theta$ gives the observed fraction of flat-spectrum quasars. A slowly increasing $\Gamma(z)$ is found as a solution of eq. 3, after giving appropriate values to the parameters. $L(\theta, z)$ can then be compared to the luminosity of the host galaxy, finding a portion of the θ-z plane (at $z \lesssim 0.3$ and $\theta \gtrsim 10°$), where quasar remnants are expected to be outshined by their hosts and observed as radiogalaxies.
B) Distribution of Γ. Following the suggestion of previous authors (Cohen 1989, Impey et al. 1991, Padovani and Urry 1992), we assume a distribution of Lorentz factors of the form $n(\Gamma) \propto \Gamma^{-x}$. We choose $x = 2$ and $\Gamma_{min} = 1$, while $\Gamma_{max}(z)$ is determined by eq. 3, after replacement of $p(\theta)$ (eq. 2) with $p^*(\theta) \propto \int_1^{\Gamma_{max}} p(\theta)\Gamma^{-x}d\Gamma$. A critical angle $\theta_G(\Gamma, z)$ separating quasars from radiogalaxies is now found for each Γ in the distribution. We find that low redshift objects with $\Gamma \lesssim 5$ should be observed as radiogalaxies for all the orientation angles.
C) Luminosity Functions of Quasars and Host Galaxies. We convolve the previous results through appropriate luminosity functions. Following the analysis and discussion by Yee (1992), we adopt for the host galaxies a gaussian distribution with $\langle M_B \rangle = -21.8$, $\sigma = 0.45$. We then need the (optical) luminosity function of steep-spectrum quasars at $z = 0$, which has to be computed through the conditional distribution function and the radio luminosity function (cf. VGC): $\Psi_{opt}(L_{opt}) = \int \Psi_R(L_R)\Phi\left(L_{opt}/L_R; L_R\right) dL_R$. The result, following again data and analysis of VGC, is a gaussian distribution with $\langle M_B \rangle \simeq -20$, $\sigma = 1.2$. Also within this model, very few quasars are expected at $z \lesssim 0.3$, except that for small angles.

The distribution of sources in the θ-z plane can be investigated observationally by means of those properties which depend strongly on the viewing angle, e.g. the superluminal velocity and the asimmetry (flux ratio between the jet and the counterjet). The expected distributions for quasars and for radiogalaxies will be computed. We expect a substantial population of low-z radiogalaxies with high asimmetry and superluminal velocity. Quantitative results could allow a direct comparison of the evolutionary unified scheme with the more conventional case of constant Γ.

REFERENCES

Barthel, P.D. 1989, *ApJ*, **336**, 606

Cohen, M.H. 1989, in *BL Lac Objects*, L. Maraschi et al., Berlin: Springer, 13

Impey, C.D., Lawrence, C.R., and Tapia, S. 1991, *ApJ*, **375**, 46

Padovani, P. and Urry, C.M. 1992, *ApJ*, **387**, 449

Vagnetti, F., Giallongo, E., and Cavaliere, A. 1991, *ApJ*, **368**, 366 (VGC)

Yee, H.K.C. 1992, in *Relationships Between Active Galactic Nuclei and Starburst Galaxies*, A.V. Filippenko, *ASP Conf.Ser.*, **31**, 417

The First Stromlo Symposium: The Physics of Active Galaxies
ASP Conference Series, Vol. 54, 1994
G. V. Bicknell, M. A. Dopita, and P. J. Quinn (eds.)

INFRARED SPECTROSCOPY OF SEYFERT 2S: A LOOK THROUGH THE OBSCURING TORUS?

S. VEILLEUX[1]
Kitt Peak National Observatory, NOAO, Tucson, AZ 85726-6732

R. W. GOODRICH
Dept. of Astronomy, Caltech, Pasadena, CA 91125

G. J. HILL
Dept. of Astronomy, University of Texas, Austin, TX 78712

INTRODUCTION

Seyfert unification theories propose that all Seyfert galaxies are fundamentally the same, differing mostly in viewing angle. Strong support for this comes from spectropolarimetry of a number of Seyfert 2s, which shows that in polarized light they have the spectra of Seyfert 1s. Detection of the broad-line regions (BLRs) in Seyfert 2s by spectropolarimetry depends critically on the existence of a scattering region with a large enough covering factor and optical depth that a significant amount of light is scattered into our line-of-sight. If an appropriate scattering region does not exist, then we will not be able to detect the BLR using spectropolarimetry. Infrared spectroscopy, on the other hand, can potentially penetrate the obscuring disk, allowing us to see the broad emission lines directly.

The present paper presents J-band spectroscopy of fifteen Seyfert 2s, including high resolution (R = 1260) spectra which can clearly resolve the broad and narrow component of Paβ and neighboring emission lines. The higher resolution of this study allows us to accurately and unambiguously remove the narrow line flux from the total line flux, and hence get a better idea of the reddening of the BLR in some cases.

RESULTS

Of the fifteen objects studied here, three have broad Paβ lines: MCG –05.23.16, Mrk 463E, and NGC 2992. Mrk 176 and NGC 5728 may also have weak broad lines – somewhat improved data should prove definitive. In NGC 5506, previously reported to have broad Paβ (Blanco et al. 1990; Rix et al. 1990), we find that the Paβ line profile is continuous and has the same shape as the nearby [Fe II] λ1.2567 line, which should not have a broad component. We interpret these observations as gas from the narrow-line region (NLR) with *no* broad component. In NGC 5506, however, the NLR profiles become *broader* with increasing

[1]Hubble Fellow

wavelength, indicating that highly reddened wings are becoming more readily visible at the longer wavelengths.

Comparison of the infrared optical depths with column depths determined from X-ray data show a general tendency for the objects with detected broad Paβ to have lower X-ray columns, although this should be confirmed with IR spectroscopy of other lines and better X-ray spectra. Given the large X-ray column depths often inferred in Seyfert 2s (e.g., Awaki 1990), and the relatively modest gain in τ we achieve from Hα to Paβ (a factor of about three), it is perhaps surprising that we find even as many as 3 obscured BLRs out of 15 objects studied in our sample. In the standard unification theory of Seyferts, this high frequency of partially obscured Seyfert 1s implies that the transition zone located between the optically-thin "throat" of the torus and the optically-thick core is rather extended. The existence of "intermediate" Seyferts (1.8s and 1.9s; cf. Osterbrock & Martel 1993) in which faint optically-detected BLRs present large reddenings (e.g., Goodrich 1990) brings support to this scenario (however, see Rudy, Cohen, & Ake 1988).

Our wavelength range also includes [Fe II] λ1.2567, thought to come from either supernova remnants in a starburst or from high-optical depth NLR gas heated by X-rays. We confirm the correlation of [O I] λ6300/Hα and [Fe II] λ1.644/Brγ (the latter transformed to [Fe II] λ1.2567/Paβ to compare to our data) found by previous authors when comparing AGN, supernova remnants, starbursts, and H II regions.

FUTURE WORK

Our long-term goal is to use Paβ and other IR lines to determine the reddenings along the lines of sight to the hidden BLRs. With this information for a statistically large sample we will then be able to recreate the profile of the dust optical depth as a function of polar angle in the obscuring torus.

ACKNOWLEDGMENTS

This work has been supported by NSF through grant AST-9121889 (RWG), and by NASA through grant HF-1039.01-92A awarded by the Space Telescope Science Institute, which is operated by the AURA, Inc. for NASA under contract NAS 5-26555 (SV).

REFERENCES

Awaki, H. 1990, ISAS Research Note, No. 473

Blanco, P. R., Ward, M. J., & Wright, G. S. 1990, *MNRAS*, **242**, 4p

Goodrich, R. W. 1990, *ApJ*, **355**, 88

Osterbrock, D. E., & Martel, A. 1993, *ApJ*, in press

Rix, H.-W., Carleton, N. P., Rieke, G., & Rieke, M. 1990, *ApJ*, **363**, 480

Rudy, R. J., Cohen, R. D., & Ake, T. B. 1988, *ApJ*, **332**, 172

The First Stromlo Symposium: The Physics of Active Galaxies
ASP Conference Series, Vol. 54, 1994
G. V. Bicknell, M. A. Dopita, and P. J. Quinn (eds.)

HST OBSERVATIONS OF THE SEYFERT 2 GALAXY NGC 5728

A.S. WILSON, G.A. BOWER, T.M. HECKMAN AND J.S. MULCHAEY
Space Telescope Science Institute, Baltimore, MD 21218

J.A. BRAATZ
University of Maryland, College Park, MD 20742

J.H. KROLIK
The Johns Hopkins University, Baltimore, MD 21218

G.K. MILEY
Leiden, The Netherlands

Images of the Seyfert galaxy NGC 5728 have been obtained with HST in the light of [OIII]$\lambda\lambda$4959,5007, Hα + [NII]$\lambda\lambda$6548,6583, and green and red continua. The emission-line images (Figs Ia,b) and their ratio (Fig If) show a spectacular bi-conical structure with overall extent 1.8 kpc. The two cones share a common axis and apex. The cones' axis coincides to within 3° with that of the nuclear radio continuum emission, but is not aligned with the rotation axis of the galaxy disk. Both the line and continuum emission (Figs Ic,d) are weak at the apex; in fact, the apex (the cross in Figs Ic,d) is contained within a dark continuum band elongated roughly perpendicular to the cones' axis. These results suggest that the apex of the cones is obscured along our line of sight, perhaps by an inclined torus. The apex almost certainly reflects the location of the "true" nucleus, for it coincides with the radio nucleus, the gaseous kinematical center of the galaxy, and the geometrical center of the ring seen in the continuum images. It has been proposed that the ring is associated with an inner Lindblad resonance, and so is likely symmetric about the mass center of the galaxy.

The dominant feature in the continuum images is the bar-like emission (AB in Fig Ic) contained within the ring. This emission is misaligned with the large-scale stellar bar (p.a. $\simeq 33°$), but agrees with the isophotal major axis of K-band light within 3″ of the nucleus (p.a. $\simeq 86°$). The elongated continuum structure may then represent a small-scale stellar bar which could be dynamically linked with the ring. Since the emission is largely contained within the cones, another possibility is that the bar-like feature is produced by electron or dust scattering of nuclear emission.

The ionization cones, the weakness of the emission at their apex, and the continuum band perpendicular to the cones' axis are all consistent with the presence of a dusty torus which blocks our direct view of the actual nucleus.

This work will be published in ApJ letters in late 1993 and was supported by NASA HST grant GO 3724.

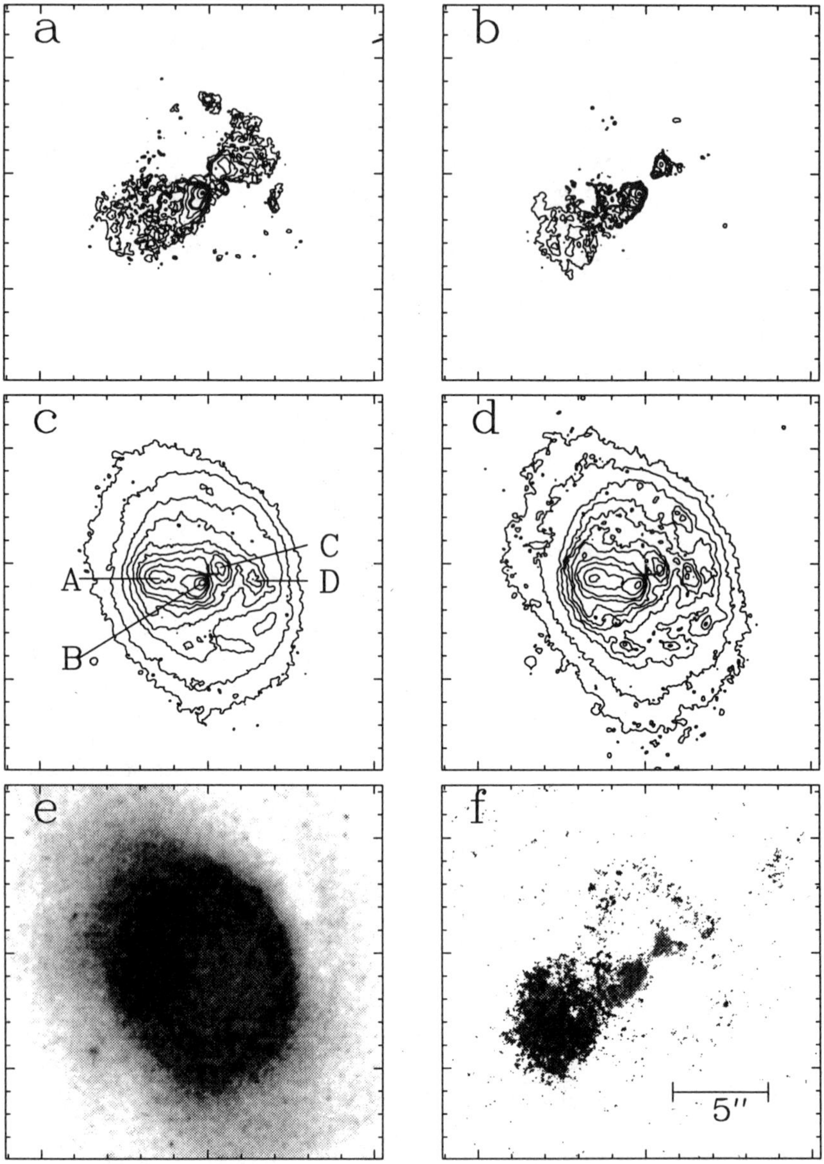

FIGURE I HST images of the central regions of NGC 5728. (a) The Hα + [NII] map and (b) the [OIII] map, each with contours plotted at 1, 2, 3, 5, 10, 20, 30, 50, and 90 % of the peaks. (c) The red and (d) green continuum images. Contours are plotted at 3, 5, 10, 15, 20, 25, 30, 40, 50, 70, and 90 % of the peak values of (c) 1.0×10^{-15} erg cm^{-2} s^{-1} Å^{-1} (arcsec)$^{-2}$ and (d) 6.6×10^{-16} erg cm^{-2} s^{-1} Å^{-1} (arcsec)$^{-2}$. (e) A grey scale plot of the green to red flux ratio map, with bluer colors as the darker shades. (f) The [OIII]/Hα + [NII] flux ratio map, which is a measure of the gaseous excitation. Here the grey scale is logarithmic, with darker shades representing larger ratios.

The First Stromlo Symposium: The Physics of Active Galaxies
ASP Conference Series, Vol. 54, 1994
G. V. Bicknell, M. A. Dopita, and P. J. Quinn (eds.)

NEAR-IR SPECTROPOLARIMETRY OF NGC 1068

S. YOUNG AND J. H. HOUGH
Division of Physical Sciences, University of Hertfordshire, Hatfield, Herts AL10 9AB

J. A. BAILEY
Anglo-Australian Observatory, PO Box 296, Epping, NSW 2121, Australia

D. J. AXON
Space Telescope Science Institute, 3700 San Martin Drive, Baltimore, USA, MD 21218

M. J. WARD
University of Oxford, Nuclear and Astrophysics Laboratory, Keeble Road, Oxford OX1 3RH

INTRODUCTION

The current unified model of Seyfert galaxies, as proposed by Antonucci and Miller (1985), has most, if not all, Seyfert 2 galaxies hosting a Seyfert 1 type nucleus in their cores. The broad lines from the type 1 nucleus can be scattered and be seen in the polarized flux spectrum, provided the geometry is correct. Whether the mechanism is electron or dust scattering has been a question of some debate. Since dust would normally have a low scattering efficiency in the infrared, near-IR spectropolarimetry should help in determining the nature of the scatterers. To this end we present the first near-IR spectropolarimetry of NGC 1068 a "Seyfert 2" galaxy in which polarized broad lines have been observed at optical wavelengths.

OBSERVATIONS

The observations were made at UKIRT, Hawaii, on the nights of 1992 November 12, 16 & 17 with the CGS4 spectrometer and IRPOL, at an effective resolution of approximately 1000 $km\ s^{-1}$. Polarization observations were made in the normal way with IRPOL, and the data were extracted using Figaro and TSP, with an effective aperture of 3×3 arcsecond^2.

RESULTS AND DISCUSSION

The polarization spectra for NGC 1068 in the region of the Paα line, is presented in figure 1. One obvious feature is the decrease in percentage polarization due to the narrow lines being less (or zero) polarized.

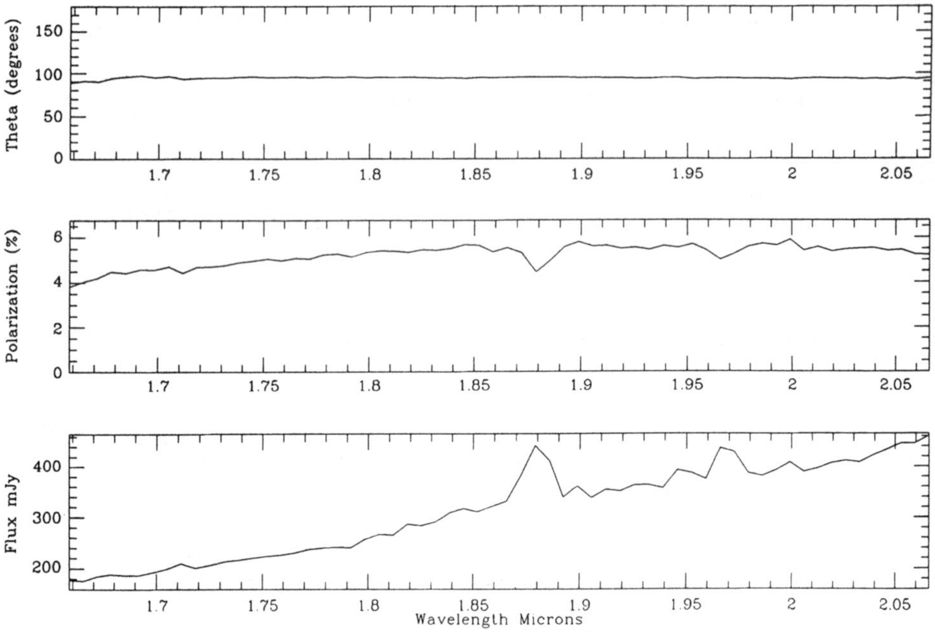

FIGURE I Spectropolarimetric data for NGC 1068

A comparison of the scattered fluxes of the broad Paα and Hα (the latter taken from our WHT observations made in 1991 February) and assuming case B recombination and the interstellar extinction law of Savage and Mathis (1979), as extended to the infrared by Landini et al. (1984) suggests that either

(a) the optical and IR broad lines are scattered by electrons, with an extinction of the scattered light of approximately 1 magnitude, which is typical of the observed values (Ward et al., 1987),

or

(b) the optical broad lines are scattered by dust and the IR broad lines are polarized by dichroic absorption as the radiation passes through a medium containing aligned grains, possibly in the nuclear torus, essentially the model proposed by Bailey et al. (1988).

If the latter is true, then we should also see the broad Paα in the direct or total flux spectrum. While there is some evidence for a broadening of the Paα line, in the total flux, our spectral resolution is insufficient to be conclusive.

REFERENCES

Antonucci R. R. J., Miller J. S., 1985, *ApJ*, **297**, 621

Bailey J., Axon D. J., Hough J. H., Ward M. J., McLean I., Heathcote S. R., 1988, *MNRAS*, **234**, 899

Hough J. H., Brindle C., Wills B. J., Wills D., Bailey J., 1991, *ApJ*, **372**, 478

Landini M., Natta A., Oliva E., Salinari P., Moorwood A. F. M., 1984, *A&A*, **134**, 284

Savage B. D., Mathis J. S., 1979, *ARA&A*, **17**, 73

Ward M. J., Geballe T., Smith M., Wade R., Williams P., 1987, *ApJ*, **316**, 138

SEYFERT GALAXIES & LINERS

Invited Papers

The First Stromlo Symposium: The Physics of Active Galaxies
ASP Conference Series, Vol. 54, 1994
G. V. Bicknell, M. A. Dopita, and P. J. Quinn (eds.)

NEW INSIGHTS INTO SEYFERT GALAXIES FROM HUT OBSERVATIONS OF NGC 1068 AND NGC 4151

G. A. KRISS, Z. TSVETANOV, AND A. F. DAVIDSEN
Department of Physics & Astronomy, Johns Hopkins University, Baltimore, MD 21218

ABSTRACT The Hopkins Ultraviolet Telescope spectra of NGC 1068 show moderately strong narrow line emission from C III $\lambda 977$ and N III $\lambda 991$, which may indicate the presence of collisionally ionized gas. This could be clouds shock-heated by their interaction with the radio jet, or material cooling from the hypothesized X-ray heated wind emanating from the nuclear region. These observations emphasize that processes in addition to photoionization should be reconsidered as important ingredients in forming the emission-line spectra of Seyfert galaxies.

HUT spectra of NGC 4151 extending below 1200 Å show strong blue-shifted absorption in the Lyman series of neutral hydrogen with a total column density of $\sim 10^{18} - 10^{21}\ cm^{-2}$. The absorbing material is dense ($> 10^9\ cm^{-3}$) and outflowing, and it has a large effective Doppler parameter ($\sim 200\ km\ s^{-1}$). The column density is insufficient to be also responsible for the X-ray absorption seen in NGC 4151, but it may play a key role in collimating the UV ionizing radiation from the nucleus. We describe a simple model in which a UV-absorbing thin atmosphere surrounding an opaque torus can collimate the radiation from NGC 4151 to produce the biconical narrow-line region seen in HST images.

INTRODUCTION

As two of the nearest and brightest Seyfert galaxies, NGC 1068 and NGC 4151 have played key roles in clarifying our ideas about the structure and workings of active galactic nuclei. The discovery of a polarized "Seyfert 1" spectrum in the light from NGC 1068 (Antonucci & Miller 1985), the prototypical Seyfert 2, invigorated the concept of a unified model for Seyfert galaxies, in which obscuration and the orientation of a single type of active nucleus modify its observational appearance.

While NGC 4151 is nominally a Seyfert 1 galaxy, it has gone through low luminosity episodes where it seems to resemble a Seyfert 2. Its spectrum is rich in optical (Anderson 1974) and UV absorption lines (Penston et al. 1981; Bromage et al. 1985; Kriss et al. 1992a) that are unusual for a Seyfert 1 galaxy, and its X-ray spectrum shows evidence for high column density absorption that only partially covers the active nucleus (Holt et al. 1980). And, unlike most Seyfert 1's, NGC 4151 has an extended biconical narrow-line region (NLR) that is more reminiscent of Seyfert 2 galaxies (Evans et al. 1993).

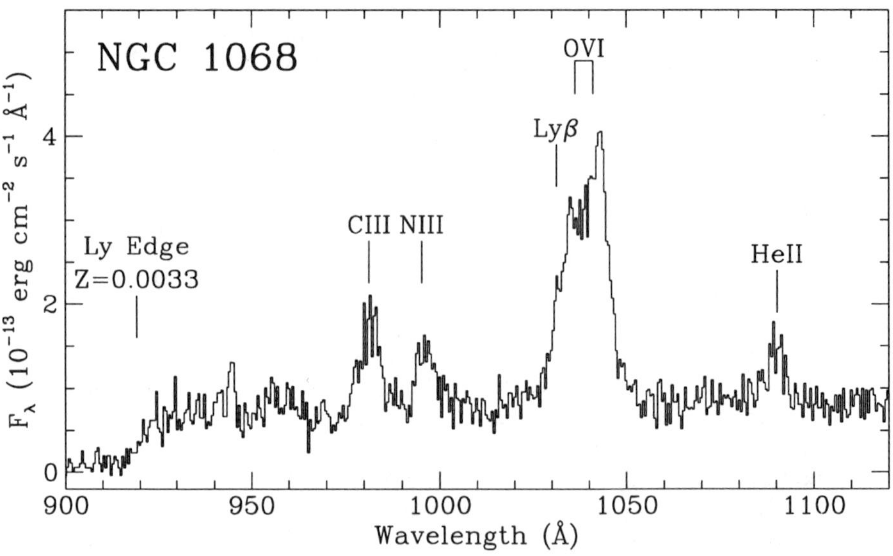

FIGURE I The 900-1120 Å section of the HUT spectrum of NGC 1068 shows moderately strong far UV emission lines plus redshifted Lyman edge absorption.

Just as observations of NGC 1068 and NGC 4151 with new instruments have often revealed new aspects of the inner workings of AGN, the unique capability of HUT to observe these Seyfert galaxies in the unexplored wavelength range between Lyα and the Lyman limit is giving us new insights into Seyfert galaxies.

EVIDENCE FOR SHOCK-HEATED GAS IN NGC 1068

Far-UV observations of NGC 1068 were obtained during the flight of the Hopkins Ultraviolet Telescope (HUT) aboard the space shuttle *Columbia* in December 1990. The spectra cover the wavelength range 830-1860 Å at a resolution of ~3 Å. Davidsen et al. (1992) describe HUT and its performance on Astro-1, and the observations of NGC 1068 are described in detail by Kriss et al. (1992b).

The unique features present in the HUT spectra below 1200 Å are shown in Figure 1, where the moderately strong emission from C III λ977 and N III λ991 is visible. He II λ1085 emission is also detected. The continuum is relatively flat in F_λ down to the redshifted Lyman edge of NGC 1068 at 918 Å. The high excitation temperature resonance lines of N III and C III are rarely seen in AGN spectra at higher redshift, and their prominence in what has long been presumed to be a photoionized emission-line spectrum is surprising. In fact, the HUT spectrum below 1200 Å bears a stronger resemblance to the Cygnus Loop supernova remnant (compare Figure 1 to Blair et al. 1990) than to any of the other AGN observed with HUT.

At a more quantitative level, we can form line ratios that are diagnostic of the gas temperature by choosing emission lines with common ground states such as (I(λ1907)+I(λ1909))/I(λ977) for C III, and I(λ1750)/I(λ991) for N III. Since the C III] $\lambda\lambda$1907,1909 lines are outside the HUT wavelength range, we use IUE measurements (Snijders et al. 1986). The observed C III line ratio of 3.15 ± 0.51 gives a temperature of $26,700^{+1900}_{-1400}$ K (90% confidence errors). The N III line ratio of 1.46 ± 0.34 implies $T = 32,700^{+5200}_{-3200}$ K. Corrections for extinction or for resonance line scattering will lead to *higher* temperatures. For $E_{B-V} = 0.4$, we deduce temperatures $> 110,000$ K. Such high temperatures are difficult to produce in photoionized gas. High ionization parameters can produce gas temperatures approaching 30,000 K, but the ratio of C IV λ1549 to the C III lines is then much higher than we observe in NGC 1068.

A likely source for shock heating is the interaction of the radio jets with the NLR and ISM of NGC 1068. The strong 1.64μm [Fe II] emission which follows the morphology of the radio jet in NGC 1068 (Cameron et al. 1994) supports this interpretation. While the clouds in the NLR are bathed in the powerful ionizing radiation field of the active nucleus in NGC 1068, the mechanical interaction with the radio jet is apparently producing additional heat input and compression that generates some of the brightest knots of [O III] emission visible in the HST image of Evans et al. (1991). Shocks generated by the radio jet are also capable of producing their own ionizing radiation field which may contribute to the line emission as shown in the models of auto-ionizing shocks by Dopita (1994). We suggest that line emission in other Seyfert galaxies that suggest interactions of radio jets with the NLR (e.g., Mrk 3, Tsvetanov et al. 1993) may have substantial contributions from shocks in addition to photoionization.

COLLIMATING THE IONIZING RADIATION IN NGC 4151

The HUT spectrum of NGC 4151 shown in Figure 2 was also obtained during the Astro-1 mission, and it is described in detail by Kriss et al. (1992a).

The type 1 Seyfert galaxy NGC 4151 presents a conundrum for the simplest unified models of AGN. If the radiation field is collimated by an obscuring torus, then simple geometry says that the NLR of Seyfert 2's seen in projection should appear biconical since we are in the shadow of the torus, while Seyfert 1 NRL's should appear isotropic. We have a fairly clear view of the optical and UV continuum and of the broad-line region (BLR) in NGC 4151, yet the high resolution HST [O III] images of its NLR show a beautiful biconical structure (Evans et al. 1993). Thus the collimation *cannot* be provided by an optically thick torus. The low column density, UV-absorbing gas visible in the HUT spectrum, however, can effectively collimate the ionizing radiation.

A slight modification of the unified model which surrounds the obscuring torus with a thin atmosphere can explain the special features of NGC 4151 (Tsvetanov & Kriss 1993). For simplicity, as shown in Figure 3, we assume uniform density and rectangular cross section, but the essential feature is that some lower density structure extend above and below the torus (e.g., the magnetically driven wind suggested by Königl 1994). This gas will be photoionized by the central continuum source, and it will remove ionizing photons from lines of sight which pass through it. We have modeled a representative cloud in the NLR which is illuminated by radiation that has passed through the thin atmosphere. At columns of $\sim 10^{21}$ cm^{-2}, the line of sight will be optically thick at all wave-

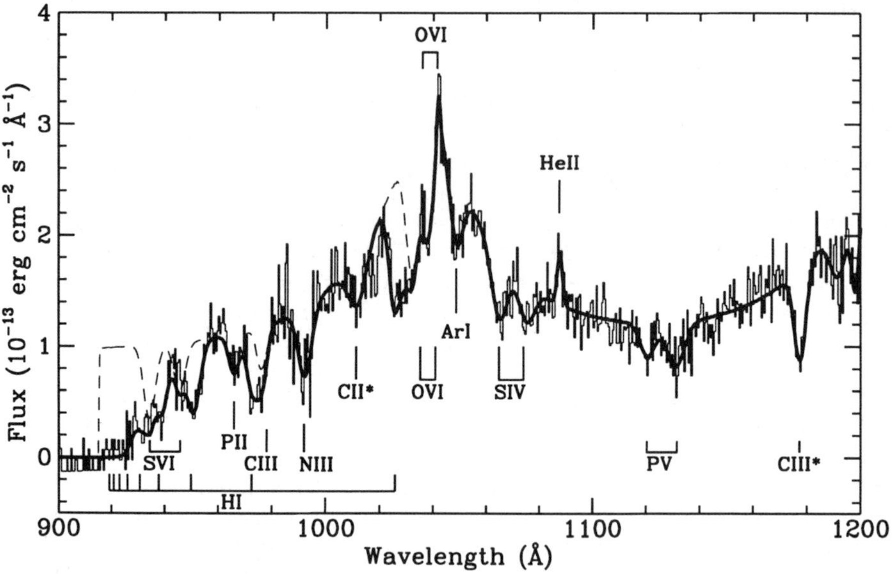

FIGURE II The far UV spectrum of NGC 4151 in the 900-1200 Å wavelength range illustrates how blended absorption in the high order lines of the Lyman series can effectively absorb the short wavelength continuum. The data are shown as a histogram, and the heavy solid line shows the best fitting power law as modified by extinction of $E_{B-V} = 0.039$, absorption by neutral hydrogen with $b = 200\ km\ s^{-1}$ and $N_H = 2.9 \times 10^{19}\ cm^{-2}$, and additional gaussian absorption lines. The dashed line shows the model *without* the hydrogen absorption.

lengths from the Lyman edge to the soft X-ray. Above the Lyman limit, and for normal dust-to-gas ratios, extinction will not have much impact until gas column densities exceed several times $10^{21}\ cm^{-2}$. Consequently, lines of sight falling within the cone defined by the shadow of the opaque torus but exterior to the ionization cone (we call this the "twilight zone"), will have a relatively clear view of the central regions at wavelengths longward of the Lyman edge.

A qualitative prediction of our model is that emission lines from ions with ionization potentials lower than the Lyman edge (e.g., S II at 10.36 eV) will be excited in clouds that fall outside the cone defined by ions with higher ionization potentials. Our photoionization calculations show that emission lines such as [S II] $\lambda\lambda 6716, 6730$ should appear as cones with larger opening angles than emission from [O III] $\lambda 5007$. This prediction can be easily tested observationally in bright, nearby Seyferts such as NGC 1068 and NGC 4151.

REFERENCES

Anderson, K. S. 1974, *ApJ*, **189**, 195

Antonucci, R. R. J., and Miller, J. S. 1985, *ApJ*, **297**, 621

Blair, W. P., et al. 1991, *ApJ*, **379**, L33

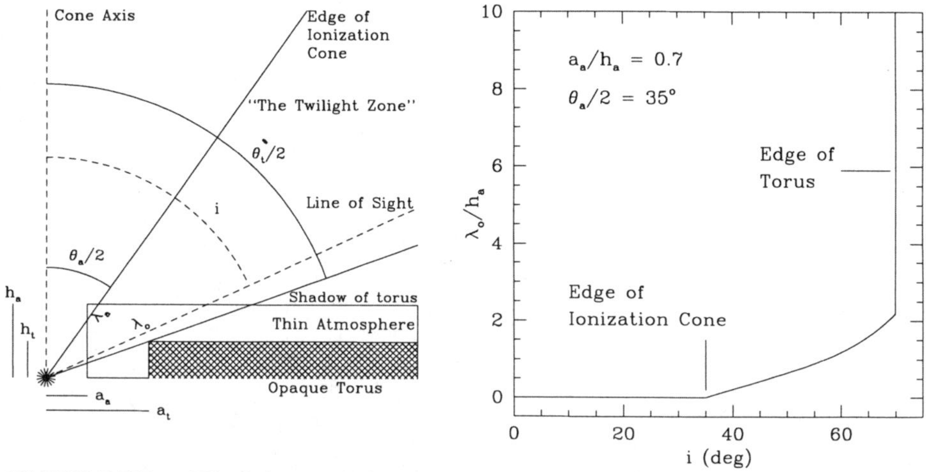

FIGURE III The left panel shows an opaque torus of rectangular cross section with a height above the central plane h_t and an inner radius a_t. It is surrounded by a uniform atmosphere with height h_a and inner radius a_a. The line of sight to the observer has a path length through the thin atmosphere of $\lambda_o = h_a/\cos i - a_a/\sin i$, where i is the inclination of the observer's line of sight from the axis of the ionization cone. As shown in the panel on the right, the relative path length through the thin atmosphere increases gradually from the edge of the ionization cone throughout the "twilight zone" until reaching the edge of the torus at $i = \tan^{-1}(a_t/h_t)$.

Bromage, G., et al. 1985, *MNRAS*, **215**, 1

Cameron, M., Blietz, M., Storey, J., & Genzel, R. 1994, this volume

Davidsen, A. F., et al. 1992, *ApJ*, **392**, 264

Dopita, M. 1994, this volume

Evans, I. N., Ford, H. C., Kinney, A. L., Antonucci, R. R. J., Armus, L., and Caganoff, S. 1991, *ApJ*, **369**, L27

Evans, I. N., Tsvetanov, Z., Kriss, G. A., Ford, H. C., and Koratkar, A. 1993, *ApJ*, in press

Holt, S. S., et al. 1980, *ApJ*, **241**, L13

Königl, A. 1994, this volume

Kriss, G. A., et al. 1992a, *ApJ*, **392**, 485

Kriss, G. A., et al. 1992b, *ApJ*, **394**, L37

reference Penston, M. V., et al. 1981, *MNRAS*, **196**, 857

Snijders, M. A. J., Netzer, H., & Boksenberg, A. 1986, *MNRAS*, **222**, 549

Tsvetanov, Z., et al. 1993, in preparation

Tsvetanov, Z., & Kriss, G. A. 1993, in preparation

The First Stromlo Symposium: The Physics of Active Galaxies
ASP Conference Series, Vol. 54, 1994
G. V. Bicknell, M. A. Dopita, and P. J. Quinn (eds.)

THE EXCITATION OF LINERS AND COOLING FLOWS

MICHAEL A. DOPITA
Mount Stromlo and Siding Spring Observatories, Canberra, Australia

ABSTRACT

A new model is proposed in which the LINER emission represents radiative bow shocks in a cool intercloud medium around dense, gravitationally confined cloudlets which are being turbulently accreted towards the nucleus. It is shown that, when proper account is taken of the ionising effects of the radiation produced in the shocked cooling region, the observed spectral characterics and velocity dispersion of the visible line emission can be accounted for by shock velocities in the range 150–400 km s^{-1}, and by intercloud densities ~ 1 cm^{-3}. The distinction between LINER emission, and other narrow-line emission in galaxies (Seyfert 2, narrow-line radio galaxies) can be understood as one of shock geometry, LINERs being spherical shocks, and others plane-parallel shocks. The luminosity of the cooling flows is consistent with the rate of change of gravitational binding energy of the dense clouds with an accretion rate equal to the mass flux from the cooling X-Ray emitting plasma. This gives a solution to the recombination problem in cooling flows, but does not address the question of the ultimate fate of the accreted gas.

INTRODUCTION

In the last fifteen years, the hypothesis that the narrow line emission features of AGN are excited by a hard power-law spectrum of UV photons has gained almost universal acceptance. In such a model, the Seyfert 2 and Narrow Emission Line Galaxies would be excited by a fairly flat UV power law with an ionisation parameter (U) of order 10^{-2} (Koski 1978). The source of these photons is assumed to be the nucleus, but the exact means whereby the non-thermal UV spectrum is generated remains obscure. The development of so called "unified models" of AGN (Barthel 1989) has been driven fairly heavily by the observation of cones of highly ionised narrow-line gas extending from the nucleus (Pogge 1988a, b, 1989; Tadhunter and Tsvetanov 1989; Fosbury 1989). In the standard picture, the nuclear radiation is assumed to be directionally beamed (Morganti et al. 1991, 1992). Despite a deal of success in describing the emission line spectrum, the weaknesses of such models are that a number of parameters remain arbitrary. In particular, the detailed physics of the nuclear source is not usually addressed, there remains a serious shortfall in the estimated number of photons based on the extrapolation of the UV spectrum (Binette, Parker, and Fosbury 1993), the model fails to give an explanation for the relatively small range of the ionisation parameter over a very large radial scale range, and it fails to even address the question of the observed velocity dispersion of the gas, typically

400–600 km s^{-1}.

The other major class of narrow-line AGNs is that of the low-ionisation nuclear emission-line regions, LINERs (Heckman 1980). These fall into a distinct region of excitation space when classified according to their emission line ratios (Phillips, Baldwin, and Terlevich 1982; Veilleux and Osterbrock 1987). If LINERs are photoionised, then again, the ionising spectrum has to be fairly flat, and the ionisation parameter is low, $U \sim 10^{-4}$ (Ferland and Netzer 1983; Stasinska 1984). LINERs, or objects having LINER-like spectra, are found in the nuclei of spiral galaxies lacking nuclear starbursts (Keel 1983a, b), in as many as 50% of elliptical galaxy nuclei (Phillips et al. 1986), and in the "cooling flows" associated with first-ranked elliptical galaxies in clusters (Heckman 1981; Cowie et al. 1983; Johnstone, Fabian, and Nulsen 1987; Heckman et al. 1989). Again, typical velocity dispersions observed in all these objects are a few hundred km s^{-1}.

A promising new model is based on mechanical energy input, rather than photoionisation from a central source. The source of the mechanical energy may be either turbulent motions of clouds in the potential of the galaxy, or direct energy deposition from an outflowing jet of relativistic electrons. The latter model is suggested by the hydrodynamic calculations of Bicknell et al. (1990). In these, a mildly supersonic jet which is confined by a denser surrounding medium will strongly interact with it through the operation of the Kelvin–Helmoltz instability at the boundary. This generates shocks in the surrounding medium, and turbulent vortices pull off dense clouds of gas embedded in a lower-density phase which is partially entrained in the jet. The turbulent collisions of clouds generate fast shocks, and slower shocks are driven into them. This physical situation appears to apply to the nearest narrow-line AGN, Cen A, where direct dynamical evidence for entrainment has been discovered in the filaments associated with the NW radio lobe (Bicknell, Dopita, and Sutherland, this conference). The optical spectrum of these filaments has been self-consistently modelled by Sutherland, Bicknell, and Dopita (1993), using our code MAPPINGS2. Our model emphasises the importance of the UV and soft X-ray emission produced in the cooling regions of high-velocity shocks, which is available to photoionise both the dense cloud and the pre-shock material. This idea was originally suggested by Binette, Dopita, and Tuohy (1985). For this reason, high-velocity shocks should be regarded as "autoionising", in that the photoionised precursor regions contribute significantly to the total line flux, and these regions are certainly much more spatially extensive than the radiative region of the fast shock.

In this paper we will investigate the hypothesis that all classes of narrow-line AGN are excited by "autoionising" shocks generated by mechanical energy deposition rather than by photoionisation by a nuclear source.

LINER GALAXIES

A number of lines of evidence suggest that the relationship between LINERs and other types of narrow-line AGN is quite intimate. LINERs are characterised by large [O II]/[O III], strong [O I], [S II] and [N II] lines in the red, and typically display velocity dispersions of 100–400 km s^{-1}. They continue the trend of decreasing velocity dispersion and Balmer luminosities towards lower excitation.

A survey of "normal" elliptical galaxies by Phillips et al. (1986) showed that low-level LINER activity is found in an appreciable fraction of these. In the spiral galaxies, LINER activity appears to be found in virtually all nuclei which do not show nuclear starburst activity (Keel 1983a, b). Elliptical LINER galaxies cover a range of radio properties, including radio quiet objects, weak nuclear sources, and Fanaroff-Riley (FR) Class I sources. However, LINERs are never found in the high-power FR II radio sources. In terms of the correlation between the optical and the radio luminosities, a similar trend emerges. There is a good correlation for the FR II class, but this disappears smoothly at the FR I–II transition. These correlations suggest that the optical emission makes a transition from predominantly jet-excited to some other excitation mechanism at the FR I–II transition.

LINERs show a variety of internal motions. Generally speaking, the lowest excitation objects show a more ordered rotation, whereas the higher excitation AGN are more chaotic systems. This strongly suggests that mechanical energy input and excitation are correlated. This is consistent with the predictions of our "autoionising" shock model. In this, dissipation of mechanical energy is fluorescently converted through photoionisation to photons at optical wavelengths, and the optical luminosity should scale as the energy flux through the shock. Therefore, shock luminosity should scale as the area of the shock, and as the cube of the shock velocity. Globally, this expectation is borne out in the models. This scaling also holds for individual emission lines which are mainly collisionally excited such as [N II], [S II] or the UV resonance and intercombination lines. However, in the case of the recombination lines of hydrogen, the decrease in recombination coefficient at high temperature tends to decrease the efficiency of production of Balmer photons at higher shock velocities. For Hβ, the luminosity is given by:

$$L_{\mathrm{H}\beta} = 6.9 \times 10^{38}\, n_0\, (V_{200})^{2.44}\, A_6 \ \mathrm{erg\ s^{-1}} \qquad (1)$$

where n_0 is the pre-shock density (cm^{-3}), V_{200} is the velocity of the fast shock in units of 200 km s^{-1}, and A_6 is the shock area in units of kpc^2. In the case of lines which are predominantly generated in the photoionised precursor zone, such as [O III] or [S III], the hardening of the UV spectrum at high velocities ensures that these lines become stronger relative to either the recombination lines or to the collisionally-excited lines.

In this model, therefore, excitation, surface brightness, and velocity dispersion are predicted all to be correlated quantities in the sense that high excitation is correlated with high shock velocity, and therefore high energy flux across the shock. For the narrow emission-line regions (NLERs) of galaxies studied by Baum et al. (1988), equation (1) was found by Koekemoer (private communication) to describe the relationship between the velocity dispersion, area, and Balmer line flux. Furthermore, such a relationship would explain the otherwise puzzling relationship between the surface brightness and level of excitation of NLER discussed by Fosbury (1989).

Do the LINERs conform to such a relation? Unfortunately, few detailed velocity and surface brightness maps of sufficient quality currently exist. However, in the case of one system displaying more orderly rotation, NGC 5077, Bertola et al. (1991) find, once again, a very good correlation between the Balmer surface brightness and the local velocity dispersion, consistent with equation (1) above, suggesting that LINERs, like the other NLERs, are also powered by radiative, autoionising shocks.

The relative strength of the lines produced in the photoionised precursors is strongly dependent upon the geometry of the shock. The photoionised region has a considerable thickness compared with that of the radiative shock region. For example, a shock having a velocity of 300 km s^{-1} has a thickness of $\sim (2.5/n_0)$ pc, whereas its photoionised precursor has a thickness of $\sim (366/n_0^2)$ pc in the case of plane-parallel geometry. Thus, in the case that the high-velocity shock is produced as a bow-shock around a dense, gravitationally-confined cloud, the precursor H II region will be characterised by spherical geometry rather than plane parallel geometry provided that the cloud is smaller than about $\sim (100/n_0^2)$ pc. The consequence of this is that the mean ionisation parameter in the H II region drops considerably, from $U \sim 10^{-2}$ in the plane-parallel case to $\sim 10^{-4}$ for bow shocks around clouds which are small in comparison to their shock precursor H II regions. Note that this is exactly the range of ionisation parameters needed in pure photoionisation models to model the Seyfert 2 and NLER galaxies on the one hand, and LINERs on the other. However, the shock excitation hypothesis has the advantage of producing these numbers naturally as a consequence of geometry within the context of a one-parameter family (of shock velocity).

The diagnostic diagrams of Phillips, Baldwin, and Terlevich (1981) or Veilleux and Osterbrock (1987) provide a powerful means of distinguishing between various excitation mechanisms. In Figure I, we show the observational results compared with the computed shock and H II region sequences in the [O I]/Hα *vs* [O III]/Hβ plane. It is clear that spherical shocks provide a good description of the LINER spectra. Furthermore, the well-known correlations of the [O I], [N II], and the [S II] flux ratios with respect to Hα (Keel 1983b; Veilleux and Osterbrock 1987), are also found to be reproduced by a simple sequence of shock velocities.

The condition that a dense cloud can survive in its passage through the shock-heated intercloud medium is that the gravitational pressure in the cloud should exceed the ram pressure of the intercloud medium. Thus;

$$\sigma_c > \left(\frac{\rho V_s^2}{G}\right)^{1/2} = 580\, P_{-9}^{1/2}\ \mathrm{M_\odot\ pc^{-2}} \tag{2}$$

where σ_c is the surface density of the cloud, V_s is the velocity of the cloud relative to the intercloud medium, and P_{-9} is the ram pressure measured in units of 10^{-9} dynes cm^{-2}. This pressure should correspond fairly closely to the pressure given by the [S II] λ6717/6731 Å line ratio. Since many LINERs have $P_{-9} \sim 1$, the surface density of the clouds which can survive is somewhat higher than that which characterises the molecular clouds in our solar neighbourhood. If we identify the clouds with dense molecular clouds being turbulently accreted to the nucleus, then presumably the intercloud medium through which they are moving derives from those (less bound) portions of the molecular clouds which have been ablated in previous cloud-intercloud or cloud-cloud encounters. The luminosity of the LINER will be determined by the rate of change of gravitational binding energy of the system of orbiting clouds in this model.

Finally, let us consider what determines the X-ray luminosity of LINERs. X-Ray emission could arise through one of three separate routes. First, it could

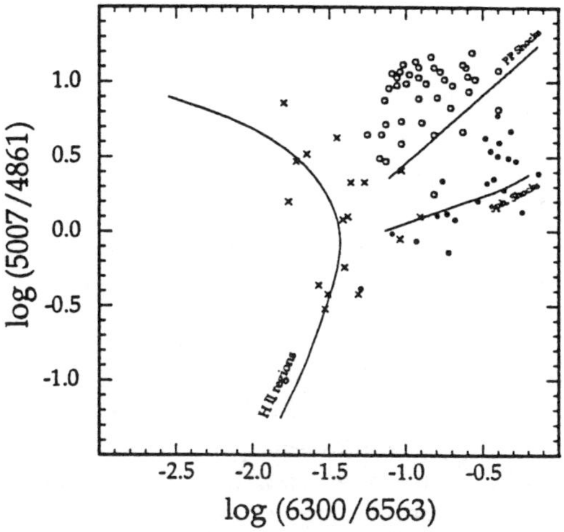

FIGURE I The observational results for Seyfert 2, NLRGs (open circles) and other objects with narrow emission line nuclei (crosses) from Veilleux and Osterbrock (1987); and for spiral nuclei with LINER emission from Keel (1983b) plotted on the oxygen excitation diagram. The H II region sequence is shown, as well as the sequence for plane-parallel shocks between 150–450 km s^{-1}, and for spherical shocks between 150–350 km s^{-1}. It is clear that these three sequences can provide an adequate description of the excitation of all classes of narrow-line emission nuclei.

arise in a cooling flow; second, it could arise by accretion onto a dense and compact central object; and third, it could be internally generated in the accretion shock itself. These can be distinguished observationally, because they produce different scaling relationships with velocity dispersion, σ, and mass flux, $\dot{\mathrm{M}}$:

First case, $$L_{\mathrm{x}} = (5\mathrm{kT}/\mu\ m_{\mathrm{p}})\dot{\mathrm{M}} \propto \sigma^2\dot{\mathrm{M}} \quad (3\ \mathrm{i})$$

Second case, $$L_{\mathrm{x}} \propto \dot{\mathrm{M}} \quad (3\ \mathrm{ii})$$

and in the third case the X-rays will scale as the mechanical flux, weighted by an emission efficiency, ε_{x}, the fraction of the total shock energy which is emitted in the soft X-ray band;

$$L_{\mathrm{x}} \propto \varepsilon_{\mathrm{x}}\sigma^2\dot{\mathrm{M}}/2 \quad (3\ \mathrm{iii})$$

The efficiency is a very strong function of shock velocity. A fit to the theoretical results gives $\varepsilon_{\mathrm{x}} \propto \sigma^{(7.5\pm1.5)}$. On the other hand, since the optical emission is (if we believe the model) generated in the accretion shocks, then optical line emission will follow a relation like (3 iii);

$$L_{\mathrm{opt}} = \varepsilon_{\mathrm{opt}}\sigma^2\frac{\dot{\mathrm{M}}}{2} \quad (4)$$

with $\varepsilon_{\mathrm{opt}} \sim$ const from shock theory, (*cf.* eqn. (1)). Using these equations we can relate the optical and the X-ray emissions in LINERs. If the galaxy has mass M and blue luminosity L_{B}, then we can take $\mathrm{M} \propto L_{\mathrm{B}} \propto \sigma^4$. Hence cases (i–iii) give:

$$\log L_x = \begin{cases} \text{const} + \log L_{opt} - 0.5 \log L_B & \text{(5 i)} \\ \text{const} + \log L_{opt} & \text{(5 ii)} \\ \text{const} + \log L_{opt} + 1.9 \log L_B & \text{(5 iii)} \end{cases}$$

Observationally we have, using the Phillips et al. (1986) data set, and that of the *Einstein* and *ROSAT* archives:

$$\log L_x = \text{const} + (0.99 \pm 0.10) \log L_{opt} + (1.68 \pm 0.20) \log L_B \quad (6)$$

This is much more like the form (5 iii), suggesting that the X-rays are produced by shocks in the accretion flow. The different models can also be distinguished by their X-ray spectra, and by the spatial distribution of the emission. Cases (i) and (iii) will display very soft thermal emission whereas Case (ii) would show a relatively hard spectrum, the X-rays being derived by accretion onto a compact object. On the other hand, Case (i) would have a source extended at least as much as the emission line extent, Case (ii) would be a point source, and Case (iii) would give a very strongly centrally peaked source, since the X-rays are only emitted with appreciable efficiency where the shock velocities are highest, by virtue of eqn. (3 iii). For the lower mass galaxies, the X-ray emission would be effectively point-like, but may become somewhat extended in the most massive objects.

COOLING FLOWS

Since their discovery, the emission line regions associated with the first-ranked ellipticals in X-ray bright clusters have usually been interpreted as being the result of cooling of the X-ray plasma in the deepest part of the potential well, where cooling times are shortest (Ford and Butcher 1979). However, this simple interpretation encountered a serious problem as a result of recognition of the "recombination problem" (Johnstone, Fabian and Nulsen 1987; Heckman et al. 1989; hereafter HBvM). It was shown that each hydrogen atom would have to make between 10 and 1000 recombinations in order to account for the observed luminosity assuming that the cooling flow has a mass flux equal to that derived from consideration of the X-ray properties (e.g. Arnaud 1988). Two suggestions have emerged which attempt to address this question. First, that repressurising shocks occur as large parcels of gas cool isochorically, and drop out of pressure equilibrium, and second, that emission is generated in mixing layers between the hot and the cold medium (Voit and Donahue 1990; Donahue and Voit 1991). Neither of these models gives sufficient additional recombinations to solve the energy problem. Crawford and Fabian (1992; hereafter CF) have pointed out that the EUV photons in the mixing-layer hypothesis determine the emission spectrum, and have argued that emission from self-absorbed mixing layers can explain the observed line ratios of the HBvM Class I sources, such as the M87 filaments associated with the radio lobes in M87 (Ford and Butcher 1979; Sparks, Ford and Kinney 1993). These are distinguished from the Class II sources by having stronger [N II] and [S II] lines relative to their [O I].

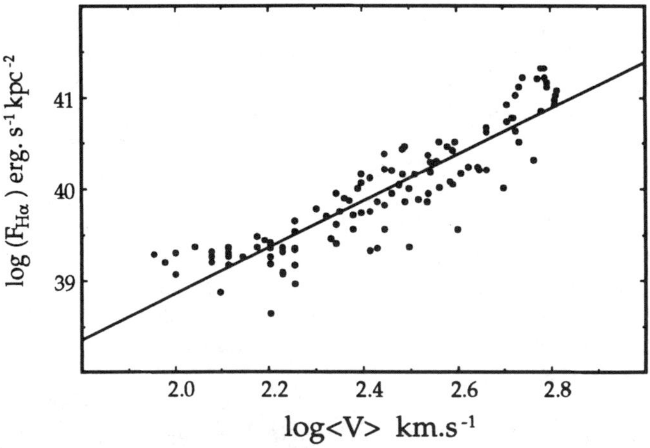

FIGURE II The velocity dispersion: surface brightness relation for the cooling flows studied by Crawford and Fabian (1992), normalised to A496. The line is a least squares fit to the data with the slope constrained at 2.4, the value given by shock theory (*cf.* eqn. 1).

The Class II sources have line ratios which put them in exactly the same part of the diagnostic plots as the LINERs, and this strongly suggests that the same excitation mechanism, radiative autoionising shocks around small, dense clouds applies to Class II cooling flows as well. The idea that these are shock-excited was argued by CF, who presented very direct dynamical evidence that this is indeed the case. They found a relationship between the local velocity dispersion and the local surface brightness of the emission which follows that expected from equation (1), but applied to Hα. This is shown in Figure II, where we have normalised each galaxy to the mean of A496 to reduce the scatter caused by the different fn_0, the product of the area covering factor of shocks and the pre-shock density. The values derived for this quantity are given in Table I.

TABLE I Core properties of Cooling Flows

Cooling Flow	$L_{H\alpha}$ erg s^{-1}	σ_{mean} km s^{-1}	σ_{cen} km s^{-1}	fn_0 cm^{-3}	L_{tot} erg s^{-1}	$\dot{M}_x$ M$_\odot$ yr^{-1}	L_{cf} erg s^{-1}
NGC1275	4.5×10^{41}	177	265	1.7	1.7×10^{43}	76	2.2×10^{42}
PKS 0745–191	7.5×10^{41}	180	210	7.1	3.0×10^{43}	310	1.0×10^{43}
3C 218	5.0×10^{40}	210	265	8.0	2.2×10^{42}	267	1.1×10^{43}
A 496	1.5×10^{40}	140	190	2.0	5.2×10^{41}	63	1.2×10^{42}
A 1795	2.0×10^{41}	200	265	0.8	8.5×10^{42}	144	5.7×10^{42}
A 2597	4.4×10^{41}	250	270	4–8	2.2×10^{43}	164	1.0×10^{43}

The excitation line diagnostics given in HBvM have been used to derive the shock parameters. It appears clear that, in cooling flows, the N/O ratio must be enhanced over solar values to between $0.15 < N/O < 0.35$ by number if we are to successfully account for the absolute strength of the [N II] $\lambda 6584$ Å line. This result should not be too unexpected, since we believe that the gas in elliptical galaxies is derived mainly from planetary nebulae, which are both theoretically expected to be (Becker and Iben 1980), and observed to be (Dopita and Meatheringham 1991) enhanced in N and He as a consequence of the dredge-up processes. In the particular excitation plot from HBvM shown in Figure III, it can be seen that individual cooling flows are characterised by a unique value of N/O, as they should be, but that different points in the cooling flows lie along a shock sequence for this value of the abundance ratio. The shock velocities indicated lie beween ~ 270 km s^{-1} in the inner regions, falling to as little as 160 km s^{-1} in the outer parts. This range agrees well with the shock velocity estimated from the dynamics (taken as 0.42 times the velocity width in the HBvM and the CF objects).

The two Type II cooling flows of CF appear to have shock velocities in the same range, again judged from both their line diagnostics and their velocity widths. However, the Type I flows (Sersic 159–03 in CF, and M87, A262 and A2052 in HBvM) do not fall on these diagnostic plots. Not only do the line ratios fail to give consistent shock velocities, but their line widths are unusually small, and inconsistent with the autoionising shock hypothesis. In the case of M87, it is clear that the emission line material is intimately associated with the radio lobes (Sparks, Ford and Kinney 1993), so it seems likely that an additional excitation mechanism is at work in Type I flows, possibly the self-absorbed mixing layers advocated by CF.

The energy dissipation rate in the Type II cooling flow emission regions is given by the energy flux across the shock:

$$\frac{dE}{dt} = 8.6 \times 10^{40} f n_0 (V_{200})^3 A_6 \text{ erg s}^{-1} \tag{7}$$

where the symbols have the same meaning as equation (1) and f is the area covering factor of shocks. All these parameters can be estimated from observation, as has been done above. Ideally, we would like to relate the energy dissipation rate to the mass inflow rate of dense clouds, assuming that the luminosity represents the turbulent dissipation of gravitational potential energy. We can do this (to order of magnitude) by assuming that the dense clouds are derived from the cooling X-ray gas, and that these release potential energy at the rate $L_{\rm cf} = 3\dot{\rm M}_{\rm x}\sigma_{\rm mean}^2/2$, where $\dot{\rm M}_{\rm x}$ is the cooling flow mass flux, and $\sigma_{\rm mean}$ is the mean line of sight velocity dispersion.

The properties of the well-observed Type II flows are summarised in Table I. We find, as did HvBM, that these mechanical luminosities agree to order of magnitude. Interestingly enough, the turbulent pressure of the intercloud gas, given by $P_{\rm IC} = 3\mu m_{\rm H} f n_0 \sigma_{\rm mean}^2/2$ is in the same ratio to the X-ray gas pressure as $L_{\rm tot}/L_{\rm cf}$ for the cases in which we can estimate it. This would then imply that the intercloud medium is kept in turbulent support with the X-ray plasma by being repeatedly shocked and partially entrained in the passage of the dense clouds.

These estimates of the properties of cooling flows allow us to construct a self consistent scenario of the way in which the cooling flow feeds the nuclear regions. In the outer regions of the flow, clouds which form by cooling are moving at subsonic velocities with respect to the surrounding hot medium, and coalesce

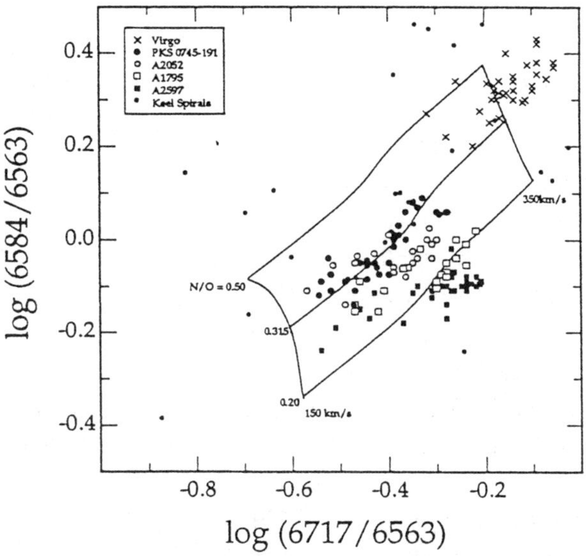

FIGURE III The [N II]/Hα *vs* [S II]/Hα diagnostic plot for cooling flows from HBvM. A grid of theoretical models are shown having shock velocities between 150 and 350 km s^{-1}, and N/O ratios between 0.2 and 0.5. The Type II cooling flows appear to be characterised by shock velocities in the range 140–270 km s^{-1}.

into dense clouds. How exactly self-gravitating clouds can form in this process is unclear, but the implication of the observations is that they exist. Falling in under gravity, these clouds enter the core region where the cooling time of the X-ray plasma is short compared with the dynamical time. Here they move at supersonic velocities with respect to the intercloud, generating a bow shock, entraining shocked intercloud in their wake, and losing energy as a result. If the X-ray halo has little net rotational support, then the infalling clouds will plunge in on radial orbits. This would ensure the increase in velocity dispersion seen in the cores of these systems, and the corresponding increase in surface brightness in the flow, without the need to pack the intercloud medium near the center of the flow, or necessarily have a greater density of clouds near the center of the galaxy. As long as this accretion process is chaotic, the intercloud medium will not aquire any net rotational support, but will be maintained as a turbulent "froth" presumably embedded in a hotter coronal phase. The existence of high velocity bow shocks ensures that a radio non-thermal component will exist in close association with the optical emission. This would provide a qualitative explanation for the observed correlation between the radio power and $\dot{M}_x$ found by HvBM.

REFERENCES

Arnaud, K. A. 1988, in *NATO Advanced Research Workshop on Cooling Flows in Galaxies and Clusters*, ed. A. C. Fabian (Dordrecht: Reidel), 31

Barthel, P. D. 1989, *ApJ*, **336**, 606

Baldwin, J. A., Phillips, M. M., and Terlevich, R. 1981, *PASP*, **93**, 5

Baum, S. A., Heckman, T., Bridle, A., van Breugel, W., and Miley, G. 1988, *ApJS*, **68**, 643

Bertola, F., et al. 1991, *ApJ*, **373**, 369

Bicknell, G. V., et al., 1990, *ApJ*, **354**, 98

Bicknell, G. V. 1993, (this conference)

Becker, S. A., and Iben, I. Jr. 1980, *ApJ*, **237**, 111

Binette, L., Dopita, M. A., and Tuohy, I. R. 1985, *ApJ*, **297**, 476

Binette, L. 1985, *A&A*, **143**, 334

Binette, L, Parker, D., and Fosbury, R. A. E. 1993 (preprint)

Cowie, L. L., Hu, E. M., Jenkins, E. B., and York, D. G. 1983, *ApJ*, **272**, 29

Crawford, C. S., and Fabian, A. C. 1992, *MNRAS*, **259**, 262 (CF)

Donahue, M., and Voit, G. M. 1991, *ApJ*, **381**, 361

Dopita, M. A., and Meatheringham, S. J. 1991, *ApJ*, **377**, 480

Ferland, G. J., and Netzer, H. 1983, *ApJ*, **264**, 105

Ford, H. C., and Butcher, H. 1978, *ApJS*, **41**, 147

Fosbury, R. A. E. 1989, in workshop on *Extranuclear Activity in Galaxies*, ed. E. J. A. Meurs and R. A. E. Fosbury (ESO: Garching), 169

Halpern, J. P., and Steiner, J. E. 1983, *ApJ*, **24**, L43

Heckman, T. M. 1980, *A&A*, **87**, 142

Heckman, T. M. 1981, *ApJ*, **250**, L59

Heckman, T. M., Baum, S. A., van Breugel, W. J. M., and McCarthy, P. 1989, *ApJ*, **338**, 48 (HBvM)

Johnstone, R. M., Fabian, A. C., and Nulsen, P. E. J. 1987, *MNRAS*, **224**, 74

Keel, W. C. 1983a, *ApJS*, **52**, 229

Keel, W. C. 1983b, *ApJ*, **269**, 466

Koski, A. T. 1978, *ApJ*, **223**, 56

Morganti, R., et al. 1991, *MNRAS*, **249**, 91

Morganti, R., et al. 1992, *MNRAS*, **256**, 1P

Phillips, M. M., et al. 1986, *AJ*, **91**, 1062

Pogge, R. W. 1988a, *ApJ*, **328**, 519

Pogge, R. W. 1988b, *ApJ*, **332**, 702

Pogge, R. W. 1989, *ApJ*, **345**, 730

Sparks, W. B., Ford, H. C., and Kinney, A. L. 1993, *ApJ*, (in press)

Sutherland, R. S., Bicknell, G. V., and Dopita, M. A. 1993, *ApJ*, **414**, 510

Tadhunter, C. N., and Tsvetanov, Z. 1989, *Nature*, **341**, 422

Veilleux, S., and Osterbrock, D. E. 1987, *ApJS*, **63**, 295

Voit, G. M., and Donahue, M. 1990, *ApJ*, **360**, L15

The First Stromlo Symposium: The Physics of Active Galaxies
ASP Conference Series, Vol. 54, 1994
G. V. Bicknell, M. A. Dopita, and P. J. Quinn (eds.)

KILOPARSEC-SCALE RADIO EMISSION IN SEYFERTS: EVIDENCE FOR STARBURST DRIVEN SUPERWINDS?

S. A. BAUM, C. P. O'DEA
STScI, 3700 San Martin Dr., Baltimore, MD 21218

D. DALLACASA
Instituto di Radioastronomia del CNR, via Irnerio n. 46, I-40126, Bologna, Italy

A. G. DE BRUYN
NFRA, Postbus 2, 7990 AA Dwingeloo, NL

A. PEDLAR
Nuffield Radio Astronomy Labs, Macclesfield, Cheshire SK11 9DL, England

ABSTRACT We report here on WSRT 6 and 20 cm observations of 13 Seyfert galaxies (see also Baum *et al.* 1993a). These observation reveal the presence of diffuse, sometimes bubble-like, radio emission in 12 of these sources on the $\sim 10''$ (kpc) scale. We show that this emission tends to align with the minor axis of the host galaxy and suggest that it is the manifestation of a superwind which sweeps the radio emitting plasma out along the pressure gradient in the ISM. We present evidence which suggests that this superwind arises from a circumnuclear starburst surrounding the AGN. Alternately, we may be witnessing large scale signs of an AGN-driven wind.

INTRODUCTION

It has been known for some time now that the nuclei of Seyfert galaxies are the sites of small, sub–kpc scale radio sources, some of which which reveal double, triple, or even jet–like structure when mapped at high ($< 1''$) resolution (e.g., Ulvestad and Wilson 1984a,b and references therein). It is generally believed that, at least the 'linear' radio sources in Seyfert nuclei result from the outflow of radio plasma from the active nucleus in the form of jets, bubbles or plasmoids (e.g., Wilson and Ulvestad 1982b; Pedlar *et al.* 1985). Inspection of the literature reveals that larger scale (kpc to tens of kpc) radio sources have been observed in some Seyfert galaxies (e.g., Wilson and Willis 1980; Condon *et al.* 1982; Wrobel and Heeschen 1988). However, it is unclear from the literature how rare (or common) a phenomenon this is or whether there is a common origin for the large scale radio emission seen in these Seyfert galaxies. In this paper, we report the results of a pilot program to search for extra-nuclear radio emission on the kpc to tens of kpc scale in 13 Seyfert galaxies.

OBSERVATIONS

We used the Westerbork Synthesis Radio Telescope in the period 1989 through 1991 to obtain $\sim$ 12 hour integrations for each of 13 Seyfert galaxies using the Digital Continuum Backend (O'Sullivan, 1993) with a bandwidth of 8 x 10 MHz at 6 cm and 8 X 5 MHz at 21 cm. We stress that this sample is not a statistical sample in any way; these sources were chosen either because they were known to have kiloparsec scale extended optical emission line regions or because there was evidence based on a comparison of synthesis and single dish fluxes that they might contain large scale radio emission. When the signal-to-noise was sufficient we used the NEWSTAR package to self-calibrate the data using the redundant baselines provided by WSRT. The typical rms noise in the 6 cm observations is about 0.08 mJy per sythesized beam. The beam has a declination (δ) dependent FWHM of $3.5'' \times 3.5/\sin\delta''$ at 6 cm (and 3.3 times that at 20 cm).

RESULTS

We detected extended radio emission in 12 of our 13 sample sources. The radio emission we have detected is in general diffuse, has a low surface brightness, and is typically not highly collimated. In a few cases where the extended radio emission has been well resolved, the radio emission appears edge brightened or 'bubble-like'. Figure I shows 4 examples. We have compared our radio images with optical images of the host galaxies. We find no clear evidence for a correspondence between optical features and the low surface brightness radio emission in our WSRT images.

DISCUSSION

We seek to understand the origin of the extra-nuclear radio emission revealed by our WSRT images. We start with the premise that there are three likely origins for this extended radio emission. These are: (1) the emission results from the directed outflow of radio emitting plasma from the active nucleus, in direct analogy to the lobes and large scale jets seen in elliptical radio galaxies and quasars, (2) the emission arises in the spiral arms and disks of the Seyfert host galaxy, (3) the emission results from the convection of cosmic rays and magnetic fields along the pressure gradient in the ISM, either in a bouyant flow or in a pressure-driven wind ('superwind'). In the case of (3), the wind could arise either from the AGN itself, or from a circumnuclear starburst and likewise, the cosmic ray electrons responsible for the radio emission could originate in a starburst or the AGN.

Orientation of the Extra-nuclear Radio Emission

Each of the three scenarios for the origin of the extra-nuclear radio emission makes specific and unique predictions for the orientation of this emission relative to the disk of the host galaxy and the axis of the nuclear radio source. Thus the

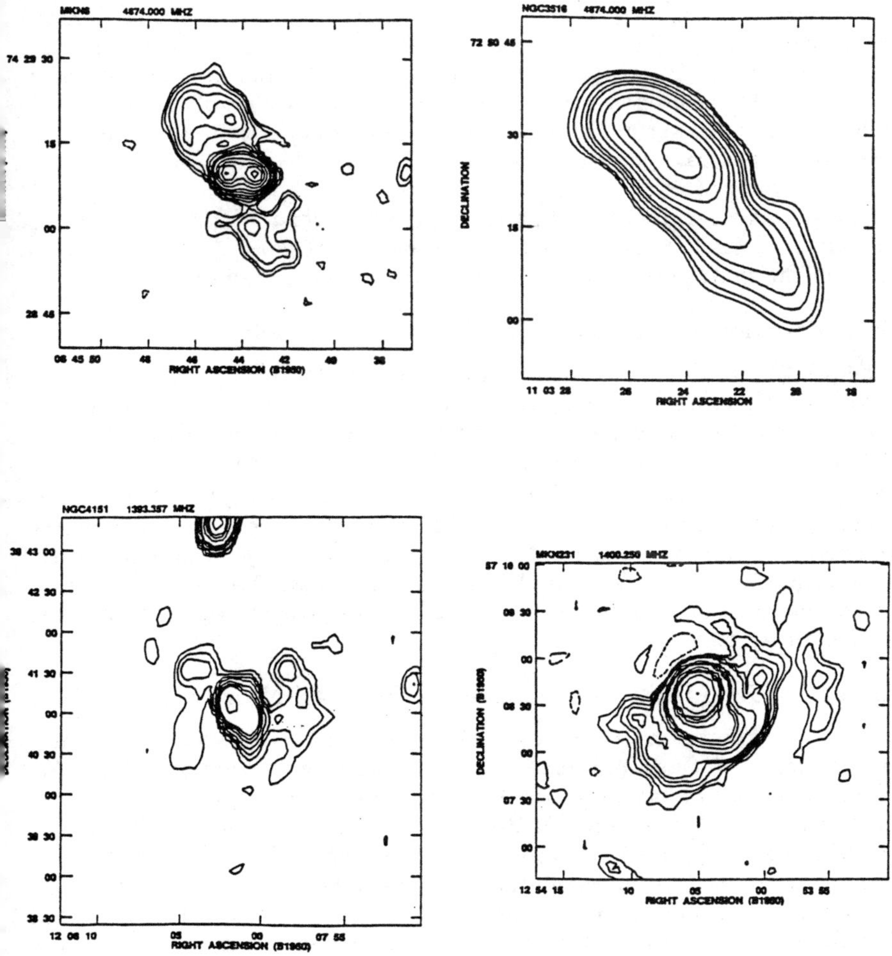

FIGURE I   Contour plots of WSRT images showing extended, diffuse, minor-axis radio emission.

orientation of the extra-nuclear radio emission can be used in a straight-forward way to try to distinguish between them.

In Figure II, we plot histograms of (a) the position angle differences between the nuclear radio source and the extra-nuclear radio source axes, (b) the position angle differences between the nuclear radio source and the major axis of the host galaxy, and (c) the position angle differences between the extra-nuclear radio source axis and the major axis of the host galaxy.

We find that, for the sample as a whole, (1) there is no tendancy for the nuclear and extra-nuclear radio source axes to align, (2) there is no tendancy for the nuclear and galaxy axes to align, (3) there *is* a tendancy for the extra-nuclear radio emission to align with the projected minor axis of the host galaxy; this tendancy becomes very strong if only Seyferts with clearly defined galaxy disks are considered. The misalignment between the nuclear radio structure and the extended radio structure suggests that, in general, the 'extranuclear' radio emission found around these Seyferts is not the analog of the large scale radio lobes seen in powerful radio galaxies and quasars (i.e., the extended radio emission is not generally a simple extension of the small scale radio jets). Moreover, the tendancy of 'extra-nuclear' radio emission to align with the (projected) minor galaxy axis suggests that this emission arises when cosmic ray electrons and magnetic fields flow outwards from the nuclear regions along the pressure gradient in the ISM, perpendicular to the plane of the galaxy disk.

Basic Properties of the Extra-nuclear Radio Emission

We find that typically, the *extra-nuclear* flux density falls in the range 1-10 mJy, and the corresponding radio powers at 1.4 GHz are $\sim 10^{20-22}$ Watts Hz^{-1} (for comparison the radio halo of M 82 has a power of roughly 10^{22} Watts Hz^{-1} [e.g., Seaquist *et al.* 1985]). The ratio of extra-nuclear (i.e., $\gtrsim$ 1 sec) to nuclear (i.e., $\lesssim$ 1 sec) radio flux density varies widely, ranging over roughly two orders of magnitude from 1/20 to 10/1. Thus, in some sources the extra-nuclear radio emission *dominates* and in others it represents a *neglible fraction* of the total radio luminosity. Based on the lack of correlation between the nuclear and extra-nuclear radio luminosities, we suggest that the nuclear and extended components may have independent origins.

Origin of the Extra-nuclear Radio Emission - AGN or Starburst?

There is a well known, extremely tight linear correlation between non-thermal radio and thermal far-infrared fluxes in extragalactic star forming regions. We calculate μ, the log of the ratio of 60 micron to 4.8 GHz flux density, for (a) the total radio flux density and (b) the extra-nuclear radio flux density for our sample of Seyferts (Figure III). We find that (1) roughly half of the Seyferts in our sample have an order of magnitude too much radio emission for their far infra-red emission compared with normal spiral galaxies, indicating that their total radio emission is dominated by emission associated with the *AGN* itself, (2) the flux density of the **extra-nuclear** radio emission agrees roughly with the flux density predicted from the 60 micron flux density, using the empirical relationship between radio and far-ir flux densities found for star-forming regions. This may be evidence that even in those sources in which the **total** radio luminosity is dominated by the AGN, the **extra-nuclear** radio luminosity may

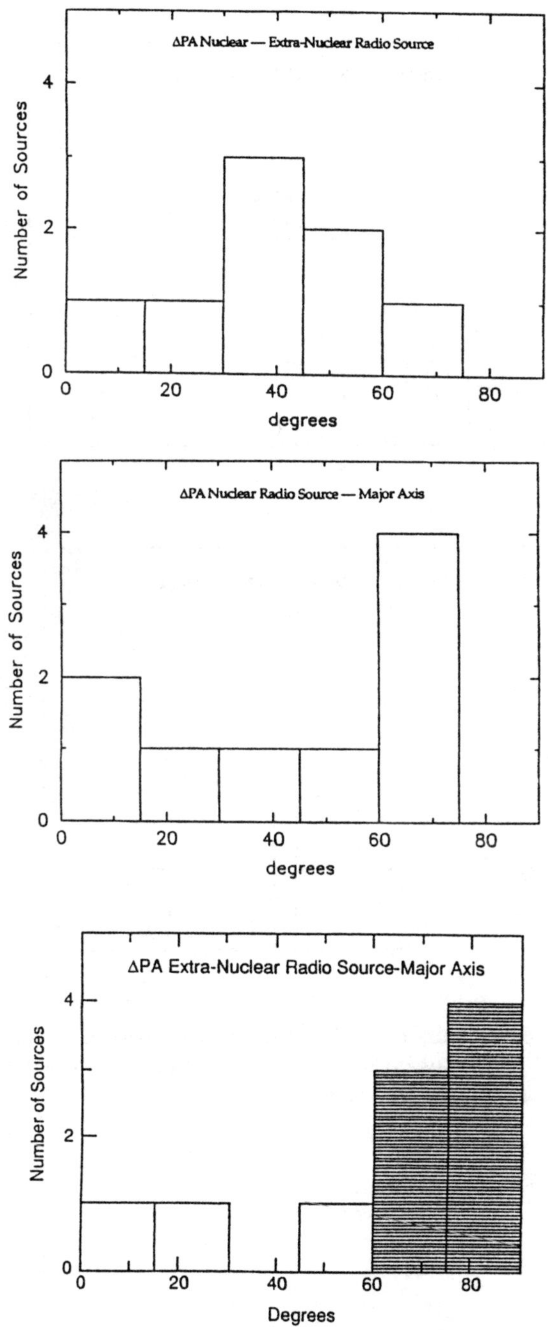

FIGURE II Histograms of position angle differences.

originate in a starburst.

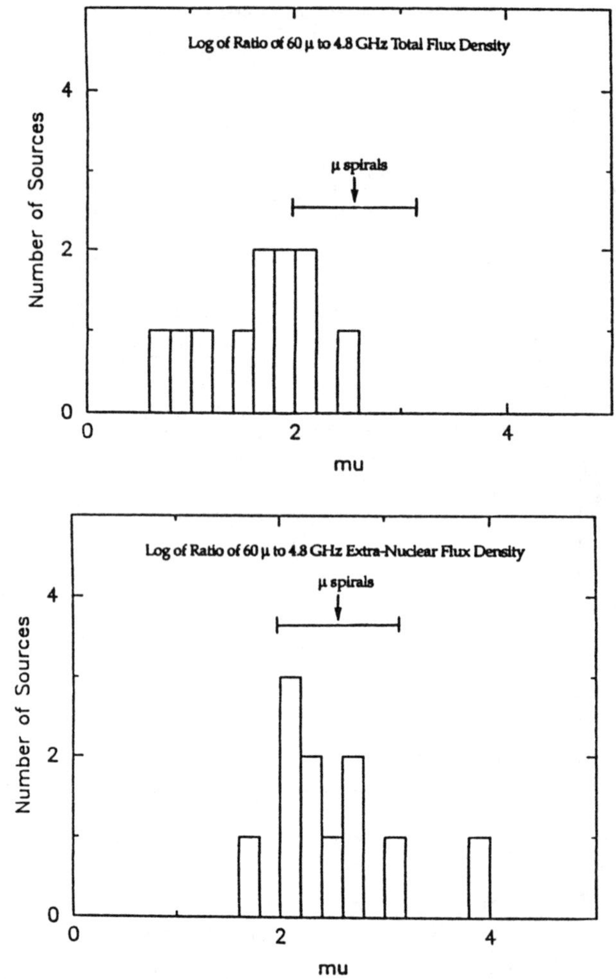

FIGURE III Histograms of μ, the log of the ratio of 60 micron to 4.8 GI flux density, for (a) the total radio flux density and (b) the extra-nuclear rad flux density for our sample of Seyferts.

Generality of the Results

Evidence for superwinds is most easily detectable in edge-on galaxies where minor axis emission is not projected on the disk of the galaxy itself. Based on optical observations of minor axis emission line gas, (Baum *et al.* 1993b, in preparation), we find evidence for superwinds in at least 1/3 of a complete sample of nearby, edge on Seyfert galaxies. Likewise, examination of low resolution radio observations of a complete sample of spiral galaxies (Condon 1987), reveals that roughly 1/2 of Seyfert galaxies with a discernible minor axis show evidence for extended radio emission along that axis (Baum *et al.* 1993a). This indicates that the phenomenon we have detected in the our WSRT observations, it in fact a general property of Seyfert galaxies.

Interpretation

The observations presented here reveal the presence of diffuse, bubble-like radio emission in some Seyfert galaxies which is oriented perpendicular to the axis of the galaxy disk. The radio morphology and orientation of this emission, its radio luminosity and the correlation of that luminosity with the far infrared luminosity of the galaxy, all suggest that this radio emission may be the manifestation of a circumnuclear starburst which generates the cosmics ray electrons responsible for the radio emission and sweeps the radio emitting plasma out along the minor axis of the galaxy disk in a starburst driven wind. The high apparent frequency of superwinds in Seyferts, would then suggest that circum-nuclear starburst are a common feature of Seyferts galaxies.

Though we currently favour the hypothesis that the energy source for the minor axis superwinds in Seyfert galaxies is a circumnuclear starburst, the possibility is still open that the source of the energy for the winds may be the AGN itself. If so, then a careful comparison of the properties of superwinds in Seyfert galaxies and starbursts may reveal differences (in the energetics and spatial distribution) of the winds in the two types of sources.

REFERENCES

Baum, S. A., O'Dea, C. P., Dallacassa, D. de Bruyn, A. G., and Pedlar, A. 1993a, *ApJ*, in press.

Baum, S. A., O'Dea, C. P., Gallimore, J., Lehnert, M., Tsetanov, Z., Mulchaey, J, Caganoff, S., 1993b, in prepartion.

Condon, J. J., 1987, *ApJS*, 65, 485.

Condon J. J., Condon, M. A., Gisler, G., and Puschell, J. J., 1982, *ApJ*, 252, 102.

O'Sullivan, 1983 in "Indirect imaging", Ed. J.A. Roberts, Cambridge University Press, Sydney, 405.

Pedlar, A., Dyson, J. E., and Unger, S. W., 1985, *MNRAS*, 214, 463.

Seaquist, E. R., Bell, M. B., and Bignell, R. C., 1985, *ApJ*, 294, 546.

Ulvestad, J. S. and Wilson, A. S., 1984a, *ApJ*, 278, 544.

Ulvestad, J. S. and Wilson, A. S., 1984b, *ApJ*, 285, 439.
Wilson, A. S. and Ulvestad, J. S., 1982b, *ApJ*, 263, 576.
Wilson, A. S. and Willis, A. G., 1980, *ApJ*, 240, 429.
Wrobel, J. M., and Heeschen, D. S., 1988, *ApJ*, 335, 677.

SEYFERT GALAXIES & LINERS

Contributed Papers

The First Stromlo Symposium: The Physics of Active Galaxies
ASP Conference Series, Vol. 54, 1994
G. V. Bicknell, M. A. Dopita, and P. J. Quinn (eds.)

RADIO OBSERVATIONS OF IC 4329A

D.L. BLANK
Dept. of Astrophysics, School of Physics, University of Sydney, Sydney, N.S.W., 2006, Australia and Australia Telescope National Facility, CSIRO, P.O. Box 76, Epping, N.S.W., 2121, Australia

R.P. NORRIS
Australia Telescope National Facility, CSIRO, P.O. Box 76, Epping, N.S.W., 2121, Australia

INTRODUCTION

IC 4329A is a nearby (z=0.01605) type 1 Seyfert galaxy only two arcmin from the centre of Abell cluster A3574 (Disney 1973). This is unusual since Seyfert galaxies, like most spiral galaxies, are rarely found inside clusters (e.g. Osterbrock 1991). Penston & Wilson (1979) discovered that IC 4329A has the steepest Balmer decrement of any known Seyfert ($H_\alpha/H_\beta \simeq 12$), a very steep optical spectral index ($\alpha = 4.4$) and that its spectra show NaI in absorption. These give values for the extinction that indicates that the absolute visual magnitude M_v is at least -23 and possibly as high as -25. Despite this, the optical image of IC 4329A is that of a relatively undisturbed edge-on Sa type spiral with a prominent dust lane.

This object is of particular interest to us because in a Molonglo Synthesis Telescope (MOST) 843 MHz survey of radio sources in southern Abell clusters, IC 4329A was discovered to have an integrated flux of 90 mJy and an apparent radio tail ~6 arcmin long (Unewisse 1993). This corresponds to a linear size of about 60 kpc assuming $H_o = 50$, implying a radio structure larger than any other known Seyfert. But the 43-arcsec resolution of the MOST observations was insufficient to determine whether this structure is physically connected to the galaxy or the effect of nearby confusing sources.

Previous radio investigations did not look at any possible large scale radio structure. Wilson & Meurs (1982) used the Westerbork array and found the integrated 1.490 GHz flux density to be 38.5 mJy. Unger et al. (1987), using the VLA, found the flux densities within a few arcsec of the nucleus to be 65 mJy at 1.490 and 24 mJy at 4.860 GHz. They attributed the discrepancy between their 1.490 GHz measurement and the earlier Westerbork result to variability in the core. The fairly flat core spectral index of $\alpha = 0.3$ is consistent with this.

OBSERVATIONS

We observed IC 4329A with the Australia Telescope in a full 12 hr synthesis on 1993 May 6 to May 7, and made images using the five antennas in the 1.5 km configuration with a resolution of 16 and 24 arcsec at 2.3 and 1.4 GHz respectively. We also made an image at 2.3 GHz which included the 6 km antenna, giving a resolution of 4 arcsec (see Fig. 1).

DISCUSSION

The radio structure seen in the MOST map turns out to be two point sources ~1 and 2 arcmin north of the nucleus. In the two images made without the 6 km antenna, the core of IC 4329A is unresolved at 1.378 GHz and has a flux of 66.7 mJy. The core is slightly resolved at 2.376 GHz and has a flux of 56.6 mJy. This gives a flat spectral index of $\alpha = 0.16$ confirming the results of Unger et al. (1987).

The 2.376 GHz map made with the 6 km antenna shows two extensions from the nucleus each about 10 arcsec long. The south eastward extension is in the plane of the galactic disc, and is also present in the 1.490 GHz and 4.860 GHz images by Unger et al. where its measured lengths are ~6 and 2 arcsec respectively. This extension may be associated with star formation.

The south westward extension is perpendicular to the plane of the galaxy, and is not seen in the earlier images. This may be an example of a nuclear-driven superwind (Baum, these proceedings). We plan to investigate both possibilities using narrowband imaging and long-slit spectroscopy.

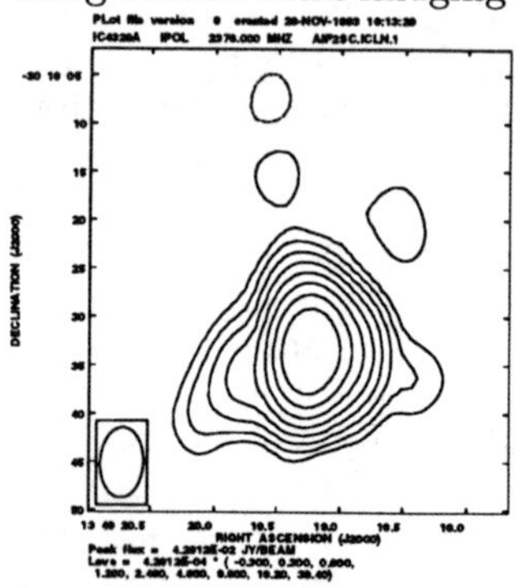

FIGURE I AT image of IC 4329A at 2.3 GHz. Contour levels are -0.30, 0.30, 0.6, 1.2, 2.4, 4.8, 9.6, 19.2, 38.4 mJy beam^{-1}. The peak flux is 42.6 mJy beam^{-1}

REFERENCES

Disney, M.J. 1973 *ApJ*, **181**, L55.

Osterbrock, D.E. 1991 in *Reports on Progress in Physics*, **54**, 579.

Penston, M.V. and Wilson, A.S. 1979 *ApJ*, **232**, 389.

Unewisse, A. 1993 *Ph.D. Thesis*, University of Sydney, Sydney, New South Wales, Australia.

Unger, S.W., Lawrence, A., Wilson, A.S., Elvis, M. and Wright, A.E. 1987 *MNRAS*, **228**, 521.

Wilson, A.S. and Meurs, E.J.A. 1982 *A&AS*, **50**, 217.

The First Stromlo Symposium: The Physics of Active Galaxies
ASP Conference Series, Vol. 54, 1994
G. V. Bicknell, M. A. Dopita, and P. J. Quinn (eds.)

HIGH RESOLUTION NEAR–INFRARED OBSERVATIONS OF THE SEYFERT 1 GALAXY NGC 7469

M. BLIETZ, L. WEITZEL, R. GENZEL AND L.E. TACCONI–GARMAN
Max-Planck-Institut für extraterrestrische Physik, Giessenbachstraße, Postfach 1603, D-85740 Garching bei München, Germany

ABSTRACT We present the results of high–resolution near–infrared (NIR) observations of the circum–nuclear environment of the Seyfert 1 galaxy NGC 7469. The maps have been obtained using our NIR Fabry-Pérot spectrometer FAST and our NIR speckle-camera SHARP. The combination of these instruments, offering both high spatial (FWHM≥0.4″) and spectral (R=1000) resolution, have enabled us to reveal the detailed distribution of the stars and of the various components of the interstellar medium (ISM) in the immediate vicinity of the active nucleus. These results have been used as input parameters for starburst model calculations which indicate a high starforming rate of ~90 $M_\odot$ yr^{-1} in the ~ 800 pc diameter circum–nuclear starburst ring.

INTRODUCTION

A major problem in the understanding of the extremely high luminosities observed towards the nuclei of Seyfert galaxies are the physical processes which give rise to the excitation of the interstellar medium (ISM) in the immediate vicinity of the active nucleus. Two models have been extensively discussed in the last decade assuming either a central, mass accreting black hole or a vigorous and short lived period of star formation. Detailed observations of the interaction of the ISM with stars and/or the active nucleus are therefore an extremely useful tool to distinguish between both scenarios.
Here we present the results of a combined effort of our MPE NIR Fabry-Pérot spectrometer FAST and our MPE NIR speckle camera SHARP to study in detail the circum–nuclear environment of the relatively nearby Seyfert 1 galaxy NGC 7469 (d=66 Mpc for H_0=75 km s^{-1} Mpc^{-1}).

THE CIRCUM–NUCLEAR ENVIRONMENT OF NGC 7469

Fig. 1a shows the 0.4″ resolution K–band continuum map obtained with our NIR speckle camera SHARP. The K–band emission reveals, in addition to the bright central peak which is taken to be the location of the nucleus, a ~800 pc diameter circum–nuclear starburst ring with several embedded emission knots indicating individual starforming regions. This result is in agreement with observations at other wavelengths, in particular with recent optical speckle (Mauder et al. 1993) and 6cm radio observations (Wilson et al. 1991).

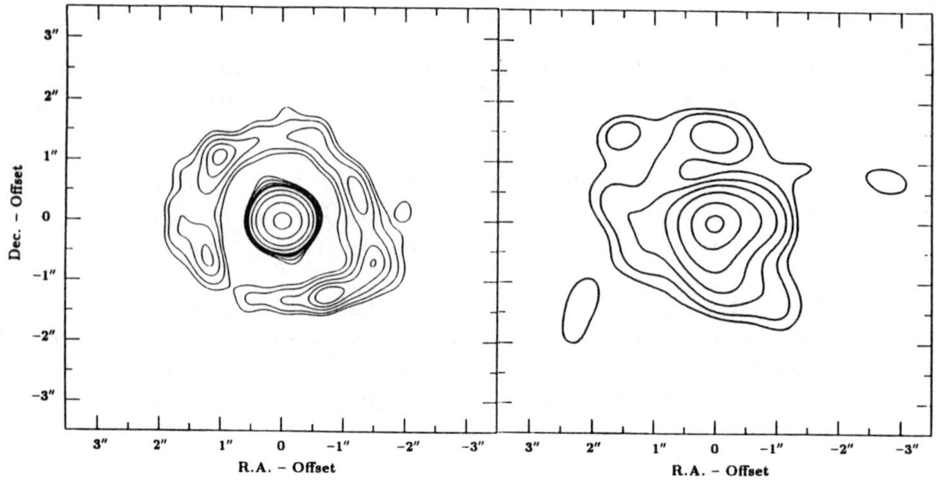

FIGURE I Contour plots of the inner ~1 kpc of NGC 7469:
a) (left) – 0.4″ resolution map of the K–band continuum
b) (right) – 0.9″ resolution map of the 2.12μm H_2 emission line

Fig. 1b represents a contour plot of the 2.12μm H_2 line emission taken with our NIR Fabry-Pérot spectrometer FAST. The emission in the H_2 line is clearly concentrated at the position of the nucleus. This indicates the presence of large concentrations of warm ($T \leq 1500$ K) molecular gas in the line-of-sight to the AGN: From the integrated H_2 line flux we determine a total mass of warm H_2 contained in the inner 300 pc of $\sim 2\times10^3$ $M_\odot$. In addition to the central component, excited molecular gas is extended over ~800 pc, spatially coincident with the region *enclosed* by the circum–nuclear starburst ring and probably acting as fueling agent for the activity in the AGN and/or the circum–nuclear starburst.

Results of starburst model calculations, taking into account the observed line intensities in our H_2 (Fig. 1), Brγ and HeI maps (not shown), indicate a burst age of $\sim 10^7$ yrs and a high starforming rate of ~90 $M_\odot$ yr^{-1} for the circum–nuclear starburst. Our calculations also indicate a stellar mass content of $\sim 1.5\times10^9$ $M_\odot$ for the starburst ring, comparable to the total molecular mass of 1.8×10^9 $M_\odot$ derived from CO line observations (Meixner et al. 1990).

REFERENCES

Mauder, W., Weigelt, G., Appenzeller, I., and Wagner, S. J. 1993, A&A, submitted

Meixner, M., Puchalsky, R., Blitz, L., Wright, M. C. H., and Heckman, T. M. 1990, *ApJ*, **354**, 158

Wilson, A. S., Helfer, T. T., Haniff, C. A., and Ward, M. J. 1991, *ApJ*, **381**, 79

The First Stromlo Symposium: The Physics of Active Galaxies
ASP Conference Series, Vol. 54, 1994
G. V. Bicknell, M. A. Dopita, and P. J. Quinn (eds.)

MEPSICRON OBSERVATIONS OF SEYFERT GALAXIES

J. GUICHARD
Instituto Nacional de Astrofísica, Optica y Electrónica. Apdo. Postal 51 y 216, Puebla, México.

I. CRUZ-GONZALEZ
Instituto de Astronomía, UNAM, Apdo. Postal 70-264, México D.F., México.

L. CARRASCO
Instituto de Astronomía, UNAM, Apdo. Postal 70-264, México D.F., México.

A. SERRANO.
Instituto Nacional de Astrofísica, Optica y Electrónica. Apdo. Postal 51 y 216, Puebla, México.

ABSTRACT Based on observations of 52 Seyfert galaxies (26 Sy1 & 26 Sy2), the narrow line region (NLR) and the broad line region (BLR) were studied. We found that the density of the [O III] zone is high ($N_e = 10^5$ to $10^6\ cm^{-3}$). The comparison with the Stasinska (1984) models suggests that the photoionization mechanism for the line formation is dominant in the NLR. We have identified FeII features in the Seyfert 2 galaxies, and found a correlation between the total optical emission and the blue emission (multiplets 37 and 38) as is known for for Seyfert 1 galaxies.

OBSERVATIONS AND DATA REDUCTION.

The observations were made with the 2.12 m telescope of the Observatorio Astronómico Nacional at San Pedro Mártir, B. C. (México), using a Mepsicron detector coupled to an echelle spectrograph in the low dispersion mode; 200 and 300 l/mm gratings were used, yielding a 12 - 15 Å resolution. The spectral coverage was 3500 - 6700 Å. The data reduction was carried out with a software package developed at the IAUNAM (Serrano & Biro 1990) and the IRAF package. Reddening corrections were applied using the extinction curve by Seaton (1979) and the H_γ/H_β ratio assuming case B recombination.

PHYSICAL CONDITIONS IN THE NLR

To derive the physical conditions of the NLR we have used the usual expressions for nebular analysis (Osterbrock 1989) of relative intensities of the [O III] lines. It is found that in the [O III] zone $N_e = 10^5 - 10^6\ cm^{-3}$ for any reasonable temperature in the context of photoionization models; such high densities have been found in the NLR of AGNs (e.g. Bonatto & Pastoriza 1990). This result would indicate the existence of a density stratification in the NLR, since the low ionization lines as [S II] must be formed in a lower density regime.

Stasinska (1984) presents a grid of model nebulae photoinized by a power-law continuum, for diferent values of electronic density ($N_e = 10^2 - 10^6 cm^{-3}$), spectral index ($\alpha = 0.7 - 2$), ionization parameter ($U = 10^{-4} - 1$), and oxygen abundances ($z = 0.02 - 2\ z_\odot$). Using the quantities E = [O II]λ3727/[O III]λ5007, M = ([O II]λ3727 + [OIII]λ5007)/H_β, we found that the photoionization models are in close agreement with the observations if subsolar abundances were present in the NLR.

FEII EMISSION

We studied the FeII emission in the Seyfert 1 galaxies to derive the total optical FeII emission, adding the contribution of several multiplets present in the spectra, being the most conspicuous those around $\lambda\lambda 4570, 5190, 5320$ Å. A correlation was found between $FeII_{op}$ and the blue emission of multiplets 37 and 38 ($\simeq 4570$Å):

$$FeII_{op} = (3.1 \pm 0.2)\ FeII_{\lambda 4570} + (0.8 \pm 0.2), \qquad r = 0.95$$

This result confirms earlier work by Phillips (1978).

We have explored the possibility of FeII emission in Seyfert 2 galaxies. Although the S/N ratio is low for these weak lines ($\simeq 5$), the spectral features found for several FeII multiplets are quite similar to those of Seyfert 1 galaxies. For Seyfert 2 galaxies, a similar correlation between $FeII_{op}$ and the blue FeII emission also exists:

$$FeII_{op} = (6.3 \pm 0.7)\ FeII_{\lambda 4570} - (2.3 \pm 1.2) \qquad r = 0.95$$

These results suggest that the FeII emission is probably produced in the outermost region of the BLR.

REFERENCES

Bonatto, C.J., and Pastoriza, M.G. 1990, *ApJ*, **353**, 445

Osterbrock, D.E. 1989, *Astrophysics of Gaseous Nebulae and Active Galactic Nuclei*, University Sciences Books, Mill Valley, CA.

Phillips, M.M. 1978, *ApJS*, **38**, 187

Seaton, M.J. 1979, *MNRAS*, **187**, 73P

Serrano, A, and Biro, S. 1990, Technical Report No. 77, IAUNAM

Stasinska, G. 1984, *A&AS*, **55**, 15

The First Stromlo Symposium: The Physics of Active Galaxies
ASP Conference Series, Vol. 54, 1994
G. V. Bicknell, M. A. Dopita, and P. J. Quinn (eds.)

SPATIALLY RESOLVED SPECTROPOLARIMETRY OF NGC 1068

M. D. INGLIS, S. YOUNG, AND J. H. HOUGH
Division of Physical Sciences, University of Hertfordshire, Hatfield, Herts AL10 9AB

D. J. AXON
Space Telescope Science Institute, 3700 San Martin Drive Baltimore, MD 21218, USA

J. A. BAILEY
Anglo-Australian Observatory, PO Box 296, Epping, NSW 2121 Australia

M. J. WARD
University of Oxford, Astrophysics and Nuclear Physics LaboratorY, Keble Road, Oxford OX1 3RH

INTRODUCTION

A turning point in the study of AGN was the discovery of broad hydrogen lines seen in polarized light in the nucleus of NGC 1068 (Antonucci and Miller 1985). A model was developed whereby the scatterers responsible for continuum and broad line polarization are located above and below the optically and geometrically thick torus which obscures our direct view of the Seyfert 1 nucleus, and hence the object appears as a Seyfert 2 galaxy. A cone shaped region of high ionization gas aligned with the radio axis and emanating from the nucleus is seen clearly in an interference filter image (Pogge 1988a). It is now widely accepted that the gas is photoionized by nuclear radiation which only illuminates the conical zone.

OBSERVATIONS

The data were taken with the spectropolarimeter on the 3.9 m Anglo-Australian Telescope on the 24, 25 and 26 October 1990. The spectropolarimeter uses a waveplate modulator built by the University of Hertfordshire, in conjunction with the RGO spectrograph and a Thomson CCD. Using the 25 cm camera and 270R grating a spectral resolution of 8.4 Åwas obtained. A two aperture dekker, each of width 1.3 arc-seconds and length 2.7 arc-seconds was used, with a separation of 23 arc-seconds.The raw spectra were reduced to the form of wavelength and flux calibrated spectra using a Cu-Ar arc and the photometric star VMA2 respectively. The polarimetric reduction, using the two normalized linear Stokes parameters, Q and U, was performed using the STARLINK TSP (Time Series Polarimetry) package developed by J. Bailey.

RESULTS AND DISCUSSION

The aperture positions (arranged at intervals of 2.5 arcseconds in an equally spaced grid) lie within, and on either side of, the U. V. ionizing cone and are at approximately the same PA as the radio axis.

The value for the Balmer decrement Hα/Hβ of 1.16 mag. for the nucleus is similar to that measured by other workers (A_v = 1.7 mag., Ward et al 1987,A_v = 1.6 mag., Koski 1978). The A_v's are seen to rise with increasing distance from the nucleus. Although an increase in dust obscuration might occur a more likely explanation is that the Hβ photons are being lost through dust scattering.

The Broad Line redshifts do not appear to be related in a simple way to the ionization cone.

The [OIII] line profiles provide information on the relative motions and extinction within the NLR. The observed blueward symmetry in the [OIII] line is similar to that seen in several other Seyfert galaxies, and has generally been interpreted as evidence for radial cloud motion coupled with differential extinction within the NLR. The redward asymmetry observed in the spectra south east of the ionizing cone is rarley seen in Seyfert galaxies (DeRobertis and Shaw 1990).

At the nucleus position there is a decrease in PA over the [OIII] line from the blue to red. At several other positions however the PA rises from the blue wing to a maximum in the core, before falling to a minimum value in the red. At the nucleus the degree of polarization rises across the line from the blue to the red as expected if the polarization is produced by dichroism in the NLR. The changes in polarization elsewhere are complex.

CONCLUSIONS

From spatially resolved spectropolarimetry we can attribute many of the observed results to dust in the NLR. This dust is responsible for scattering and extinction along our line of sight and gives rise to the steep Balmer decrement and some of the polarization and PA changes observed in the [OIII] lines. Further work to be presented in a future paper will describe the observed intrinsic flux and polarized flux along with modelling of the continuum polarization at each of the aperture positions.

REFERENCES

Antonucci, R. R. J., and Miller. J. S. 1985, *ApJ*, **297**, 621

DeRobertis, M. M., and Shaw, R. A. 1990, *ApJ*, **301**, 727

Koski, A. T., 1978, *ApJ*, **223**, 75

Pogge, R. W., 1988a, *ApJ*, **328**, 519

Ward, M. J., Geballe, T., Smith, H., Wade, R., and Williams, P., 1987, *ApJ*, **316**, 138

The First Stromlo Symposium: The Physics of Active Galaxies
ASP Conference Series, Vol. 54, 1994
G. V. Bicknell, M. A. Dopita, and P. J. Quinn (eds.)

OPTICAL AND INFRARED COLOUR RATIO MAPS OF SEYFERT GALAXIES

J.K. KOTILAINEN
European Southern Observatory, Karl–Schwarzschild–Str. 2, D–85748 Garching, Germany

M.J. WARD
Astrophysics, Nuclear Physics Building, Keble Rd, Oxford OX1 3RH, England

INTRODUCTION

An important question in AGN research is the relative contribution of nonstellar and stellar (in some cases starburst) components within the nuclear regions. Seyfert and starburst activity is known to coexist in some galaxies (e.g. Wilson et al. 1991) and the two phenomena may be connected either by an evolutionary link (e.g. Sanders et al. 1988) or by a common fueling mechanism.

We have constructed optical and infrared (IR) broad band colour ratio maps of the Seyfert galaxies NGC 1068, 3227, 4151 and 7469. These maps reveal the existence of rings and elongations not discernible in the single filter images. In the bands used (B, I and K), the contribution of emission lines to the total flux is relatively small and hence the images predominantly trace the continuum distribution. The ratios of these images are therefore sensitive to gradients in stellar populations and reddening due to dust. Details of the observations, data reduction, construction of the colour ratio maps, and full discussion of the results are presented in Kotilainen & Ward (1993).

RESULTS

The B–I map of NGC 1068 shows a good spatial agreement with the B–R map of Schild, Tresch–Fienberg & Huchra (1985), resolving the blue star forming regions in the spiral arms. In the nuclear region, there is a blue extension of about 6 arcsec diameter parallel to the radio axis. This structure agrees well with the elongation seen in the "UV/Red" map of Pogge & DeRobertis (1993), and is probably related to the scattering processes known to operate in this Seyfert. In support of this interpretation, high optical polarisation has been detected in this region by Miller, Goodrich & Matthews(1991). In the B–K map of NGC 1068 there is a red double structure separated by $\sim$ 3.5 arcsec, which extends across the nucleus and is nearly perpendicular to the blue elongation seen in the B–I map. The scale of these features is very similar to that of the H_2 emitting regions detected by Blietz et al. (1993), but the orientation is different.

The B–I map of NGC 3227 shows two blue maxima straddling the nucleus separated by $\sim$ 2 arcsec. There is reasonable spatial agreement with the millimeter molecular CO map of Meixner et al. (1990). These maxima most likely indicate the presence of star forming regions around the nucleus.

In the B–I map of NGC 4151 there is a blue elongation $\sim$ 7 arcsec in extent with two blue maxima straddling the nucleus separated by $\sim$ 2 arcsec. This structure has a similar PA to the radio components of Johnston et al. (1982), and the V–I map of Terlevich et al. (1991). A possible interpretation is scattering of the nuclear light, and a polarised jet has been detected with a similar PA, Draper et al. (1992).

The V–I map of NGC 7469 shows an irregular ring of $\sim$ 3 arcsec diameter around the nucleus with blue maxima in S, NE and NW. This structure agrees well with the millimeter molecular CO map of Meixner et al. (1990) and the radio continuum and [OIII]/Hα map of Wilson et al. (1991). Clearly the colour ratio map identifies the ring of starformation.

CONCLUSIONS

Colour ratio maps using optical and infrared images provide a new, powerful method to reveal star formation and the geometry of obscuration within the nuclear regions of Seyfert galaxies. This approach is complementary to narrow-band imaging of emission lines, and radio mapping, and it should be applied to study the statistical properties of large samples of AGN.

REFERENCES

Blietz,M.,Cameron,M., Drapatz,S., Genzel,R., Krabbe,A., van der Werf,P., Sternberg,A. and Ward,M., 1993, *ApJ*, in press

Draper, P.W., Gledhill, T.M., Scarrott, S.M., and Tadhunter, C.N., 1992, *MNRAS*, **257**, 309

Johnston,K.J., Elvis,M., Kjer,D. and Shen,P.S.B., 1982, *ApJ*, **262**, 61

Kotilainen,J.K. and Ward,M.J. 1993, *A&A*, submitted

Meixner,M., Puchalsky,R., Blitz,L., Wright,M., and Heckman,T. 1990, *ApJ*, **354**, 158

Miller,J.S., Goodrich,R.W. and Matthews,W.G., 1991, *ApJ*, **378**, 47

Pogge,R.W. and DeRobertis,M.M., 1993, *ApJ*, **404**, 563

Sanders,D.B., Soifer,B.T., Elias,J.H., Madore,B.F., Matthews,K., Neugebauer,K. and Scoville,N.Z., 1988, *ApJ*, **325**, 74

Schild,R., Tresch-Fienberg,R., and Huchra,J. 1985, *AJ*, **90**, 441

Terlevich,R., Sanchez-Portal,M., Diaz,A.I., and Terlevich,E. 1991. *MNRAS*, **249**, 36

Wilson,A.S., Helfer,T.T., Haniff,C.A., and Ward,M.J. 1991, *ApJ*, **381**, 79

RADIO GALAXIES

Invited Papers

The First Stromlo Symposium: The Physics of Active Galaxies
ASP Conference Series, Vol. 54, 1994
G. V. Bicknell, M. A. Dopita, and P. J. Quinn (eds.)

THE FR I/II BREAK AND THE BIVARIATE LUMINOSITY FUNCTION IN ABELL CLUSTERS OF GALAXIES

FRAZER N. OWEN AND MICHAEL J. LEDLOW[1]
National Radio Astronomy Observatory, Socorro, NM 87801, USA

ABSTRACT We report progress toward understanding radio source evolution by studying the properties of a large sample of radio sources in rich clusters. As reported previously, the FR I/FR II morphology change in samples of sources not chosen to be in clusters is a function of both radio and optical luminosity. In Abell clusters we show that this change occurs at the breakpoint in the bivariate radio/optical luminosity function. This suggests a close connection between the two radio structure classes. It seems likely that sources can evolve across this boundary during their lifetime.

INTRODUCTION

The general physical picture of radio galaxies has been thought to be understood for some time in terms of a beam of energy in various forms emanating from a compact object in the center of the parent galaxy and making its way through the external medium (e.g. Blandford & Rees, 1974). This picture was initially modeled analytically and more recently numerically in great detail (e.g. Williams, 1991). It is intrinsically an evolving process as the jet makes its way further out into the external gas, changes at its origin or is deflected by changing external conditions or instabilities. However, the objective basis for the model is interpretation of images of individual sources, not on any observational evolutionary picture as is the case for stellar evolution.

In the standard picture the external medium plays a major role in the evolution of the jets. The higher the density and/or pressure in the external gas, the harder time the jet has in making progress through the medium and the brighter the source should be. Thus one might expect that radio galaxies in rich clusters would have different sizes, luminosities and life histories than sources in poorer, supposedly less dense regions.

Our long-term goal is to construct a radio H-R diagram to be compared with the predictions of the standard picture and any other models that emerge. It will be at least a three dimensional diagram with radio luminosity, optical luminosity and radio size as the axes. However, in this report we will restrict ourselves to the first two parameters. We also want to study samples inside and outside rich clusters. However, most of the paper will concern either samples

---

[1]Department of Physics & Astronomy, University of New Mexico, Albuquerque, NM 87831, USA

drawn from the literature without regard to cluster membership or a complete sample of radio galaxies found in Abell clusters.

PREVIOUS WORK

The bivariate luminosity function, i.e. the radio luminosity function calculated in bins of equal optical absolute magnitude, was first reported by Auriemma et al (1977). Drawing together data from various sources and of various accuracies they found that the radio luminosity function in each optical bin could be described by two power laws with a break at a radio luminosity, P^*. Based on their limited data P^* seemed to be a constant. They also found that they could detect no difference in the luminosity function in and out of rich clusters.

Fanti (1984) compared the luminosity function of a larger sample of sources in rich clusters with the univariate version of Auriemma et al. He found no difference in the shape of the two functions and also found that it was plausible that their normalization was the same.

Much earlier Fanaroff and Riley (1974) had reported a division of radio morphology into two classes, now called FR I and FR II. FR I radio galaxies were bright near their centers and slowly faded out further from the core. FR II's were brightest at the extremities of the source and fainter close to their centers. In more recent images, FR I's usually appear to be twin, slowly broadening jets, while FR II's have bright outer hotspots. Fanaroff and Riley noted that this classification divided the sources into high (FR II) and low (FR I) radio luminosities.

Based on these data the situation appeared to be:

1. The FR I/II transition is a function of absolute radio luminosity.
2. The bivariate luminosity function has a radio break luminosity, P^*, which is the same for all elliptical galaxy absolute magnitudes.
3. The break occurs roughly but not exactly at the transition from FR I to FR II morphology.
4. These results do not seem to depend on cluster environment.

FR I/FR II 'S AND ABSOLUTE OPTICAL MAGNITUDE

In Owen (1993) we reported that the FR I/II break was a function of both radio and optical luminosity. In figure I we show a recent version of this plot. These data are collected from the literature and transformed into the R-band $M_{24.5}$ isophotal magnitude system as described in Owen & Laing (1989). As such they are not a complete sample in any sense but only serve to populate all regions of the diagram. However, there is no obvious selection which create a sharp break between the FR classes.

This diagram shows, first, that both FR I and II galaxies in this radio luminosity range occur in a wide range of optical absolute magnitude galaxies, ranging from about one magnitude fainter than L^*, which is $M_{24.5} = -22$ in this system, up to the brightest cD galaxies. Second, the relatively sharp FR I/II break for galaxies of the same absolute magnitude suggests a physical, perhaps evolutionary connection between the two classes.

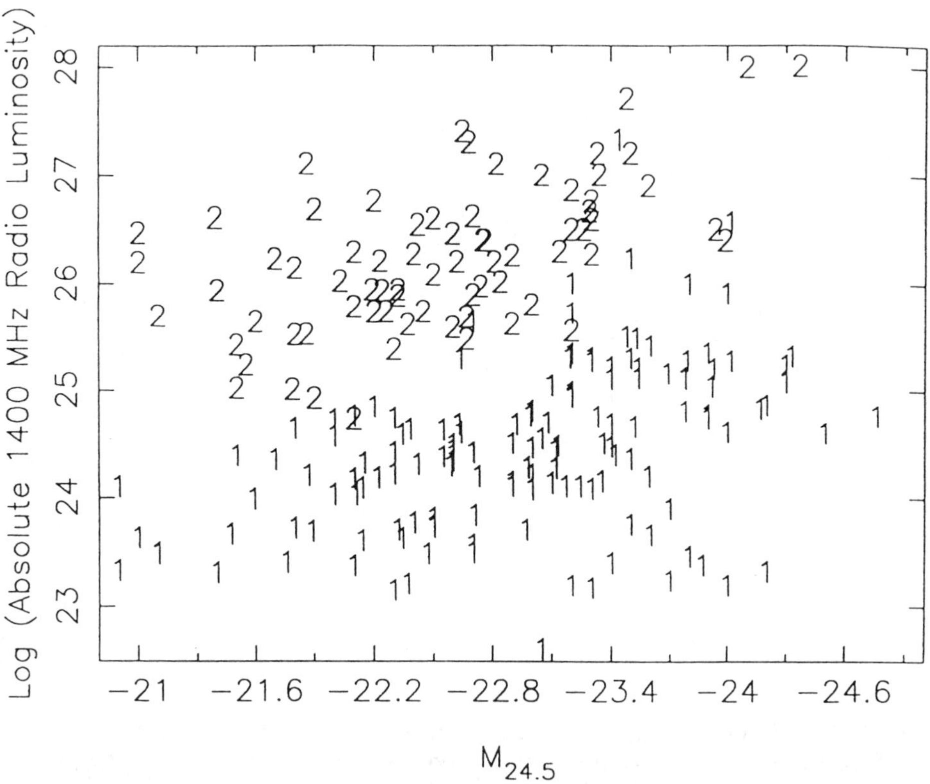

FIGURE I FR I's and II's versus radio and R-band optical luminosity. Radio units are W Hz^{-1}. The 1's and 2's refer to FR I's and FR II's, respectively.

THE BIVARIATE LUMINOSITY FUNCTION IN ABELL CLUSTERS

How does the bivariate luminosity function relate to figure I ? Unlike the data sample in figure I, we need a complete sample to study this question. Recently we completed observations of the radio and optical properties of a complete sample of Abell clusters for this purpose. 200 radio galaxies with cluster redshifts < 0.09 and 20cm flux densites > 10 mJy make up this complete sample. The sample is based on 20cm observations, mainly with the VLA, of clusters from Abell (1958), the original Abell sample, within 0.3 Abell radii of the cluster center. CCD R-band surface photometry was obtained for all the objects in the sample and redshifts have been measured to check on cluster membership.

In figure II, we show the bivariate luminosity function for this sample. Figure II shows that the bivariate luminosity function has a sharp break, P^*, as shown by Auriemma *et al*, but that it is a strong function of absolute magnitude. Furthermore the radio luminosity at the break corresponds exactly to the FR I/II break in figure I.

In comparison to the bivariate functions presented in Auriemma et al, we detect a shift in P^* of $0.4 \log P$ per 0.75 magnitude interval. Our results, how-

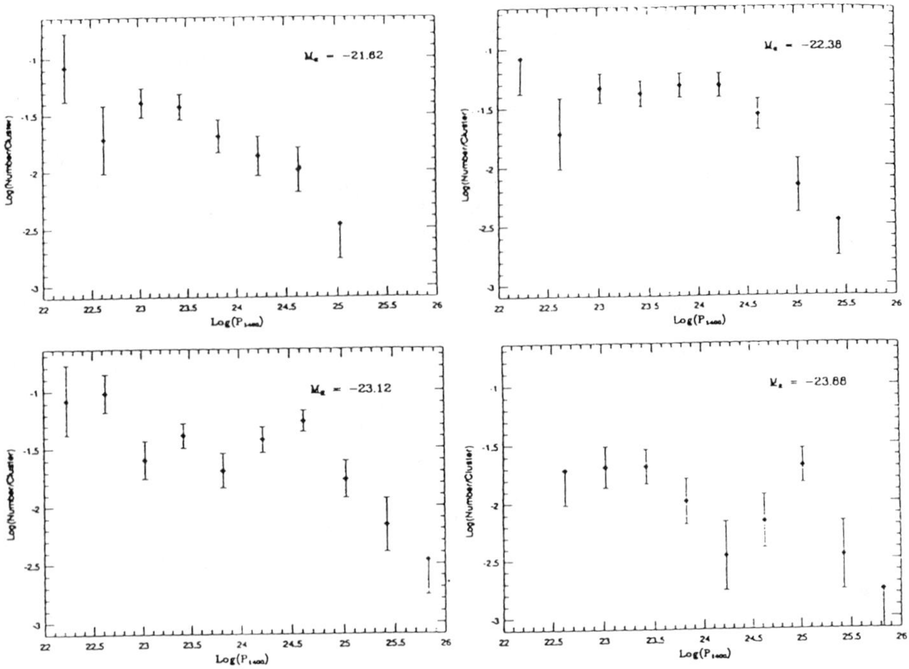

FIGURE II The bivariate luminosity function for Abell clusters of galaxies. Each of the four panels contains the radio luminosity function for a range of 0.75 in absolute magnitude. Each function is normalized to the number of radio galaxies per cluster surveyed.

ever, appear to be consistent with Auriemma et al within the errors; however, our larger sample with smaller errors allows us to draw this more detailed conclusions. The narrowness of the FR I/II transition combined with the break in the luminosity function at the same radio power suggests a close, perhaps evolutionary, connection between the two classes of sources.

Thus the picture which emerges from figures I and II is that for each range of absolute magnitude, the luminosity function has a sharp break right at the FR I/II transition. The radio luminosity of the break goes roughly like the square of the optical galaxy luminosity.

DISCUSSION

These results begin to give us an observational picture which analytic and numerical models need to explain. At least two important pieces are still missing. First, we need to include the distribution of linear sizes for these sources in the picture. de Ruiter et al (1990) have provided the best information of this subject to date for the B2 and 3CR radio samples. Their results show that the size of radio galaxies decreases with decreasing radio power and that the FR II sources are generally bigger than the FR I's. However, their results do not take into account optical luminosity and have no selection for cluster membership. We are in the process of producing such information for our Abell cluster sample.

Second, we need to understand the effect of cluster membership more clearly. So far no clear differences seem to exist which depend on cluster membership. However, rich clusters are generally thought to have higher densities and pressures than are found outside clusters. Furthermore, existing models suggest that the size and luminosity of the radio galaxies should be a strong function of the properties of the external environment. If differences between sources in and out of clusters are not observed then this suggests that either the models are wrong or our understanding of the external environment is faulty. Perhaps the external environment of almost all radio galaxies is quite similar in spite of whether or not it is included in a catalog of rich clusters ?

CONCLUSIONS

The new picture from this work is:

1. The FR I/II transition is a function of both radio and optical luminoisty and is very narrow.
2. The bivariate luminosity function has a radio break luminosity, P^* which is a function of optical absolute magnitude.
3. The FR I/II transition and P^* occur at the same radio and optical luminosities.
4. These results still do not seem to depend on cluster environment.

REFERENCES

Abell, G. O. 1958, *ApJS*, **3**, 211

Auriemma, C., Perola, G. C., Ekers, R. D., Fanti, R., Lari, C., Jaffe, W. J., and Ulrich, M. H. 1977, *A&A*, **57**, 41

Blandford, R. D. and Rees, M. J. 1974, *MNRAS*, **169**, 395

de Ruiter, H. R., Parma, P., Fanti, C., and Fanti, R. 1990, *A&A*, **227**, 351

Fanti, R. 1984, in *Clusters and Groups of Galaxies*, Dordrecht: Reidel, 1984, 185

Fanaroff, B. L. and Riley, J. M. 1974, *MNRAS*, **167**, 31

Owen, F. N. 1993, in *Jets in Extragalactic Radio Sources*, Springer-Verlag, 1993, 273

Owen, F. N. and Laing, R. 1989, *MNRAS*, **238**, 357

Williams, A. G. 1991, in *Beams and Jets in Astrophysics*, Cambridge, 1991, 342

The First Stromlo Symposium: The Physics of Active Galaxies
ASP Conference Series, Vol. 54, 1994
G. V. Bicknell, M. A. Dopita, and P. J. Quinn (eds.)

ON THE EFFECTS OF X-RAY SUBCLUMPS AND CLUSTER/SUBCLUSTER MERGERS ON EXTENDED CLUSTER RADIO SOURCES

J. BURNS, G. RHEE, K. ROETTIGER, J. PINKNEY, AND C. LOKEN
Astronomy Department, New Mexico State University, Las Cruces, NM 88003

F. OWEN
National Radio Astronomy Observatory, P.O. Box O, Socorro, NM 87801

W. VOGES
Max-Planck-Institut für Extraterrestrische Physik, D-85748, Garching, Germany

ABSTRACT We present evidence that cluster/subcluster mergers play an important role in the formation and evolution of rich galaxy clusters and radio galaxies that lie within the clusters. We have found a strong correlation between the positions of radio galaxies and X-ray subclumps in clusters. The radio galaxies appear to live within the gas residue of mergers, and this local environment appears more important in determining radio luminosities and radio morphologies than the global cluster environment. Our hybrid hydro/N-body numerical simulations of cluster mergers have led to new insights on the cluster gas environment which may solve several long-standing problems involving wide-angle tailed radio sources and cluster-wide radio halos.

INTRODUCTION

Our understanding of the dynamical and evolutionary state of clusters of galaxies is undergoing a major revision. Clusters are no longer believed to be simple, relaxed structures for the following reasons. First, substructure is now commonly observed in the optical distribution of galaxies (e.g., Rhee et al. 1991) and especially in the X-ray (e.g., Jones & Forman 1992). Recently, Mohr et al. (1993) have claimed that ≈68% of the clusters in the *Einstein* database have evidence for substructure. Second, Edge et al. (1990) and Henry et al. (1992) have found dramatic evidence for evolution in the cluster X-ray luminosity function from $z \approx 0.3$ to the present, where there appear to be fewer high luminosity clusters in the past. Third, Donahue et al. (1992) have proposed a similarly strong evolution in the numbers of X-ray cooling flow clusters with more cooling inflows in the past. Fourth, large inferred peculiar motions of 200-500 km/s have been observed for cD galaxies in rich clusters (e.g., Pinkney et al. 1993). Such high velocities appear to suggest that even clusters with large, central, dominant galaxies are far from being relaxed. Fifth, we have recently found that radio galaxies strongly correlate with the positions of X-ray subclumps in galaxy clusters (Burns et al. 1993). As we will discuss in this paper, this correlation suggests that radio galaxies live within complex gaseous environments which have not relaxed inside the broader potential well of the clusters.

As a result of the above observations, a revised picture of galaxy clusters, including their formation and evolution, is beginning to emerge. At the heart of this picture is the hierarchical growth of rich clusters via mergers of poorer groups. Such mergers are vividly seen in numerical N-body simulations using cold dark matter (e.g., Evrard et al. 1993) or hybrid hot/cold dark matter (e.g., Klypin et al. 1993) initial density perturbation spectra. This merger activity appears to be continuing today and would explain the relative abundance of substructure within Abell clusters. Because of the merging, there are more luminous X-ray clusters at present than in the past thus causing evolution in the observed X-ray luminosity function for rich clusters. Furthermore, mergers probably destroy cooling flows (e.g., Roettiger et al. 1993; Edge et al. 1992) and, thus account for the relatively higher number of cooling core clusters in the past. Finally, as described below, it appears that radio galaxies live within X-ray clumps inside clusters and these clumps probably result from the merger process.

For the remainder of this paper, we will concentrate on the relationship between X-ray substructure, cluster/subcluster mergers, and radio galaxies. The local gaseous environment and the evolutionary state of a cluster appear to strongly influence the radio properties of cluster galaxies. As a result, radio galaxies may begin to be recognized as important probes of clusters of galaxies.

RADIO GALAXIES & X-RAY SUBCLUMPS

We have undertaken several comparative studies of the X-ray and radio properties of rich and poor clusters which are revealing consistent results. The radio samples are derived from VLA 20-cm surveys of the Abell cluster catalog (Zhao et al. 1989; Owen et al. 1992, 1993) and a poor cluster catalog composed of Zwicky galaxies (Burns et al. 1987). The X-ray data consist of *Einstein* and ROSAT pointed observations, and data from the ROSAT all-sky survey (RASS; Voges 1992). These studies include the following:

Einstein/VLA Correlative Study

One of the best, albeit incomplete, databases on X-ray observations of galaxy clusters is that compiled by the *Einstein* observatory (e.g., Jones & Forman 1992). In cross-correlating the *Einstein* IPC and VLA databases, we found 41 Abell clusters that have X-ray data and radio sources that are $> 2'$ in diameter (Burns et al. 1993). Over 70% of these clusters have statistically significant multiple subcomponents.

About 75% of the radio galaxies within this sample have X-ray peaks within $5'$ of the radio galaxy positions. This fraction is still 46% when the search radius is narrowed to $1'$. This strong coincidence between the radio and X-ray peak positions is much greater than one would expect from random chance. One example of X-ray clumps around radio galaxies is in Abell 1569 which is shown

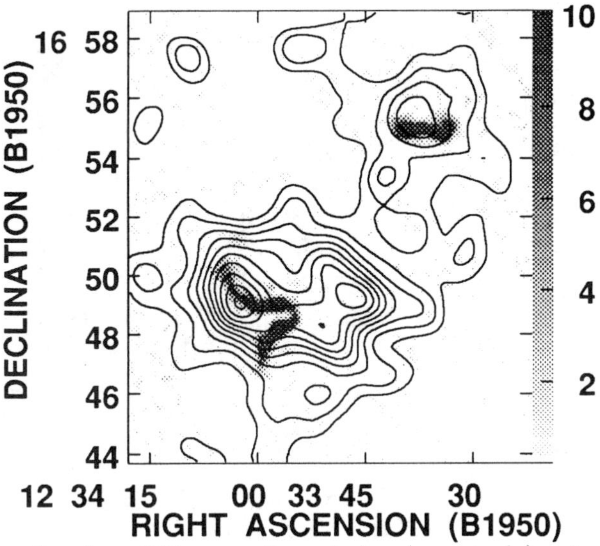

FIGURE I Overlay of 20-cm radio emission (grey-scale) onto contours of *Einstein* X-ray surface brightness for Abell 1569.

in Fig. I.

The generally extended nature of the X-ray clumps suggests that most of the X-ray peaks arise from hot gas in overdense regions emitting free-free radiation. However, we do find several examples of more compact, high surface brightness X-ray sources which could be consistent with nonthermal emission from the AGN. We are currently examining the X-ray spectra of X-ray subclumps around several radio galaxies in an effort to further differentiate between these radiation mechanisms.

ROSAT/VLA Correlative Study

Although the *Einstein* database offers significant insight into the X-ray structure of clusters, it is limited for statistical comparisons because of its incompleteness and possible biases. To remedy these problems, we have begun a detailed comparison between our statistically complete VLA survey of Abell clusters and the RASS. The RASS is the first all-sky X-ray survey with sufficient angular resolution, sensitivity, and completeness to permit such a statistical comparison.

Our initial results are for a subsample of the Abell catalog with $0.06 < z < 0.09$, since the X-ray emission from these clusters is simplest to analyze in this redshift range. We find that 76% of the radio galaxies in this sample (i.e., 90 radio sources) have RASS X-ray peaks within $5'$ of the radio positions – in excellent agreement with the *Einstein*/VLA results discussed above. The distribution of X-ray/radio offsets is strongly peaked at small separations.

We conclude from our analyses of the *Einstein* and RASS databases that radio galaxies tightly correlate with X-ray subclumps. It would appear that the local environment must play an important role in the formation and evolution of extended radio emission.

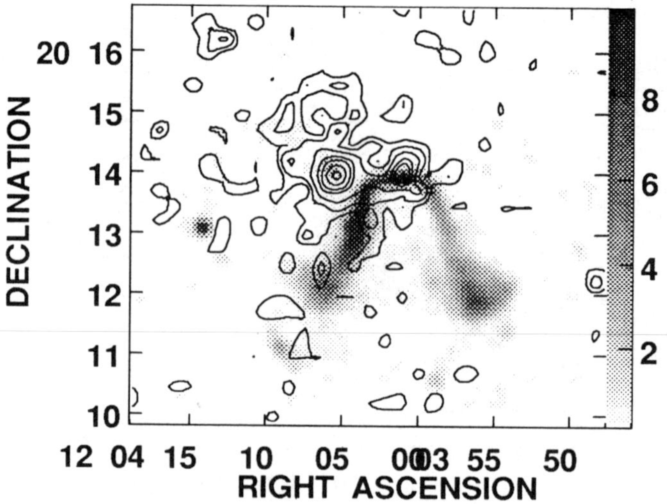

FIGURE II Overlay of 6-cm radio emission (grey-scale) onto contours of ROSAT X-ray emission for the poor group N79-299A. The eastern X-ray source is associated with NGC 4065, and the western X-ray source is associated with NGC 4061 and the HT radio source.

Poor Clusters of Galaxies

In many ways, poor clusters of galaxies may hold the key to understanding the evolution of the cluster environment and radio galaxies. There are many more poor groups than rich clusters, and poor groups are the core unit of which rich clusters are formed. Thus, by examining the radio and X-ray properties of poor clusters, we have a nearby arena to test the hierarchical merger hypothesis proposed in the Introduction.

Previously, Burns et al. (1987) presented a 20-cm VLA survey of 139 poor clusters of galaxies. These clusters are composed of Zwicky galaxies which were chosen for group membership using a percolation algorithm. Typically, these groups contain 5-10 galaxies and are relatively nearby ($z \leq 0.08$). Surprisingly, we found nine head-tail radio galaxies within these groups. This was unexpected since the average ram pressure in such poor clusters is $\approx$1% that for rich clusters, and thus the radio jets and tails in poor clusters are not expected to be bent as much as in rich clusters. However, as seen in Fig. II, poor group head-tail (HT) sources look remarkably similar to those in rich clusters. Why is this so?

The answer to this question may be found through examination of both the X-ray and optical data for these poor clusters. ROSAT images like that in Fig. II have revealed clumped X-ray emission in these poor clusters which is very similar to that in rich clusters. For N79-299A, distorted galaxy isophotes are seen on CCD images suggesting interaction between NGC 4065 and NGC 4061 (coincident with the HT source); the relative velocity between the galaxy pair is a surprisingly high 850 km/s. Relatively compact (but resolved) X-ray emission in Fig. II coincides with each galaxy, and diffuse emission surrounds this group of 10 galaxies. We suggest that this group may be in the initial stages of formation with galaxies falling into the group for the first time. NGC 4061 appears be undergoing a close passage near NGC 4065 – the relatively dense gas in the hot halos and the high encounter speed produces a large enough ram pressure to form the head-tail radio source.

Once again, it would appear that radio galaxies are particularly good pointers to clusters which are in an interesting, still evolving, dynamical state.

SOME NEW INSIGHTS ON CLUSTER RADIO GALAXIES

As a result of our initial findings on the X-ray/radio comparisons in rich and poor clusters, we have begun to develop some new insight into several long-standing problems regarding radio galaxies. The merger hypothesis is at the core of the interpretation.

First, it has been known for over 15 yrs that the radio luminosity function is virtually identical for galaxies in and out of rich clusters (e.g., Fanti 1984). This is surprising since the average higher thermal pressure in rich clusters should better confine extended radio plasma, increase the magnetic field, decrease adiabatic expansion losses, and thus increase the average radio luminosity. However, if the local environments around radio galaxies inside and outside of rich clusters are similar as suggested by our new data, then the similarity in the luminosity functions would suggest that the gas near the radio galaxies is the dominant influence on the radio properties.

Second, the lack of correlation between Abell cluster richness (a measure of the average galaxy density) and the probability of radio emission from a galaxy (e.g., Zhao et al. 1989) can be understood in a similar fashion. Once again, the global environment would seem to be relatively unimportant in comparison to the local subclump in which radio galaxies live in determining radio source properties.

Third, as noted above, poor cluster head-tail sources may be formed in regions that are still in the process of merging. Thus, radio galaxies lie at the heart of cluster merging activity from the group to the rich cluster scales.

WIDE-ANGLE TAILED RADIO GALAXIES & MERGER SIMULATIONS

In addition to the new insights on the general radio properties of clusters, the merger hypothesis has some potential for solving several other problems involving particular classes of radio sources. One such class is the wide-angle tailed (WAT) radio source. WATs are associated with D and cD galaxies at the centers of clusters. As shown in Fig. III, their radio morphology consists of narrow jets that abruptly disrupt and transform into diffuse tails which extend 100-500 kpc from the galaxy (O'Donoghue et al. 1990).

An unresolved question is what bends these V-shaped sources? The measured peculiar velocities of $>$ 200 km/s for these radio galaxies may not be enough to curve the jets according to the standard Euler's equation analysis (Eilek et al. 1984; Burns 1986).

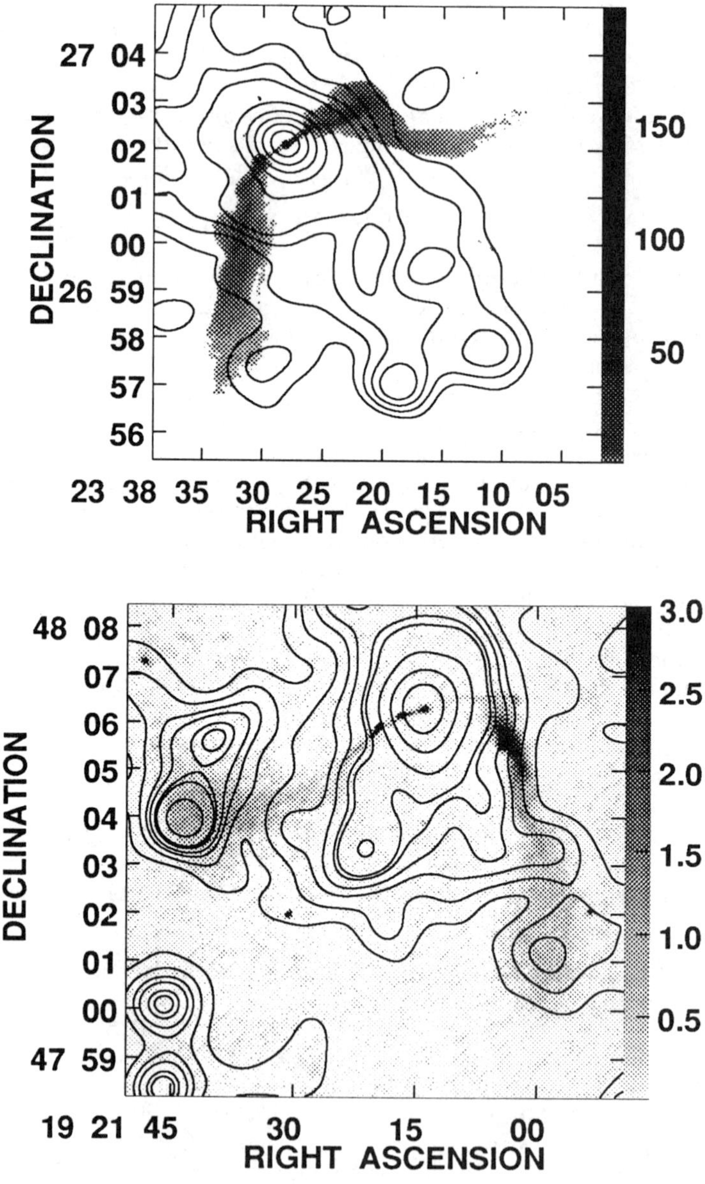

FIGURE III Overlays of VLA images of wide-angle tailed radio sources (grey-scale) onto ROSAT X-ray contour maps. A2634/3C 465 is at the top and 1919+479 is at the bottom. Note the X-ray extensions between the radio tails.

Furthermore, these clusters do not meet the conditions for a cooling flow (i.e., their central cooling time is greater than a Hubble time). According to Edge et al. (1992), 70%-90% of clusters, especially those with dominant galaxies, have cooling inflows. This would further suggest that WAT clusters are unusual.

Finally, recent ROSAT pointed observations reveal interesting asymmetries or elongations in the central X-ray emission as shown in Fig. III. Virtually all WAT clusters imaged to date with either *Einstein* or ROSAT show an X-ray extension along a line between the radio tails. This, too, is unusual since the X-ray surface brightness of most clusters with centrally dominant galaxies appears to be circularly symmetric on arcminute scales (e.g., Jones & Forman 1992).

We have begun to perform numerical simulations of cluster/subcluster mergers in an effort to better understand the gas structure and dynamics in clusters like those that contain WATs. The initial results are quite encouraging. The details of these first simulations are in Roettiger et al. (1993). In brief, we constructed a hybrid Eulerian hydrodynamics plus N-body code which allows us to model the evolution of both the gas and the dark matter as two unequal clusters (one has 8 times the mass of the second) merge. In general, we find that (1) the merger simulation reproduces central X-ray substructure like that seen in clusters, especially WAT clusters, over a significant portion of the merging timescale (> 5 Gyr). (2) Significant heating (caused by shocks as the clusters supersonically merge) and expansion of the gas occur in the core. This would suggest that a cooling flow would be destroyed by merging and would be difficult to re-establish later. This might explain why WAT clusters do not possess such cooling cores. (3) Within about 5 Gyr of the merger, the synthesized X-ray brightness profile from our simulation shows a significant excess in the center which mimics the classic signature of a cooling inflow. This suggests that one must be careful in interpreting the central X-ray surface brightness profiles of clusters and that cooling flows may not be as common as some have suggested. (4) Relatively high bulk velocities within the ICM of up to 1000 km/s are sustained for long periods (5 Gyrs) following the merger. The cluster gas continues to shock and move at relatively high speeds suggesting that the ICM is far from quiescent in clusters which have undergone merging.

How do the above properties in merging clusters help to explain the peculiar characteristics of WAT clusters? First, the N-body component of our simulation continues to show substructure and a non-Gaussian distribution of velocities well past the initial merger. So, the apparent peculiar motions of cD galaxies may reflect the unrelaxed state of the cluster rather than any significant motion by the cD within the central potential well of the cluster. Second, the mystery of what bends WATs may be solved using the high residual bulk motions of the cluster gas. We suggest that the ram pressure responsible for bending the radio jets/tails in WATs does not arise from motion of the radio galaxy through the ICM, but rather motion of the ICM past the radio plasma. This effective "wind" has a large enough value of $\rho_{ICM} v^2_{ICM}$ to explain the observed curvature of the WAT sources. Furthermore, having recognized the mechanism responsible for bending the WAT tails, one might reverse this situation and use WATs as a probe of cluster "weather". That is, observations of WATs coupled with numerical simulations could serve as the only available constraint on gas motion within clusters of galaxies since current X-ray telescopes have neither the sensitivity nor spectral resolution to measure such motions. Third, the lack of cooling flows in WAT clusters is consistent with the merger scenario. Fourth, the observed asymmetries in the central X-ray emission from WAT clusters is caused by post-merger gas which has not yet reached equilibrium in the cluster core.

Thus, the merger model has a great deal of potential to explain the many varied properties of clusters with WAT radio sources.

CLUSTER-WIDE RADIO HALOS AND MERGERS

Another long-standing problem regarding cluster radio sources is the origin of cluster-wide radio halos. How are Mpc-sized radio halos like those in Coma (e.g., Kim et al. 1990) and A2256 (e.g., Bridle et al. 1979) formed and maintained? Why do so few clusters have radio halos (Hanisch 1982)? And, why do clusters like Perseus, which have many radio galaxies that could serve as distributed sources of particles and fields, only have a relatively small diffuse source at the core and no large-scale halo (Burns et al. 1992)?

Some insight into the solution to the above questions may be found from the following additional observations. Clusters such as Coma and A2256 with large-scale radio halos do NOT have X-ray cooling inflows, but do show evidence for X-ray substructure (see White et al. 1993 for Coma and Briel et al. 1991 for A2256). We recently found that a third rich cluster with a purported radio halo, A2255, also has evidence for X-ray substructure on RASS images and a possible merger. On the other hand, clusters such as Perseus (Burns et al. 1992) and A2052 (Zhao et al. 1993) with only small ($<$ 200 kpc radius), central, amorphous radio sources have extensive cooling flows (e.g., Edge et al. 1992) and comparatively little X-ray substructure (see Böhringer et al. 1993 for Perseus). These observations coupled with the above hybrid numerical modelling suggest that clusters with halos have undergone mergers.

De Young (1992) has shown that there is insufficient energy in galaxy wakes to power radio halos via a conversion of kinetic energy. However, there is abundant kinetic energy during the merger of two clusters. The resulting turbulence in the ICM could be tapped to amplify initial seed magnetic fields (e.g., Ruzmaikan et al. 1989) and to reaccelerate relativistic electrons (e.g., Eilek 1993). Furthermore, the lifetime of such halos is governed by the decay time of the turbulence (Tribble 1993); if this decay time is short, this would imply that only a relatively few clusters which are "caught in the act" of merging are seen to have radio halos.

On the other hand, the "mini-halos" found in cooling flow clusters such as Perseus and A2052 are probably confined by the high central gas pressure in the cluster core (Loken et al. 1993). Such radio sources would have steep radio spectra and might be expected to live a relatively long time in comparison to the radio halos since they are being continuously powered.

CONCLUSIONS

It appears that clusters are more complex, dynamically evolving structures than we believed just a few years ago. If cluster/subcluster merging is as common as suggested by large-scale structure simulations and substructure observations, then some assumptions used to estimate cluster masses (hydrostatic equilibrium for X-ray gas), fraction of cluster mass in gas, the mass influx in cooling flows, etc. may not strictly hold everywhere in the clusters. Care must be taken in applying these assumptions given the dynamic nature of the cluster potential and gas.

It is now clear that radio galaxies live within X-ray subclumps which may have been produced by cluster/subcluster mergers. The radio luminosities and radio morphologies appear to be strongly influenced by this local environment surrounding the radio sources. If the local environment is indeed similar for radio galaxies in and out of rich clusters, this would explain why the radio properties are so similar for sources in rich and poor clusters.

This cluster merger scenario also has the potential for explaining many of the observed characteristics of individual classes of cluster radio sources. The optical, X-ray, and radio properties of clusters with wide-angle tailed radio sources can be simply understood if such clusters have undergone a merger. Similarly, the kinetic energy involved in the merger between two clusters appears sufficient to power cluster-wide radio halos; evidence for such mergers has been found in recent ROSAT X-ray maps for clusters with halos.

Finally, and most importantly, radio galaxies appear to participate in the evolving cluster environment. They may be triggered and shaped by the changing cluster potential. As a result, radio galaxies should be viewed with renewed interest as important tools for studying cluster evolution and cluster gas dynamics.

ACKNOWLEDGMENTS

This work has been supported by grants from NASA and the National Science Foundation. We wish to thank our collaborators including Richard White, Hans Böhringer, Rick Perley, Galen Gisler, Jean Eilek, and Anatoly Klypin for their input and advice.

REFERENCES

Böhringer, H., Voges, W., Edge, A., Fabian, A., Neumann, D., 1993, MNRAS, in press

Bridle, A., Fomalont, E., Miley, G., & Valentijn, E. 1979, *A&A*, **80**, 201

Briel, U., Henry, J., Schwarz, R., Böhringer, H., Ebelling, H., Edge, A., Hartner, G., Schindler, S., Trümper, J., & Voges, W. 1991, *A&A*, **246**, L10

Burns, J. 1986, Can. J. Phys., **64**, 373

Burns, J., Hanisch, R., White, R., Nelson, E., Morrisette, K., & Moody, J. 1987, *AJ*, **94**, 587

Burns, J., Sulkanen, M., Gisler, G., & Perley, R. 1992, *ApJ*, **388**, L49

Burns, J., Rhee, G., Owen, F., & Pinkney, J. 1993, *ApJ*, in press

De Young, D. 1992, *ApJ*, **386**, 464

Donahue, M., Stocke, J., & Gioia, I. 1992, *ApJ*, **385**, 49

Edge, A., Stewart, G., Fabian, A., & Arnaud, K. 1990, *MNRAS*, **245**, 559

Edge, A., Stewart, G., & Fabian, A. 1992, *MNRAS*, **258**, 177

Eilek, J., Burns, J., O'Dea, C., & Owen, F. 1984, *ApJ*, 278, 37

Eilek, J. 1993, preprint

Evrard, A., Summers, F., & Davis, M. 1993, preprint

Fanti, R. 1984 in *Clusters and Groups of Galaxies*, F. Mardirossian, G. Giuricin, & M.Mezzetti, Dordrecht: Reidel, 185

Hanisch, R., 1982, *A&A*, **116**, 137

Henry, J., Gioia, I., Maccacaro, T., Morris, S., Stocke, J., & Wolter, A. 1992, *MNRAS*, **386**, 408

Jones, C. & Forman, W. 1992 in *Clusters and Superclusters of Galaxies*, A. Fabian, Dordrecht: Kluwer, 49

Kim, K.-T., Kronberg, P., Dewdney, P., & Landecker, T. 1990, *ApJ*, **355**, 29

Klypin, A., Holtzman, J., Primack, J., & Regös, E. 1993, *ApJ*, in press

Loken, C., Burns, J., Norman, M., & Clarke, D. 1993, *ApJ*, in press

Mohr, J., Fabricant, D., & Geller, M. 1993, preprint

O'Donoghue, A., Owen, F., & Eilek, J. 1990, *ApJS*, 72, 75

Owen, F., White, R., & Burns, J. 1992, *ApJS*, **80**, 501

Owen, F., White, R., & Ge, J.-P. 1993, *ApJS*, 87, 135

Pinkney, J., Rhee, G., Burns, J., Hill, J., Oegerle, W., Batuski, D., & Hintzen, P. 1993, *ApJ*, in press

Rhee, G., van Haarlem, M., & Katgert, P. 1991, *A&A*, **246**, 301

Roettiger, K., Burns, J., & Loken, C. 1993, *ApJ*, **407**, L53

Ruzmaikin, A., Sokoloff, D., & Shukurov, A. 1989, *MNRAS*, **241**, 1

Tribble, P. 1993, *MNRAS*, **263**, 31

Voges, W. 1992 in *Proc. of the European 'International Space Year' Conf.*, ESA ISY 3, T.D. Guyenne & J.J. Hunt, Noordwijk: ESA Publ. Div., 9

White, S., Briel, U., & Henry, J., 1993, *MNRAS*, **261**, L8

Zhao, J.-H., Burns, J., & Owen, F. 1989, *AJ*, **98**, 64

Zhao, J.-H., Sumi, D., Burns, J., & Duric, N. 1993, *ApJ*, in press

The First Stromlo Symposium: The Physics of Active Galaxies
ASP Conference Series, Vol. 54, 1994
G. V. Bicknell, M. A. Dopita, and P. J. Quinn (eds.)

RADIO CORES IN EARLY–TYPE GALAXIES

ELAINE M. SADLER
School of Physics, University of Sydney, NSW 2006, Australia

O.B. SLEE, J.E. REYNOLDS AND R.D. EKERS
Australia Telescope National Facility, CSIRO, Epping, NSW 2121, Australia

ABSTRACT Many nearby elliptical galaxies have parsec–scale radio cores which are qualitatively similar to (though far less powerful than) the 'central engines' of powerful radio galaxies and quasars. Even at quite low radio powers, the radio continuum emission from these galaxies is dominated by an AGN rather than by starburst–related activity. Radio cores in active elliptical galaxies appear to be qualitatively different from those in (spiral) Seyfert galaxies, suggesting that an active nucleus somehow 'knows', on scales of a few parsec, whether it lives in an elliptical galaxy or a spiral.

INTRODUCTION

Radio surveys of optically–selected 'normal' galaxies with the VLA and other telescopes detect many low–luminosity sources in which most or all of the radio emission comes from the central few arcseconds. Are these sources 'central engines' like the ones seen in VLBI observations of powerful radio galaxies and quasars, or is the radio emission related to some other process such as a starburst? Wrobel & Heeschen (1991) note that the weak central radio sources in some S0 galaxies follow the FIR–radio relation (Wunderlich *et al.* 1987), and so probably arise from starbursts. As Norris *et al.* (1990) point out, however, we can usually distinguish 'central engines' and 'starbursts' unambiguously using long–baseline radio interferometry.

Here, we present some results from a large study of early–type galaxies carried out with a radio interferometer which has a resolution of a few parsecs for nearby galaxies (see Figure I). The study is described in more detail by Slee *et al.* (1994).

OBSERVATIONS

The resolution of the Parkes–Tidbinbilla Interferometer (PTI) is 0.1 arcsec at 2.3 GHz (similar to the Hubble Space Telescope) and 0.03 arcsec at 8.4 GHz.

To study core properties over a wide range in total radio power, we observed two well–defined samples of E and S0 galaxies. The 'low–power sample' comprised 69 E/S0 galaxies brighter than 14th magnitude with a 5 GHz flux

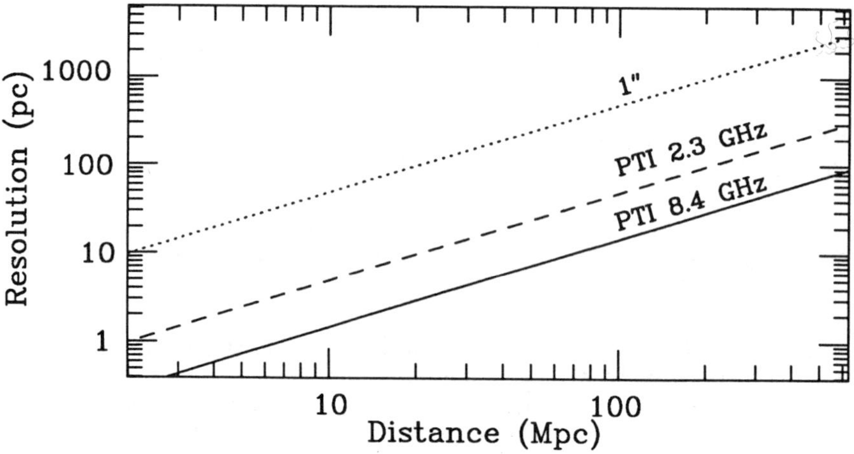

FIGURE I Resolution of the PTI at 2.3 and 8.4 GHz. The VLA at 1 arcsec resolution is shown for comparison.

density above 5 mJy, with total radio powers in the range 10^{20}–10^{23} W/Hz (H_0 = 75 km s^{-1} Mpc^{-1}). The 'medium–power sample' contained 91 galaxies originally observed by Ekers *et al.* (1989), with total powers in the range 10^{22}–10^{26} W/Hz (most of these are FR I radio galaxies). We also observed a smaller and more heterogeneous sample of radio–emitting Seyfert and spiral galaxies. The detection limit was typically 3–5 mJy at both our observing frequencies (2.3 and 8.4 GHz).

DETECTION RATE AND CORE SPECTRAL INDEX

The PTI detection rate for radio–emitting E and S0 galaxies was ~70%, and (perhaps surprisingly) was independent of both distance and total radio power. This is significantly higher than typical detection rates for Seyfert galaxies (26% for Sy1 and 46% for Sy2; Roy *et al.* 1993).

Figure II shows the detection rate as a function of total power for the two early–type galaxy samples combined. We observed only 5 galaxies with P_{tot} < 10^{21} W Hz^{-1}, so the low detection rate in the leftmost bin may not be significant. However, it is interesting that all the E/S0 galaxies observed by Wrobel & Heeschen (1991) to follow the FIR–radio relation have log P_{5GHz} < 10^{21} W Hz^{-1} and so lie below the point at which our PTI detection rate starts to drop. It is possible that this marks a real transition between radio sources where the emission is dominated by a central engine and those where other processes (perhaps related to low levels of star formation) start to become important.

The median spectral index of the PTI cores between 2.3 and 8.4 GHz is +0.3, i.e. the cores have 'inverted' radio spectra which rise towards higher frequencies. Again there is no significant variation with radio power, and the spectral index distribution for cores in the 'low–power' sample is similar to that for cores in medium–power FR I radio galaxies.

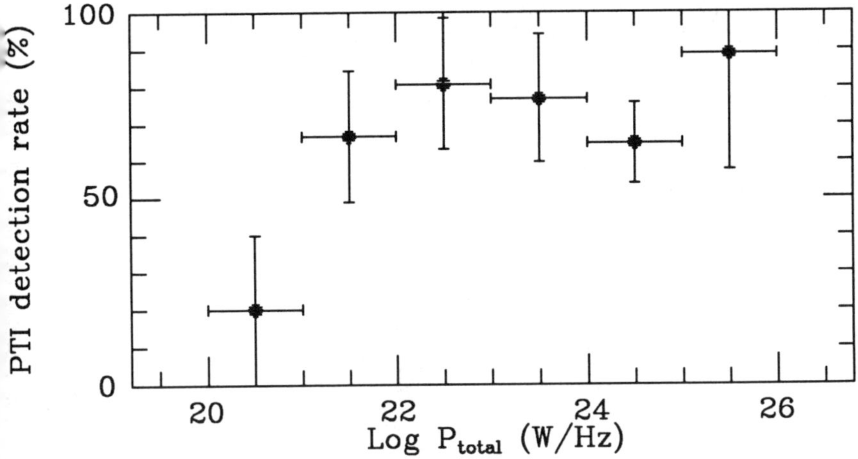

FIGURE II Detection rate of PTI cores in combined sample.

At least two observations suggest that most radio cores in the E/S0 sample are unresolved by the PTI even for the closest galaxies:
(i) The mean spectral index is independent of distance, whereas if some nearby cores were approaching resolution we should see a lower flux density at 8.4 GHz than at 2.3 GHz and hence a steepening of the spectral index.
(ii) Only 5 out of 32 sources for which we made multiple PTI observations at different hour angles showed significant (>15%) changes in flux density with position on the sky of the kind which would be expected for a partly–resolved source.
At least 80% of these cores, therefore, are probably unresolved on scales of a few parsecs. If so, they are no larger that a typical quasar core (e.g. Ghisellini *et al.* 1992).

VARIABILITY

As might be expected from their small sizes, the flux density of many of the cores is variable on timescales of a few years. In galaxies where the total radio emission is dominated by the compact core, it is possible to use lower–resolution archival data from other observers to extend the time coverage. These galaxies show typical variations of a factor of 2–3 in flux density over a 20–year period (Slee *et al.* 1994).

ARE RADIO CORES IN SEYFERT GALAXIES DIFFERENT FROM THOSE IN ELLIPTICALS?

Several observations hint at qualitative differences between the radio cores of ‘active’ elliptical and spiral galaxies, the most striking being the very different radio spectral index distributions measured with the PTI. Because of their flat core spectra, the PTI detection rate for ellipticals is roughly the same at 8.4

GHz as it is at 2.3 GHz. When we make multi–frequency observations of Seyfert galaxies, however, the core detection rate at 8.4 GHz is far lower than at 2.3 GHz. So few detections are made at the higher frequency, in fact, that only a few of the Seyfert galaxies we observed have well–defined radio core spectra (these are listed in Table I).

TABLE I PTI core spectral indices for Seyfert galaxies.

Galaxy	Type	$\alpha_{8.4}^{2.3}$
NGC 2110	E	+0.3
NGC 7213	S0	+0.5
NGC 7674	Sa	−0.7
IC 5135	Sa	< -0.5
NGC 7469	Sab	< -0.6
NGC 1068	Sb	−1.0
IC 3639	Sc	< -0.6
NGC 5506	S...	−0.7

Table I suggests that Seyfert nuclei, or at least those in spiral galaxies, have much steeper core spectra than ‘active’ ellipticals (it is interesting that the first two Seyferts in the table, which are in early–type galaxies, have spectral indices typical of the cores of E/S0 galaxies). All the spiral Seyferts have $\alpha < -0.5$, while 92±5% of the elliptical sample have $\alpha > -0.5$. Does the steeper spectral index of the Seyfert nuclei arise because the radio emission is being resolved by the PTI, or because of differences in the emission mechanism? A comparison of the PTI flux densities with VLA core fluxes on arcsec scales suggests that, unlike the ellipticals, most radio–emitting Seyferts have significant radio structure on scales between a few parsecs and a few hundred parsecs which is resolved out by the PTI. All the spiral Seyfert galaxies we observed had PTI core fluxes which were less than half their VLA core flux, while for 95% of elliptical and S0 galaxies the VLA and PTI core fluxes were roughly equal.

It appears, therefore, that the radio cores of (spiral) Seyfert galaxies are qualitatively different from those in (E/S0) radio galaxies, with radio emission from the latter being significantly more compact. This raises the possibility that an active nucleus ‘knows’, on scales of a few parsecs, what kind of galaxy it lives in. The mechanism by which this might happen is not yet clear.

HOW MANY ELLIPTICAL GALAXIES ARE ‘ACTIVE’?

If the parsec–scale cores we observe with the PTI are really ‘central engines’, as appears to be the case, we can set a lower limit on the fraction of all early-type galaxies which are currently active in this sense (since many of the ‘low–power’ galaxies were originally selected from an optically complete sample). As Table II shows, this is a strong function of optical luminosity. The fraction of ‘active’ S0 galaxies is lower than for ellipticals (except for the very brightest galaxies), but is also more uncertain because of the low PTI detection rate for S0 galaxies.

TABLE II Fraction of nearby E and S0 galaxies which are currently active.

M_B [a] (mag)	Minimum fraction 'active' E	S0
< -21.6	$> 50\%$	$> 50\%$
-19.6 to -21.6	$> 20\%$	$> 5\%$
> -19.6	very few	very few

[a] $H_0 = 75$ km s^{-1} Mpc^{-1}

CONCLUSIONS

In summary, our observations suggest that parsec–scale radio cores in early–type galaxies:

- are common
- have a flat or 'inverted' radio spectrum
- are probably very small (since few seem to be close to resolution with the PTI)
- are often variable (on timescales of a few years)

Parsec–scale radio cores appear to be less common in spiral and Seyfert galaxies than in ellipticals. When seen, they usually have steep radio spectra, in contrast to the flat or inverted spectra of the elliptical cores. This may be because they are partially resolved, and hence less compact than the 'engines' in radio galaxies.

REFERENCES

Ekers, R.D., Wall, J.V., Shaver, P.A., Goss, W.M., Fosbury, R.A.E., Danziger, I.J., Moorwood, A.F.M., Malin, D.F., Monk, A.S., and Ekers, J.A., 1989, *MNRAS*, **236**, 737

Ghisellini, G., Celotti, A., George, I.M, and Fabian, A.C., 1982, *MNRAS*, **258**, 776

Norris, R.P., Allen, D.A., Sramek, R.A., Kesteven, M.J., and Troup, E.R., 1990, *ApJ*, **359**, 291

Roy, A.L., Norris, R.P., Kesteven, M.J., Troup, E.R. , and Reynolds, J.E., 1993, this meeting.

Slee, O.B., Sadler, E.M., Reynolds, J.E., and Ekers, R.D., 1994, *MNRAS*, in press

Wrobel, J.M., and Heeschen, D.S., 1991, *AJ*, **101**, 148

Wunderlich, E., Klein, U., and Wielebinski, R., 1987, *A&AS* **69**, 487

The First Stromlo Symposium: The Physics of Active Galaxies
ASP Conference Series, Vol. 54, 1994
G. V. Bicknell, M. A. Dopita, and P. J. Quinn (eds.)

COMPACT STEEP–SPECTRUM RADIO SOURCES

CARLA AND ROBERTO FANTI
Istituto di Radioastronomia del CNR, Via Irnerio 46, 40126 Bologna, Italy
Dipartimento di Fisica, Via Irnerio 46, 40126 Bologna, Italy

ABSTRACT In this paper we review the radio properties of powerful Compact Steep-spectrum radio Sources. We also discuss some of the current interpretative models.

INTRODUCTION

A large fraction of the sources in flux limited radio samples, such as the 3CR catalogue, appear unresolved or only slightly resolved by conventional interferometry (angular sizes < 2–3 arcsec) and have normal or steep ($\alpha > 0.5$, $S \propto \nu^{-\alpha}$) high frequency spectra. These, sometimes, remain straight up to 100 GHz proving that no dominant flat spectrum core exists. These sources are named *Compact Steep–spectrum Sources* (CSS's). Their proportion is high (15–20%) especially amongst distant (z>0.2) radio sources of high power, both galaxies and quasars.

It has been argued that the majority of CSS's cannot be larger sized sources shortened by projection effects: their radio emission is therefore confined on kiloparsec scales. Which are the connections between CSS's and the larger sized radio sources? Are CSS's part of an evolutionary sequence, of which they represent the early stage (*YOUTH* scenario; e.g. Mutel and Phillips, 1988). Are they, instead, a separate class of objects, where unusual conditions of the interstellar medium (higher density and/or turbulence) inhibit the radio source from growing to large dimensions? (*FRUSTRATION* scenario).

At present, the best studied samples are those selected at 178 MHz from the 3CRR catalogue (see Fanti et al., 1990; Spencer et al., 1991 and references therein), and at 2.7 GHz from the Peacock and Wall catalogue (PW). The latter largely overlaps the 3CRR sample; however, due to its high frequency selection, it contains several sources whose spectra peak at frequencies up to a few GHz (the so called *GPS* sources; see e.g. O'Dea et al. 1991) and therefore do not belong to the 3CRR sample (Dallacasa et al., this volume). For instance, about 40% of the PW sources with $P_{2.7\mathrm{GHz}} > 10^{26}$W Hz^{-1} and a steep–straight high frequency spectrum are CSS's, but about half of them are non–3CRR.

In this review we consider the CSS's from the PW sample with $P_{2.7\mathrm{GHz}} > 10^{26}$W Hz^{-1}. We define *compact* those radio sources having projected linear size <15kpc (H_0=100km sec^{-1}Mpc^{-1}, q_0=0.5). Several other well studied CSS's, which do not belong to the complete sample defined above, are also used in the discussion of properties, when completeness is not required.

On average, we find that CSS's with higher turnover frequency have also smaller sizes. The 3CR CSS's are mostly larger than 1kpc, while those belonging to the PW catalogue *only* range mostly between less than about 100pc to ≈1kpc.

RADIO MORPHOLOGY

The *double–lobed* structure is very common (in ≈80% of radio galaxies and in >60% of quasars). A number of objects are very asymmetric in the lobe flux density ratio, or in the arm ratio, or in both. Most quasars and a good fraction of radio galaxies (especially amongst the smallest ones) have very bright jets. About 50% of quasars and 20% of radio galaxies have jets which contribute for at least 50% to the total luminosity, and in a number of cases such jets even account for most of the radio emission. The majority of the jets are asymmetric. Counter–jets are few, if any.

Core fractional luminosities at 5 GHz, evaluated from the maps with the highest available resolution, are the same, within the errors, as in large sized radio sources of similar power, both radio galaxies and quasars (median value ≈3% for quasars and $\leq$ 0.4% for radio galaxies). There is a correlation between core luminosity and jet luminosity: sources with strong jets have strong cores, and viceversa.

In a number of cases the radio structure shows remarkable distortions. In the sources with double–lobed structure and visible core, the lobes can be misaligned with respect to the core. This occurs frequently in quasars, where in about 40% of cases the misalignment angle is >20°, reaching up to 90°. In most radio galaxies, on the contrary, the lobes are aligned to within ≈20°. Distortions are commonly seen in the jets themselves. From an initially misaligned direction they can bring the jet orientation in the direction of a distant lobe, but they can also deviate a jet which, close to the core, is pointing towards the lobe. In this case, it is likely that, further out, where it is too weak to be seen, the jet will undergo a new bend in order to feed the lobe.

JET PROMINENCE, DISTORTIONS & ORIENTATION EFFECTS

How are these sources oriented with respect to the line of sight? Can we explain the presence of very bright jets, mostly, but not only, in quasars, and the distortions seen in quasars, in the frame of unified models (e.g. Barthel, 1989), i.e. in terms of projection effects plus some boosting? If we assume that relativistic motions occur, to some extent, in all radio sources close to the 'central engine', then parameters to be checked for orientation biases are: *a*) the ratio of core luminosity to extended luminosity, R_c; *b*) the ratio of jet luminosity to lobe luminosity, R_j; *c*) the asymmetry ratio, A_j, of the luminosity of main jet to that of the counter–jet (on assumption of an intrinsically symmetric ejection).

As far as A_j is concerned, we actually have too many upper limits to draw firm conclusions.

The parameter R_c in both CSS's radio galaxies and quasars is very similar to that of the larger size (> 15kpc) sources. This might lead, naively, to assume that CSS's are oriented in the sky in a way similar to their larger size cousins.

Jets, instead, are much brighter in CSS's than in the larger size sources of similar luminosity. This could be explained in terms of Doppler boosting, larger close to the central engine, coupled to an appropriate orientation distribution.

We have tried to model the R_c and R_j distributions in terms of a unique population of objects which appear either galaxies or quasars depending on the orientation (Barthel, 1989). We assumed that in both core and kpc jet bulk relativistic motions occur, possibly with different Lorentz factors. We find that, while it may be possible to reproduce the distributions of core and jet fractional luminosities (R_c and R_j) for the *whole* sample of CSS's, regardless of their optical identification, we are unable to reproduce the jet fractional luminosity distributions for radio galaxies and quasars separately. Briefly, the reason is that jet luminosities in radio galaxies (which, on the basis of unified theories, are thought to be oriented at large angles to the line of sight, and therefore should be less affected by Doppler boosting) span too a broad range of values, with some galaxies having jets as bright as the brightest jets in quasars.

We stress that we do not mean that a unified model doesn't apply to CSS's, but we think that the high jet luminosity in CSS's may not simply be due to boosting. An additional possibility, in fact, is to assume that interaction effects with the ambient medium play an important role (van Breugel et al. 1984). Such an interaction has been postulated, e.g., to explain the structure of 3C48 (Wilkinson et al., 1991): a collimated jet hits a dense interstellar cloud, it is partially decollimated and bent from its initial direction, and then brightens considerably.

It is likely that both orientation/beaming effects and interaction with the ambient medium contribute to generate the bright jets and the distortions seen.

PHYSICAL PARAMETERS AND THE FATE OF CSS'S

We have computed equipartition parameters for the lobes of CSS's. The use of equipartition is (mildly) justified since the equipartition magnetic fields allow to account, more or less, for the low frequency turnover in the overall source radio spectrum (Nan Rendong et al., 1991; Dallacasa, 1993). We limited our computation to the radio lobes, where we believe that no relativistic effects are present, while in jets Doppler boosting might play some role.

Very roughly, we get the following figures (with a dispersion of about half an order of magnitude) for the equipartition energy, pressure and magnetic field:

$$\begin{array}{llll} U_{eq} & \approx & 10^{56} \; D_{kpc}^{1.3} & \text{erg} \\ p_{eq} & \approx & 2\cdot10^{-7} \; D_{kpc}^{-1.7} & \text{dyne cm}^{-2} \\ H_{eq} & > & 10^{-4} & \text{Gauss} \end{array}$$

where D is the source total size.

If we assume that the lobe pressure is balanced by the ram–pressure due to the motion through the external hot and uniform medium, we can estimate the lobe velocity and therefore the source life time. The assumption of a central average number density of the interstellar diffuse hot gas of ≈ 0.1 cm^{-3} and a core radius of $\leq$5kpc, as found in nearby elliptical galaxies (Forman et al., 1985) leads to velocities of $\approx 3 \times 10^8$ cm sec^{-1} and to time scales of $\approx 2 \times 10^6$ years for a source 10kpc in size. If the central engine of CSS's remained active for 10–100 million years, as estimated for the Large Size sources, they would inevitably grow to hundred kpc sizes during their life time. Therefore CSS's seem to be the early stage of radio source evolution (*YOUTH* scenario).

In order that CSS's remain confined to the small size range for their whole lifetime (*FRUSTRATION* scenario), the density of the external medium, or the core radius, or both, should be at least one order of magnitude larger. While it has been suggested that at earlier epochs galaxies may have had larger and denser halos, we think that quantitative figures are still uncertain. We notice that densities >1 cm^{-3}, and/or core radii larger than 5kpc for the diffuse hot medium could lead to unacceptably high X–ray luminosities. Similar arguments have been presented extensively by De Young (1993), by means of hydrodynamical numerical simulations.

If CSS's are young objects, with ages $< 2\times10^6$ years, and the typical activity time scale for the central engine is $\approx 5 \times 10^7$ years, then CSS's should be a small fraction ($\approx$5%) of the population of which they are the young phase. Since in the high luminosity source population CSS's are $\approx$30–40%, there seems to be a contradiction with the time scales estimated above. We may overcome this difficulty if we assume that the radio luminosity of a typical source decreases with increasing time (and size). In this case we should compare proportions of sources *not at a fixed luminosity*, but at the luminosities corresponding to the different linear sizes along the evolutionary track. The dynamical life times estimated above and the fraction of radio sources which appear as CSS's, can be reconciled if the radio source luminosity is made to change approximately as size$^{-0.5}$. In the framework of the continuous energy supply models derived from Scheuer (1974), as in Baldwin (1982), the radio luminosity decreases as the source grows, due to lobe expansion in the external medium. In this class of models we obtain the required decrease in luminosity for an external density changing approximately as the inverse square of the distance from the parent object. This seems to us a plausible situation.

Above, we have considered the interaction of the source with the diffuse hot interstellar medium only, since the tepid denser and clumpy gas, which produces the narrow emission lines, is generally assumed to have a small filling factor, so that the probability of an interaction of the jets with this component is expected to be small. However such an interaction might play an additional role. For instance, if interactions are limited to few clouds, significant asymmetries could arise on the two sides of the radio sources, accounting for the strong flux density/arm asymmetries often visible.

A better knowledge of the properties of the ambient medium, from X–ray and narrow line observations, is the crucial step to choose between the *YOUTH* and the *FRUSTRATION* scenario.

REFERENCES

Baldwin, J., in *Extragalactic Radio Sources*, IAU Symp. 97, D.S. Heeschen and C.M. Wade, Reidel, 1982, 21

Barthel, P.D., 1989, *ApJ*, **336**, 606

van Breugel, W.J.M., Miley, G.K., Heckman, T.A., 1984, *AJ*, **89**, 5

Dallacasa, D., 1993, PhD thesis, University of Bologna

De Young, D.S., 1993, *ApJ*, **402**, 95

Fanti, R., Fanti, C., Schilizzi, R.T., Spencer, R.E., Nan Rendong, Parma, P., van Breugel, W.J.M., Venturi, T., 1990, *A&A*, **231**, 333

Forman, W., Jones, C., Tucker, W., 1985, *ApJ*, **293**, 102

Mutel, R.L. and Phillips, R.B., in *The impact of VLBI on Astrophysics and Geophysics*, IAU Symp. 129, M.J. Reid and J.M. Moran, Reidel, 1988, 73

Nan Rendong, Schilizzi, R.T., Fanti, C., Fanti, R., 1991, *A&A*, **252**, 513

O'Dea, C.P., Baum, S.A., Stanghellini, C., 1991, *ApJ*, **380**, 66

Scheuer, P.A.G., 1974, *MNRAS*, **166**,513

Spencer, R.E., Schilizzi, R.T., Fanti, C., Fanti, R., Parma, P., van Breugel, W.J.M., Venturi, T., Muxlow, T.W.B., Nan Rendong, 1991, *MNRAS*, **250**, 225

Wilkinson, P.N., Tzioumis, A.K., Benson, J.M., Walker, R.C., Simon, R.S., Kahn, F.D., 1990, *Nature*, **352**, 313

The First Stromlo Symposium: The Physics of Active Galaxies
ASP Conference Series, Vol. 54, 1994
G. V. Bicknell, M. A. Dopita, and P. J. Quinn (eds.)

THE STRUCTURE AND KINEMATICS OF PICTOR A

S. M. SIMKIN
Michigan State University, East Lansing, MI 48824-1116, USA

E. M. SADLER
Anglo–Australian Observatory, Epping, NSW 2121 Australia

R. SAULT
Australia Telescope National Facility, CSIRO, Epping, NSW 2121 Australia

P. ROBINSON
Michigan State University, East Lansing, MI 48824-1116, USA

ABSTRACT We have combined high–resolution images of Pictor A from the Hubble Space Telescope (HST) with 3 cm radio synthesis images from the Australia Telescope and ground–based optical images and kinematical measurements. We find a kiloparsec–scale radio jet located between the parsec–sized optical jet seen in our HST images and the radio–optical hotspot which lies more than 100 kpc from the central galaxy. The gas kinematics in the host galaxy suggest that the jet impinges strongly on the interstellar gas which surrounds the galaxy nucleus. Because it is relatively close, Pictor A may be one of the best objects in which to study the structure of the relativistic beams which produce the strong radio hotspots seen in FR II sources.

INTRODUCTION

Pictor A is the seventh brightest radio source in the sky at 408 MHz and the most powerful radio galaxy in the redshift range $0 < z < 0.04$. It is also the nearest broad–lined radio galaxy (FW0I $> 14,000\,\mathrm{km\,s^{-1}}$ from our CTIO measurements, but reported to be as high as $23{,}000\,\mathrm{km\,s^{-1}}$ by Filippenko 1985), with a highly excited nuclear spectrum. It has a compact radio core (Christiansen *et al.* 1977). Our CTIO measurements yield a lower limit to the total flux in the [O III] 5007/4958 lines of $10^{41}h^{-2}$ erg s^{-1}. Using this and standard parameters for the emitting gas clouds, we estimate the scale of the transition region between narrow–line and broad–line emission to be 80–200 pc. At the redshift of Pictor A ($z = 0.035$), $0.1''$ corresponds to a linear scale of 50 h^{-1} pc.

Because it is both nearby and an extreme example of a powerful, almost quasar–like, radio galaxy, we regard Pictor A as a natural laboratory for identifying those features which differentiate radio–loud AGN from their radio–quiet counterparts. Thus we have made a broad instrumental study of this object,

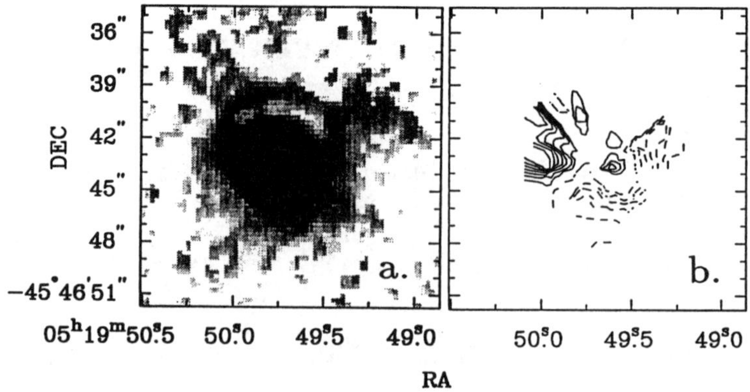

FIGURE I a) Hα image from CTIO, b) Ionized gas velocities

combining HST images with ground–based measurements from CTIO and the Australia Telescope. We summarize the results below.

GROUND–BASED OBSERVATIONS

Our ground–based images have 1–1.2″ resolution (Simkin and Sadler 1989), sampled at 0.3″/pixel. They show a wealth of arc–like, detailed structure in the ionized gas throughout the galaxy (Fig. Ia), and particularly complex structures within 5″ (2.5 to 3.0 h^{-1} kpc) of the nucleus.

Such structures are usually attributed to "mergers" or "interactions". In addition, we have used spectroscopy to measure the kinematics of this ionized gas, combining velocity measurements from several emission lines (Hα, β, [O I], [O III], [N II], and [S II]) at eight different position angles to produce a velocity "map" (Fig. Ib). There are peculiar velocity reversals in the inner 3″, which we originally suspected might be caused by a small–scale, bar–like or triaxial structure in the nucleus (de Zeeuw and Franx 1989).

HST OBSERVATIONS

Using the Planetary Camera (PC) on the Hubble Space Telescope (HST) in Cycle 1, we observed Pictor A in a narrow–band filter at 517 nm (which includes the [O III] 4958/5007 lines) and a medium–width filter at 547 nm, (which covers both [O III] and continuum.)

These observations were made up of multiple images which were merged to provide effective pixel sizes of 0.022″. They were carefully deconvolved using N. Weir's (1991a and b) MEM image restoration system and a series of expanded PSFs (Simkin *et al.* 1992). The restored images show convincing (and reproducible) extended structures on the scale of 0.1 to 0.2″ (Fig. II).

There is evidence for a small continuum jet in the direction of the large–scale radio structure (Fig. IIa) with a peak flux 4 % of that in the unresolved nucleus

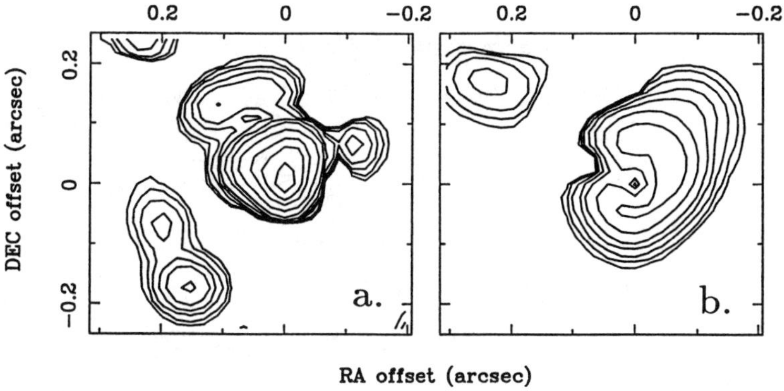

FIGURE II HST PC images, a) 547 nm continuum b) [O III] emission

and an integrated flux which is 9 % of the nuclear flux. The [O III] emission flux shows extended structure which is shifted to the north of the continuum peak by ≈ 60 mas (Fig. IIb). The continuum image also shows a second, non-stellar "knot" or secondary nucleus (which is not a point source) 0.2″ to the SE of the Seyfert nucleus.

We obtained two further sets of images with the HST PC in Cycle 2, using medium-width filters centered at 648 and 718 nm. These cover the continuum and include the emission lines of [O I] and [S II]. Reduction of these new images is still in progress, since the necessary flat-field and PSF images are not yet available. Our only confident result to date comes from convolving the "raw" images down to a resolution of 0.3″ per pixel. These clearly show that the "secondary" nucleus found in the Cycle 1 continuum image is also present at 650–720 nm.

The small-scale feature pointing to the NW radio hot-spot, seen in the HST 547 nm-continuum image (Fig. IIa), motivated us to carry out new radio observations which suggest a more complex interpretation of the inner structure of this object than our original ideas about triaxial structure implied. These radio data are discussed below.

RECENT RADIO IMAGING

Images of Pictor A at 3 cm were made with the 6 km Compact Array of the CSIRO's Australia Telescope (ATCA). Pictor A has no discernible optical continuum jet on arcsecond scales (Fraix-Burnet *et al.* 1991), and though earlier observations with the VLA (Perley, unpublished) suggested the possibility of a kpc-scale jet pointing to the nucleus from the NW hot-spot, they did not reveal a small-scale jet. However, the feature shown by our HST Cycle 1 observations was so well-aligned with the 130 kpc-distant NW hotspot that we made two 12-hour 3 cm synthesis observations with the ATCA. Figures III and IV show our preliminary 3 cm images (two more 12-hour data sets are still needed for a full synthesis).

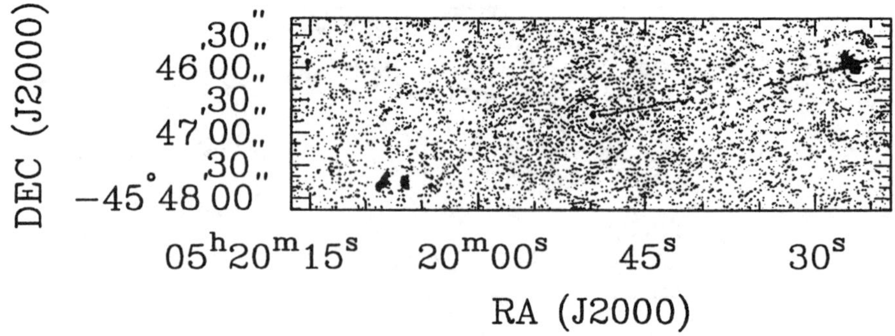

FIGURE III Grey–scale, 3 cm radio image showing NW jet

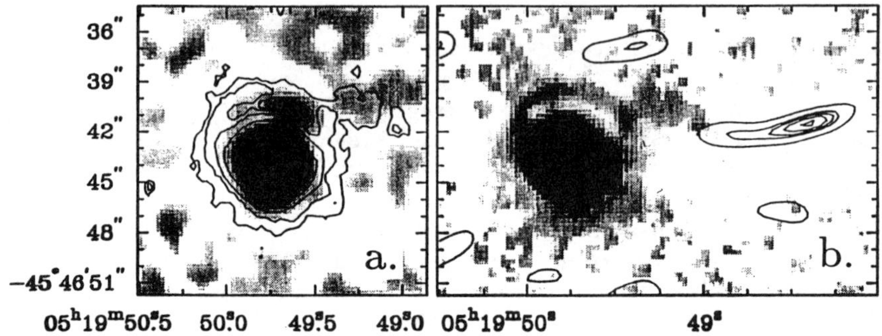

FIGURE IV a) Grey = 3 cm, lines = $H\alpha$, b) Grey = Hα, lines = inner 3 cm jet

Fig. III shows the full image including the outer hot spots (which are severely attenuated because they lie outside the AT primary beam). A jet appears to run all the way from the nucleus to the NW hot–spot. This jet is not completely straight, but changes direction between RA 5h 20m 35s and 5h 20m 40s. The eastern segment points towards the central source while the western segment, if extended to the east, intersects the SE hotspot, which is not symmetrical with the NW hotspot.

Fig. IV shows enlargements of the nuclear region. Fig. IVa shows an arc or loop in the 3 cm emission at a peak brightness which is 10% of the core source. This appears to lie **inside the gap** between the $H\alpha$ loop to the north and the main body of ionized H II around the nucleus. Fig. IVb shows the faint, inner part of the radio continuum jet, at PA 281°, overlaid on our ground–based Hα image. The ionized gas filaments to the NW appear to fill the gap between the core source and the start of the radio jet.

DISCUSSION

We can draw several conclusions from the data reported above:

1. Pictor A has a weak, secondary, diffuse, continuum nucleus which lies 0.3″ (150 pc) to the SE of the unresolved core.
2. The [O III] NLR is resolved with a size of 0.4″ × 0.2″ (roughly 100×200 pc) extending in a direction opposite to that of the second nucleus.
3. The ionized gas is strongly disturbed along the direction of the radio jet on scales of 3 to 4″ (1.5 to 2.0 kpc) with velocities in excess of $75 \sin^{-1}(i)$ km s^{-1}.
4. The loops of H II gas (on a 2 to 4 kpc scale) seem to be closely related to the 3 cm continuum, both the jet and the arc to the N.

Although radio jets are fairly common in FR II galaxies, their optical counterparts are rarely seen. This object, with its parsec–sized optical jet and kiloparsec–sized radio jet, provides a unique opportunity to investigate the mechanisms which degrade the high energy synchrotron electrons needed for optical emission. In addition, if the radio jet is indeed strongly interacting with gas in the galaxy but remains well enough focused to sustain the radio lobes, this may well provide significant constraints on the structure of the relativistic beams which produce the lobes.

ACKNOWLEDGMENTS

We thank both CTIO (operated by AURA under contract to the US NSF) and the Australia Telescope National Facility for observing time and support. Part of this work was done with financial support from NSF (AST-8914567) and HST (GO-245601-87A).

REFERENCES

Christiansen, W.N., *et al.*,1977, *MNRAS*, **181**, 183

Filippenko, A., 1985, *ApJ*, **289**, 475

Fraix–Burnet D., *et al.*,1991, *AJ*, **101**, 88

Simkin, S.M., and Sadler, E.M., 1989, *IAU Symposium* **134**, D. Osterbrock and J. Miller, eds., 484

Simkin, S.M., Sadler, E.M., and Robinson, P., 1992, *BAAS*, **24**, 727

Weir, N., 1991a, *Data Analysis, ESO Workshop* **38**, P.J. Grosbol and R.H. Warmels eds., pp 115-129

Weir, N., 1991b, *10th International Workshop on Maximum Entropy and Bayesian Methods*, W.T. Grandy and L.H. Schick (eds), pp 275-283, Kluwer Acad. Pub., Dordrecht

de Zeeuw, P.T., and Franx, M., 1989, *ApJ*, **343**, 617.

RADIO GALAXIES

Contributed Papers

The First Stromlo Symposium: The Physics of Active Galaxies
ASP Conference Series, Vol. 54, 1994
G. V. Bicknell, M. A. Dopita, and P. J. Quinn (eds.)

PRELIMINARY RADIO AND OPTICAL RESULTS FOR ROSAT SOURCES IN THE ESO KEY PROGRAM

MARTIN W. B. ANDERSON[1] AND G. L. WHITE
UWS, Nepean, P.O. Box 10, Kingswood, NSW 2747, Australia

R. D. EKERS
ATNF, P.O. Box 76, Epping, NSW 2121, Australia

I. J. DANZIGER
European Southern Observatory, Garching

INTRODUCTION

We report the current status of an extensive Radio/Optical investigation of a flux-limited sample of newly discovered soft X-ray sources selected from the ROSAT All Sky Survey (RASS; Voges, 1992) as part of an ESO Key program (Danziger et al., 1990). The sample (Table I) is divided into independent fields ($|b| > 25°$) covering 575 square degrees with all sources $>5\sigma$ (0.07-2.4 keV band).

TABLE I Summary of Key Field properties.

Field	RA	Dec	No. Src.	Size
I	01h	-40°	261	275□
II	05h	-57°	225	100□
III	11h	-27°	76	100□
IV	13h	-05°	126	100□

RADIO PROGRAM

We have observed (June 93) all sources at 4.8 GHz in fields I to III with the Australia Telescope National Facility (ATNF) Compact Array (CA) and are scheduled to observe field IV with the VLA. ATNF CA snapshot observations (5 cuts of three minutes) reach a 5σ detection level of 2-3 mJy, with all positional errors <2 arc-seconds. The detection rate within 30 arc-seconds of the ROSAT position is $\approx$18$\pm$2% (chance rate <1%). Figure I shows the fraction of flat spectrum sources ($\alpha > -0.5; S \propto v^{\alpha}$) is 58$\pm$15% for S>3mJy, α(4.8-8.6 GHz) and is consistent with the 60% estimated from Neumann et al. (1991) (S>20 mJy, α(5-10.7 GHz)). There is no evidence for radio quiet BL-Lac's as all optical BL-Lac candidates have radio associations.

---

[1]Ph.D. student at ATNF. The Australia Telescope National Facility is operated in association with the Division of Radiophysics by CSIRO.

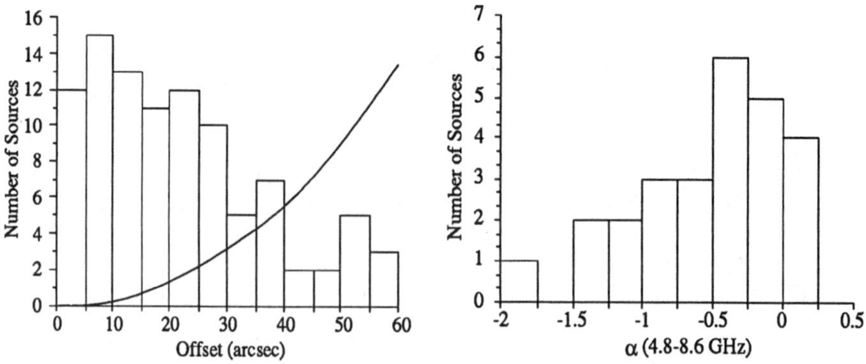

FIGURE I Offset of radio and ROSAT positions for all radio associations (S>3mJy) in a sample of 436 RASS sources overlayed with a curve giving the estimated number of false detection's for a given offset, and a distribution of α(4.8-8.6 GHz) for all field II sources within 30 arcsec of the ROSAT position.

OPTICAL PROGRAM

At present (April 93) 480 objects out of 688 X-Ray sources have been spectroscopically identified by the collaboration of Danziger et al. (1990), and redshifts have been obtained for most of the identifications. Table II gives the distribution of identifications for the Key fields and the EMSS (Einstein Extended Medium Sensitivity Survey; Stocke et al. 1991) for comparison.

TABLE II Summary of Optical Identifications.

Source Type	Stars	AGN's	BL-Lac's	Gal.	Gal. Clu.	No ID's.	Radio Detection
Key Field	23%	36%	2%	5%	6%	28%	18% S>3mJy
EMSS	26%	51%	4%	13%	2%	4%	27% S>1mJy

REFERENCES

Danziger, I. J., Trümper, J., Beuermann, H., Böhringer, H., Fleming, T., Gottwald, M., Hasinger, G., Krautter, J., Macgillivary, H., Miller, L., Pakull, M., Parker, Q., Pasquini, L., Reinsch, K., Thomas, H. C., Ulrich, M. H., Voges, W. and Zimmermann, H. U. 1990, *ESO Messenger*, **62**, 4.

Neumann, M., Reich, W., Furst, E., Brinkmann, W., Reich, P., Wielebinski, R., Fink, H.H. and Trümper, J. 1992, *MPE-Report Series*, **235**, 163.

Stocke, J. T., Simon, L. M., Gioia, I. M., Maccacaro, T., Schild, R., Wolter, A., Fleming, T. A., and Henry, J. P. 1991, *ApJS*, **76**, 813.

Voges, W. 1992, *The ROSAT All-Sky X-Ray Survey, Proc. European ISY Meeting: Space Sci. with a Particular Emphasis on High Energy Astrophysics.*

The First Stromlo Symposium: The Physics of Active Galaxies
ASP Conference Series, Vol. 54, 1994
G. V. Bicknell, M. A. Dopita, and P. J. Quinn (eds.)

A MODEL FOR DISSIPATIVE TURBULENT JETS IN FRI RADIO SOURCES

GEOFFREY V. BICKNELL
Mt. Stromlo & Siding Spring Observatories, Australian National University

MATHEMATICAL DESCRIPTION OF TURBULENT JETS

In earlier models for turbulent jets in radio sources (e.g. Bicknell, 1986), an adiabatic approximation was used for the evolution of jet surface brightness. This may be improved by taking into account the dissipation of energy from the bulk flow, mediated by turbulence. In similar fashion to the previous models integral conservation laws are used to infer the axial values of dynamical variables [details in Bicknell (1994)].

The integral equations for mass flux, momentum flux and energy flux are:

$$\begin{aligned}
\frac{dF_M}{dz} &= -2\pi \lim_{r\to\infty} r\rho v_r \\
\frac{d}{dz}\left\{2\pi \int_0^\infty \rho\, v_z^2\, r\, dr\right\} &= -2\pi \frac{dp}{dz} \int_0^\infty \left(\frac{1}{\rho} - \frac{1}{\rho_e}\right) r\, dr \\
\frac{d}{dz}\left\{2\pi \int_0^\infty \left[(h - h_e) + \frac{1}{2}v^2\right] v_z\, \rho\, r\, dr\right\} &= \frac{d}{dz}(h_e + \psi)\, F_M
\end{aligned}$$

where $F_M = 2\pi \int_0^\infty \rho\, v_z\, r\, dr$ is the mass flux, p_e the external pressure, ψ the gravitational potential and dynamical variables are statistically averaged. For light, relativistically dominated jets, the enthalpy $h = 4p/\rho$ and the difference in specific volumes, $\rho^{-1} - \rho_e^{-1} = \Delta h/4p$ where $\Delta h = h - h_e$ is the difference between jet and ISM enthalpy. Numerical implementation of the above conservation laws is facilitated by the use of a Lagrangian coordinate $u(r,z) = 2\pi \int_0^r \rho\, r\, dr$, the mass per unit length within a radius r. When pseudo-self-similar expressions are used for dynamical quantities, *e.g.* the velocity, $v(r,z) = v_c(z)V(\xi)$ where $\xi = u/\bar{u}$ and $\bar{u}$ is the value of u corresponding to the half-velocity point, the above conservation laws for momentum and energy reduce to the following ordinary differential equations:

$$\begin{aligned}
I_{V2}\frac{d}{dz}(v_c^2\, \bar{u}) + \frac{I_H}{4}\frac{dp}{dz}(\Delta h\, \bar{u}) &= 0 \\
I_{VH}\frac{d}{dz}(\Delta h_c\, v_c\, \bar{u}) + \frac{1}{2} I_{V3}\frac{d}{dz}(v_c^3 \bar{u}) &= I_V \frac{d}{dz}(h_e + \psi)\, F_M
\end{aligned}$$

where the mass flux $F_M = I_V v_c \bar{u}$, $\bar{u}(z) = \left[4\pi\, p(z)\, \bar{r}^2(z)/\Delta h_c\right] \left[\int_0^1 \left[\frac{h_e}{\Delta h_c} + H(\xi)\right] d\xi\right]^{-1}$, $\bar{r}$ is the value of r corresponding to the half-velocity point and the parameters, I_V, I_{VH} etc. represent profile integrals.

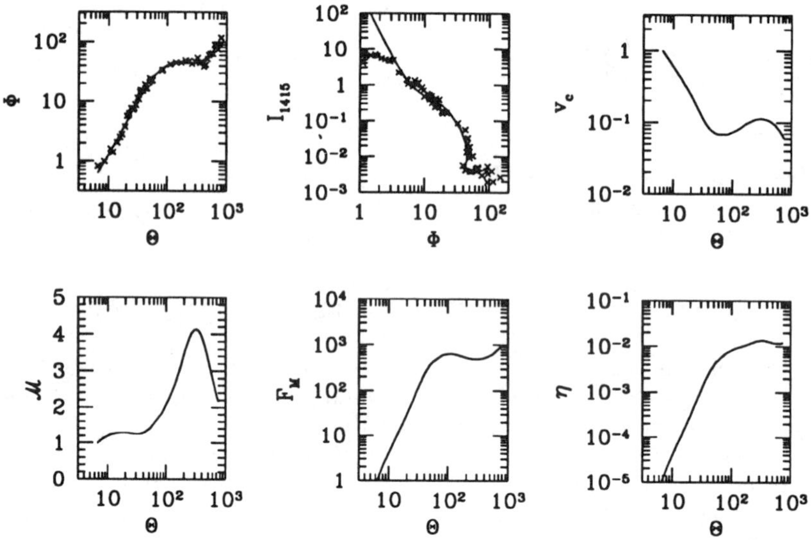

FIGURE I The FWHM of the NGC 315 jet (upper left panel) is used to model the surface brightness data (top middle panel). Other panels show the variation of velocity, Mach number, mass flux and density ratio.

MODEL FITTING

The pressure law used here is equivalent to that used by X-ray astronomers for isothermal models of elliptical galaxy X-ray surface brightness, *i.e.* $p/p_c = (1 + r^2/a^2)^{-\beta} + p_\infty/p_c$ with p_c the central pressure, a the core radius, β the ratio of virial to atmosphere temperature and p_∞ the background pressure. For a jet dominated by relativistic particles, the surface brightness, $I_\nu \propto p\, B^{(s-1)/2}\, \bar{r}$ where $s = 2\alpha + 3$ and α is the spectral index. In a region of perpendicular field, $B_{\mathbf{perp}} \propto v_c \bar{r}^{-1}$. For the parallel field region, $B_{||} \propto \bar{r}^{-2}$, although this cannot be the full story. A model fit to the NGC 315 data (Willis *et al.*, 1981; Bridle, 1982) is shown in figure I. This was obtained using Gaussian jet profiles together with $\beta = 1.15$, $p_\infty/p_c = 8 \times 10^{-6}$ and $\eta << 1$. The value of β compares favourably with that inferred from the X-Ray data ($\beta_X = 1.1$; Thomas *et al.*, 1986).

REFERENCES

Bicknell G.V., 1986, *ApJ*, **305**, 109

Bicknell G.V., 1994, in preparation

Bridle A.H., 1982, in *IAU Symposium 97: Extragalactic Radio Sources*, ed. D.S. Heeschen and C.M. Wade, 121

A.G. Willis *et al.*, 1981, *A&A*, **95**, 250

Thomas *at al.*, 1986, *MNRAS*, **222**, 655

The First Stromlo Symposium: The Physics of Active Galaxies
ASP Conference Series, Vol. 54, 1994
G. V. Bicknell, M. A. Dopita, and P. J. Quinn (eds.)

THE MOLONGLO SOUTHERN 4 JY SAMPLE (MS4)

A. M. BURGESS AND R. W. HUNSTEAD
Astrophysics Department, University of Sydney, NSW 2006, Australia

INTRODUCTION

A complete sample of 232 sources has been compiled from the Molonglo Reference Catalogue (Large *et al.* 1981) according to the following criteria: (i) declination $-85° < \delta < -30°$, (ii) integrated flux density $S_{408\ \mathrm{MHz}} \geq 4$ Jy, and (iii) galactic latitude $|b| > 10°$. About 20 per cent of the sources are quasars, with the remainder being galaxies with redshifts ranging from $z < 0.1$ to an upper limit expected to be $z > 2.5$. A substantial database is being assembled as part of this project.

Although the MS4 sample is intended in part as a southern hemisphere complement to the northern 3C survey, our main purpose is to study the morphological relationship between the radio and optical emission from radio galaxies, in particular the so-called 'alignment effect', in which the rest-frame UV continuum is aligned closely with the radio axis in galaxies with $z > 0.7$. The flux density limit has been chosen, therefore, to give sufficient numbers for a useful statistical survey; the source density is 96 sr^{-1}, as compared with 41 sr^{-1} for the 3CRR catalogue (Laing *et al.* 1983). A sky plot of the MS4 sample is shown in Figure I.

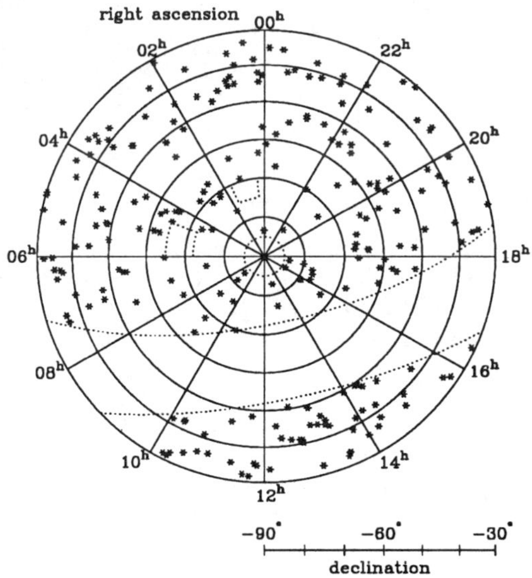

FIGURE I Sky distribution of the MS4 sources.

STATUS OF THE SURVEY

All the MS4 sources have been imaged at 843 MHz with the Molonglo Observatory Synthesis Telescope (MOST; resolution 44″) in a time-sharing mode in which each source is observed at ~8 hour angles (Subrahmanya & Hunstead 1993, in preparation). In addition to providing accurate positions, the MOST images give useful estimates of angular sizes down to ~ 10″.

The 131 sources with angular size ≤ 35″ have been imaged in snapshot mode with the Australia Telescope (AT) at 4.8 GHz (resolution 2″ or 4″), to obtain more detailed information about their structure. The sources have also been observed at 2.3 GHz with the Parkes-Tidbinbilla Interferometer (PTI) to estimate the fraction of emission coming from compact components < 0″.1.

Optical identifications were sought initially on the plates of the UK-Schmidt southern sky survey. Magnitudes and finding charts are being obtained from the COSMOS database (Yentis 1991), derived from the digitised Schmidt plates. About half the optical counterparts are fainter than the COSMOS completeness limit of $b_j = 21$, many being in blank fields. In order to complete the identifications and examine optical morphologies, we are making deep R-band CCD images with the Anglo-Australian Telescope (AAT); with typical exposure times of five minutes the limiting magnitude is $R \sim 24$. At this level, all fields not obscured by foreground stars yield a clear identification. Radio and optical positions agree to better than 1″ rms for compact sources.

The CCD image for a radio source which was in a blank field on the Schmidt plate is shown in Figure II. On the AAT image an elongated object roughly aligned with the radio structure is clearly visible.

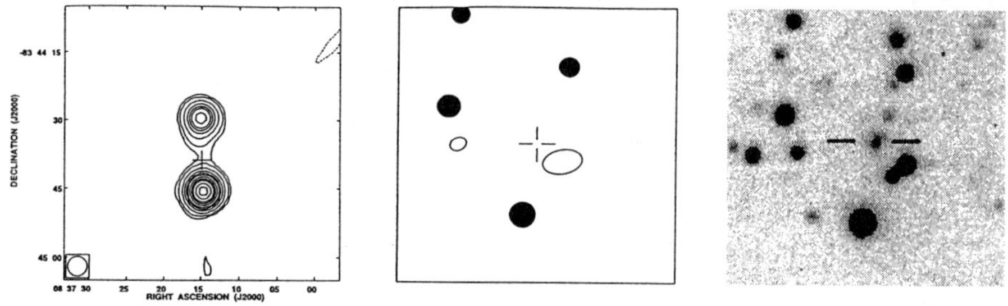

FIGURE II AT image, COSMOS finding chart, and AAT R-band CCD image of the radio source 0842−835. The crosses represent the position of the galaxy shown in the CCD image. Each panel is 1′ on a side.

REFERENCES

Large, M. I., Mills, B. Y., Little, A. G., Crawford, D. F., & Sutton, J. M. 1981, *MNRAS*, **194**, 693.

Laing, R. A., Riley, J. M., & Longair, M. S. 1983, *MNRAS*, **204**, 151.

Yentis, D. J. 1991, in *Digitised Optical Sky Surveys*, H. T. MacGillivray and E. B. Thompson, Kluwer Academic Publishers, 67.

The First Stromlo Symposium: The Physics of Active Galaxies
ASP Conference Series, Vol. 54, 1994
G. V. Bicknell, M. A. Dopita, and P. J. Quinn (eds.)

IR IMAGING POLARIMETRY OF THE NUCLEAR REGIONS OF CEN A

JEREMY BAILEY
Anglo-Australian Observatory, P.O. Box 296, Epping, NSW 2121

J.H. HOUGH AND A. CHRYSOSTOMOU
Division of Physical Sciences, Univ. of Hertfordshire, College Lane, Hatfield, Herts, AL10 9AB

INTRODUCTION

Aperture polarimetry of the nucleus of Cen A (Bailey et al. 1986) revealed a very red polarized source present in the K and L bands which was interpreted as a blazar type nucleus. To study this region in more detail we observed it on 1993 May 5 using the IRIS IR camera together with a polarimetry system developed at the University of Hertfordshire.

DISCUSSION

In the J band polarization is mostly along the dust lane, with values reaching 3% at position angles of 117 degrees near the nucleus. This polarization is likely to be due to transmission through aligned dust grains. No significant excess in polarization is seen on the IR jet, which is therefore unlikely to be synchrotron radiation or scattering of nuclear radiation.

In the K_n band the polarization is generally lower than that at J, about 1% over most of the field. This is consistent with transmission through aligned grains which should follow the Serkowski law and fall with increasing wavelength in the IR.

On the nuclear source itself a much larger K band polarization is seen of about 6.5% at a PA of 140 degrees, consistent with the aperture polarimetry of Bailey et al. (1986). It has a polarization of about 8% (after correction for dilution by stellar flux), at an angle roughly perpendicular to the jet direction.

The polarized radiation could be synchrotron emission from a blazar type nucleus, the low apparent luminosity being a consequence of beaming along an axis away from the line of sight. The very red colour would be largely intrinsic. An alternative explanation is that the red colour and polarization could both be due to absorption in a thick inner dust disk. If A_v is about 40 or greater the observed polarization can easily be produced by dichroic absorption. The observed angle would then indicate the orientation of the inner disk.

REFERENCES

Bailey, J. Sparks, W.B., Hough, J.H. and Axon, D.J., 1986, Nature, 322, 150.

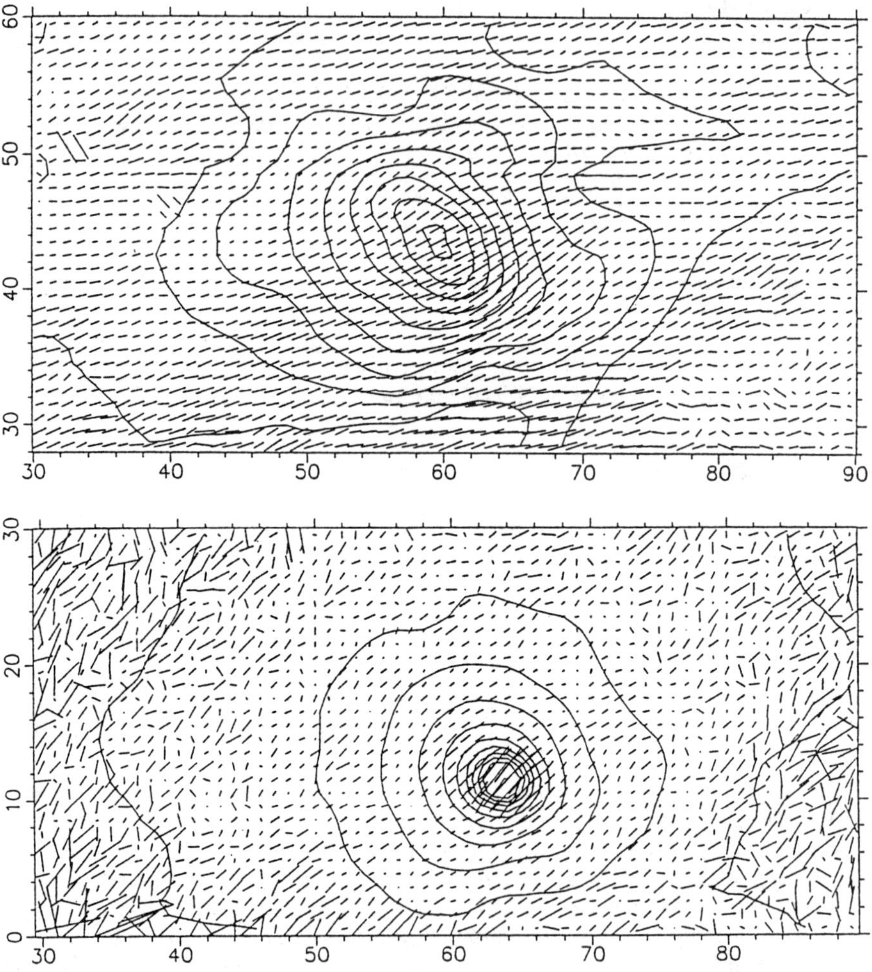

FIGURE I Polarization images of the central region of Cen A in the J (upper panel) and K_n (2.0-2.32μm, lower panel) bands. The images cover about 19 by 36 arc seconds with polarization vectors plotted for every pixel. The contours show the intensity.

The First Stromlo Symposium: The Physics of Active Galaxies
ASP Conference Series, Vol. 54, 1994
G. V. Bicknell, M. A. Dopita, and P. J. Quinn (eds.)

THE SPECTRUM OF OLD RADIO GALAXIES

J. A. EILEK AND P. A. ARENDT
New Mexico Tech, Socorro NM 87801, USA

J. E. BOROVSKY
Los Alamos National Laboratory, Los Alamos NM 87545, USA

ABSTRACT High-frequency steepening in synchrotron spectra may not be simply due to synchrotron aging in a uniform magnetic field. Old sources with ongoing particle reacceleration and a distribution of magnetic field strengths can produce synchrotron spectra very similar to those in the standard models, but with very different underlying physics. This means care must be taken in deriving source ages and dynamics from measurements of turnover frequencies.

DO WE UNDERSTAND SPECTRAL AGING?

The answer seems to be, no. That is, high-frequency turnovers in the radio spectra of extended synchrotron sources have conventionally been interperted as due to electron aging, with no reacceleration, in a uniform magnetic field (for instance, Alexander & Leahy 1987). Applying this model leads to problems, however. Statistical arguments using dynamical models of source evolution predict far too many bright, steep spectrum sources; very few such sources exist (Eilek 1993). In addition, the velocities inferred from the surface brightness distribution of type I tailed sources are a factor of ten slower than the velocities inferred from aging analyses of the same sources (O'Donoghue, Eilek & Owen 1993).

Thus, the sources do not seem to age as fast as we think they should. It may be that our dynamical models are all wrong, and that we are overestimating the source ages by a factor of ten or more. However, the simplicity of the dynamical models, and the success of hydrodynamic codes in reproducing type II sources, suggests the dynamical models are probably robust. (It seems worth noting,however, that the sharp, two-parameter type I/II break, Owen & Ledlow 1993, suggests we are still missing something important). This leaves our synchrotron emission models to be questioned; is the spectral steepening a unique sign of particle aging? When can the standard model be used? (We note that the synchrotron spectrum of Cyg A, one of the best-studied radio galaxies, may *not* fit the standard model at all; Katz-Stone, Rudnick & Anderson 1993.) In particular, *in situ* reacceleration might be expected to offset the aging problem; can it reproduce the observed spectra?

SYNCHROTRON SPECTRA FROM OLD SOURCES

We have been studying the synchrotron spectrum produced by various distribution functions (DF's) of particle energies, $n(E)$, and field strengths, $g(B)$. The electron DF might be a power law (as from shock acceleration, with synchrotron losses). Alternatively, it might be peaked, $n(E) \propto E^2 e^{-E/E_o}$, if it results from stochastic reacceleration; in this case, E_o is the energy where turbulent acceleration balances synchrotron losses. Little work has been done on $g(B)$, but real sources *must* contain a distribution of field strengths, due to internal turbulence (which may be the cause of the filaments seen in high-quality images), and also due to large-scale plasma dynamics. The general synchrotron spectrum is

$$j_{sy}(\nu) = \int_{E_{min}}^{E_{max}} \int_{B_{min}}^{B_{max}} p_o n(E) g(B) F(\nu/c_1 E^2 B) dB dE$$

where $F(\nu/c_1 E^2 B)$ is the usual synchrotron kernel; p_o and c_1 are physical constants. We have solved this equation, analytically and numerically, for several power-law and exponential $n(E)$ and $g(B)$ forms (details are presented in Eilek, Arendt & Borovsky 1993).

In particular, we find that an electron DF arising from turbulent reacceleration, combined with a magnetic field DF which is a power law up to some B_o, produces a photon spectrum which is a power law at low frequencies, and steepens exponentially at high frequencies. The transition frequency is $\nu_c \propto E_o^2 B_o$, and will decay with time as E_o or B_o decay. (For comparison, recall the standard model has $\nu_c \propto E_c^2 B$, with B constant and $E_c \propto 1/t$). Thus, a source can exhibit a power-law spectrum, with high-frequency spectral steepening, for quite different physical reasons than are usually assumed.

CONSEQUENCES FOR SOURCE MODELLING

This work shows that the sources may be fooling us. They may be much older than their standard "synchrotron lifetime", and still have a spectral break. This break may evolve only slowly, as the mean field and characteristic particle energy evolve. We need other means – for instance dynamical evolution (Eilek 1993) – to determine the age of diffuse radio sources.

REFERENCES

Alexander, P. and Leahy, J. P. 1987, *MNRAS*, **225**, 1

Eilek, J. A. 1993, submitted to *ApJ*

Eilek, J. A., Arendt, P. A. and Borovsky, J. E. 1993, submitted to *ApJ*

Katz-Stone, D. M., Rudnick, L. and Anderson, M. C. 1993, *ApJ*, **407**, 549

O'Donoghue, A. A., Eilek, J. A. and Owen, F. N. 1993, *ApJ*, **408**, 428

Owen, F. N. and Ledlow, M. J. 1993, these proceedings.

The First Stromlo Symposium: The Physics of Active Galaxies
ASP Conference Series, Vol. 54, 1994
G. V. Bicknell, M. A. Dopita, and P. J. Quinn (eds.)

TWO FREQUENCY VLBI OBSERVATIONS OF SMALL-SIZED CSS'S

D. DALLACASA
NFRA, Postbus 2, 7990 AA, Dwingeloo, The Netherlands

C. FANTI AND R. FANTI
Istituto di Radioastronomia del CNR, Via Irnerio 46,
40126 Bologna, Italy
Dipartimento di Fisica, Via Irnerio 46, 40126 Bologna, Italy

R.T. SCHILIZZI
NFRA, Postbus 2, 7990 AA, Dwingeloo, The Netherlands

R.E. SPENCER
NRAL, University of Manchester, Macclesfield Cheshire SK11 9DL,
Jodrell Bank - Great Britain

We have observed 16 *Compact Steep-spectrum radio Sources* (CSS's) from the Peacock and Wall (1982) catalogue, at 1.7 GHz using, in Mk2 mode, a VLBI global array of 10 antennas which provides a resolution of $\approx 2.5 mas$. Ten of them were also observed at 5 GHz with a global array of 15 antennas and a resolution of $\approx 1 mas$.

These sources have radio powers at 2.7 GHz comparable to those in the 3CRR sample (see Spencer et al., 1991 and references therein), but do not belong to that catalogue mostly because they have a turnover in their radio spectrum at frequencies around or above 0.5 GHz.

The source list is given in the table, together with some general information, including an indication of whether a core or a jet has been detected and a rough morphological classification. Images will be presented elsewhere. For all the sources Merlin maps at 1.7 GHz and VLA maps at 5 and 15 GHz already exist (Spencer et al., 1989). These sources, together with those selected from the 3CRR catalogue, constitute a complete sample of CSS's, unbiased with respect to any turnover in their radio spectra up to ≈ 2 GHz, which is suitable for statistical considerations on the nature of CSS's (see C&R Fanti, this volume).

The objects presented here mostly have projected linear sizes smaller than about 1kpc (roughly one order of magnitude smaller than the 3CRR CSS's). Their structures are mostly straight on the 1kpc scale, although in a few cases the orientation on this scale can be very different from that of the larger scale structure (when present). In many cases we see linear collimated features which we call *jets*, even in galaxies. This is in contrast to the larger CSS's of the 3CRR sample, where such jets only show up in quasars. In several cases we see a flat or inverted spectrum compact component that we identify with the source *core*.

On the assumption that the observed spectral turnover is due to synchrotron self–absorption in the individual sub–components, we computed, for all the sources, the expected total radio spectrum using the component sizes and flux densities and the equipartition magnetic field. The comparison of the 'composite' spectrum with the observed one is generally satisfactory, and the differences can usually be attributed to missing extended low brightness features, not detectable at 1.7 GHz, or to uncertainties in the components parameters.

A discussion of the characteristics of this *sub–sample alone* has little meaning, since it contains mostly radio sources with high frequency turnover in the spectrum, which are not representative of the CSS's as a class. These objects have to be examined together with the other CSS's as a whole. We do this in another paper presented at this Symposium.

TABLE I Source list

Source	Id.	m	z^a	S_{18cm} (Jy)	S_{6cm} (Jy)	LAS arcsec	LLS^b kpc	ν_{turn} MHz	Core	Jet	$Morph^c$
0223+341	G?	21.3	[1.3]	2.64	1.75	0.6	2.56	250	Y?	–	*Da*
0316+16	G?	22.0	[1.7]	6.93	2.89	0.3	1.27	900	–	–	*Da*
0319+121	Q	19	2.67	1.65	1.37	0.05	0.19	400	Y	Y	*cJ*
0345+337	G	19	0.244	1.96		0.5	1.21	60	–	–	??
0404+769	G	22	0.597	4.96	2.95	0.15	0.57	400	Y	Y	*D*
0428+205	G	20	0.219	3.58	2.38	0.25	0.56	1000	Y	Y?	*Da*
1225+368	G?	21.7	1.974	1.96	0.76	0.055	0.23	1200	Y?	Y	*D*
1323+321	G	19.0	0.37	4.28		0.06	0.19	400	–	–	*D*
1358+624	G	19.9	0.431	3.79	1.77	0.07	0.23	600	–	Y?	*D*
1413+349	EF	> 22		1.63	0.92	0.06	0.26	1000	Y	Y	*cJ*
1442+101	Q	17.7	3.544	2.32		0.02	0.07	950	Y?	J?	*cJ?*
1600+335	Q??			3.03	2.59	0.05	0.21	950	Y	–	*??*
1819+39	G	19	0.4	2.94		0.6	1.93	100	–	Y	*D?*
1829+29	G	20.2	0.842	2.61		2.5	10.39	100	–	Y	*D*
2230+114	Q	17.3	1.037	6.27		2.6	11.12	900	Y	Y	*D*
2342+821	G	20.5	0.735	3.21	1.74	0.18	0.73	400	–	–	*Da?*

[a] values in brackets are estimated

[b] computed with H_0=100 km $s^{-1}Mpc^{-1}$ and $q_0 = 0.5$

[c] *D*: two outer lobes; *Da*: very asymmetric lobes; *cJ*: core–jet

REFERENCES

Peacock and Wall, 1982, *MNRAS*, **198**, 843

Spencer, R.E., McDowell, J.,C., Charlesworth, M., Fanti, C., Parma, P., Peacock, J.A., 1989, *MNRAS*, **240**, 657

Spencer, R.E., Schilizzi, R.T., Fanti, C., Fanti, R., Parma, P., van Breugel, W.J.M., Venturi, T., Muxlow, T.W.B., Nan Rendong, 1991, *MNRAS*, **250**, 225

The First Stromlo Symposium: The Physics of Active Galaxies
ASP Conference Series, Vol. 54, 1994
G. V. Bicknell, M. A. Dopita, and P. J. Quinn (eds.)

EMISSION LINE PROPERTIES OF INTERMEDIATE RED-SHIFT RADIO GALAXIES AND QUASARS

R. HES
Kapteyn Astronomical Institute, P.O. Box 800, 9700 AV Groningen, The Netherlands

R. A. E. FOSBURY
ST-ECF, ESO, Karl-Schwarzschild-Straße 2, 8046 Garching, Germany

P.D. BARTHEL
Kapteyn Astronomical Institute, P.O. Box 800, 9700 AV Groningen, The Netherlands

INTRODUCTION

The diversity of AGN has been reduced by the concept of the unified schemes, where orientation and central obscuration account for the differences among the classes. As is expressed in the contributions to this conference, several observations provide support for the unification of the low radio power Seyfert galaxies. The controversial unification of high radio power objects, on the contrary, is not founded on a firm observational basis. This is due to the difficulty in studying these distant objects in detail. A statistical approach on complete samples of quasars and FRII radio galaxies is needed, necessarily more difficult because of the need of ample observing time on large telescopes.

TESTING THE RADIO LOUD UNIFIED SCHEME

A possible approach to test whether powerful radio galaxies are indeed obscured quasars is to compare the energy output of both classes of objects in radiation that is emitted isotropically. One such way is by comparison of the far infrared emission as done by Heckman et al. (1992), who found that quasars as a class are somewhat brighter than radio galaxies. Another approach was taken by Jackson and Browne (1990) who measured the luminosities in the [OIII]5007 emission line of matched samples of low-frequency selected quasars and radio galaxies, assuming this line to be emitted isotropically. Their result was that quasars are on average 5-10 times brighter in this line than otherwise similar radio galaxies. This observation has been considered a serious argument against the radio loud unified scheme. It is arguable, however, whether the [OIII]5007 line is best suited for this comparison, because it might originate partly from the nuclear regions which still suffer from dust obscuration.

Using the ESO 3.6-meter telescope we have obtained broad and narrow band images of matched samples of intermediate redshift quasars and radio galaxies from the 3CR compilation. The sources have been imaged in the light of [OII]3727, which is thought to be a better suited line for this comparison, due to its lower ionization potential and critical density. Complementing our new data with published values, we found that the difference in luminosity between both classes is considerably smaller in this line (Hes, Barthel and Fosbury, 1993): see Figure 1. This appears to support the similar intrinsic luminosities of radio galaxies and quasars in the redshift interval $0.3 < z < 0.8$, as expected in the context of the unified scheme. Note also that the stronger anticorrelation between radio core dominance and [OII] equivalent width, as compared to the relation with [OIII], reported by Baker et al. in this proceedings could be in agreement with the [OII] emission being isotropic.

As a next step we will, using newly obtained VLA data, try to correlate the structures of the [OII]3727 nebulae with the morphologies and depolarization properties of the radio sources.

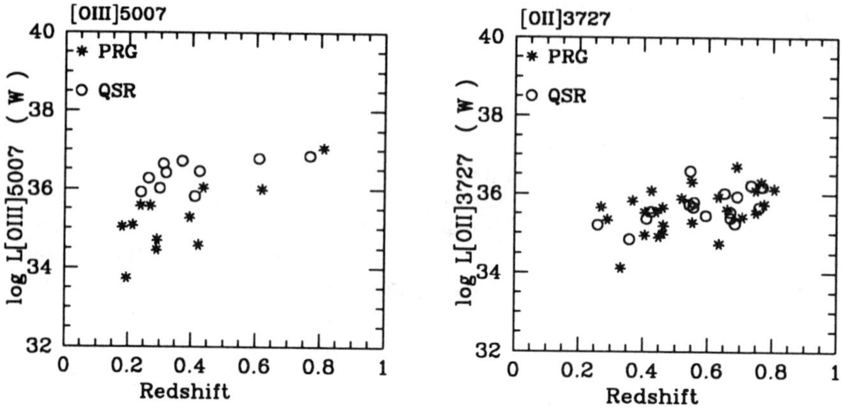

FIGURE I Luminosity of intermediate redshift powerful radio galaxies and quasars in the [OIII]5007 line (*left*) and in the [OII]3727 line (*right*) versus redshift. The data in the left panel were taken from Jackson and Browne (1990).

REFERENCES

Heckman, T.M., Chambers, K.C., Postman, M., 1992, *ApJ*, **391**, 39

Hes, R., Barthel, P.D., Fosbury, R. A. E. 1993, *Nature*, **326**, 362

Jackson, N., Browne, I.W.A., 1990, *Nature*, **343**, 43

The First Stromlo Symposium: The Physics of Active Galaxies
ASP Conference Series, Vol. 54, 1994
G. V. Bicknell, M. A. Dopita, and P. J. Quinn (eds.)

THE JETS IN RADIO GALAXY B1308-441

PAUL A. JONES
Department of Physics, University of Western Sydney, Nepean, PO Box 10, Kingwood NSW, 2747, Australia

INTRODUCTION

Radio galaxies show a wide range of large-scale structures, but they may be divided into two classes (Fanaroff and Riley 1974) based on power and morphology - the low-power FR class I sources have edge-darkened lobes and the high-power FR class II sources have bright hotspots near the ends of the lobes. There are also differences in the jets leading from the galaxy to the lobes: class I sources usually have prominent two-sided jets and class II sources have weaker one-sided jets (Bridle 1984), probably due to different Mach numbers of the jet flow (Bicknell 1985). More detailed studies of such jets are required to qualify these differences.

I argue that there are many radio galaxies, particularly of low power, that do not fit easily into either of the two simple classes of structure (Jones 1990, Jones and McAdam 1992). The radio source B1308-441 is one of these, with low power (a factor of 3.5 below the Fanaroff-Riley cutoff) but with features of both FR classes. The image observed with the Molonglo Observatory Synthesis Telescope (MOST) with a resolution of 44 arcsec (Jones 1987, Jones and McAdam 1992) shows prominent two-sided jets from the core (FR I) leading to bright spots near the end of diffuse lobes (FR II) with possible backflow. There is a prominent bright spot 2.2 arcmin from the core along the south-east jet. The source was identified with a 14.5 m elliptical galaxy having a redshift of 0.051. The angular size is 14.4 arcmin between the lobe peaks, corresponding to a projected size of 590/h kpc (using $H_0 = 100\,h\ \mathrm{km\,s^{-1}\ Mpc^{-1}}$) making it one of the largest radio galaxies known.

The MOST image does not resolve the jets across their axis. An image made with the Fleurs Synthesis Telescope (Jones 1987, Jones 1992), with a resolution of 22 arcsec, shows the four main peaks but the jets were not detected. A detailed study of the core, inner jets and and the nearby bright spot has now been made with the Compact Array of the Australia Telescope (ATCA).

ATCA OBSERVATIONS

The observation was made with the ATCA using the six antennas in a 6 km array, at two simultaneous frequencies in the 20 cm and 13 cm bands. The data were calibrated and CLEANed using AIPS and then details of the jets were determined at intervals of 15 arcsec along the axis by fitting one-dimensional gaussians perpendicular to the axis (AIPS task SLICE). The position of the ridge, the integral of flux across the jet and the deconvolved half-width were found.

The short 2 hour observation was at position angles that gave good resolution perpendicular to the jet. The results presented here are mostly from the 1.38 GHz data which have resolution 40 arcsec along the jet (similar to the MOST data), but 10 arcsec across the jet, enabling its width and position to be determined more precisely. At 2.37 GHz the clean restoring beam was 25 x 6 arcsec, but for the spectral index measurements a lower resolution image was made at 2.37 GHz to match the beamwidth at 1.38 GHz.

RESULTS AND DISCUSSION

The plot of the flux density along the source axis shows a strong, central peak which is partly due to the true core and partly the inner jet which is not adequately resolved. A flat spectrum core has been detected with the Parkes–Tidbinbilla Interferometer with flux density of 30 mJy at 2.3 GHz and 35 mJy at 8.4 GHz indicating a core size less than the fringe spacings of 0.1 and 0.03 arcsec respectively (Jones, McAdam and Reynolds, in preparation). The flux of the jets falls rapidly within the first 1 arcmin of the core with a plateau at 1.5 arcmin on the NW side. The SE jet has a bright spot 2.2 arcmin (90/h kpc) from the core where the flux rises sharply and then declines.

The spectral index distribution has a general trend to steeper spectra away from the core along both jets, but changes in the SE jet from a steep spectrum ($\alpha = -0.8$) at 1.8 arcmin from the core, to a flat spectrum ($\alpha = -0.2$) at 2.2 arcmin, as the flux increases to the bright spot.

The width of the jets increases from 3.4/h kpc (5 arcsec) near the core to over 17/h kpc (where the jets become difficult to fit as they fade), with some evidence of knottiness in the SE jet. The jets may be expanding rapidly within the first 20/h kpc, but this cannot be confirmed due the low resolution along the jet. The corresponding opening angles of the jets are 7° on the SE and 4° on the NW side out to 80/h kpc; beyond this they expand more rapidly.

The jets are not straight. There are kinks to the NW at 2 arcmin from the core where the jet expands more rapidly and fades, and along the SE jet at 1.8 to 2.2 arcmin from the core where the flux increases in the bright spot. The flatter spectrum in this region may be due to electron re-acceleration in a shock front as the jet is redirected. If the cause is a collision of the jet with a cloud in the cluster medium, then there may be some extended optical emission from the region.

REFERENCES

Bicknell, G.V. 1985, *Proc. Astron. Soc. Australia*, **6**, 130

Bridle, A.H. 1984, *AJ*, **89**, 979

Fanaroff, B.L., and Riley, J.M. 1974, *MNRAS*, **167**, 31P

Jones, P.A. 1987, *Proc. Astron. Soc. Australia*, **7**, 208

Jones, P.A. 1990, *Proc. Astron. Soc. Australia*, **8**, 254

Jones, P.A. 1992, *Australian J. Phys.*, **45**, 797

Jones, P.A., and McAdam, W.B. 1992, *ApJS*, **80**, 137

The First Stromlo Symposium: The Physics of Active Galaxies
ASP Conference Series, Vol. 54, 1994
G. V. Bicknell, M. A. Dopita, and P. J. Quinn (eds.)

EXCITATION OF EMISSION LINE REGIONS IN RADIO GALAXIES

A. M. KOEKEMOER, G. V. BICKNELL AND M. A. DOPITA
Mt. Stromlo and Siding Spring Observatories, Australian National University, Weston Creek P.O., A.C.T. 2611, Australia

INTRODUCTION

We investigate the viability of strong shocks as an excitation mechanism for extended emission-line regions (EELRs) in radio galaxies. Detailed spectroscopy is presented for EELRs in two FR II sources, PKS 0349-27 and PKS 2356-61. We show that the EELR luminosities *and* linewidths can be consistently described by the shock ionization model presented in Bicknell (1991) and Sutherland *et al.* (1993). In addition, it is difficult to reconcile the EELR morphologies and implied large opening angles with the observed nuclear flux, if the EELRs were photoionized by the nucleus as required in unified schemes.

Each source is a "classical" FR II radio double; PKS 0349-27 has a power $\log P_{1.4\mathrm{GHz}} = 26.6$ W Hz^{-1} and has been mapped with the VLA by Baum (1987) and Caganoff (1989). PKS 2356-61 has $\log P_{1.4\mathrm{GHz}} = 26.9$ W Hz^{-1}; it has been imaged in detail with the Australia Telescope (Koekemoer *et al.* 1993).

OBSERVATIONS

We observed each EELR using the AAT RGO spectrograph with a 1$''$ slit. PKS 0349-27 was observed in Dec 1991 with the IPCS in "ASPECT" mode, producing 9 parallel longslit spectra fully sampling the EELR. PKS 2356-61 was observed in Aug 1992 with the Tek CCD, producing 14 spectra across the EELR. This yielded data-cubes from [OII]λ3727 to [SII]λ6731 with 4 Å resolution.

IONIZATION MECHANISMS FOR THE EELRS

The high EELR excitation ([OIII]/H$\beta \sim 5-10$) rules out ionization by stars; two alternatives are ionization by a nuclear power-law or local autoionizing shocks.

Shocks are likely to be present in the EELRs; typical FWHM values are $500-600$ km s^{-1} and our recent high-resolution observations (50 km s^{-1}) reveal multiple components (FWHM $200-400$ km s^{-1}) in many lines. Two reasons for investigating shock ionization are: (1) the ionization parameter does not decrease with radius, requiring strong density profile constraints for nuclear ionization and (2) an ionizing radiation cone requires such large opening angles that the nucleus should not be obscured towards us. However there is no evidence for broad lines and the nuclear flux is a factor of 20 too small to ionize the EELR.

The shock model is described in Sutherland *et al.* (1993). Two clouds colliding with velocity Δv produce a shock (temperature $T \propto \Delta v^2$), radiating

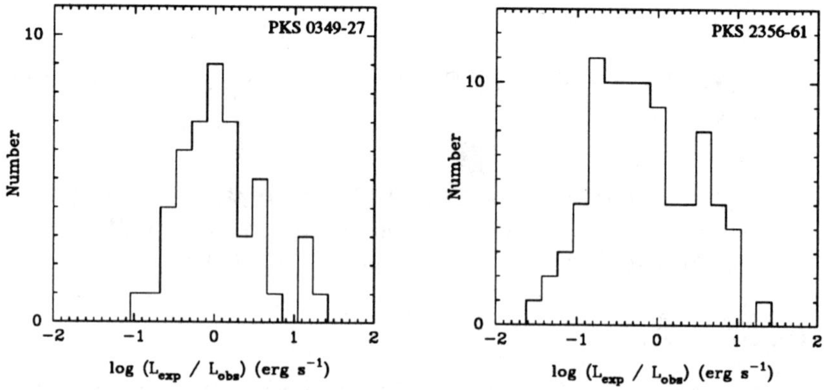

FIGURE I Ratio of Expected to Observed luminosity, for each pixel

soft X-rays/UV for $\Delta v \sim 200 - 400$ km s $^{-1}$. The ionized Balmer luminosity is:

$$L_{H\beta} \propto C\, f\, n_0 \Delta v^{2.44} A_{\rm sh} \tag{1}$$

with covering fraction C, shock cooling parameter f, pre-shock H^+ density n_0 and shock area A. If the EELRs are mainly optically thin to Balmer lines then

$$L_{H\beta} \propto \epsilon\, n_e n_p a d\, h\, \nu_{H\beta}\, \alpha^{eff}_{H\beta} \tag{2}$$

with filling factor ϵ, $n_e = n_p = n_0$, line-of-sight depth d, and cross-sectional area a of an emitting volume. Combining (1) and (2) eliminates the density:

$$L_{H\beta} = 1.46 \times 10^{35} \frac{C\, f}{\epsilon \left(\frac{d}{\rm kpc}\right) \left(\frac{a}{\rm kpc^2}\right)} \left(\frac{\Delta v}{200 \rm km\, s^{-1}}\right)^{4.88} \left(\frac{A_{\rm sh}}{\rm kpc^2}\right)^2 \rm erg\ s^{-1}$$

In Figure I we compare the expected Hβ pixel luminosities with those observed. The area a is pixel size, Δv is FWHM, and f is 0.8. Assumptions are that $d \times \epsilon$ is the mean EELR width, $C = 1$, $A = a$, and line profiles are not blended.

CONCLUSION

The EELR luminosities can readily be accounted for in terms of ionizing shocks; the scatter is mainly due to the assumptions made. Work is currently in progress to measure the variation of line ratios across the EELRs and as a function of line-of-sight velocity in order to provide better constraints.

REFERENCES

Baum, S. 1987, Ph.D. Thesis, University of Maryland

Caganoff, S. 1989, Ph.D. Thesis, Australian National University

Bicknell, G. 1991, *Proc.A.S.A.*, **9**, 93

Koekemoer, A., Schilizzi, R., Ekers, R. and Bicknell, G. 1993, in preparation

Sutherland, R., Bicknell, G. and Dopita, M. 1993, *ApJ*, **414**, 510

The First Stromlo Symposium: The Physics of Active Galaxies
ASP Conference Series, Vol. 54, 1994
G. V. Bicknell, M. A. Dopita, and P. J. Quinn (eds.)

NUMERICAL MODELS OF BENT, DISRUPTING JETS

C. LOKEN AND J. BURNS
Astronomy Dept., New Mexico State University, Las Cruces, NM 88003

D. CLARKE
Center for Astrophysics, 60 Garden St., MS 51, Cambridge MA 02138

INTRODUCTION

Wide-Angle Tailed (WAT) radio sources are unlikely to be bent by ram-pressure on the jet due to movement of the host galaxy through the intracluster medium (ICM) (e.g., Eilek et al. 1984). However, most clusters have likely undergone mergers (e.g., Burns et al. 1993) and hybrid hydro/N-body simulations of cluster/subcluster mergers (Roettiger et al. 1993) have revealed large-scale, supersonic gas motions near the center of the merged clusters. Thus, WATs may be disrupted and bent by winds in the ICM. This motivated us to simulate a jet crossing a contact discontinuity separating a static ISM from a windy ICM.

Another possible mechanism for WAT formation was studied, in 2-D, by Norman, Burns, & Sulkanen (1988) who showed that a jet would flare and disrupt upon crossing a shock. Such shocks might arise naturally at the termination of a galactic wind. In the case of an oblique shock, the gas streamlines bend towards the shock and so a disrupted jet would be expected to form an extended tail such as is seen in WAT sources. This scenario is also investigated here.

NUMERICAL SIMULATIONS

Our three-dimensional simulations were carried out with ZEUS-3D (Clarke, 1990; Stone & Norman 1992). In the first case, a jet with internal Mach number 3 propagates initially in a static ISM which is pressure-matched with the jet but 50 times denser. The jet then passes through a contact discontinuity into the ICM which is a factor of 10 less dense than the ISM. A Mach 1.3 wind blows parallel to the contact discontinuity in the ICM.

In the second model, a Mach 2.9 wind blows outwards from the galaxy and encounters an obliqe shock oriented at 70° to the initial wind direction. The ambient density and pressure increase at the shock in accordance with the jump conditions. The jet has an initial internal Mach number of 2.5 and is 100 times less dense than the wind which it initially parallels.

RESULTS

As the first jet crosses the contact discontinuity, the impact of the wind on the jet drives a strong shock into the ISM. At the same time, the jet begins to flare and bend, and rapidly becomes "naked" on the windward side. The bending

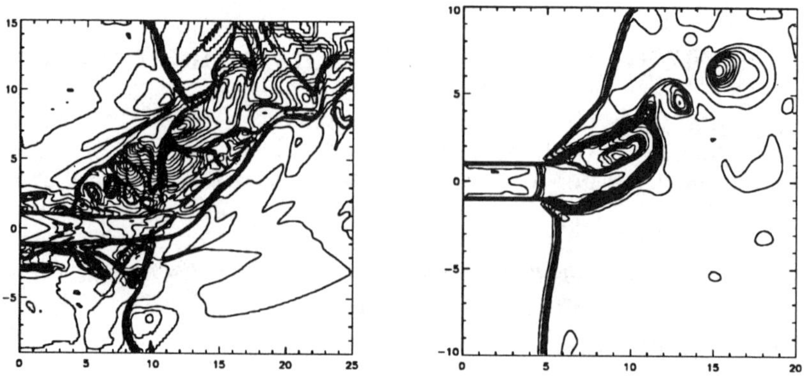

FIGURE I Log (density) contours are shown for the jet crossing a contact discontinuity (on the left) and for a jet crossing an oblique shock (right).

jet drags ISM material with it on the leeward side creating a high-density, low-pressure "wall" which the jet propagates along (Fig. I). The buffeting of the jet at the ISM/ICM interface excites instabilities which lead to the jet breaking and reforming in a series of "disconnection events".

The jet which crosses the oblique shock undergoes the flaring seen by Norman et al. (1988) as it decelerates in order to match the increased ambient pressure. A turbulent tail is subsequently dragged downstream (Fig. I) by the component of the wind velocity tangential to the shock. Note that the ISM/ICM interface is much more stable than in the case of the contact discontinuity.

CONCLUSIONS

Two possible scenarios for the formation of WATs have been investigated numerically. The passage of a jet through either a contact discontinuity or an oblique shock was found to result in the disruption of the jet and the creation of an extended tail. However, the nature of the disruption is very different in the two cases. In the case of the shock, the jet decelerates and flares as it attempts to adjust to the changed ambient pressure. In the case of the contact discontinuity, the jet is disrupted as a result of the dynamic and complex interaction at the ISM/ICM interface.

REFERENCES

Burns, J., et al. 1993, *these proceedings*
Clarke, D. A. 1992, *BAAS*, **23**, 1334
Eilek, J. A., Burns, J. O., O'Dea, C. P., & Owen, F. N. 1984, *ApJ*, **278**, 37
Norman, M. L., Burns, J. O., & Sulkanen, M. E. 1988, *Nature*, **335**, 146
Roettiger, K., Burns, J., & Loken, C. 1993, *ApJ*, **407**, L53
Stone, J. & Norman, M. 1992, *ApJS*, **80**, 791

The First Stromlo Symposium: The Physics of Active Galaxies
ASP Conference Series, Vol. 54, 1994
G. V. Bicknell, M. A. Dopita, and P. J. Quinn (eds.)

RADIO/FAR-INFRARED CORRELATION TOWARDS SOUTHERN CLUSTERS OF GALAXIES

A. M. UNEWISSE
Astrophysics Department, University of Sydney, NSW, Australia, 2006

INTRODUCTION

A tight correlation is known to exist between radio and far-infrared (FIR) emission for a variety of disk dominated field galaxies (eg: Unger *et al.*, 1989). This paper looks for evidence of a radio/FIR correlation in the direction of rich southern clusters of galaxies.

THE SAMPLE

The basis for this study is an 843 MHz radio survey of 39 nearby (distance class ≤ 3) southern galaxy clusters from the Abell, Corwin and Olowin rich cluster catalogue (Abell *et al.*, 1989) using the Molonglo Observatory Synthesis Telescope. The radio/FIR sample was obtained by correlating radio sources lying within a projected distance (r) of 500 kpc ($H_0 = 100\,\mathrm{km\,s^{-1}\,Mpc^{-1}}$) of a cluster centre against 60μm sources in the *IRAS* point source catalogue. The final sample is described more fully in Unewisse (1993).

The properties of the 15 sources in the sample associated with galaxies are summarized in Table I. In general, the galaxies have either spiral (sp) or peculiar (pec) morphology. Twelve galaxies have measured redshifts; 7 are probably cluster members, 3 are background galaxies, and 2 are foreground galaxies.

TABLE I Non-stellar radio/FIR sources in the direction of sample clusters

Abell No	z (cluster)	*IRAS* No	r (kpc)	S_{843} (mJy)	S_{60} (Jy)	id	Mag B_J	z (galaxy)	Morphology	Other Name
A2877	.0230	01059−4621	401	9	0.43	sp	13.7	.0206	SB(s)b	IC 1627
A2877		01082−4626	348	10	0.44	pec	15.3	.0239		
A2877		01091−4611	304	9	0.60	sp	14.2	.0221	SAa	ESO 243-G051
A3193	.0340	03560−5223	269	8	0.47	pec	16.0	.0334	Sa	SGC 0356.1-5223
A3381	.0350	06081−3338	94	9	0.64	pec	14.4	.0291	S?	ESO 364-G036
A3526	.0101	12435−4125	328	48	2.44	sp	14.1	.0107		
A3537	.0161	12585−3207	57	14	0.65	sp	14.6	.0319	SA(r)cd	ESO 443-G029
A3537		12593−3230	337	17	1.17	sp	17.5	.0312	Sc	HCG 063D
A3564	.0492	13322−3500	359	63	5.36	sp	14.5	.0149	S0	ESO 383-G027
A3565	.0112	13317−3312	382	30	2.08	pec	15.9			
A3574	.0142	13454−2956	148	25	2.32	gal	18.6	.1293	Liner	
A3574		13464−3003	20	90	2.04	sp	14.0	.0162	SA0,sy1	IC 4329A
A3656	.0185	19578−3847	174	20	0.77	sp?	16.9			
A3656		19582−3833	218	38	3.38	pec	13.5	.0166	SA(rs)ab	ESO 339-G025
A3747	.0306	21062−4339	255	12	0.93	gal?	16.9			

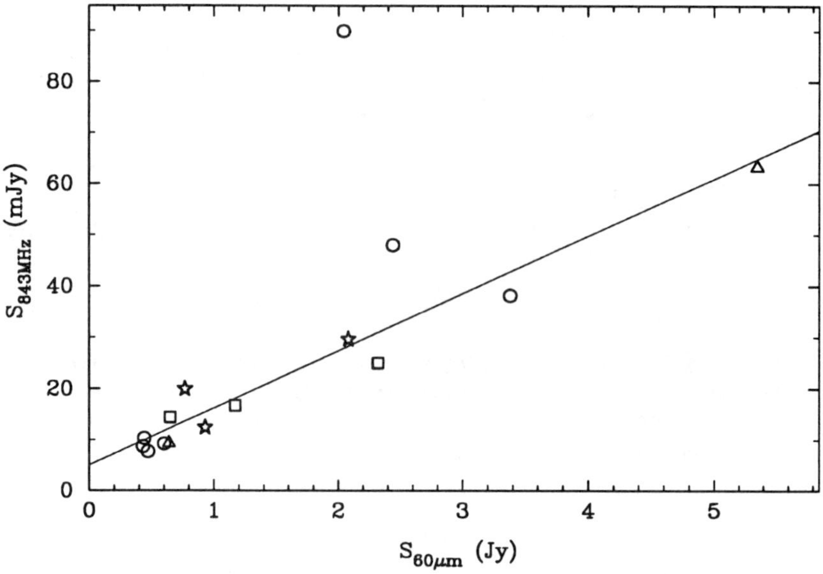

FIGURE I Radio *vs* infrared flux density for the galaxies in the sample. Symbol notation: ∘ = cluster member, △ = foreground galaxy, □ = background galaxy, and ⋆ = unknown redshift. The line is a least squares fit to all points except the one corresponding to the Seyfert which is radio loud.

RESULTS

Figure I compares the radio and infrared flux density for each galaxy in the sample. The different symbols indicate redshift information and are defined in the caption. A tight correlation is seen for all galaxies except IC 4329A (13464–3003), the only Seyfert in the sample. Such galaxies are known to be radio loud on the radio/FIR diagram (Roy *et al.*, 1992). When the units are matched, the slope shown on Figure I corresponds to a ratio of $S_{843\mathrm{MHz}}/S_{60\mu\mathrm{m}} = 0.011 \pm 0.001$.

CONCLUSION

Cluster membership appears to have no bearing on the radio/FIR correlation which may have interesting implications for a starburst origin for radio/FIR correlations.

REFERENCES

Abell, G. O., Corwin, H. G. and Olowin, R. P., 1989, *ApJS*, **70**, 1

Roy, A. *et al.*, 1992, *ASP Conf Ser*, **31**, in press

Unewisse, A. M. 1993, *PhD Thesis*, University of Sydney

Unger, S. W. *et al.*, 1989, *MNRAS*, **236**, 425

The First Stromlo Symposium: The Physics of Active Galaxies
ASP Conference Series, Vol. 54, 1994
G. V. Bicknell, M. A. Dopita, and P. J. Quinn (eds.)

HOT ELECTRONS IN THE M87 JET

MARK WALKER, DON MELROSE & LEWIS BALL
Research Centre for Theoretical Astrophysics, University of Sydney, NSW 2006, Australia

INTRODUCTION

Hubble Space Telescope images of the M87 jet (Boksenberg et al. 1992) reveal that the optical/UV morphology is strikingly similar to that of the radio (Owen, Hardee and Cornwell 1989). This is at odds with naive considerations of synchrotron cooling, whether the radiating electrons are assumed to be accelerated in the nucleus or at the leading edges of the emitting knots. We discuss possible resolutions of this puzzle, and propose a definitive observational test.

MORPHOLOGY

As noted first from the radio data (Owen, Hardee and Cornwell, 1989), the M87 jet is limb-brightened in places and the coherent structure visible between knots 'A' and 'B' is suggestive of a magnetic filament which is wrapped around the jet. These authors also noted that the interior of the jet might be largely free of magnetic field and so could function as a low-loss conduit for a flow of relativistic particles. We suggest a modification of these ideas.

It is well known that magnetized jets tend to become dominated by internal magnetic stresses as adiabatic expansion proceeds, motivating the study of "force-free" configurations (e.g. Konigl and Choudhuri 1985). Such structures have little resistance to bending and are intrinsically floppy; bending occurs as a consequence of excess pressure. We therefore suggest a model of the M87 jet as a twisted-up, quasi force-free flux rope which carries all of the power emanating from the nucleus. The spatial structure which can be resolved in the jet is then interpreted as segments of the coiled rope, these being identified with the radio-optical filaments. The observed limb-brightening of the jet indicates that the rope tends to coil into a hollow configuration.

LOCAL OR NUCLEAR ACCELERATION

Estimates of the magnetic field strength in the M87 jet imply that the synchrotron lifetime of optically-emitting electrons is short. Local acceleration in the jet has therefore been invoked to account for the observed optical synchrotron emission, but it is difficult to reconcile the detailed optical/radio correlation with existing local acceleration models. If our proposed morphology is correct it may be that electrons are accelerated locally at shocks in the flux rope, and then radiate as they continue to stream out along the rope. The cooling length is then measured along the flux rope, not along the axis of the jet.

An alternative follows from the point that almost any scenario for generating a rope-like flow of energetic particles and magnetic field incorporates axial structure within the rope. For example, in the massive black hole plus accretion disk picture most of the power flows down the center of the rope, while most of the angular momentum is carried off by the outer parts of the rope. In particular there is no reason to suppose that the magnetic field is uniform across the rope: pressure balance merely requires that the total pressure vary slowly, while the magnetic and particle contributions may anticorrelate, giving rise to a magnetically confined low-loss flow. This means that particle acceleration in the nucleus is viable.

SPECTRA

The issue of local versus nuclear electron acceleration can be decided by measuring the high-energy spectrum of the jet. At energies above 10 keV the flux from knot A should be dominated by synchrotron self-Compton emission with spectral flux $F_\nu = C\,\nu^{-0.64}$ Jy (ν is in Hz), with $C \sim 100 - 4000$ for the case of nuclear acceleration, and $C \simeq 2$ for local acceleration. This difference between the models arises as follows. If the electrons are accelerated in the nucleus their radiating lifetime must be much longer than if they are accelerated locally in the jet, in order to account for the observed optical synchrotron emission. It follows that the average synchrotron power per electron is lower in the nuclear than in the local acceleration model. (The average synchrotron emissivity may be reduced by allowing a lower uniform magnetic field, or by having a strong field with a small filling factor.) Since the average synchrotron emissivity is lower for the nuclear acceleration model, more electrons are required to produce the observed synchrotron power and the flux due to inverse-Compton scattering is therefore much larger.

Some support for the nuclear acceleration model already exists. The observed X-ray/optical flux ratio decreases with increasing distance from the nucleus (Biretta, Stern and Harris 1991) and this is attributable to synchrotron spectral ageing. However, measurement of the broad-band spectral properties at higher energies would be the decisive test. For the most extreme nuclear acceleration model, which corresponds to a jet speed of $\beta_j \sim 0.1$, the expected flux from knot A at GeV energies is on the limit of detectability of the EGRET instrument on the Gamma-Ray Observatory; other models give lower fluxes. It might even be possible to detect knot A at TeV energies (via atmospheric Cerenkov) if the electrons are accelerated in the nucleus.

REFERENCES

Biretta, J. A., Stern, C. P., and Harris, D. E., 1991, *AJ* **101**, 1632

Boksenberg, A., et al., 1992, *A&A* **261**, 393

Konigl, A., and Choudhuri, A. R., 1985, *ApJ* **289**, 173

Owen, F. N., Hardee, P. E., and Cornwell, T. J., 1989, *ApJ* **340**, 698

The First Stromlo Symposium: The Physics of Active Galaxies
ASP Conference Series, Vol. 54, 1994
G. V. Bicknell, M. A. Dopita, and P. J. Quinn (eds.)

ARE FRI'S AND FRII'S DIFFERENT?

ESTHER L. ZIRBEL AND STEFI A. BAUM
STScI, 3700 San Martin Dr., Baltimore, MD 21218

ABSTRACT We analyze the relationship between the emission line luminosity and radio power separately for edge brightened radio sources (alias FRII, Faranoff and Riley 1979) and edge darkened and more diffuse radio sources (alias FRI). We find that the FRI's and the FRII's independently show a correlation between the narrow emission line luminosity and the total radio power and/or core power where the correlations are offset in the sense that FRII's have about 10 times as much emission line activity. This suggests that either the engines of FRI's and FRII's are fundamentally different in the UV to jet thrust which they produce, or that the efficiency of converting jet kinetic energy to radio luminosity is higher in FRI's, particularly close to the radio cores.

RESULTS

The Correlation of total radio power to emission line luminosity

In Figure 2 we plot the total radio power versus emission line luminosity and we find the following:

- For the entire sample of radio galaxies we confirm the correlation between the emission line luminosity and the radio power. This relation has a slope of 0.885 ± 0.039 and a correlation probability which is better than 99.99%.
- Differentiating between FRI's and FRII's, we see clearly that each FR type has its own correlation between the line and the radio luminosity. The independent correlation probabilities are both 99.99%.
- Stronger FRII sources also have stronger emission line luminosities; i.e. powerful low emission line sources do not exist and the relationship does not level off at high radio power. This implies that we are not limited by the amount of cold gas which is seen in emission.
- The slope for the correlation of the FRII's is close to unity.
- The FRI's show a different slope. This is particularly the case if we include the lower power elliptical galaxies (the box in figure 2) obtained from Phillips *et al* (1986). It is however possible that the fit to the FRI's is bimodial, consisting of a flat baseline (i.e., a minimum line luminosity present in giant elliptical galaxies independent of the nuclear activity) followed by a slope of close to unity for the more powerful FRI's. More data for powerful FRI's are needed.

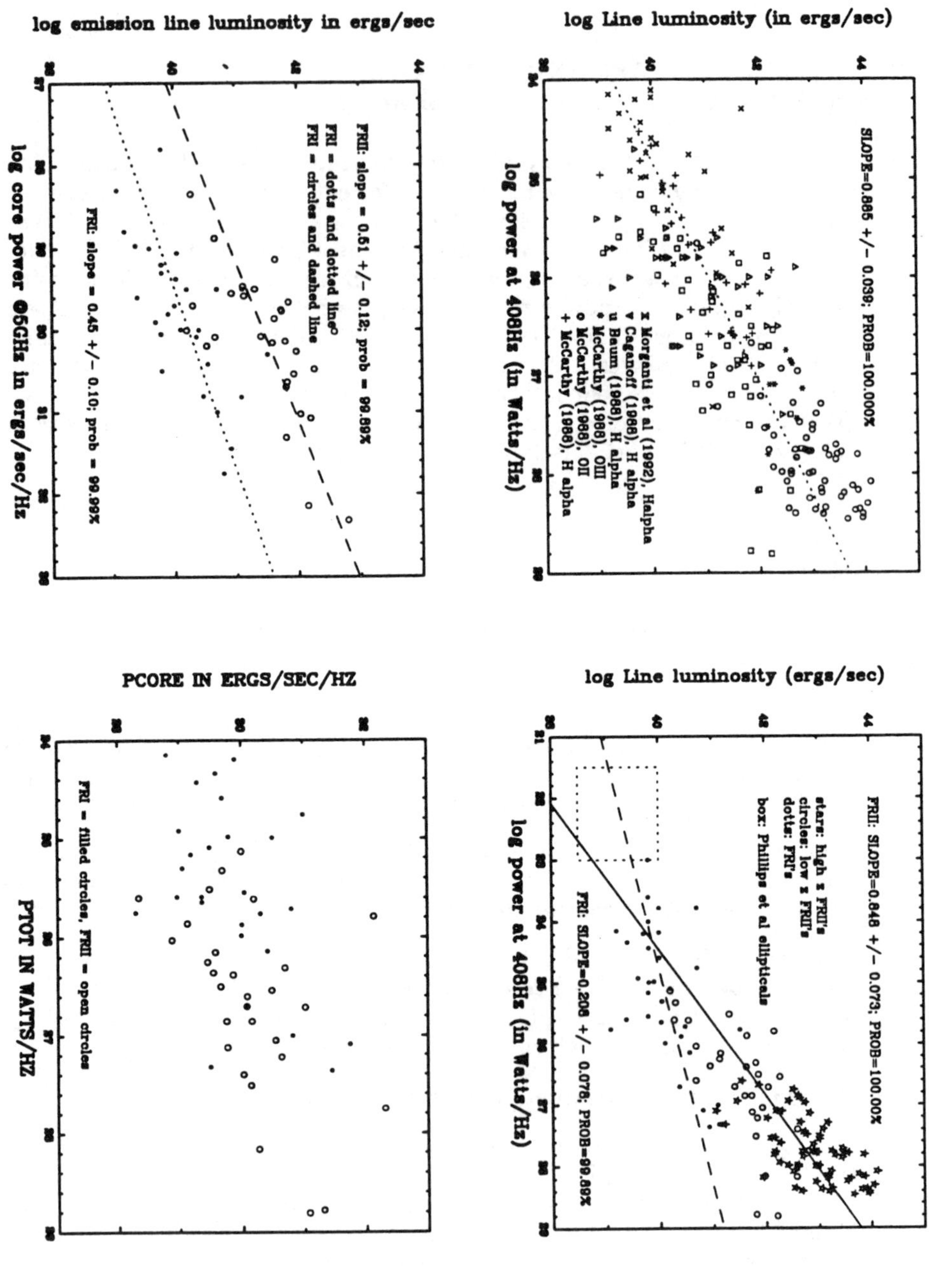
log Line luminosity (in ergs/sec)
SLOPE=0.865 +/- 0.039; PROB=100.000%
x Morganti et al (1992), Halpha
v Caganoff (1988), H alpha
u Baum (1988), H alpha
* McCarthy (1988), OIII
o McCarthy (1988), OII
+ McCarthy (1988), H alpha
log power at 408Hz (in Watts/Hz)
log emission line luminosity in ergs/sec
FRII: slope = 0.51 +/- 0.12; prob = 99.89%
FRII = dots and dotted line
FRI = circles and dashed line
FRI: slope = 0.45 +/- 0.10; prob = 99.96%
log core power @5GHz in ergs/sec/Hz
log Line luminosity (ergs/sec)
FRII: SLOPE=0.848 +/- 0.073; PROB=100.00%
stars: high z FRII's
circles: low z FRII's
dots: FRI's
box: Phillips et al ellipticals
FRI: SLOPE=0.208 +/- 0.078; PROB=99.89%
log power at 408Hz (in Watts/Hz)
PCORE IN ERGS/SEC/HZ
FRI - filled circles, FRII - open circles
PTOT IN WATTS/HZ

The Correlation Core Radio Power to Total Power

In Figure 4 we find the following:

- The output of the core radio powers of FRI's and FRII's are comparable, however the extended powers are higher for the FRII's.
- The separation between FRI's and FRII's is weaker than in either of the line to core or the line to total power plots.

Correlations of the nuclear activities

Since the extended radio power represents the output of the central engine integrated over tens of millions of years, it is important to distinguish between the radio energies deposited in the lobes and those that are directly arising from the active nucleus. Consequentially, we shall correlate the core radio powers to the emission line luminosities. In Figure 3 we find:

- While the correlation for the combined sample between the core radio power and the emissison line luminosity is weaker than that between the total power and line luminosity (also noted by Rawlings 1987), it is equally strong if we separate the two radio morphologies.
- The radio core power correlates strongly with the central narrow emission line luminosity, but the two FR-types separate out even more clearly than in the previous plots. For the same core power, the FRII's produce consistently about 10 times as much emission line luminosity as the FRI's. The independent correlations for FRI's and FRII's are significant at the 99.99% and the 99.89% confidence levels.
- The slopes for both correlations are 0.5 ± 0.12. This implies that the radio core power is proportional to the square of the narrow emission line luminosity.

REFERENCES

Baum and Heckman (1989) *ApJ*, 336,681; *ApJ*, 336, 702

Caganoff, S. 1989, PhD dissertation, The Australian National University

Faranoff and Riley (1987) *MNRAS*, 167, 31p

McCarthy (1988) PhD thesis, University of California

Morganti *et al* (1992) *MNRAS*, 254,546

Phillips *et al* (1986) *AJ*, 91, 1061

Rawlings *et al* (1989) *MNRAS*, 240, 701

HIGH REDSHIFT RADIO GALAXIES

Invited Papers

The First Stromlo Symposium: The Physics of Active Galaxies
ASP Conference Series, Vol. 54, 1994
G. V. Bicknell, M. A. Dopita, and P. J. Quinn (eds.)

HIGH–REDSHIFT RADIO GALAXIES

GEORGE MILEY
Leiden Observatory, Postbus 9513, 2300 RA Leiden, The Netherlands

ABSTRACT I give a personally-biassed review of our present knowledge about high–redshift radio galaxies.

INTRODUCTION

During the next forty minutes I shall give you my personal view of the present state of play in the high-redshift galaxy game. After a general introduction to the subject, summarizing the main differences between distant radio galaxies and their nearby counterparts, I shall concentrate on the intriguing question of what makes the UV/optical/IR continuum radiation in these objects. I shall then describe some recent work which sheds light on this question. We shall see that our knowledge of the responsible radiation mechanism is (to paraphrase Sean O' Casey) "in a state of chassis". I shall conclude with some words about the environment in which high-redshift radio galaxies are located. For those of you who are interested in pursuing some of the topics in more detail I commend you to the recent excellent review by Pat McCarthy and the references therein (McCarthy 1993).

Radio galaxies are unique cosmological probes. As with radio-loud quasars, the presence of luminous radio continuum and optical line emission enable radio galaxies to be observed and recognized at large distances, up to z = 3.8. However, unlike the situation for most quasars, their optical emission can be spatially resolved from the ground and studied in detail.

Progress in detecting distant radio galaxies has been rapid in recent years. If someone had told me only a decade ago that I would be standing here discussing galaxies having z > 2, I would not have believed them. However, the use of CCDs and the exploitation of new selection criteria has reaped dramatic rewards. Now, more than 60 radio galaxies are known with z > 2. More than half of these have been found by our group by concentrating on radio sources with the steepest spectra, most of these in a "Key Programme" of the European Southern Observatory. Although several people contributed to this Key Programme, most of the work was done by Huub Röttgering, who presented his Ph.D thesis in January and Rob van Ojik, who succeeded him.

Redshifts of 1.5 to 4 correspond to a time when the Universe was 10% - 20% of its present age. This was a crucial period in history when galaxy formation must have been rampant. It corresponds to the AGN era, a two-billion year "delta function" in the population evolution of luminous quasars and radio galaxies, when their space-density rose to a value several hundred times larger than the present density before the species mysteriously and suddenly became almost extinct.

Radio galaxies are extremely valuable for studying this period of the Universe because they emit **three components**, (IR-optical-UV continuum, emission lines and radio continuum) that are all **highly luminous and spatially extended**. Not only can different and complementary sets of diagnostics be derived from each of these components, but studies of the relationships between these various diagnostics and the interaction between the different components provides yet additional knowledge about the physical conditions. Such data are relevant for understanding the formation and evolution of galaxies, active nuclei and radio sources.

GENERAL PROPERTIES

The various spectral regimes of high-redshift radio galaxies have several properties that differ from the corresponding components in low-redshift radio galaxies. Although these changes give us information about the high-redshift Universe, we must be cautious of interpreting them as pure evolution. High-redshift radio galaxies are among the most luminous objects in the Universe at almost all wavebands and (as yet) Malmquist bias makes it difficult to disentangle the effects of redshift from those of luminosity.

The Radio Continuum.

High-redshift radio sources have smaller linear sizes than low-redshift ones, indicating the effect of a denser confining medium in the early Universe. There is conflicting evidence that distant radio sources are more bent than nearby ones. Although this effect appears to be present for the radio counterparts of quasars with normal radio spectra (Barthel and Miley 1988; Bogers et al. 1993), it is absent for radio galaxies with the steepest radio spectra (Röttgering et al. 1993).

UV/Optical/IR Emission Lines.

The associated galaxies have a rich emission-line spectrum, with enormous luminosities (Ly α typically $10^{43.5}$ - $10^{45.5}$ erg/s). The emitting gas is spatially extended in a giant halo, up to 100 kpc, with a characteristic velocity range of 10^3 km/s (FWHM). The usual rituals of interpreting spectroscopy suggest that the source of ionization is anisotropic photoionization from a hidden quasar nucleus and yield characteristic densities of 10^{-2} cm^{-3} and ionized gas masses of $\sim 10^{10}\ M_{\odot}$. As was pointed out by McCarthy in his thesis, the ionized gas is morphologically related to the radio sources for $z > 0.1$. In some cases, this can be explained as alignment due to the effects of anisotropic photoionization along the radio axes. In other cases, there is clear evidence for physical interaction between the radio jet and the ionized gas as is seen at lower redshifts most prominently in 3C 277.3 (Coma A) (van Breugel et al. 1985).

UV/Optical/IR Continuum.

The identified objects have R-magnitudes which are typically between 21.5 and 24. Until a few years ago, this radiation was believed to be produced by an old ($\sim 10^9$ y) population of stars in a "normal" giant elliptical galaxy. The main evidence for this was the remarkably low scatter and continuity in the infrared Hubble diagram (K vs. log z). A cottage industry was devoted to interpreting the spectral energy distributions of such objects in terms of evolving stellar population mixes and using the Hubble diagram to determine H_0 and q_0. The field was as elegant as the beautiful attempts to unify active galaxies on the basis of orientation that we heard so much about earlier today. Unfortunately, it was a theology built on a house of cards! (I hope that the same will not be true for the Unification Church!). Even though we now know that high-redshift galaxies are not simply elliptical galaxies whose optical properties are unaffected by their radio sources, the continuity and low scatter in the IR Hubble diagram pointed out by Lilly and Longair are important results that must be explained by any viable model for high-redshift radio galaxies.

A spanner was thrown in the works by the discovery that not only are the giant Ly α halos related to the radio sources, but the UV/optical/IR continua are also preferentially aligned along the radio sources for objects having $z > 0.7$ (Chambers et al. 1987; McCarthy et al. 1987; Chambers et al. 1988; Eisenhardt and Choski 1990; Rigler et al. 1992). We are clearly dealing with more complex systems than giant elliptical galaxies, objects that are significantly different from low-redshift galaxies.

MODELS OF THE OPTICAL CONTINUUM

Several classes of models have been proposed or considered to account for the alignment effect. Starburst models, interpret the alignment as due to star formation stimulated by the radio jet as it propagates outward from the nucleus (Rees 1989; De Young 1989; Begelman and Cioffi 1989; Bithell and Rees 1990). Two distinct scenarios are invoked. The first postulates a 10^7 y starburst superimposed on an old 10^9 y conventional elliptical galaxy. The second explains the spectral energy distribution in terms of a single relatively young (few x 10^7 y) stellar population whose formation is triggered by the jet.

Another class of models was stimulated by the discovery that, at least in a few cases, the extended emission is polarized in the UV. This prompted the suggestion that the aligned component is due to scattering of light from a hidden quasar by electrons (Fabian 1989) or dust (Tadhunter et al. 1989).

I should mention two additional possibilities that have been proposed to explain the alignment effect. Daly (1992) suggested that inverse Compton scattering of the cosmic background photons by the radio-emitting relativistic electrons might well be the culprit. However, such a model had previously been considered by Chambers et al. (1988) and discounted because of the non-coincidence of the optical and radio morphologies and the huge departures from equipartition of the radio source energetics that would be needed (up to nine orders of magnitude).

The final model I draw to your attention ia a recent proposal by Eales (1992) that the alignment is primarily a statistical effect caused by enhancement of radio emission during interaction of the jet with the medium in the surrounding galaxy. This postulates (i) the radio luminosity increases with ambient density and (ii) distant galaxies have an anisotropic density distribution, elongated along their major axes. I shall come back to this later.

RELEVANT NEW RESULTS

Contamination of Continuum by Emission Lines.

Most of the evidence for alignment of the UV/optical/IR continuum with the radio emission is based on broad-band images. IR spectroscopy by Eales and Rawlings (1993) indicate that a substantial part of the broad-band K flux is affected by the presence of emission by Hα or [OIII] in the band. This casts doubt on the reality of the reported radio/K-band alignments. In the next talk Klaus Meisenheimer will show that in some cases the alignment is suspect, even for some relatively low-redshift objects in the optical. Table 1 illustrates the problem. This shows the suspect redshift ranges for the four strong lines that dominate the visible spectra. All other lines blueward of Ly α have equivalent widths much smaller than 100Åand therefore contribute less than a few percent to the broad-band fluxes. It is clear that the strong lines are indeed a serious contaminant in the IR, but for z > 1.2, they have a negligible effect in the R-band images. Even discounting the lowest redshift objects, there is no doubt in my mind that a strong radio/optical alignment effect exists in the large remaining sample. A similar conclusion is reached by McCarthy (1993). The radio/IR alignment effect is more suspect. However, a beautiful recent image of 4C 41.17 at z = 3.8 taken by the Keck telescope in K' band excludes emission lines, but shows that the light emitted near 5000$\AA$ is clearly aligned along the radio axis (Graham et al. 1993).

TABLE I Redshifts of Contamination by Bright Emission Lines

Line	λ (A)	Optical (R-Band)	Infrared (K-Band)
$Ly\alpha$	1216	$3.8 < z < 5.5$	
$[OII]$	3727	$0.6 < z < 1.2$	
$[OIII]$	5007		$2.8 < z < 3.8$
$H\alpha$	6563		$1.7 < z < 2.7$

HST Observations.

Because the HST can provide information about sub-kiloparsec structure in such objects, Ken Chambers, Wil van Breugel, Duccio Macchetto and I therefore instigated a pilot project to observe two of the most distant galaxies known, 4C 41.17 at z = 3.8 and 4C 28.58 at z = 2.9. (Miley et al. 1992; Miley 1993). Despite the HST aberration, the pilot project was successful. The rest frame ultraviolet continua of both galaxies were well detected and exhibit clumpiness on a scale of a few hundred parsec, optimally suited to mapping with the HST and the observed morphologies were used to constrain models of distant radio galaxies. I will here mention some of the implications:

(i). Clumpiness. Both galaxies observed emit a large fraction of their fluxes (30% to 50%) in clumps. Independent of detailed models, the clumpiness implies that considerable inhomogeneities exist in the inner regions of distant radio galaxies. It is possible that the same inhomogeneities which are responsible for the HST structure are also connected with the clouds responsible for the heavy element absorption systems that sometimes occur close to the emission redshifts of some quasars. If the clumps are stellar, stellar population models would imply a mass in stars of $> 10^{10}$ M_0 contained in each 500 pc clump.

(ii). Connection with Radio. For 4C 41.17 the HST data show clearly that the alignment effect can occur strongly on the sub-kiloparsec scale. The detailed results have implications for models of the radio/optical alignment. The similar radio optical clumpiness on the circumnuclear scale is difficult to reconcile with the model of Eales (1992), for which the radio/optical alignment is a statistical effect whereby the radio emission is enhanced when the jet is emitted along the major axis of the galaxy. The similarity in the radio and optical curvature is suggestive. If the radio/optical bending is observed to be a common property in radio galaxies, it would be an argument against simple scattering models (Miley 1992).

Evidence for Young Stars.

There are two recent results which provide evidence for the presence of a young stellar population in high-redshift radio galaxies. The first is work by Hammer et al. (1993), who coadded deredshifted spectra of ten 3C radio galaxies with $0.75 < z < 1.1$. Although the resulting spectrum *shows clear absorption features indicating stars*, there is no sign of the well-known 4000Å break that would be expected from an old stellar population. Hammer et al. conclude that any old population of stars cannot be responsible for more than 15% of the flux from these objects. This really dents the old conservative viewpoint that we are observing elliptical galaxies.

The other relevant observation is the poster presented at this conference by van Ojik et al. (1993) of a spectrum of TX0211-122, a z=2.3 galaxy discovered during our ESO Key Programme. The spectrum is highly anomalous with a flux ratio of Lyα to NV λ1240 which is a factor of 30 smaller than for a typical high-redshift radio galaxy. The spectrum is unlike that of any other extragalactic object except that of the z = 2.3 IRAS galaxy F10214+4724, which is also a substantial radio source. In both objects **depression of Lyα appears to be accompanied by simultaneous enhancement of NV**. The relatively large NV/CIV ratio implies that the line-emitting gas is overabundant in nitrogen. The most feasible explanation for the spectra is that both objects have undergone a period of intense star formation. The most massive stars would evolve quickly, producing (i) the dust responsible for attenuating the Lyα and (ii) a relative overabundance of nitrogen.

Evidence for Hot Electrons.

Although there are energetic problems in explaining the alignment effect purely in terms of electron scattering (Eales and Rawlings, 1990; McCarthy 1993), it is significant that ROSAT has detected X-rays from 3C 356 at z = 1.079 (Crawford and Fabian 1993). the X-Ray source has a luminosity of approximately 2.5 x 10^{44} erg s^{-1}, implying the existence of a hot approx. (10^7K) dense ($\sim 10^{-2}$cm^{-3}) gas cooling at a rate of about 500 - 1500 $M_\odot$ per year.

Evidence for Dust.

The optical polarization work discussed by Cimatti et al. (1993) shows that high-redshift aligned radio galaxies have polarizations with a strong tendency to be oriented perpendicular to the radio axes. The wavelength dependance of the polarization fraction is unclear, although there is a slight tendency for the polarization fractions to be larger in the UV. Cimatti et al. interpret the data in terms of dust scattering. A dust mass of approximately 10^8 M_0 is required. Additional evidence for the presence of dust in at least one object is provided by the small Lyα/NVλ1140 flux ratio in TX0211-122 described in 4.3.

NATURE OF THE OPTICAL CONTINUUM

In the light of the various recent observations, it is useful to reconsider the present state of the game regarding mechanisms for producing the optical continuum. After discarding the Compton scattering and anisotropically enhanced radio emission mechanisms for the reasons discussed above, let us examine problems which we encounter if we take each of the remaining models as sole explanations for the observed IR/optical/UV continua.

Dust Scattering.

Objections to this model are:
(i) Absence of broad emission lines in the scattered light. Although one might speculate that the hidden "god" responsible for all the light is a BLLac object rather than a quasar, given the luminosities required, this would be very contrived.
(ii) Radio-optical Alignment persists into the red and IR. Because of the wavelength dependance of dust scattering, the alignment should only be present in the UV rest frame. Although the line contamination has rendered the IR imaging results unclear, the Keck continuum picture of 4C 41.17 suggests that the alignment effect persists out to at least 5000Å.
(iii) The IR Hubble Diagram. It is difficult to explain the continuity and small scatter in the K vs log z diagram.
(iv) The luminosities of the "hidden gods" would have to be typically at least 16th magnitude at z = 2, which is brighter than most known quasars.

Electron Scattering

Objections to this model are:
(i) The increase of polarization to the UV, if confirmed, would create problems. More data are needed here.
(ii) It is difficult to create enough scattered light without swamping it by stars formed in the cooling flows.
(iii) The IR Hubble Diagram (see *Dust Scattering* above).
(iv) High quasar luminosities are required (see *Dust Scattering*).

Old Stars with Young Burst.

Objections to this model are:
(i) Where is the 4000Å break?
(ii) Cannot explain the polarization.

Young Stars.

An objections to this model is that it cannot explain the polarization.

As I warned you at the beginning of my talk, the situation is in one big mess. In my view it is unlikely that any one of the proposed explanations, by itself, can account for the UV/optical/IR continuum or the alignment effect. If I were a betting man, I would back the young star model **together with** one or both of the scattering hypotheses as being **jointly** responsible.

THE ENVIRONMENT

There are several recent results which provide intriguing information about the region environment of z > 2 radio galaxies. I wish to devote the remainder of my talk to discussing these.

Companions and Clusters.

A few years ago I was asked to speak on interactions at high redshifts at a symposium in honour of Aller Toomre. I said that I felt like an atheist who had been asked to preach at a Christian revival meeting. But since then I have been born again! The reason is that a large number of high-redshift radio galaxies appear to be double, with the axis of the optical double preferentially oriented along the radio axes. Examples are 3C 65, 68.2, 194, 356, 368, TX 0828 +193 and 1436+ 157. A statistical analysis of this effect was made by Röttgering (1993), who considered 31 high-redshift radio galaxies with radio sizes of between 4" and 20". He found that the tendency of "companions" to lie close to the radio axes is significant at the 5% level. Moreover, of the two high-redshift galaxies observed in the HST pilot study, one of them 4C 28.58 is double on the sub-kiloparsec scale.

How can the doubleness of the optical images be explained? One possibility is that we are witnessing galaxy merging. An initial scenario proposed to explain the alignment effect postulated that the axis of the radio source was oriented along the direction of galaxy-galaxy interactions (Djorgovski et al. 1987), but no good mechanism was proposed to explain why this should be so. However, a connection between the radio axis and the merger axis is unnecessary. Suppose that mergers were common during the z=2 epoch and that either the starburst or scattering mechanism for enhancing luminosity occurs. When the merging galaxies are located along the beam, their luminosity would be enhanced compared with other merging galaxies and their fluxes become comparable with the parent radio galaxies.

According to such a composite starburst/merger picture a gas-rich companion is in the late stages of merging with the host galaxy of 4C 28.58. The jet responsible for the outer radio emission encounters the nuclear region of this companion, located about 3 kpc from the nucleus of the main galaxy. Shocks from the jet compress the gas, triggering an enormous episode of star formation. Mechanisms similar to those proposed to be occurring in the host galaxy by e.g. Rees (1989) and Begelman and Cioffi (1989) would operate in the companion. Scattering by dust and/or gas in the companion galaxy could also contribute to the observed flux.

Some evidence that the highest redshift galaxies are indeed surrounded by a cluster or group of less luminous galaxies is provided by the Keck picture of 4C 41.17 (Graham et al. 1993). A speculative person might conjecture on this basis that high-redshift radio galaxies are the progenitors of cD galaxies at the centres of still-forming rich clusters.

Ly α Absorption.

For the last 25 years one of the most important methods for investigating the early Universe has been to study quasar absorption lines. There are two limitations of such studies due to the fact that the light being absorbed originates in quasars. First, the quasars are unresolved, so in general no information can be obtained about the spatial scale of the absorbers. Secondly, the absorption lines having $z_{em} \sim z_{abs}$, i.e. those close to the quasar, are attenuated (presumably ionized) due to their proximity to the QSO. Absorption studies of radio galaxies would be hampered by neither of these effects. We (Hunstead, Röttgering, Miley, van Ojik and Wieringa) therefore recently instigated a project to search for absorption lines in high-redshift radio galaxies. Our initial targets were selected from galaxies having the brightest Ly α. It is in the region of this strong line that we would have the most chance of seeing absorption. Evidence for possible absorption in 4C 41.17 has been given by Hippelein and Meisenheimer (1993) and Klaus Meisenheimer will talk about that in the next presentation.

We have observed two objects with the AAT at a resolution of 1.4Å(FWHM) and are excited about the results. There appears to be several absorption features in both targets. In 0943-242, which is the subject of a poster paper in this symposium, we see a sharp deep absorption trough that reaches zero intensity and has a half-width of more than 400 km/s. The probability of obtaining such a system by a chance on the basis of the statistics of quasar absorption lines is less than 1 in 1000. The absorbers must therefore be intimately connected with the galaxy itself, or the environment of the galaxy, e.g. a protocluster. The depth of the absorption trough implies that the absorbing cloud at least covers the complete Lyα emission region (> 30 kpc). Although the exact column density involved is dependant on the number of absorbing clouds involved, the value will lie between 10^{18} and 10^{20} cm^{-2}, implying a total mass in HI of approximately 10^7 $M_{\odot}$. There appears to be a velocity shift along the slit of roughly 10 km/s which, if interpreted as rotation would imply a dynamic mass of 10^9 $M_{\odot}$. A radio jet impacting on an object of such a mass could easily be deflected (van Breugel et al. 1984). It is tempting to speculate that the culprit responsible for the Lyα absorption is similar to one of the companion objects visible in the Keck picture of 4C 41.17. This first high-resolution absorption study is certainly enough to whet the appetite for more!

CONCLUSIONS AND THE FUTURE

Taking all the various observations mentioned above, I believe that a little imagination coupled with a whiff of speculation and might lead a reasonable astronomer to the following conclusions:

(i) High-redshift radio galaxies are the progenitors of giant elliptical galaxies, located near the centres of protoclusters, maybe at the centres of giant cooling flows.

(ii) They are undergoing vigorous star formation which is producing dust.

(iii) The radio/optical alignments are produced by a combination of jet-induced star formation and scattering of light from a hidden quasar, both in the parent galaxy and in nearby companions.

(iv) The galaxies are located in highly inhomogeneous regions of the $z = 2$ Universe.

High-redshift radio galaxies will provide one of the most important tools for studying the early Universe, using the present and next generations of large telescopes. Their scales are particularly well-suited to study with the HST. Selecting objects with extreme radio spectra from the large-sky surveys at present underway with the Westerbork, VLA and Molonglo telescope will provide a rich harvest for many decades in the future.

ACKNOWLEDGMENTS

I thank my collaborators, particularly Huub Röttgering, Richard Hunstead and Rob van Ojik for many useful chats. My travel has been funded by the Mount Stromlo Symposium organization and by the Kerkhoven Bosscha Fonds.

REFERENCES

Barthel, P.D., and Miley, G.K., 1988, Nature, 333, 319.

Begelman, M.C., and Cioffi, D.F., 1989, Ap.J. (Letters), 345, L21.

Bithell, M. and Rees, M.J., 1990, Mon. Not. Roy. Astro. Soc. 242, 570.

Bogers, W.J., Barthel, P.D., Lonsdale, C.J. and Miley, G.K., 1993, in preparation.

Chambers, Miley and van Breugel, 1987, Nature, 329, 604.

Chambers, K.C., Miley, G. K., and Joyce, R.R., 1988, Ap. J. 329, L75.

Chambers,K.C. Miley,G.K. and van Breugel,W.J.M., 1990, Ap. J. 363, 21.

Chambers, K.C. and Charlot, S., 1990, Ap. J. (Letters) 348, L1.

Cimatti, A., di Serego Alighieri, S., Fosbury, R.A.E., Salvati, M. and Taylor, D., 1993, Mon. Not. Roy Astro. Soc., in press.

Crawford, C.S., and Fabian, A.C., 1993, Mon. Not. Roy. Astro. Soc., in press.

Daly, R., 1992, Ap. J. 399, 426.

De Young, D. S., 1989, Ap. J. (Letters), 342, L59.

De Young, D. S., 1981, Nature 293, 43.

De Young, D. S., 1981, Nature 293, 43.

Djorgovski, S., 1987, in Proc. Santa Cruz Astrophysics Workshop on "Nearly Normal
Galaxies" ed. S. Faber, Springer Verlag.

Eales, S.A., and Rawlings, S., 1990, Mon.Not.Roy.Astro.Soc., 243, 1p.

Eales, S.A., 1992, Ap. J., 397, 49.

Eales, S. A. and Rawlings, S., 1993, Ap. J., in press.
Eisenhardt,' and Chokshi, A., 1990, Ap.J. (Letters), 351, L9.

Fabian, 1989, M. N. R. A. S., 243, 1p.

Graham, J.,et al., 1993, Ap.J., in press.

Hammer, F., Le Fèvre, O., and Angonin, M.C., 1993, Nature, 362, 324.

Hippelein, H. and Meisenheimer, K. 1993, Nature 362, 224.

Hunstead, R., Röttgering, H., Miley, G. K., van Ojik, R., and Wieringa, M., 1993, Poster Paper, These Proceedings.

Lilly, S. J., Longair, M. S. 1984, Monthly Notices Royal Astron. Soc. 211, 833

McCarthy, P.J, van Breugel, W.J.M., Spinrad,H., & Djorgovski, S. 1987, Ap. J., (Letters), 321, L29.

McCarthy, P.J., 1993, Ann. Rev. Astron. Astrophys., in press.

Miley, G. K., Chambers, K. C., van Breugel, W. and Macchetto, D., 1992, Ap. J. (Lett), 40, L69.

Miley, G. K. 1993, Distant Galaxies in Proc. ST-ECF / STScI Workshop "Science with the Space Telescope" eds. P. Benevenuti and E. Schreier, ESO Conf. and Workshop Proc. No. 44, p1.

Rees, M.J., 1989, M. N. R. A. S., 239, 1p.

Rigler, M.A., Lilly, S.J., Stockton, A., Hammer, F., and Le Fèvre, O.,1992, Ap. J., 385, 61.

Röttgering, H., 1993, Ph.D Thesis, University of Leiden.

Röttgering, H., Miley, G.K., van Ojik, R., 1993, in preparation.

Tadhunter, C.N., Fosbury, R.A.E., and di Sergio Aligheri, S., 1989, "Proc. of the Como Conference on BL Lac Objects: 10 Years After", ed. L. Marasci

van Breugel, W.J.M., Miley, G. K., Heckman, T., Butcher, H. and Bridle, A., 1985, Ap. J. 290, 496.

van Ojik, R., Miley, G.K., Röttgering, 1993, Poster Paper, These Proceedings.

The First Stromlo Symposium: The Physics of Active Galaxies
ASP Conference Series, Vol. 54, 1994
G. V. Bicknell, M. A. Dopita, and P. J. Quinn (eds.)

A CLOSE LOOK AT HIGH-REDSHIFT RADIO GALAXIES

KLAUS MEISENHEIMER, HANS HIPPELEIN AND MARK NEESER
Planck-Institute für Astronomie Königstuhl 17, D-69117 Heidelberg, Germany

ABSTRACT We report the first results of an ongoing investigation of the extended emission line regions (EELRs) and continuum light distributions around more than a dozen high redshift galaxies. Special emphasis is placed on a detailed study of the kinematics of the EELRs and their relation to the radio source, through the utilisation of an imaging Fabry-Pérot-spectrometer. Our results show that the EELRs do not form a homogeneous class of objects. An expanding radio source, violent galaxy-galaxy interactions or a combination of both are all viable mechanisms for exiting an EELR. At $z < 1.2$ we find little support for the reported alignment of the optical continuum light distribution with the radio axis. Previous statistical evidence for this "alignment effect" seems to be strongly biased by the aligned EELRs whose line emission leaks into most broad-band filters. At the highest redshifts, our observations of 4C 41.17 ($z = 3.8$) indicate that the formation of a galactic core – marked by star formation and a radio-loud AGN – could be followed by ongoing gravitational collapse.

INTRODUCTION

As emphasized by George Miley in the previous talk, many of the strong radio galaxies at intermediate ($0.3 < z < 1.2$) and high redshift ($1.6 < z < 4$) are surrounded by huge emission line regions which in many cases are aligned with the axis of the radio source. In addition there is some evidence that the distribution of the continuum light from these galaxies is also correlated with the radio axis. In order to understand the origin of this "alignment effect" we investigated the detailed structure and kinematics of the extended emission line regions (EELRs) around a dozen radio galaxies between redshifts of 0.4 and 4. This is supplemented by a careful study of the optical continuum morphologies and their relation to both the radio sources and the EELRs.

Specifically we aim to answer the following questions concerning the origin of the EELR in each object:

Morphology: How is the EELR associated with the radio source and what is its exact shape and three-dimensional structure?

Kinematic: Are there systematic motions with respect to the (radio) core (outflow, infall, rotation, red and blue shift asymmetries)? What are the internal motions (e.g. expansion, turbulence) and are there indications for interactions between the radio source and the EELR?

Excitation: Are the EELRs excited by young stars, a central AGN, or shocks caused by the interaction between the expanding radio jet and the surrounding medium?

Our study of the continuum light distribution aims to answer an analogous set of questions: What is the continuum morphology and its relation to the radio source in individual objects? Can one detect spectroscopic evidence that the extended and aligned optical continuum (emitted in the UV) is starlight or scattered light from an obscured AGN or some other (non-thermal) source?

This approach differs from the more general, statistical studies carried out previously (McCarthy *et al.* 1987, Chambers *et al.* 1987, Rigler *et al.* 1992) because we try to disentangle the specific contributions to the overall EELR and continuum light distribution for each individual source separately. When the various constituents have been isolated we will be able to judge which components tend to bias the overall statistics towards the apparent "alignment effects". This talk is a progress report based upon analysis of about 1/3 of our data. Since space is very limited and the kinematic data show up poorly in black and white I can only display a tiny fraction of the maps presented in Canberra.

OBSERVATIONAL METHODS

We investigated the morphology and kinematics of the EELRs by employing an imaging Fabry-Perot-interferometer (FPI) in the focal reducer of the 3.5m telescope on Calar Alto (Meisenheimer & Hippelein 1992). Its spectral resolution has been optimized for this kind of study: $350\,\mathrm{km\,s^{-1}} < \Delta\lambda < 700\,\mathrm{kms^{-1}}$. With an accessible wavelength range of $440\,\mathrm{nm} < \lambda < 830\,\mathrm{nm}$ we can measure *e.g.* the $[OII]$372.7nm line at redshifts $0.2 < z < 1.2$ and Ly$-\alpha$ at $2.6 < z < 5.8$. The field of view covers $5'$ (until 1991) and $8'$ (currently).

The continuum light distribution is imaged through the same instrument by removing the FPI and using medium band interface filters ($\lambda/\Delta\lambda \cong 40$) which are selected for each individual source in such a way that contamination by emission lines is avoided.

After convolving each data set ($\gtrsim 10$ wavelength settings with the FPI and a blue, $\lambda_0 \cong 320\,\mathrm{nm}$, and red, $\lambda_0 \cong 430\,\mathrm{nm}$ continuum image) to a common beam we typically reach a spatial resolution of $1''.3$ to $1''.8$. The relative registration of the optical images has to be accurate to $\lesssim 0''.04$, the astrometric alignment with respect to the radio maps is accurate to $< 0''.5$.

RESULTS FOR INDIVIDUAL SOURCES

Table I provides an overview of the radio galaxies which have been observed and the completeness of our data. Our sample contains a dozen strong 3CR sources at $0.4 < z < 1.2$, most of which have been selected by having the most extended [OII] emission (McCarthy 1988). We also observed the first two identified radio galaxies at $z > 3$ which are surrounded by huge Ly-α halos. Only six data sets have been fully analyzed (marked by an asterisk in table I).

Our results on 3C 368 and 3C 352 have been published (Meisenheimer & Hippelein 1992, Hippelein & Meisenheimer 1992). Therefore, we will only briefly

TABLE I Observed sample.

	Source	z	completeness: FPI data	continuum
*	3C 34	0.689	yes	b,r
	3C 44	0.660	2/3	b,r
	3C 54	0.827	1/2	-
	3C 124	1.08	2/3	b
*	3C 265	0.811	yes	b
	3C 295	0.461	yes	b,r,...
	3C 337	0.635	yes	r
*	3C 352	0.806	yes	b
	3C 356	1.078	yes	b,r
*	3C 368	1.131	yes	b
*	3C 435A	0.471	yes	r
	3C 441	0.707	yes	b,r,...
*	4C 41.17	3.80	yes	r
	B2 0902+34	3.39	yes	r

summarize our study of 3C368 which contains the prototype for a well-aligned EELR (Fig. I) and may thus serve as reference for the discussion of other galaxies.

The double-lobed EELR of 3C 368 is not only perfectly aligned with the FR II radio source but also shows a similar north to south arm-length ratio. Together with the clear and undisturbed appearance of a blue-shifted southern and red-shifted northern lobe this is the most impressive evidence for a close correlation between a radio source and an EELR. A careful analysis of the southern EL lobe reveals a conical and expanding line emitting volume which lags about 10kpc behind the (presumed) bow shock of the southern radio hot spot. Interpreting this lag as cooling length through which bow shock-heated external material must pass before its densest parts reach some 10^4 K one is able to model the observations very well[1]. The model requires a bow shock velocity of 1200 kms^{-1} and an external density of $10^5 m^{-3}$. A mild north to south density asymmetry (0.8 : 1) is sufficient to account for the observed arm length ratio.

Our "line-free" continuum images do not confirm the well-aligned continuum morphology seen on broad band images (Djorgovski et al. 1987, Le Fevre et al. 1988): There is a slightly elongated central source which is severely confused by a foreground M star and three barely resolved objects around the northern lobe, the faintest of which may coincide with the northern hot spot. Obviously most of the aligned light distribution has been caused by emission lines contaminating the broadband filters. Since the southern emission line region seems to

[1]The excitation has to be provided by fast shocks which are driven by the supersonic cocoon into the surrounding EELR.

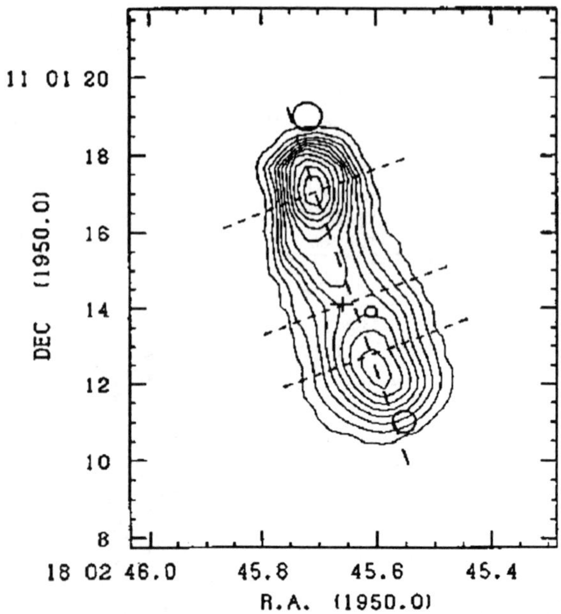

FIGURE I Contour map of the [OII] emission around 3C 368 (linearly spaced contours). The core and hot spots of the radio source are indicated by circles, a cross marks the position of the foreground star.

intersect our line-of-sight towards the core the excitation of the EELR by beamed radiation from a strong AGN core is virtually ruled out. In addition, the good correlation between the amount of linear polarization (Tadhunter et al. 1991) and the [OII] line equivalent width does not support the idea that scattered light from an aligned radiation cone produces an aligned optical *continuum.*

With the perfect radio-EELR alignment of 3C 368 in mind all the subsequently observed radio galaxies proved somewhat disappointing: 3C 352 shows a clear red-blue asymmetry in its kinematic structure, as well as the lag of the EELR boundary behind the radio hot spots. The emission line "lobes" of this source are confused, however, by a huge (> 100 kpc) emission line cloud which has no obvious connection to the radio source. Moreover, the central intensity maximum of the EELR does not fit into a purely radio source generated picture but seems to require an additional nuclear component. Thus it is impossible to separate the radio jet-triggered part of the EELR.

In 3C 435 A (Fig. II) we are faced with another extreme: although there exists a global alignment between EELR and radio axis, our imaging spectroscopy and the comparison with the continuum morphology provides little evidence for the radio source being the prime mover of the EELR. On the contrary, both the morphology and the absence of systematic motions argue in favor of a pure "interaction scenario" in which the fly-by of the northern companion galaxy has caused most of the emission line excitation. This is supported by the existence of a northern "spur", most naturally explained as tidal tail. However, this scenario fails to explain the alignment between the core-companion orientation and the radio axis (cf. Fig. II, left panel). In 3C 265 (a huge radio source, $D_{rad} = 700$ kpc with a very extended ELR, $D_{ELR} > 250$ kpc) and 3C 34 we envisage a "mixed" situation. Although the central part of both EELRs seems to be caused by violent galaxy–galaxy interactions (as indicated by the association with the companion galaxies and the confused kinematics) the very outer parts of their EELRs show good alignment with the radio axis and the red-blueshift asymmetry which characterizes the prototype 3C 368. However, the outermost EELRs lag so much behind the extent of the radio source ($D_{EL} < 1/2 D_{rad}$) that it would seem rather cavalier to identify them with cooling material behind the bow shock - although the low densities expected so far from the central mass concentration do indeed predict cooling lengths of 100 kpc or more. Anyway, the large distance between the hotspot – i.e. the current working surface of the radio jet – and the EELR is too large to draw further conclusions about the radio source expansion beyond the mere notation that the blue-shifted side is most likely that of the approaching radio lobe. The continuum images of 3C 265 and 3C 34 are dominated by several companion galaxies which show no obvious correlation with the radio source axis.

GENERAL RESULTS FOR RADIO GALAXIES

From the results of the five above-mentioned radio galaxies and the inspection of the other data sets (Table I) we can draw some preliminary conclusions about the general properties of the EELRs and continuum light distribution:

Taxonomy of the EELRs around radio galaxies ($z < 1.2$)

In our sample, we find 4 types of EELRs':

I. Large, well-aligned EELRs (e.g. in 3C 368, 3C 352, 3C 34). They are characterized by large ($D > 30$ kpc) emission line lobes clearly separated kinematically by their red and blueshift w.r.t the mean velocity ($\delta \gtrsim 500 \text{ km s}^{-1}$). In all cases the EELR exhibits large turbulent motions ($\Delta v \gtrsim 1000 \text{ km s}^{-1}$ FWHM). The maximum extent of the EELR is always less than that of the radio source[2], indicating that cooling behind the bow shock plays an important role in their formation. There exist both edge-brightened and core-brightened EELRs in this group.

II. Small, aligned EELRs (e.g. 3C 54). Although they show red- and blue-shifted emission line components, their diameters are too small to determine whether the observed material is infalling (backflow) or outflowing.

III. EELRs caused by galaxy interactions (e.g. 3C 435A, 3C 265, 3C 34). These EELRs are characterized by their complex kinematics which do not allow identification of a red or blueshifted lobe. Alignment with the radio axis is poor - if present at all. The EELRs correlate much better with the positions of companion galaxies, detected on the continuum images.

IV. Combined type (e.g. 3C 34, 3C 265). In several objects we find a mixture of a large, aligned EELR (type I) with a strong, complex EELR whose origin obviously lies in the violent interaction with a companion galaxy (type III). The combination of types II and III may also exist but is hard to detect since the interaction triggered part outshines the tiny type II EELR.

In our sample which is selected by the existence of a very large EELR ($\gtrsim 3$"), we find an equal fraction of type I and type III dominated emission line regions.

Is there an aligned continuum component?

In our entire sample of 13 radio galaxies at $0.4 < z < 1.2$ we find at most marginal evidence for the reported continuum alignment (McCarthy *et al.* 1987, Chambers *et al.* 1987). On our "line-free" continuum images all host galaxies look rather round with some fuzzy fringes ($D < 50$ kpc). Frequently we detect companions with emission lines at the redshift of the radio galaxy. Although in some cases they roughly line up with the radio axis there is no recognizable deviation from a random 2π–distribution. We conclude, therefore, that the strong "continuum alignment effect" reported in earlier studies has been mainly caused by emission lines from the aligned EELR leaking into the broad-band filters.

---

[2]Note that the northern "spur" of 3C 435A which has provided the chief evidence for line emission beyond the radio hotspot (van Breugel & McCarthy 1989) is most likely a tidal tail.

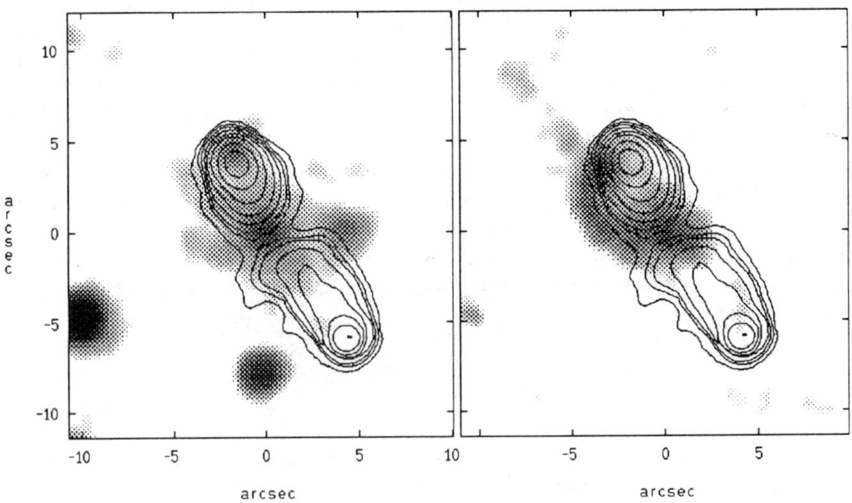

FIGURE II Greyscale images of the continuum light distribution (left) and [OII] emission around 3C 435 A with radio contours overlaid.

Excitation of the EELRs and the $L(EL) \sim L(rad)$ relation.

In several EELRs of type I (3C 368, 3C 352) we find convincing evidence that the gas is ionized by fast shocks (*cf.* Bicknell, Dopita, these proceedings) which are driven by the cocoon surrounding the radio jet. This excitation process provides a natural explanation for the observed proportionality between the radio luminosity $L(rad)$ and the (extended) line luminosity $L(EL)$. On the other hand, the large fraction of interaction triggered EELRs found in our sample and the occurrence of centrally enhanced EELRs, casts some doubt on the idea that the kinetic power of the radio jets alone controls both the non-thermal radio and the emission line luminosity.

4C41.17 - A RADIO GALAXY AT VERY HIGH REDSHIFT

So far, we have observed only those two radio galaxies at $z > 3$ with the highest (extended) Ly-α luminosities: B2 0902+34 ($z = 3.39$, Lilly 1989) and 4C 41.17 ($z = 3.80$, Chambers *et al.* 1991). Here we restrict ourselves to a discussion of 4C 41.17. In a recent letter to Nature (Hippelein & Meisenheimer 1993) we reported the detection of an absorbing hydrogen cloud in the line-of-sight towards the huge Ly-α EELR around 4C41.17 (Fig. III). In that letter we mainly discussed the possibility that the absorption could be caused by foreground absorbers ("Lyman forest"). A recent HST observation by George Miley *et al.* (1992) as well as our improved astrometry shows that both the optical continuum peak and the likely core of the radio source (B2 in Chambers *et al.* 1991) coincides with the location of the absorber. Thus the second alternative, in which the absorbing cloud belongs to the mass aggregation around 4C41.17 appears to be more favorable.

The blue shift of the absorption line w.r.t. the mean emission line redshift then indicates an infall of the emission line gas towards the dense absorbing HI cloud near the continuum peak. This interpretation is supported by the details of the kinematics measured with our imaging FPI: In addition to a large turbulent velocity spread, which well covers the relative velocity of the absorption (a necessary condition for its detection), we find that the central part of the EELR is somewhat redshifted w.r.t. its surroundings (Fig. III, see Hippelein & Meisenheimer 1994 for further details). We interpret this as being caused by ongoing accretion onto the center of mass of the 4C 41.17 complex, lying somewhat eastwards of the present continuum peak. In the densest parts of the accreted cloud, star formation has already set in and causes the continuum peak. The continuum elongation in the east-west direction (HST image!) may indicate that the surface of star formation proceeds eastwards from the "core" marked by both the Ly-α absorption and a sharp boundary in the continuum distribution.

In this scenario the apparent Ly-α absorption is caused by the densest parts of the accreted gas. Overall the ionization of hydrogen is still dominated by accretion shocks making the eastern half of the cloud most prominent in Ly-α. The "gap" in both the Ly-α and the HST image seems to be caused by gas

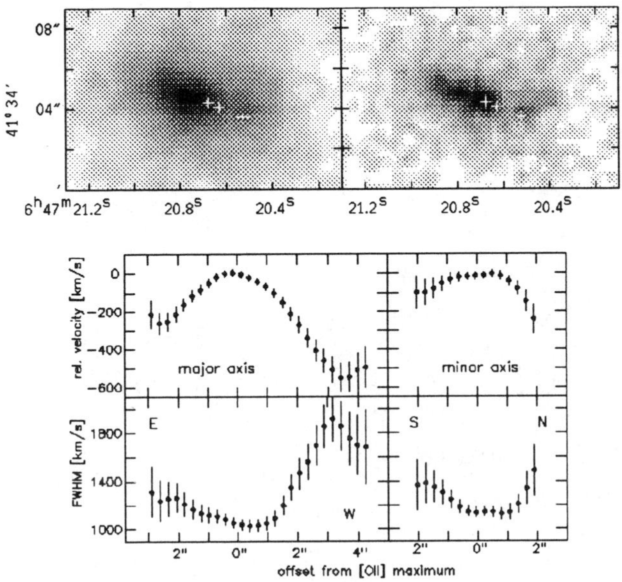

FIGURE III The Ly-α cloud around 4C 41.17 (upper left) with the absorbing hydrogen cloud detected on a monochromatic image (λ = 583 nm, upper right). Crosses mark the triple radio source B1,B2,B3. In the lower panels the kinematics across the peak of the emission line cloud is shown.

which has already some dust mixed in. If the sketched scenario is correct it would imply that the completed formation of a galactic core could be accompanied by an ongoing gravitational collapse in the outskirts of the cloud. Even if the first burst of star formation produces heavy elements (and as a consequence: dust) very rapidly, the delayed outer collapse continues to generate a huge amount of (unobscured) Ly-α luminosity.

REFERENCES

Chambers, K.C., Miley, G.K. & van Breugel, W.J.M. 1987, Nature **329**, 604

Chambers, K.C., Miley, G.K. & van Breugel, W.J.M. 1990, *ApJ* **363**, 21

Djorgovski, S. *et al.* 1987, *AJ* **93**, 1307

Hippelein, H. & Meisenheimer, K. 1992, *A&A* **264**, 472

Hippelein, H. & Meisenheimer, K. 1993, Nature **362**, 224

Hippelein, H. and Meisenheimer K. 1994 *A&A* in preparation.

Le Fèvre, O., Hammer, F. and Jones, J. 1988, *ApJ* **331**, L73

Lilly, S.J. 1988, *ApJ* **333**, 161

McCarthy, P.J. 1988, *PhD Thesis*, Berkeley.

McCarthy, P.J. *et al.* 1987, *ApJ* **321**, L29

van Breugel, W.J.M. & McCarthy, P.J. 1989, in E. Meurs & R. Fosbury (eds.): *Extranuclear Activity in Galaxies*, ESO proceedings No. 32, p. 227.

Meisenheimer, K. & Hippelein, H. 1992, *A&A* **264**, 455

Miley, G.K., *et al.* 1992, *ApJ* **401**, L69

Rigler, M.A. *et al.* 1992, *ApJ* **358**, 61

Tadhunter *et al.* 1992, *MNRAS* **256**, 53p

STARBURSTS, MERGERS & ACTIVE GALAXIES

Invited Papers

The First Stromlo Symposium: The Physics of Active Galaxies
ASP Conference Series, Vol. 54, 1994
G. V. Bicknell, M. A. Dopita, and P. J. Quinn (eds.)

STARBURST GALAXIES AT HIGH REDSHIFT

DANIEL W. WEEDMAN
Department of Astronomy and Astrophysics, The Pennsylvania State University, State College, PA 16802

ABSTRACT It is demonstrated that, in the absence of extensive dust obscuration, star forming galaxies should already be observable at high redshifts, while existing observations do not find them. Proceeding with the assumption that such galaxies are dusty, prototype objects are described and suggestions are made for an observational strategy to locate highly redshifted analogues with ground based observations. Choosing the redshift interval $2 < z < 2.5$, because so much is already known about quasars in this interval, it is demonstrated that starburst galaxies should be found there within samples of optically red objects with K band magnitudes about 20th magnitude and 20 cm radio fluxes of fractions of a mJy or less.

INTRODUCTION

While there are numerous things which I could describe that we don't understand about the relations among AGN, starbursts, and primordial galaxies, this presentation will focus on those things which we would like to know within the redshift interval $2 < z < 2.5$. I choose this because it remains the interval in which we know most about the quasar luminosity function; the reason is simply that the photographically based survey techniques find quasars most completely when their Ly α line is readily observable and of such observed wavelength to make the quasar appear blue in color. Consequently, a great deal of data are available on the luminosity function of quasars in this interval and the nature of quasar evolution from then until the present (Boyle, Shanks and Peterson 1988; Schmidt, Schneider and Gunn 1991; Hewett, Foltz and Chaffee 1993). This is, therefore, an obvious interval in which to consider other phenomena, such as starbursts and AGN, which relate to the quasars.

Unlike the situation for quasar surveys, there has been a notable lack of success in finding high redshift, star forming galaxies based upon their optical luminosities (corresponding to ultraviolet in the galaxy rest frames). Even though very faint, blue galaxies are well known, (e.g. Tyson 1988, review in Koo and Kron 1992), none have been found at redshifts $2 < z < 2.5$ in unbiased optical surveys. Suspected regions of star formation in this interval identified by optical characteristics have only been located by association with radio galaxies (Miley et al. 1992; Windhorst, Mathis and Keel 1992), or with intervening damped Ly α absorption lines from galaxy disks (Wolfe et al. 1992; Lowenthal et al. 1991), or with quasars at comparable redshifts (Steidel, Sargent and Dickinson 1991; Macchetto et al. 1993). Why isn't it easy to see high redshift, starburst galaxies? It is easy to show that, in the absence of severe dust absorption within the starburst, *such galaxies should be visible.*

Consider the luminosity function of local starbursts. The most complete luminosity function is that determined by IRAS based upon the infrared reradiation of starlight by dust. From comparisons of this infrared luminosity function to star forming galaxies found in other ways, it is clear that the majority of star formation in local galaxies is obscured by dust. This is even the case for those starburst episodes that were detected in the Markarian survey, which deliberately selected on the basis of visibly blue galaxies. Even within the Markarian galaxies, it seems that only 10% of the ultraviolet luminosity from hot stars escapes (Weedman 1991). Even though dusty starbursts are not easily visible in the rest frame ultraviolet, we can estimate how they would appear if the dust weren't there.

Starting with the luminosity function in the IRAS bright galaxy sample (Soifer et al. 1987), we know the bolometric luminosity function of these star forming galaxies because virtually all of the starlight is reradiated in the IRAS wavelengths. Soifer et al. give a luminosity function in units of L_{bol}. Furthermore, it is straightforward to estimate that, for a wide range of star forming initial mass functions, the bolometric luminosity of the collection of hot stars is approximately 1000 times the monochromatic luminosity L_λ at 128.5 nm (Kunth and Weedman 1987). This means that we can "clean" the IRAS luminosity function to determine what the ultraviolet luminosity function would be if there were no dust to absorb the ultraviolet in star forming regions. I wish for the current exercise to determine the luminosity function at a rest wavelength of 215 nm. This is chosen as being the rest frame wavelength observed in R band for a starburst galaxy at z = 2.25, in the center of the interval of interest. Taking the translation from 128.5 nm to 215 nm from the median starburst spectra in the IUE atlas of Kinney et al. (1993) - "SB" and "BCDG" in their entries - yields $L_\lambda(215)/L_\lambda(128.5) = 0.46$. Combining with the infrared-derived L_{bol} and transforming into frequency units, $L_\nu(215) = 7.1 \times 10^{-16} L_{bol}$ for $L_\nu(215)$ the rest frame luminosity in ergs s^{-1} Hz^{-1} at a frequency corresponding to 215 nm and L_{bol} in ergs s^{-1}.

The next step is to estimate how many galaxies within this assumed ultraviolet luminosity function should be seen within $2 < z < 2.5$. An R mag. observing limit of 25 is adopted as that is reasonable for the faint galaxy surveys available. Using the appropriate cosmological relations (for $H_o = 75$ km s^{-1} Mpc^{-1} and $\Omega = 0.2$), I find that there should be 7000 blue galaxies deg^{-2} within $2 < z < 2.5$, even with no evolution of luminosity or density. How the starburst galaxy luminosity function actually evolves is an issue currently receiving intense scrutiny; for now, the best estimate is that parameterizing the evolution by pure luminosity evolution would indicate that luminosity of a starburst galaxy increases with $(1 + z)^2$ (Lonsdale and Chokshi 1993). Adopting this form of evolution would predict 60,000 blue galaxies deg^{-2} to R = 25 within $2 < z < 2.5$. Observed R counts of faint galaxies yield 10^5 deg^{-2} (Koo and Kron 1992), so the high redshift ones should consist of 7% to 58% of the total. They are not there. Unless we wish to conclude that star formation in the early universe was less common than at present, the only alternative explanation is that these starbursts at high redshift are dusty and so are not easily seen via their ultraviolet light.

For the remainder of my presentation, I will assume that the starbursts are there and are dusty, so the purpose of what follows is to suggest ways of finding them anyway.

THE PROTOTYPE DUSTY STARBURSTS

It is first necessary to describe the prototypical objects that we are searching for. Most fundamentally, these objects are large, highly concentrated masses of gas and dust within which luminosity is being produced. Wrestling with the issue of how to determine whether a shrouded source is a starburst or an AGN is an ongoing theme at this Workshop (as is the issue of whether there is a meaningful difference between the two things since they are often in intimate association). My chosen prototypes of shrouded starbursts are dusty, luminous sources which have nearby regions of vigorous star formation in which the unobscured blue stars can be seen and which also have very large amounts of molecular gas associated with the luminous region. In such a case, even if one argues that the shrouded source itself is not powered by starlight, its presence is nevertheless an indicator that a starburst is ongoing.

The prototypical examples I propose for such things are the nucleus of the interacting system IC 694/NGC 3690 (Markarian 171, Arp 299 - with the specific component of interest being labelled in what follows as NGC 3690A) and the nucleus of Arp 220. These two are the best examples because data at many wavelengths exist for the small, nearly unresolved, shrouded component, and these fluxes are used to yield the luminosity as a function of frequency shown in Figure 1. (All luminosities are derived assuming $H_o = 75$ km s^{-1} Mpc^{-1} and $\Omega = 0.2$, and luminosities are shown in the rest frame of the source; references for fluxes are in the caption.) Note the close similarity in the luminosity distributions of these two objects.

At radio frequencies, these prototypes illustrate the confusion between attempting to decide what is purely a starburst and what also harbors an AGN. Both are compact radio sources, marginally resolved at sub-arcsecond scales, and so interpreted as very compact starbursts (Condon et al. 1991). Yet, both also harbor faint sources unresolved at VLBI scales with brightness temperatures > 10^{7o} (Lonsdale, Smith and Lonsdale 1993). It is suggested that the latter sources could be young supernova remnants, which would be a consequence of the starburst, but they could also be interpreted as low-luminosity AGN within the starburst knot. (Note that component NGC 3690D, the radio source having these properties as labelled in Condon et al. and Lonsdale et al., is the same as component NGC 3690A defined by Gehrz et al. and discussed herein.)

As a prototypical example of a shrouded AGN, I choose the nucleus of Cygnus A, and its luminosity distribution is also shown in Figure 1. Except for the much greater radio luminosity, it is also similar to the prototypes chosen as starbursts. For a final comparison to these condensed, luminous, dusty objects, the highly redshifted object IRAS 10214+4724 is shown. Whether the luminosity of this famous object arises primarily from a starburst is controversial (Lawrence et al. 1993), but it is the most dramatic example now known of a huge mass of molecular gas being converted into something that produces light (e.g. Solomon,

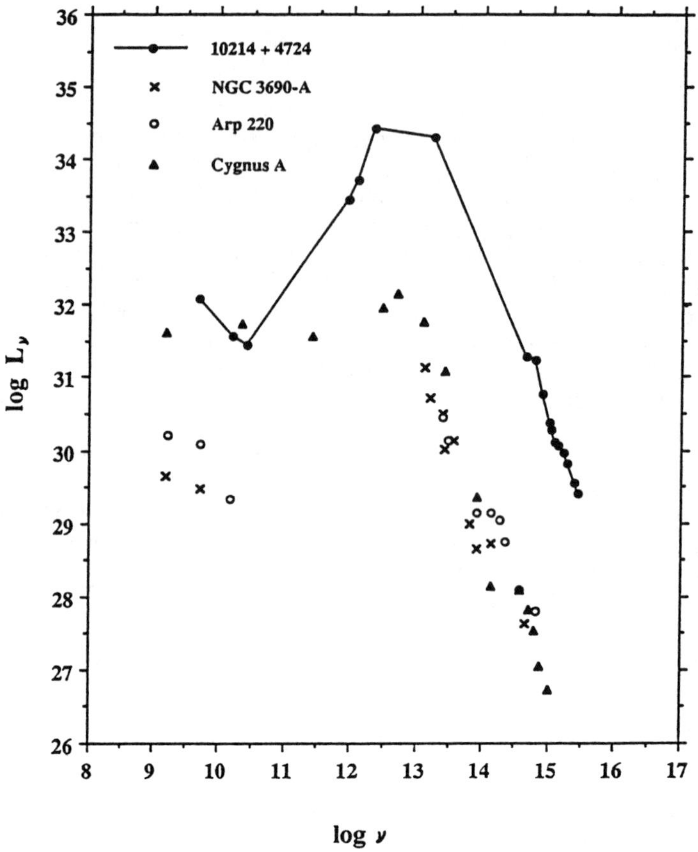

Figure 1: Spectra in the rest frames of the sources for the IRAS source 10214+4724, the dusty starburst knots NGC 3690A and Arp 220, and the dusty AGN Cygnus A. Luminosities are derived assuming $H_0 = 75$ km s^{-1} Mpc^{-1} and $\Omega = 0.2$ using fluxes from these references: a. For 10214+4724, from the compilation in Rowan-Robinson et al. (1993); b. For NGC 3690A, from the fluxes for component A in Gehrz, Sramek and Weedman (1983) with the highest frequency point for the continuum deduced using the Hα equivalent width and flux for component A from Armus, Heckman and Miley (1989); c. For Arp 220, from the fluxes for the nuclear component only (5" beam) in Sanders et al. (1988) and from radio fluxes in Norris (1988); d. For Cygnus A, from the compilation of fluxes for the nucleus in Djorgovski et al. (1991). IRAS-derived far-infrared luminosities are not shown for NGC 3690A and Arp 220 because both objects are large and extended infrared sources, and the IRAS beam is much larger than the knots described, but the IRAS results are included for Cygnus A because the infrared source seems localized to the nucleus.

Downes and Radford 1992). The most interesting point of this comparison is that the spectral shape of 10214+4724 closely resembles the prototype starburst cores, except that 10214 is about 1000 times more luminous at all wavebands.

OBSERVABLE PROPERTIES OF DUSTY STARBURSTS AT HIGH REDSHIFTS

In order to understand dusty starbursts at high redshifts, and how this phenomenon relates to the quasars already known at such redshifts, we must find more examples. What is the optimum search strategy for seeking out things with luminosities intermediate between 10214 and 3690A at $2 < z < 2.5$? What one really wants are more surveys in far infrared wavelengths; 10214 was barely detected by IRAS and only at 60μm so less luminous examples at comparable redshifts should not exist in the IRAS database. The next opportunity for far infrared surveys will be with ISO and then SIRTF, although neither will provide all-sky surveys. In the meantime, we have to try with ground based facilities to locate similar objects, which means using the optical, near-infrared, and radio fluxes as selection criteria.

Obviously, the sources of interest are very red. The spectra in Figure 1 are in the source rest frame; in the observer's frame, observed R band arises at 215 nm in the source and observed K band arises at 677 nm for $z = 2.25$. Looking at these wavelengths in Figure 1, the observed flux ratio $\log [f_\nu(R)/f_\nu(K)] = 1.4$. Using the flux calibration that at 15th mag., $f_\nu(R) = 3.08$ mJy and $f_\nu(K) = 0.67$ mJy, the color in magnitudes (R-K) = 4.3. Initial selection of a highly redshifted dusty starburst requires, therefore, that an object be this red. Even though such red galaxies are already known (e.g. Elston, Rieke and Rieke 1988), they are generally thought to be distant ellipticals, red because of red stars.

The actual value of achievable magnitude limits is such that substantially less luminous objects than 10214 should exist at comparable redshifts in deep K band samples (10214 has K magnitude of 16). An object just like NGC 3690A or Arp 220 at $z = 2.25$ would have $K = 25$ and $R > 29$, so we would not expect to find similar things with available observing techniques. Because a magnitude of $K = 20$ is readily achievable, however, detecting objects with approximately 1% the luminosity of 10214 is quite feasible.

If the $(1 + z)^2$ luminosity evolution deduced from local IRAS starburst galaxies continues to high redshift, the analogues of NGC 3690A or Arp 220 at $z = 2.25$ would have K magnitude approximately 22. Because such objects even then would be 250 times less luminous than 10214, many of them should exist at comparable redshifts. At the bright end of the starburst galaxy luminosity function (Soifer et al. 1987), the number of objects increases by a factor of 10^5 for a factor of 250 decrease in luminosity. Even if there is only one object in the entire sky like 10214, this means we can expect another 99, 999 waiting to be found at the limits of ground based telescopes.

Getting on with it then, consider another selection criterion that can help to find dusty starbursts. This is the ratio of radio to near-infrared flux. Once again, note the similarity among NGC 3690A, Arp 220, and 10214. For all three sources, $1 < \log [f_\nu(20 \text{ cm})/f_\nu(K)] < 2$, in the observer's frame for $2 < z < 2.5$. This means that such a source at $K = 20$ would have $0.07 \text{ mJy} < f_\nu(20 \text{ cm}) < 0.7$

mJy. While faint, the VLA detection of such a source is possible. Faint radio sources are common, but it is known that the radio source population below 1 mJy is dominated by faint, blue galaxies at redshifts < 1 (Windhorst, Fomalont and Partridge 1993). This means that faint, red radio sources - of the kind we are seeking - are not common and might thereby be distinctive.

The radio spectra also provide a selection constraint. The radio spectra of starbursts have $-0.3 > \alpha > -0.9$, for $f_\nu \propto \nu^\alpha$, with the shape of the power law determined by the proportion of thermal relative to non-thermal radiation, the latter from supernova remnants (Sramek and Weedman 1986, Condon and Yin 1990). The flatter spectra tend to be associated with the denser knots, with the spectra being steeper in more extended regions (Stine 1992). Some starburst knots are so dense that an intrinsically steeper, primarily non-thermal spectrum can also be flattened by free-free absorption (Condon et al. 1991). The presence of a flat spectrum, compact source does not necessarily mean, therefore, that the synchrotron self-absorption characteristic of an AGN is present. Empirically, the spectra in Figure 1 indicate that in the observer's frame at 6 cm and 20 cm for a high redshift source, the spectrum is steep, as seen for both Arp 220 and 10214. The faint red radio sources we are seeking could be expected, therefore, to be sources which are compact - of arcsecond scale - having nominally steep spectra ($-0.3 > \alpha > -1.0$).

SUMMARY

Even though far-infrared imagery, the most important search technique for new examples of highly redshifted starbursts, is not currently available, seemingly viable suggestions are made for how such objects might be found with ground based observations alone. The detection of only a few examples at high redshifts of objects similar in spectra and a factor of 100 more luminous than the prototype cores of NGC 3690A and Arp 220 would greatly enhance our understanding of the luminosity function of star forming galaxies in the crucial early phases of the universe. Such detections would confirm that previous searches for primordial galaxies have failed because these galaxies are dusty. Knowing how many starburst galaxies accompany quasars in the early universe is crucial to understanding the relationship among these galaxies, quasars, and the progenitors of galaxies now present locally.

REFERENCES

Armus, L., Heckman, T.M., and Miley, G.K. 1989, *ApJ*, **347**, 727
Boyle, B.J., Shanks, T., and Peterson, B.A. 1988, *MNRAS*, **238**, 957
Condon, J.J., and Yin, Q.F. 1990, *ApJ*, **357**, 97
Condon, J.J., Huang, Z.-P., Yin, Q.F., and Thuan, T.X. 1991, *ApJ*, **378**, 65
Djorgovski, S., Weir, N., Matthews, K., and Graham, J.R. 1991, *ApJL*, **372**, L67
Elston, R., Rieke, G.H., and Rieke, M.J. 1988, *ApJL*, **331**, L77
Gehrz, R.D., Sramek, R.A., and Weedman, D.W. 1983, *ApJ*, **267**, 551
Hewett, P.C., Foltz, C.B., and Chaffee, F.H. 1993, *ApJL*, **406**, L43

Kinney, A.L., Bohlin, R.C., Calzetti, D., Panagia, N., and Wyse, R.F.G. 1993, *ApJS*, **86**, 5

Koo, D.C., and Kron, R.G. 1992, *ARAA*, **30**, 613

Kunth, D., and Weedman, D.W. 1987, in *Exploring the Universe with the IUE Satellite,* ed. Y. Kondo, (Dordrecht:Reidel), p623

Lawrence, A. et al. 1993, *MNRAS*, **260**, 28

Lonsdale, C.J., Smith, H.E., and Lonsdale, C.J. 1993, *ApJL*, **405**, L9

Lonsdale, C.J., and Chokshi, A. 1993, *AJ*, in press.

Lowenthal, J.D., Hogan, C.J., Green, R., Caulet, A., Woodgate, B., Brown, L., and Foltz, C. 1991, *ApJL*, **377**, L73

Macchetto, F., Lipari, S., Giavalisco, M., Turnshek, D.A., and Sparks, W.B. 1993, *ApJ*, **404**, 511

Miley, G.K., Chambers, K.C., van Breugel, W.J.M., Macchetto, F., Windhorst, R., Mathis, D.F., and Keel, W.C. 1992, *ApJL*, **400**, L1

Norris, R.P. 1988, MNRAS, **230**, 345

Rowan-Robinson, M. et al., 1993, *MNRAS*, **261**, 513

Sanders, D.B., Soifer, B.T., Elias, J.H., Madore, B.F., Matthews, K., Neugebauer, G., and Scoville, N.Z. 1988, *ApJ*, **325**, 74

Schmidt, M., Schneider, D.P., and Gunn, J.E. 1991, in *The Space Distribution of Quasars*, *ASP Conf. Series 21*, ed: D. Crampton, (ASP: San Francisco), p. 109

Soifer, B.T., Sanders, D.B., Madore, B.F., Neugebauer, G., Danielson, G.E., Elias, J.H., Lonsdale, C.J., and Rice, W.L. 1987, *ApJ*, **320**, 238

Solomon, P.M., Downes, D., and Radford, S.J.E. 1992, *ApJL*, **398**, L29

Sramek, R.A., and Weedman, D.W. 1986, *ApJ*, **302**, 640

Steidel, C.C., Sargent, W.L.W., and Dickinson, M. 1991, *AJ*, **101**, 1187

Stine, P.C. 1992, *ApJS*, **81**, 49

Tyson, J.A. 1988, *AJ*, **96**, 1

Weedman, D. 1991, in *Massive Stars in Starbursts*, ed: C. Leitherer, N.R. Walborn, T.M. Heckman, and C.A. Norman, (Cambridge: Cambridge U.P.), p. 317

Windhorst, R.A., Mathis, D.F., and Keel, W.C. 1992, *ApJL*, **400**, L1

Windhorst, R.A., Fomalont, E.B., and Partridge, R.B. 1993, *ApJ*, **405**, 498

Wolfe, A.M., Turnshek, D.A., Lazetta, K.M., and Oke, J.B. 1992, *ApJ*, **385**, 151

The First Stromlo Symposium: The Physics of Active Galaxies
ASP Conference Series, Vol. 54, 1994
G. V. Bicknell, M. A. Dopita, and P. J. Quinn (eds.)

ACTIVITY IN GALACTIC NUCLEI: STARBURSTS AND BLACK HOLES

JUDITH J. PERRY
Institute of Astronomy, University of Cambridge,
Madingley Road, Cambridge CB3 0HA, U.K.

ABSTRACT The abundances deduced from the observed broad lines in the spectra of active galactic nuclei (AGN) require that significant enrichment of nuclear interstellar gas precede—or be co-temporal with—the observed activity. Supernovae in a starburst stellar cluster can provide both the observed heavy elements *and* the fuel supply required to produce the observed broad-band luminosity. The symbiosis between a nuclear starburst stellar cluster and a central black hole appears able to explain most AGN observations. The mass lost during normal stellar evolution by the young dense cluster creates the nuclear interstellar medium (ISM), whose velocity field is described by a bipolar flow. The flow sustains the bolometric luminosity of the AGN because its accretion branch feds the radiating accretion disc of the BH; the accretion is accompanied by a supersonic axial outflow. The stellar mass-loss is fed into the nuclear ISM by stars whose Keplerian velocities are highly supersonic. The resulting shocks are radiative, leading to the formation of cool, line emitting gas condensations. Because of the coupling of the mass-energy source of the luminosity with the production of the line-emitting gas, the broad line equivalent widths are roughly constant over a wide range of luminosities, in agreement with observation. Similarly, the constancy of the ionization parameter of the lines is a direct consequence of the tight feed-back inherent in the hydrodynamics of cloud formation.

INTRODUCTION

The most striking feature of active galaxies—their enormous luminosity—strains 'normal' sources of 'fuel' and requires extreme conditions for its explanation, *whatever* model is invoked. The observed luminosity requires an equivalent rate of conversion of mass-energy of $\dot{M} = L_{\rm Bol}/\epsilon c^2$ (where ϵ, the efficiency of conversion of mass energy to radiation is normally taken as roughly 10% for accretion onto BHs). This requires

$$\dot{M} = 17.5 L_{47}/\epsilon_{0.1}\ M_{\odot}{\rm yr}^{-1} \qquad (1)$$

The problems finding such high rates of fuel supply have been the subject of discussion for years (Rees, 1984, Perry, 1993b). Neither normal stellar evolution in galaxies, nor accretion of the local ISM onto supermassive black holes can provide such high rates.

However, activity in galactic nuclei manifests itself not only through this 'output of energy unsustainable over the Hubble time', but also through the radiation of prominent broad emission lines from heavy elements (in all objects except some blazars). Therefore, *whatever* processes generate the broad-band continuum, it is clear that activity *cannot occur before* star formation and evolution through to supernovae, even at redshifts of 5. This fact, a direct result of the appearance of heavy element emission lines, is so basic as to often be ignored! Although abundances are difficult to determine precisely, abundances appear to be *at least* solar, and recent estimates by Hamann & Ferland (1993) range up to ten times solar. In contrast, analysis of the absorption line systems observed in most high redshift QSO spectra shows that the absorbing gas has abundances significantly lower than solar, with mean values—at redshifts of 2—of only one-tenth solar in the systems containing most of the baryonic mass—the damped Ly-α systems (Pettini, *et al.* 1993). Thus, at the same, and *later*, epochs as we observe highly enriched gas in the nuclear broad line region (BLR), the overwhelming proportion of diffuse gas in galaxies is highly under-enriched. These observed abundances are an important constraint on theories of AGN.

Historically, the first models of radio galaxies and quasars—by Shklovskii in 1960, Field in 1964, and Colgate in 1967 (see Perry, 1993b)—invoked central clusters of supernovae or pulsars to power the observed optical and radio luminosities, through their energy and relativistic particle emission.

Attention turned away from stellar cluster theories as the arguments for a gravitational origin for the energy generation of QSOs gained favour—particularly through disc accretion onto BHs (Lynden-Bell, 1969)—and because early SB models failed to account for the rapid variability of the continuum, for the broad and narrow emission lines, or for the observed radio jets. However these new theories—based solely on BHs, and their accretion of whatever gas came under their influence—have not been able to explain satisfactorily the complex behaviour of the circumnuclear region: the structure, creation and evolution of the broad-line emitting region (the BLR), for example, escapes explanation and continues to require "hand-of-God" intervention. This difficulty is in contrast to the dramatic and convincing successes of the BH accretion theory, which seems able to explain the continuum emission and its variability (Rees, 1984). The excitement and fascination of studies of the "monster" underlie continuing attempts to explain all aspects of activity solely in terms of the BH, reducing the rôle of the parent galaxy to that of an onlooker.

Attention has gradually been shifting back to the rôle the parent galaxies play in the activity at their centres and there has been a well publicised debate in the last few years between those who believe that either black holes, or starbursts, alone, explain activity in galactic nuclei. I believe this debate is futile, although often instructive and sometimes fun! Each camp can produce a long list of observed features which the other theory is strained to—or cannot—explain. Neither can produce convincing observational support, or evidence conclusively ruling out the alternative theory.

My collaborators and I have long held quite a different point of view. The very fact that both black holes and starbursts explain much of what we see, albeit different aspects of the phenomena, led us—long before the current debate began to rage—to propose that a symbiosis between starbursts and black holes was necessary to explain the observed properties of active galaxies. Originally it was the singular failure of black hole theories, alone, to provide any explanation for the ionization parameters, equivalent widths, abundances, and other properties of the broad line region—most of which are remarkably constant over more than 4 orders of magnitude change in the central luminosity—that led us to consider the role of young stars near the black holes. We believe that a cluster of young mass-losing stars surrounding the black hole naturally, and in a uniquely self-consistent manner, explains most of the salient features of the observations. In 1982, John Dyson and I suggested that a young stellar cluster *surrounding* a massive BH could be important in explaining the origin of the broad emission line clouds (Dyson & Perry, 1982). Developed in detail in 1985 (Perry & Dyson, 1985, hereafter PD) our model invokes radiative SN shocks, which result when SN explode in high density interstellar gas in galactic nuclei, to create the cool BLR "clouds"; in this theory the BLR is produced *self-consistently* with the fuelling of the continuum by the mass loss from the stellar cluster. In 1987 we further incorporated the accretion disc, and its line as well as continuum emission (Collin-Souffrin, Dyson, McDowell & Perry,1987, hereafter CDMP). Robin Williams has been modelling the evolution of the stellar cluster, and he and I and Amanda Baker have recently been studying the hydrodynamics of the nuclear interstellar medium.

In 1985, Terlevich & Melnick (Terlevich & Melnick, 1985) revived the idea that AGN could be explained by massive starbursts alone, without the need for central massive BHs. They first modelled the early stages of evolution of a *high metallicity* cluster as it passes from being a normal HII region to the development of 'warmers' when it may appear as a Seyfert 2 galaxy. More recently, Terlevich *et al.* (1992) have extended their model to include supernovae exploding into high density shells left behind by their high metallicity progenitors. In this case, the SN shocks are radiative, and broad emission lines can be radiated. In their model, the bolometric luminosity is the result of the integrated stellar light from the starburst.

What are the implications of the observed BLR abundances for these three broad categories of AGN theory? BH accretion theories, if they do not involve starbursts, must account for the existence of highly enriched gas in the accretion region at a cosmic time when most of the available reservoir of gas is under-enriched. Currently it is widely believed that the mass supply must come from the distributed ISM of the parent galaxy, or from the IGM of the local cluster, since mass-loss from a normal, old stellar population is known to be inadequate. A spate of recent observations has implicated galaxy collisions in the AGN phenomenon (see Perry, 1993 for a recent review), and suggests that the diffuse ISM of the host galaxy, or the IGM of a local cluster, can be provoked to flow into the nucleus by an encounter or bar. However, at least by redshifts of 2 and greater, this gas must be processed through stars if it is to create the observed enriched BLR gas as well as the accretion luminosity. It is then not illogical to assume that it may first form a starburst. Might we then not observe activity when those stars are in their active mass loss phase, within the first 10^8 yrs after the initial formation of the cluster? If so, we must consider a BH/starburst theory. On the other hand, the starburst-only theory requires a pre-existing high metallicity environment, because the only progenitors of isolated radiative SN are high-metallicity stars. Although these may be common in galactic nuclei at present, it is not obvious that at high redshift enough highly enriched gas exists to create a coeval starburst of $10^{11} M_{\odot}$ or more.

Nonetheless, the association of nuclear activity with the starburst phenomenon receives substantial circumstantial support in the local (low redshift) universe and a wealth of observational evidence and theoretical interpretation has accumulated in the past few years which suggests that such a configuration may be the result of, for example, gravitational encounters between galaxies which trigger the rapid infall of matter (in a free-fall time) to the nucleus where it can then form a coeval stellar cluster (see Perry, 1993).

If disturbances to the galactic potential, either through bars or encounters, trigger infall on a free-fall time scale the formation of nuclear starbursts seems a likely outcome. However, the scale of the resulting starburst is unclear: observed "classical" nuclear starbursts (of the sort invoked by Terlevich *et al.* as parent AGN) form on spatial scales of hundreds of parsecs, not parsecs. It is tempting to speculate that the presence of a supermassive black hole may cause the gas to continue to flow into the nucleus to the smaller scales required to form a starburst/BH symbiotic system. If gas does flow down into the central parsec, then until the local density builds up to high enough densities to overcome the tidal shear induced by the black hole (which in other contexts has been invoked to destroy stars) star formation will be inhibited. It is only when the local integrated mass exceeds that of the black hole that these tidal shear forces become insignificant, and that star formation can occur. It thus seems plausible that if a starburst system is to form in the immediate neighbourhood of the black hole, that the total stellar mass must equal or exceed that of the BH. Indeed, quite independantly of this argument, our simulations of the mass and energy budget of a symbiotic starburst/BH system indicate that just such a mass ratio is required to explain the observations.

Unfortunately, current models are limited by resolution problems. Models of interacting galaxies which show that inflow of the ISM to the "nucleus" occurs define the nucleus as their smallest resolution element: which in turn is larger than the total size of interest to those who observe and model nuclear starbursts. In turn, their smallest resolution element is larger than the BLR. So, until we can patch the theoretical models together at their boundaries, it will be difficult to progress beyond plausibility arguments when we discuss the cause and effect chain linking galaxy collisions to activity in the centres of galaxies.

Interestingly, the supernovae associated with starbursts play a central rôle in both theories for starburst AGN—those with, and without, a central engine. However, there are important differences both in the numbers, required progenitors, and dynamics in the two theories, some of which are highlighted below.

AGN AS NUCLEAR STARBURST STELLAR CLUSTERS AND CENTRAL BLACK HOLES

If a compact starburst stellar cluster surrounds a BH which is accreting gas, then:

• The cluster acts as a reservoir of mass, which it releases in a steady and controlled fashion—through normal processes of stellar winds and supernovae—into the local interstellar medium (ISM).

In the first 10^7 to 10^8 yrs of evolution of a starburst stellar cluster, mass-loss from massive stars can easily meet the requirements of a mass supply for an accreting black hole sufficient to fuel even a very luminous QSO (Williams & Perry, 1992; 1994, in preparation). For example, for a Salpeter IMF with upper and lower mass limits of 1 to 50 $M_\odot$ an "instantaneous burst" cluster of 10^9 $M_\odot$ will generate mass loss sufficient to fuel bolometric luminosities of 10^{47} erg s^{-1} at a few times 10^6 yrs, decaying to 10^{46} erg s^{-1} by several 10^7 yrs. Results for a wide variety of IMFs and mass limits show similar features; the main differences are in the relative length of the SN epoch and the effective L/M of the cluster. In all cases, in contrast, starburst only models require cluster masses more than 100 times as great in order to generate these luminosities; typically, luminosities of 10^{47} erg s^{-1} require coeval starbursts of over 10^{11} $M_\odot$.

- The BH, through accretion of much of this ISM material, 'generates' the observed broad band luminosity, $L(\nu)$. The level of the bolometric luminosity, L_{Bol} is controlled, in turn, by the rate of mass loss by the cluster. This mass loss is fed to the BH through the accretion branch of the bipolar flow of the nuclear ISM created by the stellar mass loss. There is an accompanying supersonic axial outflow.
- The distributed ISM is kept hot and ionized by this central radiation field to whose generation it has contributed by feeding the BH. For the broad-band spectra typical of AGN, this means that it is in equilibrium at the Compton temperature, typically of order 10^7 K.
- The supernovae (SN) in the cluster act as pistons, compressing the ISM and creating BLR clouds in a self-regulating feedback loop. These continuously created, evanescent clouds are photoionized by the central continuum, and radiate the high ionization lines (HIL).
- The accretion disc generates, in addition to continuum radiation, low ionization lines (LIL) from its outer regions.

Self-regulating Feedback

The feedback loop between the creation of the continuum luminosity and the line emitting gas is a straightforward result of the coupling of hydrodynamics and radiation processes. In galactic nuclei with massive BHs at their centres both the stellar Keplerian velocities and the ISM wind and accretion flow velocities are supersonic, ensuring that *all* stellar mass-loss is shocked before it can mix into the local ISM and subsequently get to the accreting BH. The ram pressure of these shocks is so high that they are radiative (Perry, 1993a), (Dyson, Perry & Williams, 1992). The only equilibrium state available to the high pressure shocked gas is photoionization equilibrium, at roughly 10^4 K. The special rôle of SN arises because the radiative shocks they generate are spatially so large that the radiative cooling time behind the shocks is shorter than the time the gas is trapped in the high pressure region. Thus in these large shocks the gas has sufficient time to cool into equilibrium with the radiation field (PD, Perry, 1993b) and it forms BLR "clouds". These are kept photoionized by the central continuum—a continuum generated by accretion of the ISM, itself formed from cluster mass loss dominated by those SN which create the BLR clouds—closing the loop.

It is this self-regulating feedback loop, controlled by the coupling of hydrodynamic and radiative processes, which gives rise to the remarkable similarity of many of the observed features of nuclear activity in galaxies.

Some Physical Details of the BH/Starburst Model

We assume that a dense, coeval nuclear starburst stellar cluster surrounds a massive BH in the nucleus of a galaxy. *If* so, then there are a number of direct, and inevitable, consequences, a few of which I now review.

- *Shocks in the Nuclei of Active Galaxies*

One of the important linch-pins of the symbiotic feedback described above is the presence of radiative shocks in the stellar cluster. What are the physical conditions that lead us to conclude that they are inevitable? If a BH exists at the centre of a galaxy then the Keplerian orbital velocities are

$$v_{\rm Kepler} \approx 660\ (M_8/r_{\rm pc})^{1/2}\ {\rm km\ s}^{-1}. \tag{2}$$

The sound speed in the ISM depends on temperature; assuming that it is at the Compton temperature of the observed non-stellar central continuum, $T_{\rm ISM} = 10^7 T_7$ K, then

$$v_{\rm sound} \approx 380\ T_7^{1/2}\ {\rm km\ s}^{-1}. \tag{3}$$

We can see immediately that almost *all* stars within $r < 3(M_8/T_7)$ pc move through the ISM supersonically (Perry, 1993b). They drive strong bow-shocks into the ISM, and their atmospheres are likewise subjected to strong shocks. Therefore, all mass injection into the ISM—whether through stellar winds, supernovae, tidal stripping or stellar collisions—involves shocks. This would be true even in a static ISM; however, the nuclear ISM is characterised by a supersonic bipolar flow (see below).

Shocks convert kinetic energy into thermal energy: the post-shock gas is heated and compressed, and in the limit of strong shocks the post-shock temperature, $T_{\rm S} \approx 10^8\ \omega^2$ K (where the shock velocity, $v_{\rm shock} \equiv 3000\ \omega$ km s^{-1}, is the relative velocity between the flowing ISM and the stellar wind or SN ejecta and is the vector sum of $v_{\rm Kepler}$, $v_{\rm ISM}$ and $v_{\rm injection}$). The post shock density is 4 times the preshock density. The effect of such shocks on the ISM is to raise the pressure—and thereby to lower the ionization parameter—by a factor $4T_{\rm S}/T_{\rm Compton} = 40\omega^2/T_7$.

- *Radiative Shocks*

The effect of the shocks discussed above is to create local regions whose gas pressure is typically forty times as high as that of the ambient, unshocked, ISM. Because the shocked gas is 'superheated' and out of equilibrium with the incident radiation field, it cools rapidly by radiative processes—inverse Compton scattering, Bremsstrahlung, and finally by recombination. If the cooling is isobaric, then when the temperature falls to the Compton temperature the gas is typically forty times as dense as the ambient ISM. At this density the gas is no longer at equilibrium at the Compton temperature , and the gas continues to cool in an attempt to reach equilibrium (PD). How low a temperature is ultimately achieved depends only on the size of the shock—it is the size of the shock which determines if the cooling is isobaric. The size of the shocked region in any global flow is determined solely by the 'obstacle' causing the shock: *e.g.* by the star or supernovae. Ordinary stars are surrounded by spatially small shocked regions, and the shocked gas flows out of the high pressure isobaric region before it can cool. Supernovae, however, inject typically 10^{51} erg into their ejecta: the resulting shocks are large enough for the gas to cool before it flows out of the isobaric region. The SN expands until the ram pressure at the shock front is balanced by the pressure in the remnant:

$$P_{\mathbf{stagnation}} = \text{Energy Density of SN Bubble.} \tag{4}$$

$$\rho_{\rm ISM} v_{\rm ISM}^2 \approx 3E_{51}/(4\pi r_{\rm shock}^3). \tag{5}$$

The shocked gas remains in the subsonic, isobaric post-shocked region for a time, $t_{\rm flow}$, given by the size of the shock, $r_{\rm shock}$ and $v_{\rm flow}$. In those shocks which are large enough so that $t_{\rm flow}$ exceeds the cooling time, $t_{\rm cool}$ the gas reaches radiative equilibrium at approximately 10^4 K.

Although the immediate post-shock gas is shock ionized, and the ionization structure of the cooling gas is dominated by time dependent recombination, the broad lines are emitted primarily by gas which has cooled into equilibrium with the central continuum, and thus the line spectra are—as is observed—characteristic of photoionized, and not of shock ionized, gas. Davina Innes and I have modelled the shocks in detail; preliminary results have been reported in (Perry, 1992).

• The BLR created by radiative shocks in the symbiotic starburst/BH or starburst only models have many similarities, and several important differences. In the starburst/BH model, the supernovae ejecta shock against a diffuse nuclear ISM created by the integrated mass-loss of the whole cluster; in the starburst only model, the ejecta shock against a shell of dense gas created by the pre-supernova progenitor star. In the latter case, only high-metallicity clusters have progenitors which can create the required high density shells; furthermore, unless essentially *all* the supernovae are of the required radiative type, the cluster mass becomes extremely large if it is to account for the observed luminosities of bright AGN. An additional difficulty is that known starbursts have superwinds, which could be expected to disrupt many of the pre-supernovae shells. In the symbiotic case, on the other hand, the cluster need not be an initially high-metallicity one, since the it is the global ISM which interacts with the supernovae, not a pre-existing local dense shell. It is this difference, as well as the conversion of the mass- energy of the stellar mass loss by the BH that leads to the orders of magnitude difference in the size of the required starburst stellar clusters between the two theories.

• *Size of Shock Created BLRs*

Because the Compton cooling depends on the local radiative flux, $t_{\rm cool}$ depends inversely on the distance from the central source: this dependence on distance introduces a 'critical' distance within which cooling occurs, and beyond which it does not. We find that supernovae can create BLR clouds by shock compression within a distance from the central source given by (PD):

$$r_{\rm BLR,\ \mathit{max}} \approx L_{47}^{1/2} E_{51}^{1/4} \ \rm pc \tag{6}$$

where $L_{47} \equiv L_{\rm Bol}/(10^{47}{\rm erg/s})$ and $E_{51} \equiv E_{\rm SN}/(10^{51}{\rm erg})$. This theoretical result is in striking agreement with estimates based on photoionization models which yield $L_{47}/r^2_{\rm BLR} \approx 1$ for the bulk of the BLR. There is also a very weak dependence on shock velocity and pressure (see (PD) as well as on the energy of the supernovae explosions; these introduce only small variations about this critical value. The fact that cooling—and concomitantly continuous 'cloud' creation—occurs *within* a critical distance from the central source means that there will be an extended BLR, with cool gas also at very small distances from the nucleus. Observational support for the predicted extended region is now accumulating as the results of reverberation mapping and complex photoionization modelling are combined. In addition, evidence for temporal changes in the distribution of BLR gas is now emerging from analysis of monitoring campaigns (Perry, van Groningen & Wanders, 1994).

• In the starburst only model the size scale of the BLR is at least as large as is the nuclear starburst itself, and thus a critical test of the model is that the BLR must be spatially resolvable, particularly in the brighter objects (Terlevich *et al.* 1992).

• *Flow Patterns in the Nuclear Region*

The above discussion relies not only on the physics of the radiative shocks, but also on the mass-loss properties of the stellar cluster and on the hydrodynamics of the nuclear region. Robin Williams, Amanda Baker and I have been studying the hydrodynamics of a nuclear region containing a starburst stellar cluster, a supermassive central black hole and a cold accretion disc (Williams, Baker & Perry, 1994, in preparation). Most importantly, the flows are all bipolar with supersonic axial outflow accompanied by accretion onto the disc. Narrow (often conical) stagnation regions separate the accretion and outflow regions, and global shocks often sweep through the region. Depending on the strength of the cluster mass loss and the gravitational field of the cluster and BH, the outflows from the nucleus into the surrounding galaxy are either sporadic, and in some cases explosive, or steady—in the form of narrow highly supersonic hydrodynamical jets, or broad, quasi-spherical outflows.

• *The Accretion Disc and "Remnant" BHs*

If a dense starburst is present surrounding the accreting black hole the presence of the cluster deepens the potential well into which gas accretes, and at the edge of the core significant additional shear forces increase the temperature of the disc (Perry & Williams, 1992). Whether or not the infra-red feature which is caused by this additional shear is observable or not, it is clear determine the mass of the central black hole determined by models of the "big blue bump" can be in error: the big blue bump measures the mass of the core of the stellar cluster plus that of the central black hole. Our cluster mass-loss models generally favour stellar core masses greater than that of the central black hole, by factors ranging from roughly 3 to 30. Thus estimates of the masses of "dead" QSOs in the centres of inactive galaxies are probably too high. If the active cluster is dominated by massive stars, the cluster will lose most of its mass during the active phase, both in stellar winds and in the final supernovae explosions. ¿From the hydrodynamical models of the ISM, we estimate that roughly half of this mass-loss may be lost to the core region in a supersonic wind. Until fully self-consistent models, including the stellar dynamics, of the region are available, safer guides to residual black hole masses, are the minima calculated for Eddington accretion over 3–10 $\times 10^7$ yrs.

Consequences of the Symbiotic Feedback of BH and Starburst Cluster

• The remarkable constancy of the observed equivalent widths of the emission lines from object to object over many orders of magnitude of continuum luminosity implies that the mass of radiating BLR gas increases roughly in proportion to the increase in continuum luminosity. This proportionality arises as a natural consequence of the feedback between cluster and BH, since the BLR clouds are created as a byproduct when the central luminosity is generated. The total amount of mass cooled by SN shocks depends on the rate of mass flow through the shocks: *i.e.* the mass flux in the ISM times the covering factor of the SN bubbles. The same global ISM mass flow controls the level of the continuum. These quantities are directly proportional to each other, leading to equivalent widths independent of luminosity, as required.

• The hot diffuse ISM is optically thin, as is required by observations of short time-scale X-ray variability. This optical depth requirement severely limits the density of the ISM (PD) and therefore the mass of the ISM in the region; this means not only that the local ISM cannot pressure confine the BLR clouds, but also that models in which the BH is fed over the long term by an ISM which is not continuously resupplied would be in conflict with observations. This difficulty supplying the BH is overcome because the mass supply is in the condensed form of stars in the local stellar cluster—effectively an optically thin, controlled release, local, mass storage system.

• The ionization parameter—the ratio of the local radiation pressure to the gas pressure in the radiating gas—of the HIL is remarkably constant amongst AGN. In shock created HIL clouds the gas pressure is equal to the ram pressure of the radiative shocks: this is controlled by the mass flux through the region and the shock velocity. The radiation pressure also is directly related to the mass flux, which therefore effectively drops out of the ionization parameter, leading to an ionization parameter in remarkable agreement with observation.

• The HIL clouds are created within the entire region interior to a maximum radius determined by the condition that the cooling time behind the shocks be smaller than the flow time. This maximum radius is equal to the 'typical' or mean radius found for the BLR from photoionization simulations of the emission line spectra. Furthermore, since clouds are created throughout an extended region (rather than in a shell) there is a range of densities and ionization parameters, as is inferred from detailed study of different lines.

• The emission lines are predicted to vary, on short time scales, in direct response to variations in the ionizing continuum. However, on time scales long enough for the distribution of cold clouds to change, we expect secular changes in the emission line strengths and changes in the line profiles which reflect the local creation and subsequent evaporation of line emitting material. There are suggestions in recent data that such changes are now being detected.

• The accretion disc radiates both the continuum luminosity and the low ionization lines (LIL). The HIL and LIL are intimately coupled both because both are created within and by the mass flux in the nucleus and because the LIL are excited in large part by X-rays reflected from the HIL region (CDMP).

• The HIL are observed to be systematically blue shifted with respect to the LIL. This occurs because the HIL are created predominantly in an outflowing wind above the accretion disc which radiates the LIL. The disc occults the far-side HIL, giving a systematic blue shift of the centroid of the lines CDMP).

• The dense nuclear stellar cluster partially changes the structure of accretion discs in comparison to those around isolated BHs. The resulting increase in the dissipation of the gravitational binding energy of the accreting gas, particularly in the outer regions of the disc, may well explain the small 'infra-red bump' seen in many spectra (Perry & Williams, 1993).

• The SN inject heavy elements into the local ISM (the source of the BLR clouds) explaining the appearance of heavy element emission lines at all redshifts. In contrast, direct accretion of unenriched intergalactic, or halo material at high redshift without intermediate stellar processing on its way to forming emission line clouds cannot explain the abundances seen.

• The Broad Absorption Lines (BAL) seen in the spectra of roughly 10% of (mostly high redshift) active galaxies result when a radiative SN happens to lie on the line-of-sight to the continuum source (Dyson & Perry, 1992). The covering factor of these compact SN accounts for the frequency with which BALs are observed, and the ionization state and velocities of the cold, rapidly moving ejecta are all in remarkable agreement with the observations. The systematic blue shift of the BAL is a direct result of the fact that the ejecta moving towards the central continuum source is stopped significantly earlier by interaction with the outflowing ISM than is the outward moving ejecta (Dyson & Perry, 1992).

• The model is robust to the initial mass function (IMF) of the stellar cluster, since the physics of both the feeding of the BH and the creation of the line emitting gas is dominated by high mass stars (Williams & Perry, 1993b).

• The total mass of the BH/starburst system required to explain activity is minimal. *Starburst only models require two to three orders of magnitude more mass to explain the bolometric luminosity.* This is because the integrated mass-energy of the cluster stellar mass loss (which can be converted by accretion onto the BH into luminosity) is several orders of magnitude greater than the total luminous energy radiated by the stellar cluster (even if the mechanical energy output of all its SN is all efficiently radiated) (Williams & Perry, 1992).

• The identification of activity with a nuclear starburst phase has numerous attractive features. The expected episodic activity agrees with many current interpretations of the luminosity evolution of active galaxies, which suggest that active lifetimes are of the order of several 10^7 to 10^8 years (see Perry, 1993b)—similar to the SN phase lifetimes of starbursts. Some density evolution is also expected, proportional to the changing rate of galaxy interactions.

CONCLUSIONS

If activity manifests itself as a result of the symbiosis outlined above, then clearly the fundamental parameters of the model are the mass of the BH and the total mass of the stellar cluster, its spatial distribution, and its IMF. I believe that a "one-parameter" model will not succeed in explaining nuclear activity, however attractive unified models are. In addition to viewing angle and the mass of the central engine, we must consider the range of phenomena which result from different ratios of the mass and compactness of the cluster to the mass of the central BH.

All our models assume 'standard' stellar evolution tracks are valid also for stars in the nuclei of galaxies. However, in the situation discussed here, where all stars are bathed not only in the intense radiation field of the central source, but are subjected to strong shock heating and ablation of their atmospheres, it is unclear if this is a valid assumption. This work has opened a new Pandora's Box of questions, not only about the viability of the formation of dense, coeval starburst stellar clusters, but about the structure and evolution of the stars which may be formed and forced to live out their short lives in such unusual environments. We shall be investigating some of these questions in the near future—but if the scenario I have outlined here does actually describe the centres of galaxies which are 'active' then surely many of the details will change as our understanding of the symbiosis of stars and black holes progresses.

ACKNOWLEDGMENTS

Many of the ideas reviewed in this talk with which I, personally, have been associated were developed over the past ten years in collaboration with John Dyson, and more recently in collaboration with Robin Williams, Suzy Collin and Davina Innes. I am grateful to them for their stimulating collaboration. Space limitations have meant that I was unable to refer to most contributions to the observations and theory of both starbursts and AGN and their connection; however, an extensive review with references can be found in (Perry, 1993b).

REFERENCES

Collin-Souffrin, S., Dyson, J. E., McDowell, J. C., & Perry, J. J. 1987. *A&A* **232**, 537

Dyson, J. E., & Perry, J. J. 1982. in *Third European IUE Conference* p. 595, eds Rolfe, E., Heck., A., & Batrick, B. ESA, Noordwijk

Dyson, J. E., & Perry, J. J. 1992. in *Testing the AGN Paradigm* pp. 553-555, eds Holt, S. S., Neff, S. G., & Urry, C. M., AIP Conf. Proc. 254

Dyson, J. E., Perry, J. J., & Williams, R. J. R. 1992. in *Testing the AGN Paradigm* p. 548, eds Holt, S. S., Neff, S. G., & Urry, C. M., AIP Conf. Proc. 254

Hamann, F. & Ferland, G. 1993. *ApJ* **418** 11

Lynden-Bell, D. 1969. *Nature*, **223**, 690

Perry, J. J. 1993a. in *First Light in the Universe: Stars or AGN?* pp. 25-96, eds & , Editions Frontieres

Perry, J. J. 1993b. in *Central Activity in Galaxies - From Observational Data to Astrophysical Diagnostics* pp. 25-96, eds Sandqvist, Aa. & Ray, T. P., Springer-Verlag

Perry, J. J. 1992. in *Relationships between Active Galactic Nuclei and Starburst Galaxies* pp. 169, ed. Filippenko, A. V. (Astro. Soc. Pac. Conference Series

Perry, J. J., & Dyson, J. E. 1985. *MNRAS* **213**, 665 (PD)

Perry, J. J., van Groningen, E., & Wanders, I. 1994. in preparation

Perry, J. J., & Williams, R. J. R. 1993. *MNRAS* **260** 437

Pettini, M., Smith, L. J., Hunstead, R. W, & King, D. L. 1993 *MNRAS in press.*

Rees, M. J. 1984. *Ann. Rev. Astr. Astrophys.* **22** 471

Terlevich, R., Melnick, J. 1985. *MNRAS* **213**, 841

Terlevich R.T., Tenorio-Tagle, G., Franco, J., Melnick, J. 1992. *MNRAS* **255**, 713

Williams, R. J. R. 1993. Ph.D. Thesis, University of Cambridge

Williams, R. J. R. & Perry, J. J. 1992. in *Physics of Active Galactic Nuclei* p. 697 eds. Duschl, W. J. & Wagner, S. J., Springer-Verlag, Heidelberg

Williams, R. J. R. & Perry, J. J. 1993. *MNRAS in press*

The First Stromlo Symposium: The Physics of Active Galaxies
ASP Conference Series, Vol. 54, 1994
G. V. Bicknell, M. A. Dopita, and P. J. Quinn (eds.)

NEAR–INFRARED OBSERVATIONS OF STARBURST AND SEYFERT GALAXIES

DUNCAN A. FORBES
Lick Observatory, University of California, Santa Cruz, USA

MARTIN J. WARD
Astrophysics,Department of Physics, Oxford University, Oxford, UK

ABSTRACT

Technological advances in near–infrared detectors have opened up a number of new avenues in active galaxy research. For a large sample of active galaxies we examine the correlation between infrared emission lines and radio continuum luminosity, which provides further insight into the differences and similarities between starburst and Seyfert galaxies. Here we present recent near–IR narrow–band imaging data from the AAT. In starburst galaxies there is a direct spatial coincidence between the location of near–infrared [FeII] line emission and compact steep spectrum radio sources (both appear to be tracers of SN activity). There is evidence that extragalactic supernova remnants can be considerably more luminous than their counterparts in the Galaxy.

INTRODUCTION

The relative role of stellar processes (e.g. SN activity) vs those associated with a compact nuclear source (generally accepted to be a black hole with an accretion disk), in active galaxies, are still controversial. In order to study the nuclear activity with high spatial resolution, and penetrate the dust obscuration that is often present, we must currently rely on radio and/or infrared observations. Here we combine information from radio synthesis mapping and narrow–band imaging obtained using the latest generation of IR instruments. A resolution of one arcsec corresponds to 10–100 pc for the nearest active galaxies, which is an interesting scale in which to probe the circumnuclear activity.

NEAR–IR DIAGNOSTIC LINES

Near–IR spectra of active galaxies reveal a wealth of emission and absorption lines. Many of these lines can be used as probes of the activity and physical conditions in the circumnuclear environment. They have the advantage over optical lines that the extinction is much reduced ($A_V \sim 10A_K$). Some of the strongest lines commonly observed in active galaxies are given in Table 1. Of particular note are He I and O I which give information about physical conditions of the gas, [FeII] which probably traces fast shocks (see below), [Si VI] is a high ionization line seen in some Seyfert galaxies, H_2 (1–0) is the ro–vibrational line of molecular hydrogen, the CO bands are luminosity sensitive and can therefore be used to trace the distribution of supergiants and the 3.28 μm feature which is not generally seen in the spectra of Seyfert nuclei, is probably associated with dust grain emission, and starburst activity. Here we concentrate on the emission lines of [FeII] at 1.64 μm, H_2 and the hydrogen recombination line at 2.17μm (Brγ).

TABLE I Near–IR diagnostic lines

Line	Wavelength
He I	1.08 μm
O I	1.13 μm
[FeII]	1.26 μm
P β	1.28 μm
[FeII]	1.64 μm
P α	1.88 μm
[Si VI]	1.96 μm
He I	2.06 μm
H_2 1-0	2.12 μm
Brγ	2.17 μm
CO bands	2.3-2.4 μm
PAH	3.28 μm

NEAR–IR LINES AND RADIO CONTINUUM EMISSION

Line emission from [FeII] is produced by electron collisions from fast shocks and possibly hard photoionization (P.I.). As well as the correct ionization conditions, an enhanced abundance of Iron in the gas phase seems to be required (Greenhouse *et al.* 1991). Such a grain destruction mechanism would result from a SN blast wave, or other fast shocks. It is not clear if sufficient grain destruction results from X–ray heating.

The non–thermal radio emission from starburst galaxies is due to SNRs and superwinds. In Seyferts this emission is often associated with jets or lobes powered by processes at very small scales (possibly a black hole). It can be difficult to separate these mechanisms by radio emission or spectral index alone.

Fig. 1 shows the [FeII] vs 6cm radio emission from the nuclear region ($\sim$ few hundred parsecs) of active galaxies, using data from the literature and our own work (Forbes & Ward 1993). A strong correlation (σ = 0.27) is seen for all galaxies; there is no trend with activity type. For starburst galaxies this correlation is simply explained in terms of SN activity. The [FeII] emission can therefore provide an independent estimate of the SN rate. For Seyfert galaxies the mechanism for [FeII] emission is not so clear but may have an additional component due to radio jet induced shocks. High resolution line imaging will provide an important test.

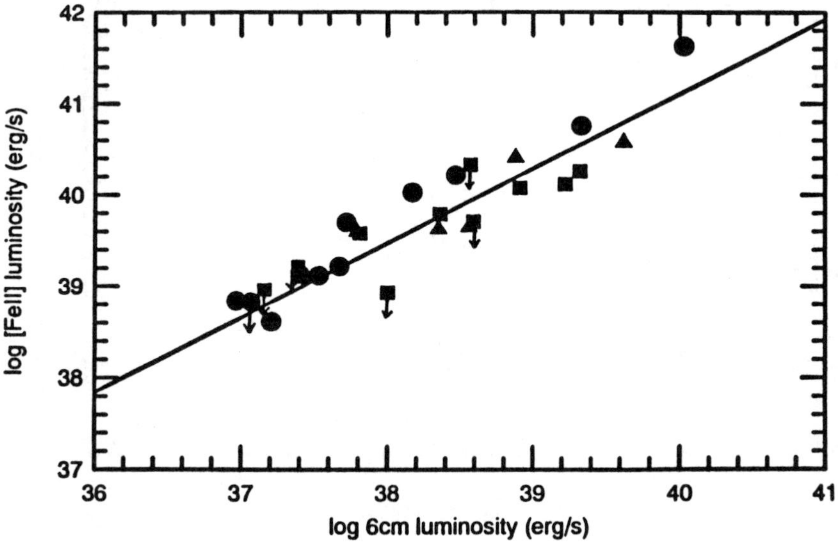

FIGURE I [FeII] 1.64μm vs 6cm radio emission for galactic nuclei. Filled circles, triangles and squares represent starburst or LINER galaxies, galaxies with composite spectrum and Seyfert galaxies respectively. Also shown is the best-fit regression line. The dispersion about this line is 0.27 in the log.

Molecular hydrogen has many possible excitation mechanisms. One such possibility is X–ray heating from SNRs. If so, we would expect similarly strong correlation with 6cm radio emission. Fig. 2 reveals that the correlation for H_2 is weaker ($\sigma = 0.40$) than for [FeII] line emission (Fig. 1). This suggests that the trend in Fig. 2 is secondary in nature and that other excitation sources for H_2 are more likely e.g., P.I. or shocks from regions of young star formation. We also note the location of the merging galaxy NGC 6240 in Fig. 2. This galaxy is extremely luminous in H_2 relative to its radio emission. Line imaging, discussed below, dramatically illustrates the source of this excess H_2 emission.

NEAR–IR LINE IMAGING

Perhaps the best example of the connection between radio and [FeII] emission is given by NGC 1808. Figure 3 shows a 200s [FeII] line image of NGC 1808 obtained recently with the IRIS imaging spectrograph on the AAT. The seeing was $\sim 0.8''$. The image reveals a bright nucleus surrounded by a kpc sized ring of '[FeII] hotspots'. These hotspots represent the sites of SNRs and show a good spatial correspondence to the non–thermal compact sources seen in the radio continuum (Saikia *et al.* 1990).

Starburst galaxies may possess SNRs that are more luminous than Galactic SNRs. One particular remnant in M82 is $\sim 100\times$ more luminous than Cas A in both radio and [FeII] emission (Greenhouse *et al.* 1991).

Near–IR imaging of the nearby starburst galaxy NGC 253 has been pre-

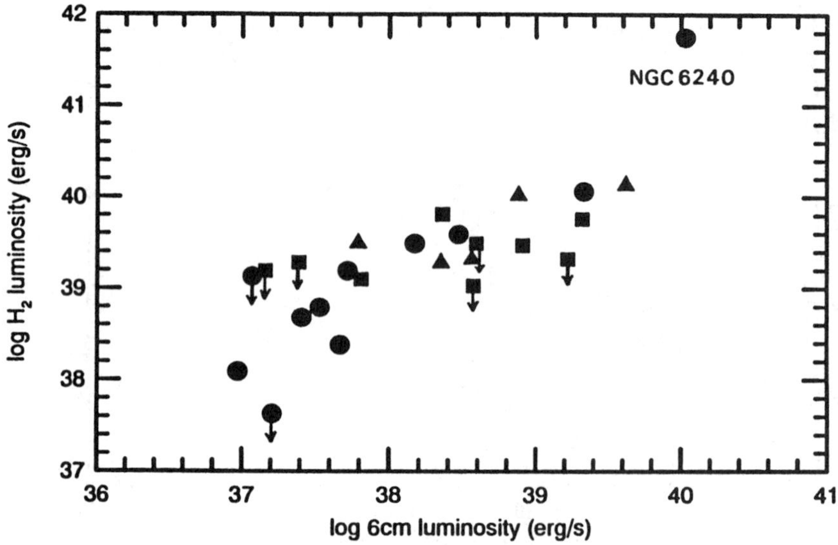

FIGURE II H_2 1-0 S(1) vs 6cm radio emission for galactic nuclei. The same symbols are used as in figure 1. The dispersion is 0.40 in the log.

sented elsewhere (Forbes *et al.* 1993). There we find that the spatial distribution of H_2 and Brγ emission is similar, indicating that the H_2 is excited in young star formation regions. The H_2 emission also reveals filamentary structure. beyond the disk which may be excited within the superwind. [FeII] does not appear to be excited by this wind, perhaps due to the low gas density present.

One of the few Seyfert galaxies imaged to date is NGC 1068 (Blietz *et al.* 1993). There the [FeII] follows the same P.A. as the radio jet with a correlated morphology. This suggests that radio induced shocks may also generate [FeII] emission but this is not the only possibility.

Finally, the H_2 line imaging of NGC 6240 by van der Werf *et al.* (1993) reveals that the majority of the H_2 emission is concentrated between the two merging nuclei. Thus the extreme H_2 luminosity seen in figure 2 for this galaxy appears to be caused by shock excitation at the interface of the ISM of each galaxy. There is also some evidence for outflowing molecular hydrogen in the galaxy superwind.

CONCLUDING REMARKS

The new generation of near–IR arrays, for both imaging and spectroscopy, provide powerful tools for the study of active galaxies. Infrared emission line diagnostics can replace the standard optical ratios, for regions of high extinction. Similarly, narrow–band IR images reveal structures which are obscured at optical wavelengths. Supernovae, and their remnants have characteristic signatures in both their radio and IR line emission. The relationship between these properties has already helped in our understanding of the AGN/starburst connection.

FIGURE III [FeII] line image of NGC 1808. Linear contours are shown. North is up and east is left. The bar represents 2″. Note the similarity to the radio maps published by Saikia *et al.* (1990).

ACKNOWLEDGMENTS

We thank Ray Norris, Andrew Phillips, Reinhard Genzel and group at MPE for help and useful discussions.

REFERENCES

Blietz, M., *et al.* 1993, *ApJ*, in press

Forbes, D. A. & Ward, M. J. 1993, *ApJ*, in press

Forbes, D. A., *et al.* 1993, *ApJ*, 406, L11

Greenhouse, M. A., *et al.* 1991, *ApJ*, 383, 164

Saikia, D. J., *et al.* 1990, MNRAS, 245, 397

van der Werf, P., *et al.* 1993, *ApJ*, 405, 522

STARBURSTS, MERGERS & ACTIVE GALAXIES

Contributed Papers

The First Stromlo Symposium: The Physics of Active Galaxies
ASP Conference Series, Vol. 54, 1994
G. V. Bicknell, M. A. Dopita, and P. J. Quinn (eds.)

CO(1 – 0) DETECTION IN A MERGING POLAR RING GALAXY

M. ARNABOLDI
Mt. Stromlo and Siding Spring Observatories, ACT

M. CAPACCIOLI
Osservatorio Astronomico di Capodimonte, Napoli, Italy

F. COMBES
Observatoire de Paris-Meudon, Meudon, France

OBSERVATIONS

Photometric and spectroscopic data on the PR galaxy ESO 603–G21 were acquired with EFOSC2 and the ESO 2.2m telescope at La Silla, Chile, in August of 1992. Surface photometry in B and R shows a stellar component with an apparent major axis at P.A.= 24° and axis ratio $c/a = 0.54$. There is a ring at P.A.= 114°, identified by a warped dust lane and a strong, clumpy, and thick Hα emission, which strongly affects the SW part of the galaxy. The Hα–continuum image shows ionized gas clouds which have not completely settled down onto the ring. The B and R luminosity profiles along the apparent major axis of the host galaxy follow a Freeman type II exponential disk indicating that the stellar component is probably an S0. Its $B - R$ *vs.* radius diagram indicates a strong reddening towards the nucleus: here the color is $B - R = 2$, while the outer regions are rather blue, with $B - R \simeq 1$. In the $B - R$ map the ring appears much bluer that the host galaxy.

Long–slit spectra were obtained in the 4650 – 6800Å range (to include Mg2, NaD lines and Hα emission), at two different position angles, P.A.= 114° and 24°, for a total integration time of three hours. When the slit is aligned with the PR, at P.A.= 114°, the emission line rotation curve shows that the velocity in the ring is still rising at the edges of the gas distribution, with $\Delta V_{max} = 300$ kms^{-1}. The absorption line rotation curve shows that stars follow the gas kinematics in the inner 8", with $\Delta V_{max} = 250$ kms^{-1}. At P.A.= 24°, the emission line rotation curve shows no motion of the gas perpendicular to the ring, confirming a standard ring kinematics for the PR. In contrast the absorption line rotation curve indicates that motion of the stellar component along this axis has a $\Delta V_{max} = 200$ kms^{-1}. This stellar kinematics suggests the presence of a galactic triaxial potential, with the neutral axis of motion most likely being at P.A.= 69°.

CO(1 – 0) emission has been detected using the 15m millimetric SEST telescope at ESO, la Silla, in May 1993. Fig. 1 shows the intensity *vs.* position diagram obtained from the single dish observations of the nucleus, position (0, 0), and of the edges of the ring, (–20, 8) and (20, –8); the integration time for each position is of 5 hrs. The spectra show the presence of molecular clouds in the

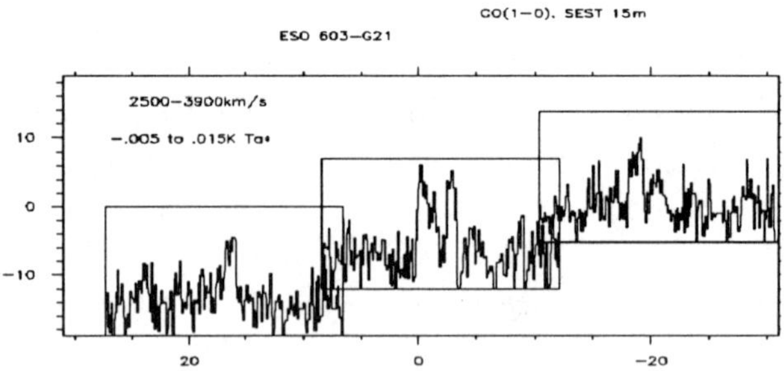

FIGURE I Intensity *vs.* position diagram of the CO(1-0) emission in the polar ring galaxy ESO 603–G21.

system, which in turn implies that there is ongoing star formation in this PR. This was previously indicated by observation in the UV band with the IUE satellite of an analogous PR system AM 2020–504 (Arnaboldi et al. 1993). The gaussian fit to the emission lines gives the following velocity components: 3114 and 3263 kms^{-1} in $(0,0)$, 3235 kms^{-1} in $(20,-8)$, and 3073 kms^{-1} in $(-20,8)$. The double horn shape of the central emission, and the presence of the lowest velocity component at $(-20,8)$ indicate a still rising rotation curve for the molecular clouds at the ring edges, demonstrating that these clouds are indeed in the ring. We derive a total H_2 mass of $7.6 \cdot 10^8 M_\odot$, assuming the standard H_2/CO conversion ratio of $N(H_2)(\mathrm{mol\ cm}^{-2}) = 2.3 \cdot 10^{20} I(CO)(\mathrm{K\ kms}^{-1})$, and adopting a distance of 42 Mpc.

DISCUSSION

The morphology of the PR in ESO 603–G21 and the presence of infalling ionized gas towards the ring indicated that this structure is young. The host galaxy is most probably an S0 with a strong reddening towards the center, which may be due to absorption by the ring. The rotation curves for the star component along two orthogonal position angles show evidence of a triaxial gravitational potential in the system. The emission line rotation curves show a standard ring kinematics for the ring.

The detection of $CO(1-0)$ emission in the ring indicates that the ring structure has formed from a strong interaction event, supporting previous evidence of accretion and star formation in PR galaxies (Arnaboldi et al. 1993, Combes et al. 1992).

REFERENCES

Arnaboldi, M., Capaccioli, M., Barbaro, G., Buson, L., Longo, G. 1993, *A&A*, **268**, 103

Combes, F., Braine, J., Casoli, F., Gerin, M., van Driel, W. 1992, *A&A*, **259**, L65

The First Stromlo Symposium: The Physics of Active Galaxies
ASP Conference Series, Vol. 54, 1994
G. V. Bicknell, M. A. Dopita, and P. J. Quinn (eds.)

MERGERS AND STARBURSTS IN MODERATE REDSHIFT FIELD GALAXIES

MATTHEW COLLESS
Mount Stromlo & Siding Spring Observatory, Australian National University, Weston, A.C.T. 2611, Australia

DAVID SCHADE, R.S.ELLIS
Institute of Astronomy, University of Cambridge, Madingley Road, Cambridge CB3 0HA, U.K.

T.J.BROADHURST
Dept of Physics & Astronomy, Johns Hopkins University, North Charles St, Baltimore, MD 21218, U.S.A.

We have used the tip-tilt correction of HRCam on the CFHT to obtain subarcsecond images of 26 galaxies from the redshift survey of Colless et al. (1990,1993). The primary sample of 17 galaxies have enhanced star formation indicated by [OII] equivalent widths greater than 20Å, while the comparison sample of 9 galaxies have equivalent widths less than 10Å. By fitting exponential disks or $r^{1/4}$-law bulges we obtain scalelengths for the blue and red stellar populations and so broadly establish the location of the star-formation (in the nucleus or the disk) for each galaxy. We have also searched for nearby faint companions in order to determine whether the star-formation might be linked to tidal interactions or mergers. We find that these moderate-redshift galaxies generally have straightforward low-redshift analogues, in that their colours, sizes and luminosities are consistent with those of various types of $z \approx 0$ galaxies. The star-forming objects have morphologies ranging from early-type spirals to irregulars, and absolute magnitudes spanning the range M^*-1 to M^*+5. Some have star-formation concentrated in their nuclei while others have star-formation occurring across the entire disk. Although none of the comparison sample have close companions, 30% of the galaxies with enhanced star-formation have companions with projected separations less than 10 h^{-1} kpc. The numerical coincidence between this fraction and the 40% excess (compared to $z \approx 0$) in the number of galaxies with enhanced star-formation found by Colless et al. (1990) is striking. Although the sample is small, this is the first direct evidence suggesting that interactions or mergers play a significant role in the increased fraction of field galaxies with enhanced star-formation at moderate redshifts.

REFERENCES

Colless M.M., Ellis R.S., Taylor K., Hook R.N., 1990, MNRAS, 244, 408

Colless M.M., Ellis R.S., Broadhurst T.J., Taylor K., Peterson B.A., 1993, MNRAS, 261, 19

The First Stromlo Symposium: The Physics of Active Galaxies
ASP Conference Series, Vol. 54, 1994
G. V. Bicknell, M. A. Dopita, and P. J. Quinn (eds.)

KINEMATICS OF THE MOLECULAR AND IONIZED HYDROGEN GAS IN NGC 3256

JAYANNE ENGLISH AND KEN FREEMAN
Mount Stromlo and Siding Spring Observatories, Private Bag, Weston Creek Post Office, ACT 2611, Australia

RAY NORRIS
Australia Telescope National Facility, Box 76, Epping, NSW 2121, Australia

INTRODUCTION

The extended arms apparent in our HI Australia Telescope Compact Array (ATCA) observations (in collaboration with S. Aalto, and R. Booth) provide evidence that starburst galaxy NGC 3256 owes its peculiar optical morphology to an interaction event between galaxies. Previous studies (such as Graham et al., 1987) show that the visible stellar light in the central region has a characteristic $r^{0.25}$ luminosity profile. This suggests that NGC 3256 is a late stage merger which will produce an elliptical galaxy as the remnant. The following is a preliminary examination of two theoretical propositions and uses this ATCA data, our Parkes HI profile, and our velocity field map of the H_α component produced using the Double Beam Spectrograph on the 2.3m MSSSO telescope.

Kinematic Decoupling of the Core

If a kinematically distinct core is observed in an elliptical galaxy, it is perceived to be a signature of the merger event which produced the elliptical. Hernquist and Barnes (Nature, 1991) propose the following scenario for the merger of gas rich disks in certain encounters. Just after first passage, gas is deposited in the inner core region of each parent galaxy. The mass of this gas, which is comparable to that observed in starbursts, is subjected to strong torques during the subsequent merger. Thus it loses its memory of its original angular and orbital momentum. However the gas in the outer region retains some memory of its momentum. Hence the gas in the inner region and the gas in the outer region of a merger remnant may be kinematically decoupled.

Globular Cluster Progenitors

In a merger, gas fragments are shocked in the tails and loops that have been torn off during the interaction process. The resulting star formation within these fragments is analogous to the formation of globular clusters within clumps surrounding collapsing protogalaxies proposed by Searle and Zinn (1978). Hence we expect to find globular cluster progenitors in loops, tails, and clumps of gas in merging galaxies.

OBSERVATIONAL RESULTS

The gas in the centre of the system is coupled to the gas in the outer regions:

In general, the H_α component of the system displays a global rotation which is consistent with the motion of the extended arms of neutral hydrogen.

The centre of the velocity range for both gas components is about 2820 km s^{-1}. (The optical velocity range is about 300 km s^{-1}; the radio velocity range is about 480 km s^{-1}).

Prominent features in both neutral and ionized gas are spatially coincident. In particular an absorption feature in the ATCA HI data is centred on the brightest H_α region and covers essentially all of the disk-like area. (This positioning indicates HI is absorbing emission from two bright sources, one associated with the H_α region, which are apparent in the 3 cm radio continuum and [Fe II] data of Norris and Forbes (1993)).

Star forming gas fragments are created during interactions:

Some H_α regions do not adhere to the projected circular motion. For example, there is a redshifted H_α feature in the south west which deviates from the mean velocity of its neighbourhood by about 180 km s^{-1}.

The HI gas in the extended arms displays a clumpy distribution in both spatial and velocity regimes.

CONCLUSIONS

Our observations of the H_α gas in the inner 100 arcsec (=20 kpc) and HI gas in outer regions spanning almost 9 arcmin (=100 kpc) of NGC 3256 provide evidence of consistent global rotation. If NGC 3256 is in a late stage of merging, it is unlikely that the process will produce a remnant elliptical with a kinematic decoupling signature. (However we note that other data indicate that the cores of the parent galaxies have not yet begun coalescing.)

We observe distinct velocity features consistent with the idea that during the interaction of the parent galaxies fragments, filaments or clumps of gas are formed. The associated H_α emission indicates that star formation occurs within them. We argue that this may be associated with globular cluster formation in merging galaxies.

REFERENCES

Graham, J.R., Wright, G.S., Joseph, R.D., Frogel, J. A., Phillips, M. M., Meikle, W.P.S., (1987), Star Formation in Galaxies, ed. C.J. Lonsdale Persson, NASA CP-2466, 517

Hernquist, L. and Barnes, J.E., (1991), Nature, **354**, 210.

Norris, R. and Forbes, D., (1993), in preparation.

Searle, L., and Zinn, R., (1978), Ap. J., **225**, 357

The First Stromlo Symposium: The Physics of Active Galaxies
ASP Conference Series, Vol. 54, 1994
G. V. Bicknell, M. A. Dopita, and P. J. Quinn (eds.)

KINEMATICS OF ARP 118

C.F. MCCAIN AND K.C.FREEMAN
Mt. Stromlo & Siding Spring Observatories, Private Bag, Weston Creek, P.O., Weston, ACT2611

INTRODUCTION

This is a study of a class of interacting galaxies in which one component is an SO galaxy and the other component is chaotic in structure, sometimes in the form of a ring or HII blobs and loops. In many examples, there is no evidence of a visible nucleus in the irregular component, and *Freeman & de Vaucouleurs (1974)* suggested that this component may be formed from interstellar matter stripped from the SO companion. An alternative explanation is an encounter between two spiral galaxies: one component which has a more tightly bound ISM strips off ISM from the other component, resulting in a stripped SO and a chaotic component. One of the nuclei might be hidden behind the ISM as to give the impression of single nucleus for the whole system.

How this mechanism actually happens, how it affects the velocity and star distributions, and what are the stellar populations in these systems are of particular interest in this study. To try to answer these questions, we are doing spectroscopy and optical imaging, and will be doing radio imaging, of a sample of these objects, with dynamical modelling using *SPH* to follow soon.

One particular system shows spectacular kinematics, with internal motions up to 1000 km/s. This is **Arp 118** (**NGC 1144/43**), which consists of a very distorted disk with two bright optical sources (**NGC 1144**), and a compact SO galaxy (**NGC 1143**). The brightest nucleus of **NGC 1144** shows a Seyfert-2 spectrum *(Hippelein, 1989)*. Infrared observation indicates extranuclear star formation activity *(Joy and Ghigo, 1988)*

We have taken velocity measurements from $H\alpha$ emission lines for about ten position-angles with **NGC 1143** as the reference point, and B, R, I and Kn (2.0-2.3 microns) images of **Arp 118** using the 2.3 m telescope and the Anglo-Australian Telescope at Siding Spring. Dynamical modelling for this system will follow soon.

OBSERVATIONS

- the velocity starts at $\sim$ 8200 km/s for most of the pa's; this emission comes from the material in the spheroidal component
- the velocities in **NGC 1144** rise rapidly by up to $\sim$ 1000 km/s with increasing distance from **NGC 1143**
- around the (brightest) nucleus of **NGC 1144**, the $H\alpha$ emission lines show splittings of up to $\sim$ 500 km/s, suggestive of concentrations of material

moving at different speeds.

- there seems to be a strong interaction between NGC 1143 and NGC 1144, as can be seen from the multiple-peaks of $H\alpha$ lines emitted by material between the two components

- the images show a stellar disk extending from **NGC 1144**; there are two major loops coming out from this component, and one is suggestive of another disk which might once belonged to the other bright nucleus.

DISCUSSION

What is the origin of these enormous velocity gradients in **NGC 1144**? An internal motion of 1000 km/s is very large for interacting pairs (which normally have velocities $\leq$ 300 km/s). It is commmonly believed that galaxy-galaxy collision with such high velocity difference will result in minimal disturbances. One possible explanation is a face on fast encounter of two galaxies, maybe an elliptical and a disk galaxy, or two disk galaxies, with the one galaxy stripping the interstellar matter from the other galaxy. A slightly different scenario is needed to explain the phenomena if the mentioned loop is actually a disk. In this case, **Arp 118** is a result of interactions between three galaxies, possibly an elliptical/SO with two spiral galaxies of different sizes.

REFERENCES

Freeman, K.C. and de Vaucouleurs, G. 1974. *ApJ*, **194**, 569

Hippelein, H.H. 1989. *A&A* **216**, 11

Joy, M. and Ghigo, F.D. 1988 *ApJ*, bf 332, 179

The First Stromlo Symposium: The Physics of Active Galaxies
ASP Conference Series, Vol. 54, 1994
G. V. Bicknell, M. A. Dopita, and P. J. Quinn (eds.)

A STARBURST RADIO GALAXY AT Z=2.34?

ROB VAN OJIK, GEORGE MILEY, HUUB RÖTTGERING, MALCOLM BREMER
Leiden Observatory, Postbus 9513, 2300 RA Leiden, The Netherlands

INTRODUCTION

We report the discovery of a distant radio galaxy at a redshift $z = 2.34$ with a peculiar emission line spectrum. the spectrum of TX0211−122 is similar to that of the IRAS galaxy F10214+4724. TX0211−122 is likely to be undergoing a massive starburst.

TX0211−122 was discovered in a survey to enlarge the sample of galaxies having $z > 2$, by studying ultra steep spectrum (USS) radio sources (Chambers *et al.*, 1990; Röttgering, 1993). It was identified with an object with an R-magnitude of 22.7. Optical spectroscopy showed that TX0211-122 has very weak Lyman α and very strong NV $\lambda1240$ emission, strikingly different from other radio galaxies at similar redshifts (McCarthy 1993).

TABLE I Line ratios of TX0211-122 and other relevant objects.

Line-Ratios	TX0211 total	TX0211 5"	Average Hi-z-RG	F10214	NGC1068	SN1987A April 1988
Lyα/CIV	1.1	0.9	6	0.2	-	-
NV/CIV	0.7	0.8	0.2	0.5	0.8	> 20
SiIV/CIV	0.1	0.1	0.1	0.2	0.2	> 3
HeII/CIV	0.7	0.6	2	0.7	0.5	> 3
CIII]/CIV	0.3	0.4	0.7	0.6	0.2	> 5

DISCUSSION

Two anomalous properties of the spectrum of TX0211−122 must be explained: 1. the deficiency of Lyman α and 2. the enhancement of NV.

The spectrum of TX0211−122 is remarkably similar to that of the IRAS galaxy F10214+4724 at $z = 2.286$ (Rowan-Robinson *et al.*, 1991). We regard the fact that the anomalies in the spectrum of TX0211−122 are so large *and* the close similarities with the spectrum of F10214 as more than a coincidence. The large IR luminosity of F10214 is evidence for the presence of considerable amounts of dust in this object, which is presumably responsible for the depression of Lyα. This may also be the case for TX0211−122. The most likely hypothesis is that the depression of Lyman α both in TX0211-211 and F10214 is due to dust.

The N/C abundance ratio of TX0211 is $\sim$10 times solar, requiring substantial chemical evolution. A vigorous, massive starburst is required to provide these large N/C abundances. (Hamman and Ferland, 1992).

An exceptionally hard nuclear continuum spectrum cannot produce the strong NV emission, without enhancing the CIV emission. However the com-

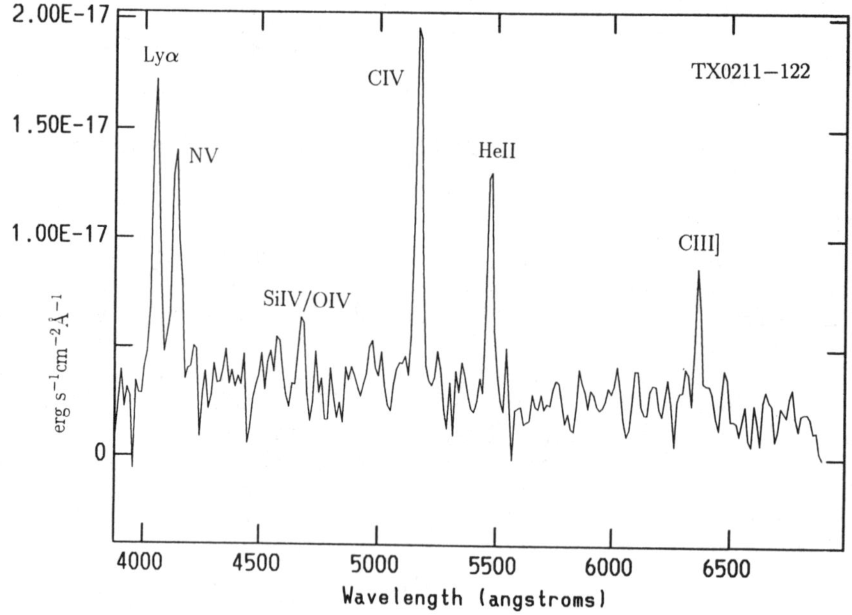

FIGURE I The optical spectrum of TX0211−122. The emission line ratios are comparable to those of IRAS F10214+4724.

bination of young hot stars and photoionization by the AGN can enhance the NV emission without enhancing CIV, as suggested for NGC1068 (Truong and Bruhweiler, 1991), which has a similar NV/CIV ratio.

The anomalous spectrum of TX0211−122 and its similarity with the IRAS galaxy is intriguing. We regard it as evidence that occasionally in high-redshift active galaxies conditions occur that simultaneously attenuate Lyman α *and* enhance NV λ1240 by large but similar amounts. The most likely culprit is a massive starburst, occurring in the central region of TX0211−122, possibly as the result of the passage of the radio jet. The most massive stars evolve quickly, producing the dust responsible for attenuating the Lyman α emission and a relative overabundance of nitrogen. Photoionization from the youngest stars and the AGN then provide a source of hard ionization, enhancing NV λ1240.

REFERENCES

Chambers, K.C. Miley, G.K. and van Breugel, W.J.M.: 1990, *ApJ*, **363**, 21.

Hamman, F., and Ferland, G.J.: 1992, *ApJ*, **391**, L53.

McCarthy, P.J., 1993, *ARA&A*, in press.

Röttgering, H., 1993, *Ph.D Thesis*, University of Leiden.

Rowan-Robinson, *et al.*: 1991, *Nature*, **351**, 719.

Truong, K.Q., and Bruhweiler, F.C.: 1991, *ApJ*, **379**, 609.

SUMMARIES

The First Stromlo Symposium: The Physics of Active Galaxies
ASP Conference Series, Vol. 54, 1994
G. V. Bicknell, M. A. Dopita, and P. J. Quinn (eds.)

SUMMARY: AN OBSERVER'S PERSPECTIVE

DANIEL W. WEEDMAN
Department of Astronomy and Astrophysics, The Pennsylvania State University, State College, PA 16802

INTRODUCTION

The organizers of this workshop served us well in assembling the group which enjoyed a week of southern hemisphere hospitality for the "first" Stromlo Symposium. The assembly was notable for the intimate mixing of theorists and observers which forced the sharing of ideas via the alternation of equations with the latest multimedia imagery. In the end, satisfaction was gained for all - because the observers have now seen what we were told all along had to be there, and the theorists were impressed by new observations which reinforce where is the real heart of the AGN puzzle.

THE TORUS AND THE DISK

Most of the ongoing attempts to explain many active galaxies which look very different rely on effects of orientation and perspective to change drastically the appearance of the same galactic nucleus, depending on the observer's direction. The Lick School pieced together the geometry from years of careful spectroscopic observations of Seyfert galaxies; the radio astronomers deduced the need for directed beams by considering core dominated vs. lobe dominated sources. Accounting for such anisotropies is usually achieved using some kind of disk. Now, we have a real picture that illustrates most of the previously hypothesized structures. This is the picture shown here by Ian Evans in the first presentation at the symposium: the remarkable Hubble Space Telescope image of NGC 4261 showing a dusty torus, from the center of which emerges a light cone and a radio jet.

To me, this image crystallized a lot of concepts that were already present in many models of AGN. It makes it seem obvious, for example, that we should differentiate between the "dusty torus" and the "disk". The torus is the 10 to 100 pc scale donut containing large amounts of dust and molecular gas which can geometrically obscure the nucleus, depending on orientation. The torus is also home to the starbursts. Whereas the "disk" is the much smaller, inner region which is responsible for collimating jets and for providing fuel to the innermost luminosity source and gravitational potential. (Joe Miller pointed out that in the picture of NGC 4261, the disk is warped relative to the torus, as judged by the direction of the radio jet, confirming even the details of the elegant deductions from modeling of other objects based on spectroscopic properties.)

This visualization explains away a lot of the confusion between whether various nuclei are powered primarily by starbursts or by non-thermal processes associated with the disk. It is no surprise that both are present and difficult to distinguish when the torus and disk cannot be spatially separated. As the torus houses the raw material for starbursts, the observable nature of the starburst depends on the molecular mass of the torus, its age, its dust content, and orientation. The disk and its luminosity generating processes are much smaller spatially and may be independent of the presence of a torus; it seems more likely that a disk could be present without a torus than conversely, because material at the inner edge of a torus would naturally feed a disk. A torus requires hundreds of millions of solar masses of material; a disk and the boil off that makes the broad line region only requires hundreds of solar masses. As a starburst proceeds within the torus, the torus dissipates on a time scale of 10^7 years. A disk can remain, continuing the non-thermal power source, which can even be enhanced if the leftovers of the starburst add to the gravitational potential. Extrapolating to an evolutionary scenario, one would conclude that the activity in M87 is older than in Centaurus A, or that in NGC 4151 older than in NGC 1068, because Cen. A and NGC 1068 still have dusty tori.

Understanding the starburst and the obscuration of nuclear luminosity requires, therefore, that we understand the torus. Understanding the ultimate source of the non-stellar luminosity requires knowledge of the energy generating processes in the disk. It does not bother me substantially that every AGN or quasar may be a little different in the details of these processes. We were reminded here by Anne Kinney of the complexity arising within stellar analogues whenever hot gas and magnetic fields mix. If our only knowledge of solar-like stars were based on flares in X-rays and gamma-rays, each star would appear unique and there would be vast differences among their luminosities. Consequently, we might well expect that the observables from an accretion disk in an AGN will differ drastically from object to object even though the fundamental astrophysics is similar. Nevertheless, there is a symptom which shows unambiguously that we are dealing with the basic mystery of a quasar

BEAMS ARE IT

Observers, having been spoiled by Hertzsprung and Russell, spend careers looking for correlations. A most astounding correlation in the AGN observables was emphasized at this meeting by Peter Michelson in his review of results from the Compton Gamma Ray Observatory. *All of the quasars or AGN seen as gamma ray sources are already known to be blazars.* The VLBI radio astronomers had already identified apparent superluminal motion in these objects, nicely interpreted as evidence for relativistic plasma motion toward the observer. The resulting "relativistic beaming" greatly amplifies the flux seen from the observer's direction, so the fact that gamma ray quasars are detected only in this way is a clear demonstration that the gamma rays arise from the same clouds as the VLBI scale radio sources. Here is a fundamental relation among photons differing by 10^{18} in energy. Many intriguing little anecdotes turn up. There is one source which has been detected, not only at Kev, Mev, and Gev energies, but at Tev energies from ground based observations of air showers! Think of it - an AGN which makes the atmosphere glow. This object is Markarian 421, a blazar but one found originally because Markarian was using a small telescope to hunt for evidence of starbursts (called "superassociations" in the Armenian school.)

Although the mystery and importance of these beams (or jets) has always driven the radio astronomers, the rest of us perhaps needed this hit from heavy gamma ray photons to wake us up. What is crucial now for observers is careful planning of correlated VLBA and GRO observations of selected sources while the gamma ray observing facility is available.

The jets, presumably collimated somehow by the disk, provide an intrinsic source of anisotropy for luminosity emerging from the nucleus. This is contrasted with the anisotropy produced by the torus, which only blocks light in certain directions. Beaming anisotropy may explain some other puzzles, such as the light cone in NGC 4151 which Kriss discussed here. Any torus which produces this light cone by shadowing is hard to accomodate, because the nuclear luminosity source and the broad line region which are so easily seen in NGC 4151 would be hidden by this torus, in the manner of NGC 1068. (Remember that, according to Miller and Antonucci, the nucleus of NGC 1068 would be so bright as to be visible to the unaided eye if not obscured by the torus!) But the VLBI-scale structure in NGC 4151 is aligned with the cone, as it should be if the cone is a consequence of intrinsic beaming from this inner jet. How can we learn more about beams and jets?

CENTAURUS A

Perhaps the southern hemisphere's major attraction, the galaxy Centaurus A, once considered so bizarre that it was ignored in a kindly way, holds many of the keys. If our current prejudices are correct, it has it all: an ongoing starburst within a dusty torus arising from an interaction, interior to which is a disk collimating the emergent radio jet. Its big advantage, of course, is that it is so much the closest example of these combined phenomena. We saw in this meeting that there is still much to be explored in probing the characteristics of the beam from the nucleus of Cen. A. Our hosts showed a fascinating analyis of emission line filaments in the outer reaches of the galaxy which they conclude are ionized by photons produced by shock waves arising when the beam ablates a molecular cloud. As observing capabilities on the ground and in space improve, Cen. A will become a renewed target. Being able to map and monitor it with the highest possible spatial resolution, for example, is a crucial motivation behind having southern hemisphere observing and tracking capabilities for the Japanese Very Long Baseline Space Observing Program (orbiting VLBI) scheduled for 1995, and a crucial contribution from NASA to this mission will be the facilities of the Tidbinbilla station, just down the road from where we speak. Given all of the local expertise and involvement in probing this magnificent specimen, I suggest that the "second" Stromlo Symposium be entitled "Centaurus A".

WHERE ARE THE DISTANT STARBURSTS?

We observers will continue to approach the integrated starburst/AGN problem from the extreme positions of probing in great detail a few nearby examples, while pushing to define the overall populations at the greatest observable redshifts. Nature has made raw quasars easy to find, primarily because their ultraviolet spectrum easily emerges to yield high redshifts quite readily from the strong emission lines. As a consequence, we already know a great deal about the evolution of the quasar population through 90% of the age of the universe. What we are missing from the distance reaches of the universe is knowledge of the star formation process. Until we have that, relating stars, galaxies, and quasars in the early universe will not be possible. This makes the search for highly redshifted starbursts a crucial goal, with the objective of determining the luminosity function of star forming galaxies as a function of redshift with as much precision as we know it for quasars. In my presentation here, I mentioned a few search techniques based on infrared and radio observations. Malcolm Longair pointed out that submillimeter detector arrays will soon be incorporated on ground based surveys (if one defines Mauna Kea as on the ground). These will allow the detection of starburst galaxies whose rest frame far infrared luminosity is redshifted into submillimeter wavelengths. Radio astronomers from NRAO reminded us that this will be extended to even higher z when the millimeter array is built. And the infrared astronomers still anticipate flying cooled space telescopes to resume the far infrared surveys where IRAS left off. It seems that the observers will continue to do their part, and we can expect them to earn their way back to future meetings which review our progress in this fundamental topic.

The First Stromlo Symposium: The Physics of Active Galaxies
ASP Conference Series, Vol. 54, 1994
G. V. Bicknell, M. A. Dopita, and P. J. Quinn (eds.)

A THEORETICAL PERSPECTIVE

GEOFFREY V. BICKNELL AND PETER J. QUINN
Mount Stromlo & Siding Spring Observatories, Australian National University

INTRODUCTION

The pace of research in the arena of Active Galaxies has been rapid over the last few years, spurred in part by the operation of the Hubble Space Telescope, in part by other new observational facilities (*e.g.* the Compton Gamma Ray Observatory and ROSAT), but also through the continuing progress made on the theoretical front and in various observational programs carried out on major ground-based optical and radio facilities throughout the world. In Australia, the completion of the Australia Telescope and the continuing program of advanced instrumentation on the Anglo-Australian Telescope have stimulated new work in Active Galaxies. This symposium has reflected the rapid international development of this field and we have benefited from numerous interesting and well prepared contributions in various areas of active galaxy research. Here we attempt to summarize and relate the major results of the symposium while at the same time highlighting what we believe to be some interesting questions raised by the various papers.

THE CENTRAL ENGINE AND ITS ENVIRONS

Hubble Space Telescope Observations of AGN

Ian Evans's paper on HST observations of AGN, fittingly the first of the symposium, has underlined the fact that ultraviolet and optical imaging together with spatially resolved spectrophotometry have contributed greatly to our knowledge of AGN. The observations summarized by Evans cover a wide range of AGN topics: Unified Schemes, Jets, and the Starburst-AGN connection. Therefore, the implications of the HST data are discussed in the relevant sections below. Other papers in these proceedings also involve HST observations and these are summarized at the appropriate point.

The Γ-Ray Emission from AGN

Less than 10% of Active Galaxies display jets or radio emission. However, the physical processes associated with jets and their production place some of the most stringent constraints on the source of energy in AGN. It is therefore fitting that some of the most spectacular recent results to be presented at this meeting were the EGRET observations of significant Γ-Ray fluxes from AGN, summarized by Peter Michelson. Clearly the most sensible interpretation of the EGRET results is that the Γ-Rays (1) arise from Compton scattering of ambient "soft" photons by relativistic electrons and/or positrons and (2) are relativistically beamed. Variability further constrains the emission region to $\lesssim 0.1\,\mathrm{pc}$.

An important concept for discussing Γ-Ray emission from jets, is that of the Γ-Ray photosphere, introduced by Roger Blandford. Within this photosphere, pairs are created off the soft photon field implying (1) that the active constituents of jets at the photospheric radius are pairs and Γ-Rays and (2) an increase in variability time scale with increasing photon energy. Both Blandford's and Begelman's reviews note that it is more attractive to consider the source of soft photons to originate from outside the jet (*e.g.* from the central accretion flow) rather than from the jet itself (the Synchrotron-Self-Compton model). Beams of electrons and/or positrons are rapidly decelerated by the ambient radiation field and Arieh Königl and colleagues have therefore modeled the spectrum produced by unneutralized proton beams propagating through an ambient X-Ray photon field. (One justification for this approach is the argument by Celotti and Fabian (1993) that VLBI jets are electron-proton dominated.) Königl shows an interesting fit to the spectrum of 3C 279 obtained using from a model wherein the initial beam is continually decelerated by an extended corona.

The initial configurations of jets are an issue. One possibility (Blandford) is that a beam essentially begins as Poynting Flux, raising questions concerning its conversion into a mechanical energy flux.

Don Melrose has reviewed IDV and considered explanations in addition to those involving high Lorentz factors, including refractive interstellar scintillations (RISS), specific models of shocks produced by jet inhomogeneities and coherent emission. As he shows, RISS are ruled out if the suggested correlated radio and optical variability is real. If beaming is the explanation then the required Lorentz factors are huge (~ 100). Another model reviewed by Melrose, the shock model of Qian *et al.* (1991) obviously has some appeal in view of the interpretation of knots in jets as moving shocks.

More on Jets

Another interesting piece of jet physics which was discussed here (again by Blandford) is whether jets are surrounded by a more slowly moving magnetically dominated wind. This would complete the analogy between extragalactic jets and jets from Young Stellar Objects. The latter appear to be surrounded by a slow-moving wind which drives the associated bipolar flows. The Seyfert super winds reviewed by Baum *et al.* may be indicative of such a morphology. Baum and her colleagues show that the extranuclear radio and IR luminosities are consistent with the standard radio-IR relation, suggesting that the origin of the winds is a circumnuclear starburst. However, they also speculate that the energy driving the winds may originate from the AGN. The concept of a wind enveloping jets in radio galaxies raises the prospect that VLBI jets are not free, as previously believed but are, in fact, pressure confined.

The HST observations of jets, reviewed by Evans, have presented new dilemmas for jet theorists. The initial impression from the HST observations of the M87 jet, imaged for the first time with a comparable resolution to the VLA observations, is that the radio and optical structures are identical. However, Sparks, Biretta and Marchetto (1992) have shown that the jet knots are more centrally condensed in the optical/UV than in the radio and this provides an important constraint on the theory of the emission from this jet. The edge brightened morphology of part of the M87 jet and the filamentary appearance of the optical jets in PKS 0521-36 and 3C66B suggests that particle acceleration is occurring at the respective surfaces of these jets, possibly mediated by the helical surface waves that are produced on supersonic jets (*cf.* Owen, Hardee and Cornwell, 1989). Another view on the M87 jet has been put forward in these proceedings by Walker *et al.* who suggest that it is a force-free flux rope advected from the nucleus and twisted into a hollow configuration. Significant γ-ray emission is a possible consequence of this model. The elucidation of the relationship between the optical (HST) and radio jets and the interaction between the jets and dense gas in Pictor A which was the subject of the paper by Simkin *et al.* also promises to bring to light some interesting new radio source physics.

A potentially significant new idea was introduced at this symposium by Gvaramadze and Machabeli. They have included the magnetic field as an important dynamical component of accretion (although not in the same way as Königl or Blandford or Hawley and colleagues) and have outlined the way in well collimated jets may be formed through the so-called "cumulative effect" of shock-waves generated by free-falling filaments. A jet envelope is also a natural consequence of this model, this being formed near the standing shock formed at the inner edge of the accretion disk. Accretion flow onto a dynamically significant magnetic field is extremely complex, involving reconnection and the distortion of the magnetic field itself by the accretion. Thus, numerous details of the model remain to be calculated. One therefore looks forward to a number of interesting theoretical papers following from this work. This is another area of contact between stellar and AGN physics, accretion onto magnetized white dwarfs being a topic of current interest. (See also the discussion of Kinney's paper, below.)

X-Ray Emission; Thermal and Nonthermal Emission

Wilkes and Begelman have given clear and complementary accounts of the disentanglement of thermal and nonthermal effects in AGN. Wilkes's account of how the differences in Seyfert 1 and QUASAR X-Ray spectral slopes on the one hand, and the differences in *Einstein* and *EXOSAT* slopes on the other, were resolved by *GINGA* observations underlines the importance of a detailed spectral coverage of AGN. The physical idea, due to Guilbert and Rees (1988), that cool gas should be abundant in the environs of AGN and would be responsible for the production of spectral features in X-Rays, is well supported by the data. Both Wilkes and Begelman conclude that the IR and X-ray emission are not related by SSC emission. Perhaps the main argument for this is that although the X-Ray and IR emission are often correlated, this is not always the case, and the far-IR turnover is too steep to be due to synchrotron self absorption. As summarized by Wilkes, the variability study of Done *et al.* supports this conclusion. On the subject of the origin of the power-law X-Ray component, Wilkes has summarised the evidence that this is due to SSC emission in radio loud objects. However, Begelman has pointed out that this elegant notion does not work for the γ-rays – hence the models involving inverse Compton off the AGN background referred to above. The X-Ray variability indicates a small source region. Wilkes's comment, that although X-Ray spectra contain a multitude of components, the data are generally consistent with one component varying more than another, is of interest. The *intrinsic absorbers* discussed by Wilkes and their association with Gigahertz Peak Spectrum sources supports the view that these sources are "frustrated jets" that have not made it out of the dense core of a cooling flow. However, the paper on the IR comparisons of radio-loud Active Galaxies by O'Dea *et al.* suggests that there are details to be addressed, such as the apparent lack of conversion of directed jet energy into IR flux, before such a view can be made secure.

The Broad Line Region

Whilst there is at least some consensus on the physical processes associated with the Γ-Ray and X-Ray fluxes, we have, recorded in these proceedings, two qualitatively different models of the broad-line region, one due to Bill Mathews and the other due to Judith Perry. Both Mathews and Kriss and Krolik have drawn attention to the fact that reverberation studies show that specific lines form over a narrow range in radius, that is, the Broad Line Region (BLR) is stratified. A consequence of this is that the line widths cannot be formed through the excursions of clouds throughout the entire BLR. Mathews's solution to this is to invoke the "bouncing cloud" model in which the line profile is formed through the oscillations of clouds about equilibrium radii determined by the balance between gravitation and radiation pressure, suggesting that this circumvents some of the problems encountered by other models involving clouds. One prediction of this model is that the required central mass ($\sim 3 \times 10^8\, M_\odot$)is about 10 times larger than "naive" estimates of the virial mass calculated from reverberation studies.

Disks

Anne Kinney's paper comparing the spectra of Cataclysmic Variables and AGN is a timely reminder that there is no evidence for disks in AGN, despite the fact that they are a popular theoretical construct. There are various ways around this, including spectral broadening by differential motions, reprocessing by a thick corona and variation of effective gravity in the spectrum formation region (see Kinney's and Begelman's papers for a discussion of this). Thus it may not be surprising that the Lyman edge signatures typical of accretion disks in CVs are not present. On the other hand, the microlensing of the QUASAR 2237+0305 (see Begelman's review) has also called into question the existence of disks in AGN and the investigation of other scenarios for energy production is certainly worthwhile. In these proceedings Blandford has suggested a mechanism involving the production, through Lense-Thirring precession, of non-closed orbits of clouds which subsequently collide and shock. Such a scenario could be modeled, at least approximately, using fairly straightforward numerical techniques.

Undeterred by the lack of evidence for disks, Hawley, Gammie and Balbus have discussed physics and simulations which in a sense represent the "state of the art" in accretion disk theory. Hawley *et al.* begin by summarizing the problems with pure hydrodynamic disks: They are stable, do not become turbulent, and therefore will not drive an accretion flow. The discovery of the Balbus-Hawley instability (see Balbus and Hawley, 1992, and references therein) driven by weak magnetic fields changes the situation dramatically, producing rapid instability growth and interchange of mass and angular momentum. The effective $\alpha \sim 0.1$ occurring in their simulations (limited to some extent by the finite size of the "shearing box") is of the order of magnitude that most workers have assumed in the theory of AGN discs. However, the details of the transport in the disk are different from the standard picture - energy is not dissipated on the spot for example, but may be locked up in the magnetic field to be released in reconnection events in the disk corona, possibly leading to coronal holes . Hawley and colleagues also note that the dragging in of interstellar magnetic field into an hour-glass configuration as required in centrifugally driven hydromagnetic wind models, is not an obviously viable process in a turbulent disk.

UNIFIED SCHEMES

Seyfert Galaxies

Perhaps the most complete set of contributions to these proceedings deal with *Unified Schemes* in their various forms. Joe Miller has neatly summarized the evidence that at least some Seyfert 2 galaxies are in fact obscured Seyfert 1s. The summary of Tran's (1993) Ph.D. thesis represents the latest chapter in the work done at Lick on polarization observations of Seyferts. Careful separation of polarized and unpolarized components has evidently identified two featureless continua, one polarized (FC1), the second (FC2), often dominant, and unpolarized. The spectra of both are indicative of free-free emission at several times 10^5 K. Miller speculates that FC2 is in fact the gas responsible for the scattering of the broad lines and FC1.

Despite being one of the major proponents the idea that at least some Seyfert 2s are obscured Seyfert 1s, Miller is reluctant to suggest that all Seyferts can be unified in this way. This reluctance mainly centers on the *radio* properties of Seyfert galaxies. One important difference is the differing relationships between continuum polarization and radio morphology: perpendicular for Seyfert 2s, parallel for type 1s. Miller notes that this may be due to differing scattering geometries and possibly this could be related to the Krolik-Begelman (1986) molecular torus (see also Pier and Krolik, 1992) or Königl's obscuring wind model discussed below. Laura Kay's paper, which complements Miller's in a number of respects, emphasizes that the polarization method for detecting obscured Seyfert 2s is only successful when the fraction of galaxy starlight is small and notes that is not clear whether this constitutes a selection effect or a real physical difference (related to luminosity). Kay also discusses the higher radio power and flatter IR spectra of the BLR Seyfert 2s (Moran *et al.*, 1992; Hutchings and Neff, 1991) concluding that the former difference may not be too large and that the latter may be related to the physics of the obscuring tori. However, the paper by Alan Roy and Ray Norris, extends the points concerning radio properties made by Miller and Kay, clearly showing that there are real radio differences between Seyfert 1s and 2s: Surprisingly (for a unification advocate) the 2s have systematically larger radio fluxes, inconsistent at first sight, with relativistic boosting of Seyfert radio emission. Roy and Norris consider two alternatives for this, (apart from an intrinsic difference) both involving preferential absorption by the narrow-line region in Seyferts and suggest that the Seyfert 1 and 2 radio dichotomy may disappear at higher frequencies. The results of their future observations obviously will be of great interest.

The subarcsecond morphology of the Narrow Line Region in NGC 1068 as revealed by HST observations (summarized by Evans), supports the Antonucci-Miller view of this object although the location of the scattering mirror remains elusive. Reassuringly, the inferred apex of the cone is near the southernmost component of the radio triple. However, this does raise the question (again) of the explanation for jet sidedness. Why is the one-sided jet pointing in the direction of the presumably isotropic OIII emission from the ionization cone? We should also note that the coincidence of the bright OIII and radio knots in the NGC 1068 ionization cone is circumstantial evidence for excitation of at least some of the optical line emission through interaction with the jet (which is required to be at least mildly relativistic). That there are OIII knots *not* coincident with radio knots requires the radio jet to "thrash about" with a time scale $\sim$ 100 yrs. This is reminiscent of Scheuer's (1982) dentist's drill model. Detailed models remain to be worked out.

The Nature of the "Obscuring Torus"

Miller offers two constraints on the properties of the object obscuring the broad line region: (1) The existence of ionization cones and the pattern of scattered light suggest that it is flattened or toroidal; (2) The required size (larger than the broad line region) precludes it being the (hypothetical) accretion disk.

Miller raises the possibility that the obscuring regions may be transitory "dumps" of large amounts of material resulting from mergers or interactions, citing the evidence that many Seyferts do in fact possess double nuclei or companions. Krolik and Begelman (1986) have published a model of a molecular torus for the obscuring region and this model is supported to some extent by the HST observations of NGC 4261 and M51, reviewed by Evans (see the HST image of NGC 4261 on the front cover).

The Krolik-Begelman torus is supported by the velocity dispersion of the constituent molecular clouds. Königl suggests a different model: a centrifugally driven hydromagnetic wind consisting of an opaque outer region of neutral gas and dust and an inner region of ionized gas. Radiation pressure from the central object "blows open" the inner region of this wind creating ionization cones. Hot gas, initially at 10^6 K, cools in the inner wind and provides the scattering "mirrors" at a temperature $\sim 10^5$ K, as inferred by Miller and his collaborators. Clearly this model has some very appealing features: It explains the difference in slopes of the near IR continua referred to by Kay and the extended mid IR emission in Seyfert 2s; a dependence of the wind opening angle on radiation pressure would introduce a luminosity dependence into Seyfert 1/2 unification (*e.g.* the lack of highly X-Ray luminous Seyfert 2s). It may also go some way to explaining the observations of NGC 4151 which Evans has argued are inconsistent with pure geometrical obscuration. This feature is possible because of the optical transparency of the region of the wind which absorbs the ionizing X-Rays. Whether the Kartje-Königl model is true in detail depends upon the presently uncertain characteristics of disk driven winds: Such winds have been discussed a great deal in various contexts (bipolar flows from young stellar objects, for example) and are currently being investigated numerically. The simulations have encountered difficulties which relate strongly to the physics of magnetized accretion discs (*e.g.* the Balbus-Hawley instability, see Balbus and Hawley, 1992 and references therein) and it may be some time before we know how centrifugally driven disk winds work in detail. Infrared polarimetric studies which are now possible with IR arrays may shed light on the magnetic field orientation in a number of objects (albeit at some distance from the central object) and tell us whether the field configurations implied by disk driven winds are possible.

FRII Galaxies and Quasars

Peter Barthel, as one of the leading pastors of the *Unification Church*, gave us a compelling sermon on why we should believe that the parent population of QUASARS are FRII Radio Galaxies. As in all religions, there is some dissension (*e.g.* Singal) which probably indicates that the proposition has not yet attained the status of dogma. However, Robert Laing and his fellow disciples have presented a good case that, while not all FRIIs are buried QUASARS, FRII galaxies with high excitation emission lines are consistent with the unification scheme as proposed by Barthel. The discovery of broad-line spectra in the infrared in a number of narrow-line radio galaxies certainly does lend support to the idea.

Whilst there now seems to be a willingness to accept the idea of relativistic beaming in at least the broadest outline, there still remain some interesting problems to be resolved. The Laing-Garrington effect (Laing, 1988; Garrington *et al.*, 1988) demonstrating a correlation between polarization and jet-sidedness in QUASARS has been one of the observational linch-pins of relativistic beaming. On the other hand Liu and Pooley (1991a,b) have shown that in radio galaxies, there is a strong correlation amongst relative lobe length, steeper radio spectrum and depolarization. In the Bridle *et al.* paper (presented by Peter Scheuer) the authors comment on the possible devastation of relativistic beaming ideas implied by the combined results. However, their high resolution observations of the lobes of a sample of QUASARS provide some reassurance for advocates of relativistic beaming. They show that there is a combination of beaming and environmental factors at play with asymmetries in lobe length responsible for spectral asymmetries in the extended emission and the jet side controlling the spectral index of the hot spots (for which there is an adequate explanation in terms of relativistic effects). The remaining worrying aspect of their observations is that it is not only the hot spots which have a flatter spectrum on the jet side. Possibly this implies that the emission neighboring hot spots is also relativistically beamed. As Bridle *et al.* remark, relativistic beaming survives ...by the skin of its teeth!

FRI Galaxies and BL-Lacs

While the FRII members of the Unification Church seem to be rejoicing in the knowledge that beaming has bestowed upon them, the FRI congregation seems to be echoing with a chorus of *Hallelujah Brother!* The summary of the work of Meg Urry and Paolo Padovani on the statistics of FRI sources and BL-Lac objects provides strong support for BL-Lacs being beamed FRIs. There had previously existed some difficulties with the concept of FRIs harboring relativistic jets. However, it now appears that relativistic core jets and kpc-scale sub-relativistic flow are plausibly dynamically consistent. Moreover, following on from the ideas relating to the polarization at the bases of FRI jets (see Robert Laing's contribution), Paola Parma and her Bologna colleagues have shown that, if the asymmetric bases of class I jets are interpreted using the Doppler beaming model, the velocities ($\sim 0.6c$) are in accord with those implied by deceleration of an initially relativistic jet to transonic velocities. Thus it appears to be coming accepted that all jets start off life as at least mildly relativistic and that, in a sense, is a type of unification in itself. One caveat is the uncertainty in the relationship between X-Ray and radio selected BL-Lac objects. Why are the required Lorentz factors different (~ 3 for the former and $\gtrsim 7$ for the latter)? The solution to this particular problem may lend insight into the production of X-Ray emission in all types of Active Galaxy.

The lower power of FRI galaxies suggests that even though their jets may be relativistic they may not be quite as relativistic as those in higher powered FRII sources and QUASARS. To date, limitations on sensitivity have precluded observations of all but a few FRI sources. However, we have seen here (Venturi *et al.*; Giovannini *et al.*) some of the first results on VLBI imaging of samples including FRI sources. There is little structural difference between FRIs and FRIIs with still the commonest structure being a one-sided core jet aligned with the strongest kpc scale structure. The *lower* limits on the jet Lorentz factor ($\sim 1.3 - 2$) are suggestive of lower flow velocities. Clearly, however, further progress must await the completion of the VLBA to see if some two-sided structure emerges. (This is possible if the Lorentz factors are as low as 2 and the jets are not too highly inclined). FRI jets do not show high apparent proper motions, however, and this is inconsistent with relativistic flow if one naively interprets proper motions as indicative of jet velocities. This situation can be remedied if the knot motions represent reverse (and possibly oblique) shocks in not too rapidly moving jets. Such shocks may be caused by variations in the jet velocity om a time scale of a few years and therefore could constrain the jet production mechanism.

Beaming of Optical Emission

The Barthel unification paradigm for QUASARS and Radio Galaxies focuses on the radio emission. The paper of Baker *et al.* complements the radio work and presents a strong case for beaming of the *optical* emission. She and her colleagues have convincingly shown the diminution of the OII equivalent width with increasing core dominance (following on earlier work on the OIII equivalent width by Jackson and Browne, 1989) attributing this to an increase in the optical continuum as the core becomes more visible. This obviously appealing interpretation of the data highlights the intriguing Roy-Norris result and it will be of interest to see what the final answer is on the interpretation of the Seyfert observations. The Baker *et al.* paper also reemphasizes a result that has been known for some time (*e.g.* Yee, 1980), namely that as a function of the core dominance parameter R, the H_β equivalent width is practically constant.

There was considerable discussion on the constancy of the H_β equivalent width in the final symposium session with Mike Dopita putting the view that, in view of the likely differences, amongst QUASARS, of the covering factor of the broad line region, the constancy of this parameter is remarkable. See the discussion of Judith Perry's paper, below, for one idea on this characteristic of

AGN.

Is Aspect Enough?

Most discussion of Unified Schemes and indeed most of the discussion above, has centered on the rôle of aspect. Two other papers, however, have indicated that other parameters are required. Steve Rawlings has suggested that accretion rate, $\dot{M}$, is another important parameter for separating radio loud and radio quite ellipticals, that in earlier epochs most ellipticals were accreting at near the Eddington limit, $\dot{M}_{\rm Edd}$ and that when $\dot{M}$ drops below $\dot{M}_{\rm Edd}$ the production of relativistic jets with a high kinetic energy flux and radio loud activity are favoured. In a sense, Rawlings's adoption of $\dot{M}$ as another parameter, is consistent with Judith Perry's advocacy for the ratio of black hole mass to stellar cluster mass because of the effect of the star cluster on the mass accretion rate. (The ideas of Perry and colleagues are discussed further below in the Starburst section.)

IR emission and Unified Schemes

Other aspects of IR emission and its relationship to Unified Schemes are covered in the paper by O'Dea *et al.*: There is a difference in the mid-IR properties of Broad-Line Radio Galaxies (BLRG) and Narrow Line Radio Galaxies (NLRG) which O'Dea *et al.* claim shows that the IR emission is anisotropic, if the unification paradigm is to hold. The difference disappears in the Far-IR. This result is broadly consistent with the notion of partially opaque obscuring tori (which become optically thin at long wavelengths) provided that one accepts the conventional interpretation that, except for objects such as Blazars and BL-Lacs, the IR emission is due to dust (see Begelman and Königl). Note however, the contribution by Laing *et al.* who show that not every NLRG has a right to be described as an obscured BLRG.

SEYFERT GALAXIES AND LINERS

A major part of the discussion of Seyfert galaxies took part in the section on Unified Schemes. Nevertheless, other important aspects of Seyferts were also reported. The ultraviolet spectra of NGC 1068 discussed by Kriss *et al.* are one set of data that support the idea that photoionization by a central power-law (or other spectrum) is not the only mechanism responsible for line emission in AGN. The strong narrow line emission from C III $\lambda 977$ and $N\ III \lambda 991$ is indicative of shocks. There are two possible interpretations: (1) There is a mixture of shock excitation and power-law photoionization or (2) Shocks are producing enough ionizing radiation to photoionize the precursor gas. Bicknell has summarized how the ionization of the Centaurus A filaments may be produced by autoionizing shocks and Dopita has modeled the spectra of LINERS and Seyferts using autoionizing shock models with somewhat different geometry. This idea "unifies" emission line luminosities and kinematics. However, probably it is mainly relevant to the Narrow Line Region. Reverberation studies of the Broad Line Region seem to confirm photoionization by a central source as the favored mechanism.

RADIO GALAXIES

The FRI/II Classification

Frazer Owen presented further results from the comprehensive cluster radio source survey which they and their colleagues have undertaken over the last few years. The structure of the diagram of radio sources in the radio-optical plane is intriguing - a clear break between class I and class II sources across a dividing line on which $L_{\rm radio} \propto L_{\rm opt}^2$. As Owen and Ledlow state, this diagram establishes, for the first time, the plausibility of an evolutionary connection between class II and class I sources. One speculates that class II activity may be initiated by a merger and that the activity lasts for approximately a dynamical time scale of the merger ($\sim 10^7$ yrs). There then may follow a more lengthy period ($\sim 10^8$ yr) of quieter class I activity.

Cluster Mergers and their Effects on Radio Sources

It is not an easy task to give the last talk of a symposium. However Jack Burns easily held the audience's attention with a paper by himself and his colleagues on the effects of subcluster mergers on radio sources. This paper represents a compelling synthesis of X-Ray, radio and optical observations with high powered N-body and hydrodynamic simulations. We have learned that Radio Galaxies coincide (more or less) with the peaks of subclumps in merging clusters and that the long-lived peculiar galaxy velocities and hydrodynamic motions can have interesting effects on the large scale structure of the radio sources. (This notion of "cluster weather" has been advocated by Ron Ekers for a number of years.) Burns *et al.* show that the dynamics of merging clusters can explain the appearance of wide-angled tails (WATs) and the absence of cooling flows around WAT galaxies and suggest that the formation and maintenance of cluster radio halos may be related to the turbulence produced by the merger. Moreover, their work now provides a reason for why the luminosity function of cluster and field galaxies are so similar - it is principally determined by the local sub-cluster environment prior to the formation of richer clusters. As Burns *et al.* remark, the elucidation of these properties of radio sources now make it feasible to use radio sources as a diagnostic of cluster mergers.

HIGH REDSHIFT RADIO GALAXIES

One of the more interesting discoveries during the last decade has been that of High Redshift Radio Galaxies, selected mainly by their steep radio spectra. We have been fortunate at this symposium in having two complementary papers on this subject, one by George Miley on the general characteristics of High Redshift Radio Galaxies and the other by Klaus Meisenheimer on some specific examples drawn from an extended sample. As Miley has explained, the physics of these galaxies is not well understood although he presents an outline of some of the processes that may be important. These include: (1) Vigorous star formation, producing dust (hence polarization) and (2) Radio-IR-Optical-UV alignments produced by jet-induced star formation. Clearly, the radio and *continuum* alignment question is an ongoing debate at present because of the potentially confusing effect of strong emission lines in optical and infrared pass bands. However, there is at least evidence in one class of high redshift source (see Meisenheimer's contribution) for alignment of the radio continuum and *line* emission. The clumpiness of the emission on scales of ~ 500 pc as revealed by HST and radio observations (Miley) is highly significant and probably crucial to a correct interpretation of these sources. As Miley shows, if the clumps are stellar then they contain $\gtrsim 10^{10}\, M_{\odot}$ of stars per clump. This is a typical size for local starburst regions and since these high redshift galaxies contain many of these clumps this underlines the extreme nature of these objects.

The taxonomy of these objects outlined by Meisenheimer is a useful one for elucidation of the essential physics. There are the well aligned Extended Emission Line Regions (EELRs) referred to above, which have been an important component of the prototype high redshift radio galaxies to date. Meisenheimer argues that this class is due to jet-induced, shock-excited emission line gas. The large inferred turbulent motions ($\sim 500\,\mathrm{km\,s^{-1}}$) means that, as we have seen in other contexts, the shocks are autoionizing. However, shocks with such high velocities have not been modeled to date.

There are also other high-z galaxies in which the dominant excitation mechanism seems to be related to mergers which must have been numerous at these epochs. In this class Meisenheimer claims that there is no preferred radio-optical-infrared alignment. Then there are examples of mixed types in which the production of jets within the environment of merging galaxies leads to a burst of emission line activity. Miley's explanation for the alignment of companions along the radio axis, as being due to strong excitation when a jet points in the direction of a companion, is an appealing one. Another feature of these galaxies which has not been modeled in detail at present is the relatively broad Ly_{α} absorption systems associated with them. As Meisenheimer suggests these may be indicative of ongoing gravitational collapse (and possibly consistent with the large number of apparent starburst clumps).

STARBURSTS, MERGERS AND AGN

It has long been held that there is a connection, either direct or evolutionary, between mergers, starbursts and AGN. The most recent manifestation of a "starburst-only" scenario for AGN has been the work of Terlevich and collaborators (*e.g.* Terlevich *et al.*, 1992 and references therein). Judith Perry has given as a clear account of hers and her collaborators' work over the last ten years, involving studies of the symbiosis between black holes and starburst clusters in AGN. The motivation for this stems from the metal rich gas that is implied by the broad line observations and the recognition that there is a problem in providing the high mass fluxes to a black hole from a hot, optically thin ISM. Perry's paper is a clear account of the ideas that she and her colleagues have developed and here we summarize some of the outstanding features of their models. These depend upon the very direct relationship between mass-loss via winds and supernovae, the fueling of the black hole, the production of the continuum luminosity and the production of broad line clouds through radiative shocks in supernovae. The coupling of these processes leads to both a constant H_{β} equivalent width (see the discussion of the Baker *et al.* paper above) and the constancy of the ionization parameter. Thus the Perry *et al.* ideas for the production of the broad line region stand in marked contrast to the "bouncing cloud" model presented by Mathews as does their mechanism for broad absorption line systems (absorption by supernova remnants) compared to that suggested by Miller (grazing lines of sight across the obscuring torus). Another interesting contrast between the Perry and Mathews models is in the inferred mass of the black hole, respectively smaller than and larger than previous estimates. We do not attempt to make any judgement on the relative merits of these ideas here; however the suggested mechanisms and parameters are so different that there must be some way of distinguishing them.

Dan Weedman's contribution highlights why there is so much speculation about the relationship between AGN (in particular quasars) and starburst galaxies – pure examples of the latter at high redshift are extremely difficult to find. As Weedman points out, such galaxies must be shrouded in dust, making their redshifted UV emission difficult to detect. Nevertheless, he gives useful ground-based search strategies which may be successful in tracking down starburst galaxies in the redshift range $2 < z < 2.5$ with luminosities $\sim 1\%$ of the highly luminous starburst galaxy 10214+4724. It is only when such galaxies are found that we will gain an observational handle on the link between QUASARS and starbursts. One should also note the result by Fillipenko, Ho and Sargent (1993) (reviewed in Evans's contribution) showing the lack of UV stellar absorption lines in the spectrum of the Seyfert galaxy, NGC 4395. This one observation tends to argue against a starburst driven AGN but does not rule out an evolutionary or symbiotic connection as implicated in the models of Perry and colleagues. On the other hand Elaine Sadler and colleagues have shown that a number of nearby ellipticals show evidence for AGN like activity without any evidence for starbursts and this needs to be reconciled with any ideas of starburst-AGN activity – symbiotic, evolutionary or otherwise. Another interesting, perhaps crucial, aspect of the Sadler *et al.* paper is the difference in the core spectral indices of the Seyfert and Elliptical galaxies in their sample, the former having steep spectra, the latter flat. As Sadler *et al.* remark, the radio sources in the respective galaxies seem to know something about the environment.

The contribution by Forbes and Ward also underlines the value of IR imaging in selected lines such as [FeII] and H_2 as important in distinguishing between AGN and starburst activity and for tracing shock excitation – induced by jets, SNRs or colliding interstellar media in merging galaxies.

Another aspect of Starbursts and Active Galaxies was presented in this symposium by Ron Ekers. The ideas discussed by Ekers relate to the initiation of massive star formation through AGN activity and production of cosmic rays. The latter partially ionize giant molecular clouds, freezing in the magnetic field, thereby inhibiting the formation of low mass stars and biasing the Initial Mass Function to high masses, as observed. Part of Ekers's purpose in promoting a connection between cosmic rays and star formation is in providing a basis for the Radio-IR relation. In the Active Galaxy area, the idea is interesting in that it places a different emphasis on the starburst - AGN connection, possibly providing part of a feedback loop tying AGN and starburst activity together.

THE X-RAY BACKGROUND

In the previous section we have discussed the possible relationships between one of the potential precursors to active galaxies. This seems to be an appropriate place, therefore, to discuss the paper by Boldt and Leiter on the constraints placed on the evolution of Active Galaxies by observations of the Cosmic X-Ray background (CXB). This fascinating paper represents the culmination of work involving the synthesis of data on the Cosmic X-Ray Background, the observed evolution of AGN in X-Rays, accretion physics and cosmology. The bottom line of this paper is that Boldt and Leiter have made a strong case for the existence of a class of Precursor Active Galaxy (PAG) which makes a substantial contribution to the CXB. Their case is based on:

1. The luminosity evolution of quasars implies that the CXB could be generated by normal AGN if the observed number density is $\sim 4\times10^3\,\mathrm{degree}^{-2}$ for $\Omega = 0$ and $\sim 6\times10^2\,\mathrm{degree}^{-2}$ for $\Omega = 1$. Surface brightness fluctuations of the CXB are consistent with the number density estimates for $\Omega = 0$ but inconsistent for $\Omega = 1$.

2. The energy density of the CXB implies a comoving density of black holes $\approx (0.6-2.1)\,M_\odot\,\mathrm{Mpc}^{-3}$. When compared with present-epoch estimates of black hole mass densities in currently active galaxies this estimate implies a small ratio, f, of active to dormant black holes consistent with estimates due to Kormendy and Richstone (1992).

3. The black hole mass per AGN implied by accretion and the implied duty-cycle averaged black hole mass and density are more consistent with the constraints on black hole masses in Seyferts and with the density implied by the CXB if $\Omega = 1$ and $f < 0.042$ (although the $\Omega = 0$ result does not differ by more than a factor of order unity).

4. When the CXB spectrum is fit by a combination of Seyfert and QUASAR spectra, there is a residual spectrum which is well described by compact Eddington-limited sources. These are the Precursor Active Galaxies proposed by Boldt and Lieter to account for the additional faint sources implied by the ROSAT observations in conjunction with $\Omega = 1$. (See point 1).

Although point 3 does not make a compelling case for $\Omega = 1$, this value of the density parameter is of course strongly favoured by cosmologists. It will be interesting to see if future work identifies these PAG and if in fact they are related to the faint starburst galaxies that Weedman has suggested should exist.

CONCLUDING REMARKS

In this summary, we have covered a lot of the material in the invited papers and attempted to relate many of the themes arising in those papers to one another. Unfortunately there is not the space to cover all of the contributed papers as well. However, we trust that the reader will find this summary useful and that it will serve as a guide to the major topics discussed at the symposium.

REFERENCES

Balbus S.A. and Hawley J.F., 1992, *ApJ*, **400**, 610

Celotti A. and Fabian A.C., 1993, *MNRAS*, **264**, 228 Fillipenko A.V., Ho L.C. and Sargent W.L.W., 1993, *ApJ*, **410**, L75

Garrington S.T. *et al.*, 1988, *Nature*, **331**, 147

Guilbert P.W. and Rees M.J., 1988, **233**, 475

Hutchings J.B. and Neff S.C., 1991, *AJ*, **101**, 434

Kormendy J. and Richstone D., 1992, *ApJ*, **393**, 559

Krolik J.H. and Begelman M.C., 1986, *ApJ*, **308**, L55

Laing R.A., 1988, *Nature*, **331**, 149

Liu R. and Pooley G.G., 1991a, *MNRAS*, **249**, 343

Liu R. and Pooley G.G., 1991b, *MNRAS*, **253**, 669

Miller, J. S., Goodrich, R. W., and Mathews, W. G. 1990, *ApJ*, **378**, 47

Moran E., Halpern J., Bothn G. and Becker R., 1992, *AJ*, **104**, 990

Owen F.N., Hardee P.E. and Cornwell T.J., 1989, *ApJ*340, 698

Pier, E., & Krolik, J. H. 1992, *ApJ*, **401**, 99

Qian S.J. *et al.*, *A&A*, **241**, 15

Scheuer P.A.G., 1982, in *IAU Symposium 97, Extragalactic Radio Sources*, ed. D. S. Heeschen and C. M. Wade, Reidel, Dordrecht, 163

Sparks W.B., Biretta J.A. and Macchetto F., 1992, *ApJ*, in press

Terlevich R.T., Tenorio-Tagle, G., Franco, J., Melnick, J. 1992. *MNRAS* **255**, 713

AUTHOR INDEX